SYSTEM
AND
SIGNAL ANALYSIS

The Oxford Series in Electrical and Computer Engineering

M.E. Van Valkenburg, *Senior Consulting Editor*
Adel S. Sedra, *Series Editor, Electrical Engineering*
Michael R. Lightner, *Series Editor, Computer Engineering*

SYSTEM AND SIGNAL ANALYSIS

Second Edition

CHI-TSONG CHEN

*State University of New York
at Stony Brook*

New York　　　　*Oxford*
OXFORD UNIVERSITY PRESS

Oxford University Press

Oxford   New York
Athens   Auckland   Bangkok   Bogota
Bombay   Buenos Aires   Calcutta   Cape Town
Dar es Salaam   Delhi   Florence   Hong Kong
Istanbul   Karachi   Kuala Lumpur   Madras
Madrid   Melbourne   Mexico City   Nairobi
Paris   Singapore   Taipei   Tokyo   Toronto

and associated companies in
Berlin   Ibadan

Copyright © 1994, 1989 by Oxford University Press, Inc.

Published by Oxford University Press, Inc.,
198 Madison Avenue, New York, New York, 10016
http://www.oup-usa.org

Cover credit: Mel Lindstrom/Tony Stone Images

Library of Congress Cataloging-in-Publication Data available upon request

ISBN: 0-19-510722-5

9 8 7 6 5 4 3 2

Printed in the United States of America
on acid-free paper

In memory of my father and grandfather

Preface

PURPOSE

This book is intended primarily as a text for a sophomore-junior course in systems and signals for electrical engineering, mechanical engineering, system engineering, and engineering science students. It can also be used as a reference book for practicing engineers. It introduces the following five major topics:

1. Fundamental concepts (causality, linearity, time-invariance, lumpedness, and their implications);
2. System analysis (the Laplace transform and z-transform);
3. Signal analysis (the Fourier transform, frequency spectrum and its computation);
4. Stabilities and their implications (filtering, frequency response, model reduction and operational amplifier circuits);
5. State-variable equations and computer simulations.

The purpose of this text is to provide a common background for subsequent courses in control, communication, electronic circuits, filter design, and digital signal processing. Thus the discussion of some topics, such as modulation, DFT, and Bode plot, is not exhaustive. Some topics such as filter design and the detailed structure of the fast Fourier transform (FFT) are not discussed. (The FFT, however, is extensively used to compute frequency spectra.) The reader is assumed to have had courses in general physics, including simple circuit analysis, and elementary linear algebra. Some knowledge of differential equations is helpful but is not essential.

CHAPTER DESCRIPTIONS

Chapter 1 introduces continuous-time, discrete-time signals and their simple manipulations. The differences between sinusoidal functions and sinusoidal sequences are extensively discussed. The frequency of sinusoidal sequences is then defined and justified. The simplest version of the sampling theorem is discussed. Chapter 2 introduces systems with and without memory. The concepts of state, zero-input response, and zero-state response are introduced. We then introduce the concepts of linearity, time-invariance, and their implications. These concepts are used in Chapter 3 to develop the convolution description for linear time-invariant (LTI) systems. Difference (differential) equations are then developed to describe LTIL (lumped) systems. Numerous examples are used to show how difference (differential) equations are developed. Chapter 4 introduces the Laplace transform, which transforms convolutions and differential equations into algebraic equations. Thus system analysis becomes simple and straightforward. We discuss general forms of the zero-input response and the zero-state response. The concept of complete characterization is introduced in using transfer functions to describe systems. Chapter 5 discusses the discrete counterpart of Chapter 4. In system analysis, only simple signals such as step functions are used, because they are sufficient to reveal characteristics of LTIL systems.

Chapter 6 studies continuous-time signal analysis. The Fourier series is first developed for periodic signals and then extended to the Fourier transform for aperiodic signals. The concept of the frequency spectrum is then introduced and its physical meaning is discussed. The counterpart of Chapter 6 for discrete-time signals is presented in Chapter 7. Computer computation of the discrete-time Fourier transform leads naturally to the discrete Fourier transform (DFT). Computation of the frequency spectrum of continuous-time signals is also discussed. Chapter 8 studies the question of stability, which is essential in designing systems to process signals. It discusses bounded-input bounded-output (BIBO) stability, its conditions (Hurwitz or Schur), and methods of checking stability (the Routh and Jury tests). For BIBO stable systems, we define the frequency response. If a system is not BIBO stable, the system is generally useless in practice and the frequency response is meaningless. Chapter 9 studies the BIBO stability of operational amplifier (op-amp) circuits. It shows that the ideal op-amp model cannot always be used to analyze actual circuits. Model reduction, which is widely used in practice, is discussed using the concepts of BIBO stability and frequency spectrum. Finally, marginal and asymptotic stabilities are introduced and then applied to study Wien-bridge oscillators. The last chapter introduces state-variable equations and their implementations using op-amp circuits and simulations on digital computers. The realization problem, namely to find state-variable equations for transfer functions, is discussed. Transfer functions can then be simulated on computers through state-variable realizations. Appendix A reviews solving linear algebraic equations; it is discussed from the computer computational point of view.

RATIONALE

This text discusses some topics that are not usually covered in signal and system texts. Its arrangement of topics is again not entirely conventional. Therefore it is appropriate to give some explanations.

Many texts discuss first the Fourier transform (signal analysis) and then the Laplace transform (system analysis). To tie in their relationship, these books discuss first two-sided Laplace transform and then reduce it to the (one-sided) Laplace transform. This text discusses first system analysis and then signal analysis for three reasons: (1) The characteristics of LTIL systems are determinable using any signal, in particular, a step function. Thus, signal analysis is not a prerequisite in system analysis. (2) The Laplace transform is simpler and more general than the Fourier transform in system analysis, therefore, there is no need to study the Fourier transform prior to system analysis. (3) Because the Fourier transform can be easily obtained from the (one-sided) Laplace transform and because Laplace transform tables are much more widely available than Fourier transform tables, it is more convenient and efficient to study first the Laplace transform. The application of two-sided Laplace transform is very limited, therefore, this text studies exclusively the one-sided Laplace transform.

The concept of stability is either briefly mentioned or completely omitted in some texts. This is justifiable if systems are built using passive circuit elements, because the systems will be automatically stable, and the discussion becomes unnecessary. However, present-day systems may be built using operational amplifiers or special digital hardware, and the issue can no longer be neglected. If a system is not stable, it may burn out (in the case of electrical systems), disintegrate (in the case of mechanical systems) or overflow (in the case of computer programs). Thus an unstable system is useless in signal processing and its frequency-domain analysis is meaningless. Therefore stability is studied before the concepts of frequency response and filtering are introduced.

The transfer function has been widely used in analysis and design. It describes, however, only the zero-state response of systems; therefore, its employment must be justified. This text introduces the concept of complete characterization. This concept is equivalent to the concepts of controllability and observability in state-variable equations. Although the concepts of controllability and observability are fairly complex, the concept of complete characterization is simple and easily understandable, making it appropriate in texts of this level.

The concept of model reduction is fundamental in engineering. For example, the gain of an amplifier is often reduced to a constant. In spite of its widespread use, the concept is rarely discussed in control, electronics, or other type of texts. In order to discuss this topic, we need the stability concept of systems and the frequency spectrum of signals. Thus, it is logical to discuss it in this text.

The concept, analysis, and design using state-variable equations are not as simple as those using transfer functions; therefore, state-variable equations are introduced in the last chapter. A complete treatment of the topic requires more sophisticated mathematics (eigenvalues, eigenvectors, and similarity transformations) than is used in this text, therefore it is not attempted. Stressed here is its employment on computer simulations and computations, which is pertinent in this computer age.

FEATURES

The following is a listing of some of the features of this text:

1. Continuous-time system and signal analysis, and discrete-time system and signal analysis are developed in parallel. This provides the opportunity of comparing their similarities and differences and enforcing the understanding of either method.

2. It answers questions that are often raised in system and signal analysis, such as

   (a) Why differential (difference) equations are preferable to convolutions in studying lumped systems.

   (b) Why, in using the Laplace transform, the region of convergence can be completely disregarded.

   (c) Why the Laplace transform is used mainly in system analysis, and the Fourier transform is used mainly in signal analysis.

   (d) Why, in signal analysis, signals are often extended to $-\infty$ in time.

   (e) Why, in design, we may use exclusively transfer functions without considering zero-input responses.

3. It discusses practical problems such as

   (a) Difficulties in obtaining impulse responses by measurement.

   (b) Automobile suspension systems and the effect of worn out shock absorbers.

   (c) Collapse of elevated highway and suspension bridge from stability, resonance, and energy points of view.

   (d) How to choose the number of data points in using the DFT (or FFT), and how to choose the integration step size in computer computation.

   (e) Computer computation of the frequency spectrum of continuous-time signals.

   (f) Stability of operational amplifier circuits, and why differentiators may not be built in practice.

   (g) Model reduction and its application to accelerometers and seismometers.

4. It discusses delayed and advanced form difference equations and discusses its use (or not to use).

5. The frequency of digital signals is defined through analog signals and is justified using a camcorder.

6. It discusses the relationships among the continuous-time and discrete-time Fourier series, Fourier transforms, Laplace transform, and z-transform. It also discusses the difference between the discrete-time Fourier transform and discrete Fourier transform (DFT).

7. It stresses basic concepts in filtering: stability and steady-state response.

8. It introduces BIBO stability, marginal stability, and asymptotic stability and their application to practical circuits.

9. It introduces the Routh test and Jury test. The Routh test presented is simpler than the conventional cross-product formulation.

10. It discusses operational amplifier (analog computer) simulation and digital computer simulation of continuous-time systems.

11. The (one-sided) Laplace transform is used to compute the (two-sided) Fourier transform. The (one-sided) z-transform is used to compute the (two-sided) discrete-time Fourier transform.

12. Every concept is illustrated using numerical examples.

THE ISSUE OF USING COMPUTERS

Most system and signal analyses can now be easily carried out using widely available digital computers and software. For example, commands *step* and *fft* in MATLAB yield immediately the unit step response of systems and the frequency spectrum of signals. Therefore, this text does not dwell on computational mechanics; instead it concentrates on fundamental concepts and general properties of signals and systems. However, we believe that solving a number of simple problems by hand will enhance the understanding of a topic and give confidence in using a digital computer. We also believe that only after mastering a topic can one interpret intelligently computer printouts. Therefore, the advent of digital computers should not alter our basic learning process. We should, however, take advantage of digital computers and learn to be proficient in their employment. At present, a large number of software packages are available commercially. This text discusses, at the end of most chapters, the use of the *Student Edition of MATLAB*, The Math Works, Inc., Englewood Cliffs, NJ: Prentice Hall, 1992. Most of the analyses discussed in this text can then be carried out using the commands in MATLAB.

NOTATIONS

This text uses a large number of variables. Some of them are dummy or intermediate variables and are introduced for mathematical necessity. Others are used to denote physical variables. These physical variables can be grouped in two ways: continuous and discrete, time domain and frequency domain. Continuous time is denoted by t, which can assume any real number in $(-\infty, \infty)$.

Discrete time is denoted by k, which can assume only integers. Continuous frequency is denoted by ω, which can assume any real value in $(-\infty, \infty)$; discrete frequency is denoted by m, which can assume only integers. Lowercase letters are used to denote time functions (signals) such as $f(t)$ or $f[k]$. The function $f(t)$ is a continuous-time function (because of t) and $f[k]$ is a discrete-time function (because of k). To stress further the difference, continuous independent variables are enclosed by parentheses and discrete independent variables by brackets. Frequency domain representations of $f(t)$ or $f[k]$ are denoted by $F(\omega)$ or $F[m]$. It is important to remember that t and k are associated with time and ω and m are associated with frequency.

Capital F is used in many ways in this text. $F(s)$ is used in Chapter 4 to denoted the Laplace transform and $F(z)$ in Chapter 5, the z-transform. In Chapter 6, $F(\omega)$ is used to denote the Fourier transform and $\overline{F}(s)$ the Laplace transform. If $F(s)$ were used to denote the Laplace transform as in Chapter 4, then we must use a different notation, such as $\overline{F}(\omega)$, to denote the Fourier transform. In Chapter 6, we encounter mostly the Fourier transform, therefore we choose the simpler $F(\omega)$ (without bar) to represent it. To avoid possible confusion, every notation is explained whenever it first arises in a chapter or a section.

COURSE ADOPTION

This text contains more material than can be taught in a one-semester course. At Stony Brook, we cover Chapters 1 through 6 and Chapters 8 and 9 by skipping asterisked sections and examples. If time permits, we cover Chapter 7, BIBO stability of discrete-time systems (asterisked sections in Chapter 8) or marginal and asymptotic stability (asterisked sections in Chapter 9). Another possibility is to discuss some topics in Chapter 10. Thus, this text can be used in many ways.

In addition, for a one-semester course, the following combinations are possible.

1. Emphasize the continuous-time case by skipping the discrete-time case in Sections 1.4–1.6; 2.7; 3.1–3.3, Chapters 5 and 7.
2. Emphasize the system part by skipping the signal part in Chapters 6 and 7.

The order of signal analysis (Chapters 6 and 7) and system analysis (Chapters 3, 4, and 5) can be reversed by skipping Sections 6.4.1 and 7.3.3. The entire book is also suitable for a two-semester sequence course on signals and systems. A solution manual is available from the publisher.

ACKNOWLEDGMENTS

This edition differs greatly from the first edition. A new chapter on signals is added. The original Chapter 2 on linear algebra is moved to the appendix. All other chapters have been completely rewritten and reorganized. More examples are added and more motivation is provided. Many people helped me in writing this and the first edition of this book. I am grateful to Professors Armen Zemanian and John Murray for the discussion of some systems concepts; to Velio Marsocci, operational amplifiers, and to Stephen Rappaport, modulation problems and general presentation. The notations in Chapter 2 were the result of a suggestion of Professor Rappaport. I am grateful to Professor Eliahu Jury for suggestions in improving the presentation of stability theory. I would also like to thank Ms. Toni Sue Bowins for preparing the photograph in Figure 1.4.

This text was reviewed by many people. The reviewers of this edition are: John Taque, Ohio University; James A. Bucklew, University of Wisconsin; Jorge L. Aravena, Louisiana State University; V. John Matthews, University of Utah; W. T. Baumann, Virginia Polytechnic and State University; and Adrian Papamarcou, University of Maryland. Their thoughtful and critical comments led me to restructure the text and to add the first chapter on signals. I am grateful to Professor Petar Djuric for many ideas of revision after using the first edition of the book, and to Professor Yuan-Hwang Chen, National Sun Yat-Sen University, Taiwan, for reading the entire manuscript of this edition. My special thanks go to Professor Da-Zhong Zheng, a visiting scholar from Tsinghua University, China, who scrutinized every word and every equation of the manuscript and made a great number of suggestions. He also helped me in galley and page proof and proofread the solutions manual. I would also like to thank Mr. Mike Lee and Mr. Cezary Purzynski who helped me in using the computer and various software. It has been a pleasure working with Ms. Emily Barrosse, Senior Acquisition Editor, Laura Shur, Michelle M. Slavin and Joanne Cassetti of Saunders College Publishing. Finally, I would like to thank Ms. Dolores Wolfe of York Production Services who did an excellent job in coordinating the production of this book.

Chi-Tsong Chen
Stony Brook
January 1994

Table of Contents

* May be omitted without loss of continuity.

Signals

1.1 INTRODUCTION

This text studies signals and systems. We encounter both of them daily almost everywhere. When using the telephone, your voice, a signal, is transformed by the microphone, a system, into an electrical signal. That electrical signal is transmitted, maybe through a satellite circulating around the earth, to the other party and then transformed back into your voice. In the process, the signal may have been processed many times by many different systems. In addition to their role in communications, signals can be used in making medical diagnoses and in detecting the presence of an object, such as an airplane or a submarine. For example, the electrocardiogram (EKG) shown in Figure 1.1(a) and the brain waves (EEG) shown in Figure 1.1(b) can be read to determine the heart condition and the state of mind of the patient. Figure 1.2 shows the total number of shares traded daily on the New York Stock Exchange. The trading volume is a signal that, together with other signals such as gross national product (GNP), unemployment rate, and interest rates, can be interpreted to determine the status of the U.S. economy. We also use signals to control our environment and to transfer energy. To control the temperature of a room, we may set the thermostat at 20°C in daytime and, to save energy, 15°C in the evening. Such a control signal is shown in Figure 1.3(a). Electricity is sent to our home in the sinusoidal waveform shown in Figure 1.3(b), which in the United States has peak values of $\pm 110 \times \sqrt{2}$ volts and repeats itself 60 times per second.

All signals in the preceding examples have one independent variable and are called *one-dimensional signals*. Pictures such as the ones in Figure 1.4 are signals with two independent variables; they depend on the horizontal and vertical positions. In this text, our study will focus on signals with one indepen-

Figure 1.1 *Continuous-time signals*

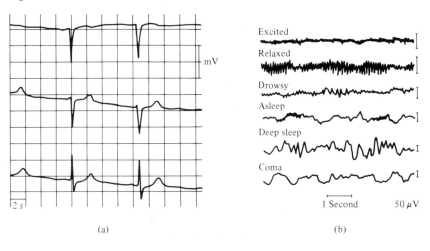

(a)   (b)

dent variable. The independent variable is, unless stated otherwise, time. Thus, signals can be considered as *single-valued functions* of time that carry information. Signals and functions will be used synonymously here.

Signals that are defined at every instant of time, such as those in Figures 1.1 and 1.3, are called *continuous-time signals*. Signals that are defined only at discrete instants of time, such as the one in Figure 1.2, are called *discrete-time signals*. This chapter first studies simple continuous-time signals and their manipulations and then studies discrete-time signals.

Figure 1.2 *Discrete-time signals. (Copyright © 1993 by the New York Times Company. Reprinted with permission.)*

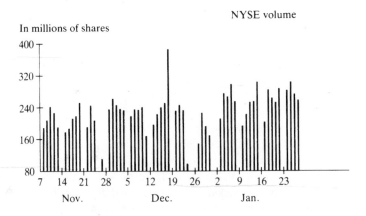

Figure 1.3 *(a) Control signal (b) Household electric voltage*

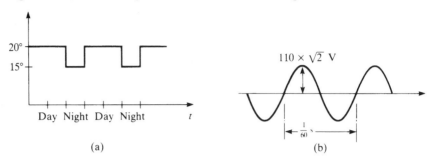

(a)   (b)

1.2   ELEMENTARY CONTINUOUS-TIME SIGNALS

A continuous-time signal will be denoted by $f(t)$ with t ranging from minus infinity to positive infinity, written as $f(t)$ for all t in $(-\infty, \infty)$. In reality, no signal starts from the negative infinite time and lasts forever. However, mathematically it is convenient to consider signals to be defined for all t. Thus, the signal

$$f(t) = 1 \tag{1.1}$$

is a constant for all t as shown in Figure 1.5(a). A trivial signal, it is nevertheless an important one. The time instant $t = 0$ is a relative one; it may be the instant we start to study the signal. Because $t = 0$ is mainly used for reference, we call it the *reference time*.

Figure 1.4 *Picture deblurring (Provided by Drs. G. W. Stroke and M. Halioud)*

Figure 1.5 *(a) Constant function (b) Unit step function (C) Ramp function*

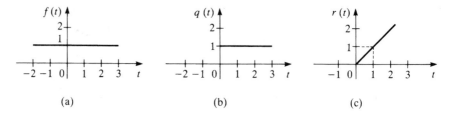

(a)   (b)   (c)

Unit step function The *unit step function*[1] is defined as[2]

$$\maltese \qquad q(t) := \begin{cases} 1 & \text{for } t \geq 0 \\ 0 & \text{for } t < 0 \end{cases} \qquad\qquad (1.2)$$

and is shown in Figure 1.5(b). The value of $q(t)$ at $t = 0$ is defined, for convenience, as 1; in fact, it can be defined as 0, 2, −3.1, or any other value. The information or energy of continuous-time signals depends on magnitude as well as *time duration*. At isolated t, the time width is zero (zero duration), and its energy is zero; thus, its value has no effect on systems. This is the reason that the value of $q(t)$ at $t = 0$ can be arbitrarily defined. For convenience, the step function is referred to the unit step function unless its magnitude is specified other than 1.

Ramp function The *ramp function* is defined as

$$\maltese \qquad r(t) := \begin{cases} t & \text{for } t \geq 0 \\ 0 & \text{for } t < 0 \end{cases} \qquad m = 1 \qquad (1.3)$$

and is shown in Figure 1.5(c). The slope of the ramp is 1, thus the function can be called more informatively as the unit ramp function. Clearly, the function $ar(t)$ is a ramp with slope a. Similarly, we may define the *acceleration function* as t^2 for $t \geq 0$ and 0 for $t < 0$ or, more generally, the *polynomial function* as

$$\maltese \qquad f(t) = \begin{cases} b_0 + b_1 t + \cdots + b_n t^n & \text{for } t \geq 0 \\ 0 & \text{for } t < 0 \end{cases}$$

Clearly, the step, ramp, and acceleration functions are special cases of the polynomial function.

---

[1]The unit step function is generally denoted by $u(t)$. Because $u(t)$ will be reserved to denote the input signal of systems, we use $q(t)$ to denote the step function.

[2]We use $A := B$ to denote that A, by definition, equals B. We use $A =: B$ to denote that B, by definition, equals A.

Sinusoidal function A continuous-time signal $f(t)$ is said to be *periodic* with period P if

$$f(t + P) = f(t) \tag{1.4}$$

for all t. If (1.4) holds, then

$$f(t) = f(t + P) = f(t + 2P) = \cdots = f(t + nP)$$

for every positive integer n. Thus if $f(t)$ is periodic with period P, then it is periodic with period $2P, 3P, \ldots$. The smallest such P is called the *fundamental period* and the function repeats itself every P seconds. The most important periodic signal is the sinusoidal function

$$f(t) = A \sin(\omega t + \theta) \tag{1.5}$$

shown in Figure 1.6, where $A, \omega,$ and θ are called, respectively, the *amplitude, frequency,* and *phase.* The sinusoid is periodic with fundamental period

$$P := \frac{2\pi}{\omega} \tag{1.6}$$

The fundamental period is the smallest time interval for the sinusoid to repeat itself or to complete a cycle; thus, its inverse $f := 1/P$ gives the number of repetitions or cycles in one second. The unit of f is Hz, or cycles per second; f is related to ω, with unit radians per second or rad/s, by $f = \omega/2\pi$. In mathematical equations such as (1.5), frequency ω must be expressed in the unit of rad/s; in writing and conversation, it is, however, much more convenient to express frequency in Hz. In this text, the notation f will not be used, and ω may be expressed in rad/s or in Hz. Thus, if we state that ω is 5 Hz, then $\omega = 5 \times 2\pi = 10\pi$ rad/s. A 1 KHz (1000 Hz) sinusoid has fundamental period 1/1000 second or 1 millisecond; a 20 MHz (20×10^6 Hz) sinusoid has

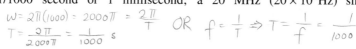

Figure 1.6 *Sinusoid*

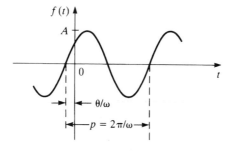

fundamental period $1/(20 \times 10^6) = 50 \cdot 10^{-9}$ second or 50 nanosecond. Note that the preceding discussion applies to cosine functions as well because

To convert to sine:

$A \cos(\omega t + \theta - 90°) = A \sin(\omega t + \varphi)$

$$A\sin(\omega t + \theta) = A\cos\omega t$$

if the phase θ equals 90 degrees or $\pi/2$ rad.

- **Real exponential function—time constant** Consider the exponential function

$$f(t) = e^{at} \tag{1.7}$$

where a is a real constant. It reduces to $f(t) = 1$ if $a = 0$. If a is positive, then e^{at} increases from 0 to infinity as t increases from $-\infty$ to ∞, as shown in Figure 1.7(a). If a is negative, then e^{at} decreases to 0 as t approaches infinity, as shown in Figure 1.7(b). The rate of increase or decrease depends on the magnitude of a. If a is negative, then

$$\frac{f\left(t + \frac{1}{|a|}\right)}{f(t)} = \frac{e^{a\left(t + \frac{1}{|a|}\right)}}{e^{at}} = e^{\frac{a}{|a|}} = e^{-1} = \frac{1}{2.7} = 0.37$$

which means that the value of e^{at} decreases to 37% of its original value whenever the time increases by $1/|a|$. The number $1/|a|$, the inverse of the magnitude of a, is called the *time constant*. Using the time constant, a rough sketch of e^{at} can be easily obtained.

- **Complex exponential functions—positive and negative frequencies** The signals discussed so far are all real-valued functions. We now discuss a complex-valued function. Consider the *complex exponential* function $e^{j\omega t}$ for all t, where ω is a real number. Using *Euler's formula*[3]

$$e^{j\omega t} = \cos\omega t + j\sin\omega t \tag{1.8}$$

where $j = \sqrt{-1}$, we see that $e^{j\omega t}$ is a complex-valued function with real part $\cos \omega t$ and imaginary part $\sin \omega t$. If we plot $e^{j\omega t}$, using time as the horizontal coordinate as in the preceding examples, then the plot will be three dimensional, as shown in Figure 1.8(a), and is difficult to visualize. Therefore, it is customary to plot $e^{j\omega t}$ on a complex plane, as shown in Figure 1.8(b), with t as a parameter on the plot. Because

$$|e^{j\omega t}| = \sqrt{(\cos\omega t)^2 + (\sin\omega t)^2} = \sqrt{1} = 1$$

[3]In honor of Leonhard Euler (1707-1783), a Swiss mathematician.

Figure 1.7 *Real exponential functions*

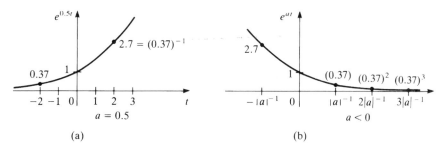

(a)   (b)

for all t, the plot of $e^{j\omega t}$ on a complex plane is a unit circle, as shown. As t
increases, $e^{j\omega t}$ with $\omega > 0$ will rotate around the unit circle counterclockwise.
The larger ω, the faster the rotation. For example, if ω is 2π rad/s or 1 Hz, then
it takes one second to complete one rotation. If ω is 10 Hz, then it takes one-
tenth of a second to complete one rotation.

Now if ω is negative, then $e^{j\omega t}$ will rotate clockwise. Thus, positive and
negative frequencies indicate only different directions of rotation. Both positive and negative frequencies are needed in signal analysis. For example,
Euler's formula in (1.8) implies

$$\cos \omega t = \frac{e^{j\omega t} + e^{-j\omega t}}{2} \qquad \sin \omega t = \frac{e^{j\omega t} - e^{-j\omega t}}{2j} \qquad (1.9)$$

If the negative frequency is not used, then the expression in (1.9) is not possible.

Figure 1.8 *Complex exponential*

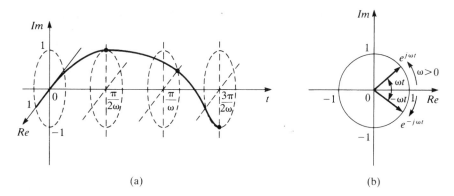

(a)   (b)

EXERCISE 1.2.1

Plot e^{j2t} on the complex plane for $t = 0$, !, $\pi/2$, π, and 4. Do the same for e^{-j2t}.   ■

EXERCISE 1.2.2

Derive (1.9) from (1.8).   ■

Positive-time and negative-time signals A signal $f(t)$ is called a *positive-time* signal if $f(t) = 0$ for $t < 0$ and a *negative-time* signal if $f(t) = 0$ for $t > 0$. Otherwise, it is called a two-sided signal. Clearly, the unit step and ramp functions are positive-time signals. The signals in Figures 1.5(a), 1.6, and 1.7 are two-sided signals. If a signal does not start from $-\infty$, and if we can choose the reference time $t = 0$, then we can choose $t = 0$ as the instant where the signal starts to appear and the signal becomes positive-time. Therefore, it is fair to say that most signals encountered in practice are positive-time signals.

1.2.1   Boundedness in magnitude and infinite time

A continuous-time signal $f(t)$ is said to be *bounded* in a time interval if there exists a constant M such that

$$|f(t)| \le M$$

for all t in the time interval. The specification of the time interval is important. For example, the signal e^{at} with $a < 0$, as shown in Figure 1.7(b), is not bounded in $(-\infty, \infty)$ but is bounded in $[0, \infty)$.[4] If a signal does not grow to positive or negative infinity, then it is bounded.

In reality, it is difficult to generate unbounded signals. For example, operational amplifier circuits can be built to generate step, ramp, sinusoidal, or real exponential functions. However, if the operational amplifier circuits are driven by ±15 V power supplies, then all signals in the circuits are limited to, roughly, 2 volts below the supplied voltage. Thus, in the case of ramp and exponentially increasing functions, the generated signals will *saturate* or remain at 13 volts or the circuits will burn out. Thus, signals encountered in practice are all bounded.

Infinity is a very convenient notion, even though a formidable one, in mathematics. Equations involving infinity often require rigorous justification. In engineering, we shall use its convenience and disregard its mathematical sub-

---

[4]We use $[a,b)$ to denote $a \le t < b$. In other words, the parenthesis does not include equality but the bracket does. Because we can never reach positive or negative infinity, we do not use brackets with infinity.

tleness. More importantly, infinity in engineering need not be infinity in mathematics. For example, 10^{100} seconds, hours, or years from now is still very far away from infinity. However, in some engineering applications, signals may be considered to have reached infinite time in 10 seconds, as is explained in the following.

Consider the positive-time part of the exponential function in Figure 1.7(b). It approaches zero as t approaches infinity. Because the function decreases by 37% whenever time increases by one time constant $1/|a|$ and because

$$(0.37)^5 = 0.007$$

the function decreases to less than 1% of its original value after *five time constants* and may be considered to have reached 0 in some applications in control and filter design. In this case, we may consider

$$\infty \approx 5 \times (\text{time constant})$$

For example, the function $e^{-0.1t}$ may be considered to have reached 0 in $5 \times 1/0.1 = 50$ seconds and the function e^{-5t} in $5 \times 1/5 = 1$ second. Thus, infinity in engineering is a relative one; it can be very long or very short.

EXERCISE 1.2.3

If we consider e^{-bt} with $b > 0$ to have reached 0 in five time constants, what are infinities for $b = 0.01$, 1, and 100?

[**ANSWERS:** $500s$, $5s$, $0.05s$] ■

1.3   MANIPULATION OF CONTINUOUS-TIME SIGNALS

In this section, we will discuss some simple manipulations of signals.

Shifting   Let $f(t)$ be a function, as shown in Figure 1.9(a), and T be a positive number. Then $f(t - T)$ shifts $f(t)$ to the right or *advances* $f(t)$ by T seconds, as shown in Figure 1.9(b). This can be verified by checking a number of t. For example, if $T = 1$, as in Figure 1.9(b), then $f(t - T) = f(-0.5) = 1$ at $t = 0.5$ and $f(t - T) = f(0) = 2$ at $t = 1$ and so forth. Similarly, we can verify that $f(t + T)$ shifts $f(t)$ to the left or *delays* $f(t)$ by T seconds, as shown in Figure 1.9(c). For example, if $f(t) = q(t)$, the unit step function, then we have

$$q(t - T) = \begin{cases} 1 & \text{for } t \geq T \\ 0 & \text{for } t < T \end{cases} \tag{1.10}$$

Figure 1.9   *Shifting*

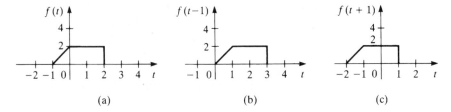

(a)   (b)   (c)

EXERCISE 1.3.1

Consider the function shown in Figure 1.10, or

$$p_a(t) := \begin{cases} 0 & \text{for } t < -a \\ 1 & \text{for } -a \le t \le a \\ 0 & \text{for } t > a \end{cases} \tag{1.11}$$

It is called a *rectangular pulse* with amplitude 1 and width $2a$. If $a = 2$, plot $p_a(t - 1)$, $p_a(t - 2)$, $p_a(t - 3)$, and $p_a(t + 3)$. Which are positive time? Which are negative time?   ■

    Time shifting, especially delay, is common in practice. Music from a compact disc (CD) or an audio tape is a delayed signal. It is not possible to advance a signal in *real time*. For example, no sound will appear before it is uttered. Taped signals, however, can be advanced or delayed with respect to the reference time $t = 0$. Operations from tapes may be called non-real-time operations. Strictly speaking, advanced signals from tapes are still delayed signals.

Flipping   Let $f(t)$ be a function as shown in Figure 1.9(a). Then $f(-t)$ is the flipping of $f(t)$, with respect to $t = 0$, to the negative time, as shown in Figure 1.11(a). If T is positive, then $f(-t - T)$ shifts $f(-t)$ to the left by T seconds, as shown in Figure 1.11(b). This can be verified by checking a number of t. Note that $f(t - T)$ shifts $f(t)$ to the right; whereas $f(-t - T)$ shifts $f(-t)$ to the left. If we write $f(-t - T) = f(-(t + T))$, then $f(-t - T)$ can also be obtained by

Figure 1.10   *Rectangular pulse*

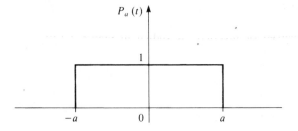

Figure 1.11 *Flipping and shifting*

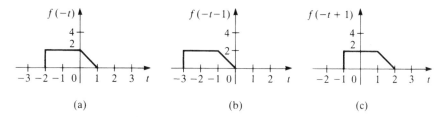

flipping $f(t + T)$ with respect to $t = -T$ to the negative time. This can be veri-fied by comparing the plots in Figures 1.9(c) and 1.11(b). Similarly, $f(-t + T)$ shifts $f(-t)$ to the right by T seconds, as shown in Figure 1.11(c). It can also be obtained by flipping $f(t - T)$ with respect to $t = T$ to the negative time. We can summarize this discussion as follows. Let $f(t)$ be an arbitrary signal and let T be a positive constant. Then we have

1. $f(-t)$ is the flipping of $f(t)$ to the negative time with respect to $t = 0$.
2. $f(t - T)$ is the shifting of $f(t)$ to the right by T seconds.
3. $f(t + T)$ is the shifting of $f(t)$ to the left by T seconds.
4. $f(-t + T)$ is the shifting of $f(-t)$ to the right by T seconds or the flipping of $f(t - T)$ to the negative time with respect to $t = T$.
5. $f(-t - T)$ is the shifting of $f(-t)$ to the left by T seconds or the flipping of $f(t + T)$ to the negative time with respect to $t = -T$.

EXERCISE 1.3.2

Verify that if $f(t)$ is a positive-time signal, then $f(-t)$ is a negative-time signal.

Multiplication Consider two signals $f(t)$ and $h(t)$ defined for all t in $(-\infty, \infty)$. Then their product $g(t) := f(t)h(t)$ forms a new signal. If $h(t)$ is constant with magnitude a and if a is positive and larger than 1, then $g(t)$ is an amplified ver-sion of $f(t)$. For example, if $f(t)$ is as shown in Figure 1.9(a) and $h(t) = 2$, then $g(t) = 2f(t)$ is as shown in Figure 1.12(a), which is obtained by multiply-ing the value of $f(t)$ at every t by 2. If a is positive and smaller than 1, then $f(t)$ will be attenuated. If a is real and negative, then the magnitude of $f(t)$ will be reversed or inverted, as shown in Figure 1.12(b). The brain waves shown in Figure 1.1(b) are the result of amplification. Every signal will be attenuated after traveling a long distance; thus, signals traveling across the Atlantic undersea cables are repeatedly amplified.

If $h(t) = q(t)$, the unit step function defined in (1.2), then $q(t)f(t)$ trun-cates the negative-time part of $f(t)$ to zero. Define

Figure 1.12   *(a) Amplification (b) Inverting*

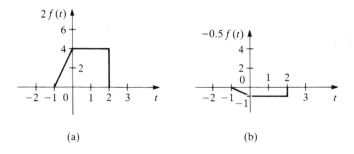

(a)   (b)

$$f_+(t) = \begin{cases} f(t) & \text{for } t \geq 0 \\ 0 & \text{for } t < 0 \end{cases} \tag{1.12a}$$

and

$$f_-(t) = \begin{cases} 0 & \text{for } t > 0 \\ f(t) & \text{for } t \leq 0 \end{cases} \tag{1.12b}$$

Clearly, $f_+(t)$ is a positive-time signal and equals the positive-time part of $f(t)$ whereas $f_-(t)$ is a negative-time signal and equals the negative-time part of $f(t)$, as shown in Figure 1.13. If we use the unit step function, then $f_+(t)$ and $f_-(t)$ can be written simply as

$$f_+(t) = f(t)q(t) \qquad f_-(t) = f(t)q(-t) \tag{1.13}$$

for all t. Thus, multiplication by $q(t)$ can achieve *truncation*.

If $h(t) = p_a(t)$, the rectangular pulse defined in (1.10), the multiplication is called *windowing*. It picks up $f(t)$ only for t in $[-a, a]$ and truncates the rest to zero. The rectangular pulse is the simplest window. More complex windows

Figure 1.13 *Truncations*

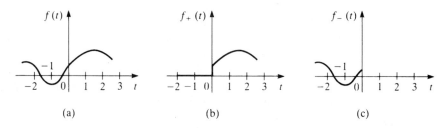

(a)   (b)   (c)

are discussed in most texts on digital signal processing. See, for example, Reference [3]. Note that the pulse can be expressed as

$$p_a(t) = q(t + a)q(-t + a) \qquad (1.14)$$

for all t, where $q(t + a)$ is $q(t)$ shifted to the left by a seconds, as shown in Figure 1.14(a), and $q(-t + a)$ is $q(-t)$ shifted to the right by a seconds, as shown in Figure 1.14(b). Their product clearly yields the rectangular pulse.

If $h(t) = \cos \omega_0 t$, then the multiplication is called *modulation* and $f(t)\cos \omega_0 t$ is called a modulated signal. Signals sent by ratio or TV stations are all modulated signals. Without modulation, human voices may interfere with each other and are difficult to send over long distances. Thus, modulation is essential in communication. This will be discussed further in Chapter 6.

Addition Consider two signals $h(t)$ and $f(t)$ defined for all t. Their sum, $h(t) + f(t)$, forms a new signal, where the summation is carried out point by point at every t. For example, if $h(t) = q(t + a)$ and $f(t) = -q(t - a)$, then their sum yields the rectangular pulse in Figure 1.10, that is,

$$p_a(t) = q(t + a) - q(t - a) \qquad (1.15)$$

Thus, the rectangular pulse can be obtained by multiplying two step functions, as in (1.14), or by subtracting two step functions, as in (1.15). Note that signals received by radios or TV are sums of a number of signals emitted by different stations.

EXAMPLE 1.3.1

Consider the function shown in Figure 1.9(b). We show that it can be decomposed as

$$2r(t) - 2r(t - 1) - 2q(t - 3)$$

where $r(t)$ is the unit ramp (with slope 1) defined in (1.3). To create a ramp with slope 2, we multiply $r(t)$ by 2. Subtracting $2r(t - 1)$ from $2r(t)$ yields a

Figure 1.14 *(a) Shifted step function (b) Flipped and shifted step function*

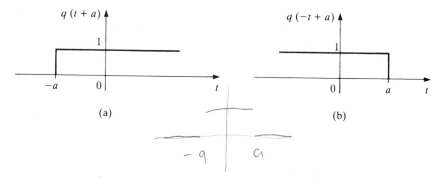

(a)

(b)

ramp with slope 2 in [0,1] and a shifted step function with magnitude 2, as shown in Figure 1.15. Thus $2r(t) - 2r(t-1) - 2q(t-3)$ equals the function in Figure 1.9(b).

EXERCISE 1.3.3

Verify that the function in Figure 1.9(c) can be expressed as

$$2r(t+2) - 2r(t+1) - 2q(t-1)$$

and the function in Figure 1.11(a) can be expressed as

$$2r(-t+1) - 2r(-t) - 2q(-t-2)$$ ■

A signal can be expressed as the sum of a positive-time and a negative-time signal. For example, a constant function $f(t) = 1$ is two-sided, the unit step $q(t)$ is positive time and its flipped version $q(-t)$ is negative time. Then we have

$$\boxed{q(t) + q(-t) = 1}$$ (1.16a)

and, by multiplying $f(t)$ to (1.16a),

$$f(t)q(t) + f(t)q(-t) = f_+(t) + f_-(t) = f(t)$$ (1.16b)

Mathematically speaking, these equations are incorrect, because the equalities do not hold at $t = 0$. Recall that $q(0)$ is defined as 1, thus we have $q(0) + q(-0) = 2 \neq 1$ and $f(0) + f(0) \neq f(0)$, unless $f(0) = 0$. If we modified the definition of $q(0)$ from $q(0) = 1$ to $q(0) = 0.5$, then the equalities hold for all t. This modification, however, is not necessary. Two continuous-time

Figure 1.15 *Subtraction of two ramp functions with slope 2*

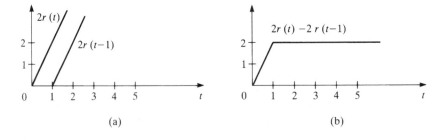

(a)   (b)

signals that differ only at isolate t (zero duration) have the same property and the same effect on systems; therefore, they are considered to be the same signal in engineering. Thus, we write (1.16). This is the same reason that the value of the unit step function $q(t)$ at $t = 0$ can be assigned as 1 or any other value.

1.4 DISCRETE-TIME AND DIGITAL SIGNALS

The signals illustrated in all the preceding figures except Figure 1.2 are functions of every instant of time or a continuum of time and are called continuous-time signals. The signal in Figure 1.2 is defined only at discrete instants of time and is called a discrete-time signal. Although some signals are inherently discrete time, most discrete-time signals are obtained from continuous-time signals by *sampling*. For example, the temperature in a room is a continuous-time signal. If it is *recorded* or *sampled* only at 2 P.M. and 2 A.M. of each day, then the data will appear as shown in Figure 1.16(a). It is a discrete-time signal, for it is defined only at discrete instants of time. The instants at which the data appear are called *sampling instants*; the time intervals between two subsequent sampling instants are called *sampling intervals*. In the discrete-time signal shown in Figure 1.16(a), the sampling intervals are all the same. But sampling intervals may not always be the same. For example, if the highest and lowest temperatures of each day were recorded as shown in Figure 1.16(b), the sampling intervals would be all different. In this text, we study only discrete-time signals with equal sampling intervals, which are called the *sampling periods*.

The amplitude of a discrete-time signal can assume any value in a continuous range. If its amplitude can only assume values from a *finite* set, then it is called a *digital signal*, as shown in Figure 1.17(a). The gap between assumable values is called the *quantization step*. For example, if the temperature is recorded with a digital thermometer with integers read out, then the amplitude can assume only integral values and the quantization step is 1. Thus, digital signals are *discretized* in time and *quantized* in amplitude; discrete-time signals are discretized in time only. The signals processed on digital computers are all digital signals.

In processing and transmission, digital signals are usually expressed in binary form, a string of zeros and ones, as shown in Figure 1.17(b). Since the

Figure 1.16 *Discrete-time signals*

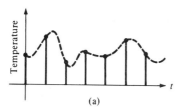

(a)

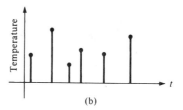

(b)

Figure 1.17 *Digital signals*

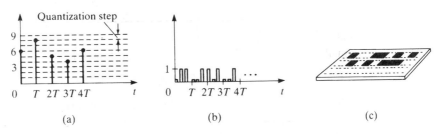

(a) (b) (c)

string of zeros and ones does not resemble, in shape, the original signal waveform, a digital signal may be called a nonanalog signal. [A continuous-time signal usually has the same waveform as the physical variable; thus, it is also called an *analog* signal.] On a notebook computer powered by 5.5 V supply, one may be represented by a voltage range from 3 to 5 volts and zero by 0 to 1 volt. On a compact disc, one and zero are represented by the presence and absence of a tiny bulb, as shown in Figure 1.17(c).

An analog or a continuous-time signal can be transformed into a digital signal using the so-called analog-to-digital (A/D) converter. Conversely, a digital signal can be transformed into an analog signal using a digital-to-analog (D/A) converter. These converters are widely available, and conversions can be easily achieved.

[Because digital signals are represented by strings of zeros and ones, their processing and transmission are much less susceptible to noise than are analog signals.] Coupled with the flexibility and versatility of digital circuits and digital computers, the processing of digital signals has become increasingly popular and important. For example, although human voices are continuous-time signals, they are now being transmitted digitally, as shown in Figure 1.18, to avoid cross-talk, distortion, or transmission noises. Music is now being recorded in digital form on compact discs because it can be reproduced with higher fidelity than with conventional analog recordings.

Because the number of bits used in practice must be finite, every discrete-time signal must be approximated by a digital signal in processing or transmission. Analysis of digital signals is not simple, however. To see why, consider digital signals with magnitude limited to three decimal digits, ranging from 00.1 to 99.9. The sum of 87.0 and 45.8, which is 132.8, and the product of 10.1 and 20.5, which is 207.05, are outside the range of three decimal digits. The prod-

Figure 1.18 *Digital voice communication*

uct $0.1 \times 0.2 = 0.0$ is indistinguishable from the product $0.0 \times 0.2 = 0.0$. Thus, manipulation of digital signals may encounter difficulties. No such problem will arise in discrete-time signals. Therefore, in analysis and design, digital signals are often considered as discrete-time signals. If the quantization step is small, say, 10^{-3}, then the error of approximating a digital signal by a discrete-time signal or vice versa is small and can simply be disregarded. If the quantization step is large and the error cannot be neglected, then the error is analyzed using statistical method. See, for example, Reference [3]. In conclusion, discrete-time signals are approximated by digital signals in implementation, and digital signals are considered as discrete-time signals in analysis and design. To misuse terminology, we sometimes refer to discrete-time signals as digital signals.

1.5 ELEMENTARY DISCRETE-TIME SIGNALS AND THEIR MANIPULATION

Consider an analog signal $f(t)$ for t in $(-\infty, \infty)$. Its sampling with sampling period $T > 0$ yields a discrete-time signal $f(kT)$, where k is an integer ranging from $-\infty$ to ∞. To simplify notations, we define

$$f[k] := f(kT) \qquad (1.17)$$

for all integer k in $(-\infty, \infty)$. When we use brackets, the sampling period T is suppressed. In a real-time operation, such as in a telephone conversation, the sampling period T cannot be disregarded. If $f(kT)$ is stored in a tape and then processed in non-real-time, then $f(kT)$ is simply a *sequence of numbers*; the sampling period T does not play any role and can be assumed simply as 1. Thus, our discussion often disregards T, unless it plays an essential role. For convenience, discrete-time signals are also called *sequences.*

Our discussions of discrete-time signals will be brief, unless one differs substantially from continuous-time signals. As in the analog case, a discrete-time signal $f[k]$ is a positive-time sequence if $f[k] = 0$ for $k < 0$. It is a negative-time sequence if $f[k] = 0$ for $k > 0$.

Consider the sequence

$$f[k] = 1 \qquad (1.18)$$

It equals 1 for all integer k in $(-\infty, \infty)$, as shown in Figure 1.19(a). It is a two-sided sequence and is the discrete counterpart of the constant function defined in (1.1).

Impulse sequence The impulse sequence is defined as

$$\delta[k] = \begin{cases} 0 & \text{for } k \neq 0 \\ 1 & \text{for } k = 0 \end{cases} \qquad (1.19a)$$

Figure 1.19 *(a) Constant sequence (b) Impulse sequence (c) Step sequence*

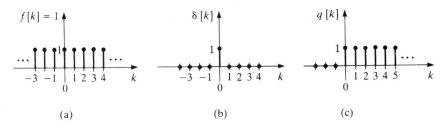

(a)   (b)   (c)

for all k. The sequence has only one nonzero entry at $k = 0$ and is zero elsewhere, as shown in Figure 1.19(b). The impulse sequence is also called the *Kronecker delta sequence.*[5] Its analog counterpart is more complex and will be discussed later, when its application arises.

Let i be a fixed integer. Then $\delta[k - i]$ shifts the nonzero entry of $\delta[k]$ to $k = i$ or

$$\delta[k - i] = \begin{cases} 0 & \text{for } k \neq i \\ 1 & \text{for } k = i \end{cases} \tag{1.19b}$$

Note that there are two integers k and i in (1.19b): i is fixed and k is a variable ranging from $-\infty$ to ∞.

The impulse sequence is useful in representing sequences. For example, consider

$$f[-2] = 4 \quad f[-1] = 0 \quad f[0] = 2 \quad f[1] = 0 \quad f[2] = -5 \quad f[3] = 1$$

This sequence has values 4, 0, 2, 0, -5, and 1, at $k = -2, -1, 0, 1, 2$, and 3, respectively. This sequence can be plotted on a graph or be represented by a table. If it is to be expressed analytically, then we must use impulse sequences. The sequence $f[k]$ equals 4 at $k = -2$ and thus contains $4 \times \delta[k + 2]$. It equals 2 at $k = 0$ and thus contains $2 \times \delta[k - 0]$. Proceeding forward, we can write the sequence[6] as

$$f[k] = 4\delta[k + 2] + 2\delta[k] - 5\delta[k - 2] + \delta[k - 3]$$

Similarly, if we want to express $f[k]$ explicitly in terms of $f[0], f[\pm 1], f[\pm 2], \ldots$, then we write

[5]Named after Leopold Kronecker (1823-1891), a German mathematician.
[6]It is incorrect to write $f[k] = 4 + 2 - 5 + 1$.

$$f[k] = \cdots + f[-2]\delta[k+2] + f[-1]\delta[k+1] + f[0]\delta[k]$$
$$+ f[1]\delta[k-1] + f[2]\delta[k-2] + \cdots$$
$$= \sum_{i=-\infty}^{\infty} f[i]\delta[k-i] \qquad (1.20)$$

This holds for every integer k. Note that, in the summation, k is fixed and i ranges from $-\infty$ to ∞. For example, if $k = 10$, then (1.20) becomes

$$f[10] = \sum_{i=-\infty}^{\infty} f[i]\delta[10-i]$$

Every $\delta[10-i]$ in the summation, as i ranges from $-\infty$ to ∞, is 0 except for $i = 10$; thus, the infinite summation reduces to

$$\cdots + f[8] \times 0 + f[9] \times 0 + f[10] \times 1 + f[11] \times 0 + \cdots = f[10]$$

This verifies the correctness of (1.20). Whenever we have two or more variables, we must keep track of which is fixed and which is running. In (1.20), it is meaningless to keep i fixed. If we vary k and i from $-\infty$ to ∞ at the same time, then confusion will arise.

EXERCISE 1.5.1

Express the following sequences in terms of impulse sequences:

    (a) $f[-5] = -5$, $f[-3] = -3$, $f[-1] = -1$, $f[1] = 1$, and $f[2] = -2$.

    (b) $f[k] = k - 2$, for $k = 0, 1, 3$, and $f[k] = 0$ otherwise.

[**ANSWERS:** (a) $f[k] = -5\delta[k+5] - 3\delta[k+3] - \delta[k+1] + \delta[k-1] - 2\delta[k-2]$. (b) $f[k] = -2\delta[k] - \delta[k-1] + \delta[k-3]$.] ■

EXERCISE 1.5.2

Verify (1.20) for $k = -5$ and 20. ■

EXERCISE 1.5.3

Verify that the <u>constant sequence in (1.18) can be expressed as</u>

$$\boxed{1 = \sum_{i=-\infty}^{\infty} \delta[k-i]} \qquad \text{✳}$$

for all k. ■

● **Unit step sequence** The unit step sequence is defined as

$$q[k] := \begin{cases} 1 & \text{for } k \geq 0 \\ 0 & \text{for } k < 0 \end{cases} \quad \text{or} \quad q[k - i] := \begin{cases} 1 & \text{for } k \geq i \\ 0 & \text{for } k < i \end{cases} \quad \text{(1.21)}$$

and is shown in Figure 1.19(c) for $i = 0$. It is a positive-time sequence. The sequence $q[-k]$ flips $q[k]$ to the negative time with respect to $k = 0$ and is a negative-time sequence. If i is a positive integer, then $q[k - i]$ shifts $q[k]$ to the right i sampling instants. The impulse sequence can be obtained from the step sequence as

impulse sequence →
$$\delta[k] = q[k] - q[k - 1] = q[k]q[-k] \quad \text{(1.22)}$$

(Verify). In the continuous-time case, the value of a function at any isolated time instant is not important. Thus, we can write (1.16) even though it does not hold at $t = 0$. In the discrete-time case, there is no time duration and signals are defined only at isolated time instants. Thus, the value at every instant cannot be disregarded and the discrete counterpart of (1.16a) is

$$q[k] + q[-k] - \delta[k] = 1 \quad \text{(1.23)}$$

for all k. If we do not include the impulse sequence in (1.23), then the equality does not hold at $k = 0$. Alternatively, we have $q[k] + q[-k - 1] = 1$, for all k.

EXERCISE 1.5.4

Let N be a positive integer. Then the sequence

$$p_N[k] := \begin{cases} 1 & \text{for } -N \leq k \leq N \\ 0 & \text{for } k < -N \text{ and } k > N \end{cases} \quad \text{(1.24)}$$

is called a *rectangular sequence* of length $2N + 1$. Such a sequence for $N = 4$ is plotted in Figure 1.20. Show

$$p_N[k] = q[k + N] - q[k - (N + 1)] = q[k + N]q[-k + N]$$

where $q[k]$ is the unit step sequence defined in (1.21). ■

As in the analog case, we may define the unit ramp sequence as

$$r[k] := \begin{cases} k & \text{for } k \geq 0 \\ 0 & \text{for } k < 0 \end{cases} \quad \text{(1.25)}$$

Figure 1.20 *Rectangular sequence*

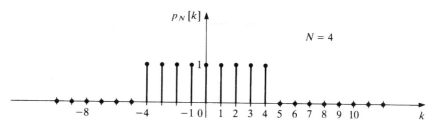

It is a positive-time sequence. The sample of the real exponential function e^{at} is $e^{akT} = (e^{aT})^k =: b^k$, where $b := e^{aT}$. Thus, the real exponential sequence can be defined as

$$f[k] = b^k \qquad (1.26)$$

for all k. The sequence reduces to a constant sequence if $b = 1$. If b is real and larger than 1, the sequence increases from zero to infinity as k increases from $-\infty$ to ∞. If b is real, positive, but smaller than 1, the sequence decreases from infinity to 0 as k increases from $-\infty$ to ∞.

EXERCISE 1.5.5

Plot b^k with $b = -1$ for $k = -1$, 0, 1, 2, and 3. Do the same for $b = 2$ and $b = -0.5$. ■

Let $f[k]$ be an arbitrary sequence and i be a positive integer. Then we have, as in the continuous-time case,

1. $f[-k]$ is the flipping of $f[k]$ to the negative time with respect to $k = 0$.
2. $f[k - i]$ is the shifting of $f[k]$ to the right by i sampling instants.
3. $f[k + i]$ is the shifting of $f[k]$ to the left by i sampling instants.
4. $f[-k + i]$ is the shifting of $f[-k]$ to the right by i sampling instants or the flipping of $f[k - i]$ to the negative time with respect to $k = i$.
5. $f[-k - i]$ is the shifting of $f[-k]$ to the left by i sampling instants or the flipping of $f[k + i]$ to the negative time with respect to $k = -i$.

1.6   SINUSOIDAL SEQUENCE AND ITS FREQUENCY

Consider the analog sinusoid $f(t) = \sin \omega_0 t$. It is periodic with fundamental period $P = 2\pi / \omega_0$, or it repeats itself every P seconds. We sample it with sampling period $T > 0$ to yield

$$f[k] := f(kT) = \sin \omega_0 kT \qquad (1.27)$$

for $k = 0, \pm1, \pm2, \ldots$. This is a discrete-time sinusoid. A discrete-time signal $f[k]$ is said to be periodic with period N, where N is a positive integer, if

$$f[k] = f[k + N] \qquad (1.28)$$

for all integers k in $(-\infty, \infty)$. This definition is similar to the one in (1.4) for the continuous-time case except that the period is now required to be an integer. If (1.28) holds, then

$$f[k] = f[k + N] = f[k + 2N] = \ldots = f[k + nN]$$

for any k and every positive integer n. Thus if $f[k]$ is periodic with period N, it is periodic with period $2N$, $3N$, ... The smallest such N is called the *fundamental period*.

Now we discuss some differences between sinusoidal functions and sinusoidal sequences. The analog sinusoid $\sin \omega_0 t$ is periodic for every ω_0. Its sampled sequence in (1.27), however, is not necessarily periodic for every ω_0 and every sampling period T. Indeed, if $\sin \omega_0 kT$ is periodic, then there exists a positive integer N such that

$$\begin{aligned} \sin \omega_0 kT = \sin \omega_0(k + N)T &= \sin (\omega_0 kT + \omega_0 NT) \\ &= \sin \omega_0 kT \cos \omega_0 NT + \cos \omega_0 kT \sin \omega_0 NT \end{aligned}$$

for all k. This holds if and only if

$$\cos \omega_0 NT = 1 \qquad \text{and} \qquad \sin \omega_0 NT = 0$$

or

$$\omega_0 NT = 2\pi n \qquad \text{or} \qquad \frac{\omega_0 T}{\pi} = \frac{2n}{N}$$

for some integer n. Thus $\sin \omega_0 kT$ is periodic if and only if $\omega_0 T / \pi$ is a rational number. In other words, $\sin \omega_0 kT$ is periodic if and only if there exists an integer n such that

$$N = \frac{2n\pi}{\omega_0 T} \qquad (1.29)$$

is a positive integer. The smallest such N is the *fundamental period* of $\sin \omega_0 kT$. Note that the preceding discussion also applies to $\cos \omega_0 kT$. Here is an example.

EXAMPLE 1.6.1

Consider the sequence $\sin 2k$ or $\cos 2k$. Clearly, we have $\omega_0 T = 2$. Because $\omega_0 T / \pi = 2 / \pi$ is not a rational number, the sequence is not periodic. In this case, there exists no integer n in $N = 2n\pi / 2$ such that N is an integer and the sequence will never repeat itself. The sequence $\sin 0.01\pi k$ is periodic because $\omega_0 T / \pi = 0.01\pi / \pi = 1/100$ is a rational number. Its period is $N = 2n\pi / \omega_0 T = 2n\pi / 0.01\pi = 200n = 200$ by choosing $n = 1$. The sequence $\cos 3\pi k$ is periodic with period $N = 2n\pi / 3\pi = 2n / 3 = 2$ by choosing $n = 3$.

EXERCISE 1.6.1

Which of the following are periodic? If any are, find their periods.

(a) $\sin 0.1k$

(b) $\cos 10.1\pi k$

(c) $\sin 0.1\pi k$

(d) $\cos 3\pi k / 7$

[**ANSWERS:** (a) No, (b) 20, (c) 20, (d) 14.]

In conclusion, $\sin \omega_0 kT$ is not necessarily periodic; it is periodic if and only if $\omega_0 T$ can be expressed as $r \times \pi$, where r is a rational number. The frequency of analog $\sin \omega_0 t$ is ω_0; it equals $2\pi / P$, where P is the fundamental period. Following this, we may define the frequency of digital $\sin \omega_0 kT$ as $2\pi / NT$, where N is the fundamental period. Although this definition is most natural, it is not appropriate, as the following example shows.

EXAMPLE 1.6.2

Consider $\sin 0.1\pi k$ with $T = 1$. If $n = 1$, then

$$N_1 = \frac{2n\pi}{\omega_0 T} = \frac{2n\pi}{0.1\pi} = 20n = 20$$

Thus, the sequence is periodic with period 20 and repeats itself every 20 sampling periods, as shown in Figure 1.21(a). Now consider $\sin 0.3\pi k$ with $T = 1$. The smallest integer n to make

$$N_2 = \frac{2n\pi}{\omega_0 T} = \frac{2n}{0.3} = \frac{20n}{3}$$

an integer is $n = 3$. Thus, $\sin 0.3\pi k$ is periodic with fundamental period 20. It repeats itself every 20 sampling period, as shown in Figure 1.21(b). Thus, both $\sin 0.1\pi k$ and $\sin 0.3\pi k$ have the same fundamental period and, consequently, the same frequency if we adopt the preceding definition. However, the rate of change of $\sin 0.3\pi k$ is three times faster than that of $\sin 0.1\pi k$. Thus, the frequency as defined is not consistent with the usual sense of frequency and should not be used.

If we cannot use the fundamental period to define the frequency of digital $\sin \omega_0 kT$, then how do we define it? We can define it through analog sinusoids. Consider a digital $\sin \omega_0 kT$. We call an analog $\sin \bar{\omega} t$ an *envelope* of $\sin \omega_0 kT$ if the sample of $\sin \bar{\omega} t$ with sampling period T equals $\sin \omega_0 kT$, that is,

$$\sin \omega_0 kT = \sin \bar{\omega} t \Big|_{t=kT} \tag{1.30}$$

Figure 1.21 *Two periodic sequences with the same fundamental period*

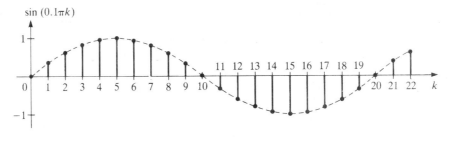

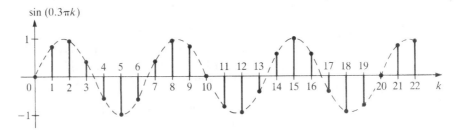

for all k. Note that if $\sin \bar{\omega}t$ is an envelope of $\sin \omega_0 kT$, then $\sin (\bar{\omega} + 2n\pi/T)t$, for every integer n is also an envelope of $\sin \omega_0 kT$. Indeed, because of

$$\sin nk\,2\pi = 0 \quad \text{and} \quad \cos nk\,2\pi = 1 \quad (1.31)$$

for any integers n and k (positive, negative, or zero), we have

$$\sin (\bar{\omega} + \frac{n2\pi}{T})t \bigg|_{t=kT} = \sin (\bar{\omega}kT + 2nk\pi) = \sin \bar{\omega}kT \cos (nk\,2\pi)$$

$$+ \cos \bar{\omega}kT \sin (nk\,2\pi) = \sin \bar{\omega}kT$$

Thus, every $\sin \omega_0 kT$ has infinitely many envelopes. The *primary envelope* of $\sin \omega_0 kT$ is defined as the envelope with the smallest frequency in magnitude or, more precisely, the analog $\sin \omega t$ with

$$-\pi < \omega T \le \pi \quad \text{or} \quad -\frac{\pi}{T} < \omega \le \frac{\pi}{T} \quad (1.32)$$

such that $\sin \omega t \,|_{t=kT} = \sin \omega_0 kT$. Then the frequency of $\sin \omega_0 kT$ is defined as the frequency of the analog $\sin \omega t$. This is stated formally as a definition.

Definition 1.1 The frequency of a digital $\sin \omega_0 kT$, periodic or not, is defined as the frequency of the analog $\sin \omega t$ with ω in (1.32) whose sample, with sampling period T, equals $\sin \omega_0 kT$.

EXAMPLE 1.6.3

The digital $\sin 1.1\pi k$ with $T = 1$ has envelopes $\sin 1.1\pi t$, $\sin 3.1\pi t$, $\sin (-0.9\pi)t$, $\sin 5.1\pi t$, $\sin (-2.9\pi)t$, ... Three of them are plotted in Figure 1.22(a). Note that their intersections yield the sequence $\sin 1.1\pi k$. Among all these envelopes, the smallest frequency in magnitude is -0.9π. Thus, the primary envelope of $\sin 1.1\pi k$ is $\sin (-0.9\pi)t$ and the frequency of $\sin 1.1\pi k$ is -0.9π radians per second.

EXERCISE 1.6.2

Consider $\sin 2k$. Figure 1.22(b) shows the plots of $\sin 8.28t$, $\sin 2t$, and $\sin (-4.28t)$. Do the samples of these three continuous-time sinusoids with sampling period $T = 1$ equal the discrete-time sinusoidal sequence $\sin 2k$? Are they envelopes of $\sin 2k$? Which one is the primary envelope? What is the frequency of $\sin 2k$?

[**ANSWERS:** Yes, yes, $\sin 2t$, 2 rad/s.]

Figure I.22 *Three envelopes of sin 1.1πk and sin 0.6πk*

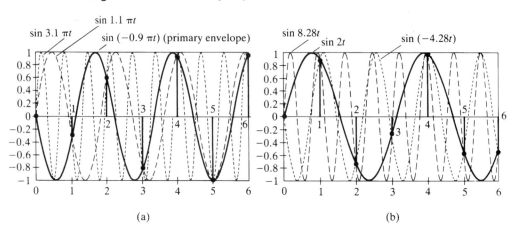

(a)   (b)

If we use Definition 1.1, then the frequency of digital $\sin \omega_0 kT$ is defined whether $\sin \omega_0 kT$ is periodic or not. Before we give a physical justification of the definition, note that the representation of digital $\sin \omega_0 kT$ is not unique. Because of

$$\sin [(\omega_0 T + n\, 2\pi)k] = \sin \omega_0 kT \cos (nk\, 2\pi) + \cos \omega_0 kT \sin (nk\, 2\pi) = \sin \omega_0 kT$$

which follows from (1.31), two different digital $\sin \omega_1 kT$ and $\sin \omega_2 kT$ in fact denote the same sequence if the difference of $\omega_1 T$ and $\omega_2 T$ equals 2π or its multiple. For example, the following digital sinusoidal sequences

$$\sin 7.1\pi k \qquad \sin 5.1\pi k \qquad \sin 3.1\pi k \qquad \sin 1.1\pi k \qquad \sin (-0.9\pi k) \qquad \sin (-3.9\pi k)$$

all denote the same sequence. This can be stated generally as

$$\sin \omega_1 kT = \sin \omega_2 kT \qquad \text{if} \qquad \omega_1 T = \omega_2 T \text{ (modulo } 2\pi) \qquad \textbf{(1.33)}$$

where $\omega_1 T = \omega_2 T$ (modulo 2π) means

$$\omega_1 T - \omega_2 T = n \times (2\pi)$$

for any negative, zero, or positive integer n. It means that if $\omega_1 T$ and $\omega_2 T$ differ by a multiple of 2π, they are still considered equal. In order to eliminate this nonuniqueness, we shall restrict $\omega_0 T$ to lie in an interval of 2π such as

$$(-\pi, \pi], \qquad [0, 2\pi), \qquad \text{or} \qquad [\pi, 3\pi) \qquad \textbf{(1.34)}$$

If we restrict $\omega_0 T$ in $(-\pi, \pi]$, or in (1.32), then the primary envelope of $\sin \omega_0 kT$ is $\sin \omega_0 t$ and the frequency of $\sin \omega_0 kT$ is simply ω_0. Therefore, whenever we are given a digital $\sin \omega_0 kT$ or $\cos \omega_0 kT$, we shall reduce ω_0 to the range in (1.32) by adding or subtracting $2\pi / T$ or its multiples.

EXAMPLE 1.6.4

To find the frequency of $\sin 27k$ with sampling period $T = 0.5$, first we write $\sin 27k$ as $\sin (54kT)$. Its frequency should lie inside $(-\pi / T, \pi / T] = (-2\pi, 2\pi] = (-6.28, 6.28]$. This is not the case; therefore, 54 should be reduced to lie inside the region by subtracting $4\pi = 12.56$ or its multiple. Therefore, we have

$$\sin 54kT = \sin 41.44kT = \sin 28.88kT = \sin 16.32kT = \sin 3.76kT$$

Thus, the frequency of $\sin 27k$ with sampling period $T = 0.5$ is 3.76 rad/*s*.

EXERCISE 1.6.3

Find the frequencies of the following sequences for $T = 1$:

(a) $\sin 4.2\pi k$

(b) $\cos 5.2\pi k$

(c) $\sin 100\pi k$

(d) $\sin 10k$

(e) $\sin (-2.1k)$

(f) $\cos 20k$

Which is not periodic?

[**ANSWERS:** (a) 0.2π rad/*s*. (b) -0.8π rad/*s*. (c) 0, (d) -2.56 rad/*s*. (e) -2.1 rad/*s*. (f) 1.16 rad/*s*. (d), (e), and (f).] ■

EXERCISE 1.6.4

Find the frequencies of $\sin 2.4\pi k$ in rad/*s* and in Hz for $T = 0.1, 0.5, 1, 2,$ and 5.

[**ANSWERS:** 4π rad/*s* = 2 Hz, 0.8π rad/*s* = 0.4 Hz, 0.4π rad/*s* = 0.2 Hz, 0.2π rad/*s* = 0.1 Hz, 0.08π rad/*s* = 0.04 Hz.] ■

1.6.1   Frequencies of analog and digital sinusoids— sampling theorem

Consider an analog sin ωt with frequency ω. The frequency of its sampled sequence is generally not ω. For example, the frequency of analog sin $1.8\pi t$ is 1.8π rad/*s*, but the frequency of its sampled sequence with sampling period $T = 1$

$$\sin 1.8\pi t\big|_{t=kT} = \sin 1.8\pi k = \sin (1.8\pi - 2\pi)k = \sin (-0.2\pi)k$$

is -0.2π rad/*s*. From the preceding discussion, we can establish that the relationship between the frequency of sin ωt and the frequency of its sampled sequence sin ωkT is as shown in Figure 1.23. This relationship can be visualized using a camcorder to shoot a wheel with a mark A on the rim. For convenience, we assume the camcorder shoots one frame per second or, equivalently, the sampling period is 1. If the frequency of the wheel is 0, 1, or 2 Hz, that is, the wheel is stationary or spins one or two complete turns per second, then the wheel on the *tape* (the sampled signal) will remain stationary and its frequency is 0 Hz. If the frequency of the wheel is 0.25 Hz, then its sampled signal or the wheel on the tape will also spin ¼ turn per second, as shown in Figure 1.24(a). If the frequency of the wheel is 0.75 Hz, or the wheel spins ¾ turn per second, then the wheel on the tape will spin ¼ turn per second but in the *opposite* direction, as shown in Figure 1.24(b). Thus, the frequency of the digital signal is −0.25 Hz. If the wheel spins 1.25 Hz or 2.25 Hz, then the wheel on the tape will spin ¼ Hz as shown in Figure 1.24(a). This is a good illustration of the relationship between the frequencies of analog and digital signals. This also shows that the frequency of digital sinusoidal sequences as defined in Definition 1.1 is consistent with our perception of frequency.

The preceding discussion has many practical applications. Consider a wheel that has an etched line, as shown in Figure 1.25. When the wheel is spin-

Figure 1.23 *Frequencies of digital and analog sinusoids*

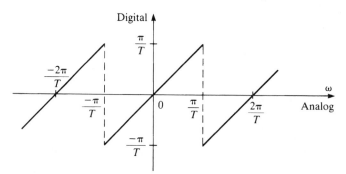

Figure 1.24 *The sampled signals of a spinning wheel, with (a) ω = 0.25, 1.25, 2.25 Hz, . . . , and (b) ω = 0.75, 1.75 Hz, ...*

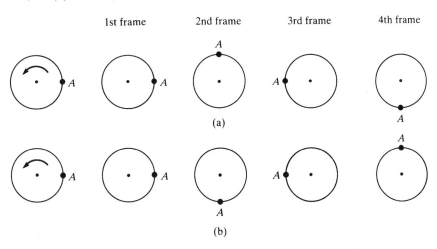

ning at a high speed, it is not possible to see the line. But suppose we aim a timing gun at the line. The timing gun periodically emits pulses of light and acts as a sampling system. If the frequency of sampling $(2\pi / T)$ equals the rotational frequency of the wheel, then the etched line becomes stationary. This technique is used to set the ignition time in automobiles. This is also the basis of the strobing effect. A *stroboscope* is a device that periodically emits flashes of light and acts as a sampling system. The duration of flashes may be in the range of microseconds, and the frequency of emitting flashes can be adjusted. If the frequency of flashing (sampling frequency) is identical to the rotational frequency of an object, the object will appear stationary; if the sampling frequency is slightly smaller than the rotational frequency, then the object will appear to rotate slowly in the opposite direction; and if the sampling frequency is slightly larger, then the object will appear to rotate slowly in the same direction. Thus the stroboscope can be used to observe the vibration or rotation of mechanical objects and to measure the speed of rotation.

Figure 1.25 *Timing gun*

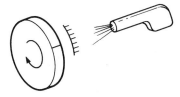

The highest frequency of analog $\sin \omega t$ is infinity; the highest frequency of digital $\sin \omega kT$ is π / T, as was shown in Figure 1.23. There can be a different derivation of this highest frequency. Consider the sinusoidal sequence

$$\cos \omega_0 kT \quad \text{with} \quad \omega_0 = \frac{2\pi}{NT} \tag{1.35}$$

where N is a positive integer. The sequence is periodic with period N because $n = 1$ is the smallest integer to make

$$\frac{2n\pi}{\omega_0 T} = \frac{2n\pi}{2\pi / N} = nN = N$$

an integer. Clearly, the smaller N, the larger ω_0 is. The smallest positive integer is $N = 1$; its corresponding sequence $\cos \omega_0 kT$, however, is a constant sequence as shown in Figure 1.26(a). Thus, its frequency should be zero. The next smallest positive integer is $N = 2$; its corresponding sequence is plotted in Figure 1.26(b). Its value changes from the largest to the smallest or vice versa at every k, thus the sequence $\cos \omega_0 kT$ with $N = 2$ has the largest rate of change and, consequently, the largest frequency that is $2\pi / NT = \pi / T$. Figure 1.26(c) shows $\cos \omega_0 kT$ with $N = 5$; its rate of change and frequency are clearly smaller than those in Figure 1.26(b).

Consider an analog $\sin \omega_0 t$ with fundamental period $P = 2\pi / \omega_0$. Referring to Figure 1.23, we see that the frequency of its sampled sequence with sampling period T may not equal ω_0. However, if

$$\frac{\pi}{T} > \omega_0 \quad \text{or} \quad T < \frac{\pi}{\omega_0} = \frac{P}{2} \tag{1.36}$$

then the frequency of the sampled sequence $\sin \omega_0 kT$ equals ω_0. In this case, it is possible to determine the frequency of $\sin \omega_0 t$ from its sampled sequence. This is the simplest version of the *sampling theorem*. It states that if the sampling period is less than half of the fundamental period of $\sin \omega_0 t$ or,

Figure 1.26 *Three sequences*

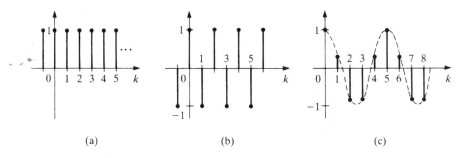

(a) (b) (c)

equivalently, more than two points[7] are sampled in one fundamental period of sin $\omega_0 t$, then the frequency of sin $\omega_0 t$ can be determined from its sampled sequence. If the sampling period is larger than or equal to half of the fundamental period, or two points or less are sampled in one fundamental period, then it is not possible to determine the frequency of sin $\omega_0 t$ from its sampled sequence.

EXERCISE 1.6.5

Find the frequencies of the sampled sequences of sin $2t$ for $T = 0.1$, 0.5, 1, 2, and 5. What is the fundamental period of sin $2t$? Is the answer consistent with the sampling theorem?

[**ANSWERS:** 2 rad/s, 2, 2, -1.14, and -0.512. Fundamental period $= \pi = 3.14$. Yes.]   ∎

A summary of the preceding discussion follows.

1. Not every sinusoidal sequence sin $\omega_0 kT$ or cos $\omega_0 kT$ is periodic. It is periodic if and only if $\omega_0 T$ is a product of π by a rational number. Analog sin $\omega_0 t$ is periodic for every ω_0.   *digital only*

2. Two sinusoidal sequences sin $\omega_1 kT$ and sin $\omega_2 kT$ denote the *same* sequence if ω_1 and ω_2 differ by $2\pi/T$ or its multiple. Analog sin $\omega_1 t$ and sin $\omega_2 t$ denote different sinusoids if $\omega_1 \neq \omega_2$.

3. Every sin $\omega_0 kT$ can be reduced to sin ωkT with

$$-\frac{\pi}{T} < \omega \le \frac{\pi}{T}$$

Then the frequency of sin $\omega_0 kT$ is defined as ω. It is in fact the frequency of analog sin ωt whose sample equals sin $\omega_0 kT$. This is defined for all sin $\omega_0 kT$ whether sin $\omega_0 kT$ is periodic or not.

4. The frequency range of discrete-time sin $\omega_0 kT$ is

$$\boxed{\left(-\frac{\pi}{T}, \frac{\pi}{T}\right] \text{ (in rad/s)} \quad \text{or} \quad \left(-\frac{1}{2T}, \frac{1}{2T}\right] \text{ (in Hz)}} \qquad (1.37)$$

or $(-0.5, 0.5]$ in Hz if $T = 1$. This range is determined solely by the sampling period. The highest frequency of sin $\omega_0 kT$ is π / T. If $T = 1$, the highest frequency is π rad/s or 0.5 Hz. The frequency range of continuous-time sin ωt is $(-\infty, \infty)$; the highest frequency of sin ωt can be as large as desired.

---

[7]For example, we may sample 2.1 points in one period or 21 points in ten periods.

5. The frequency of $\sin \omega_0 t$ can be determined from its sampled sequence if the sampling period T is less than half of the fundamental period of $\sin \omega_0 t$ or two or more points are sampled in one fundamental period.

To conclude this section, let us discuss complex exponential sequences. As with analog, the complex exponential sequence is defined as

$$f[k] = e^{j\omega_0 kT} = \cos \omega_0 kT + j\sin \omega_0 kT \tag{1.38}$$

Clearly, all discussion for sinusoidal sequences applies here. That is, the complex exponential sequence is periodic if and only if $\omega_0 T/\pi$ is a rational number. Because

$$e^{j2n\pi} = \cos 2n\pi + j\sin 2n\pi = 1 \tag{1.39}$$

for every integer n, we have

$$e^{j(\omega_0 + n\frac{2\pi}{T})kT} = e^{j\omega_0 kT} e^{j2nk\pi} = e^{j\omega_0 kT} \tag{1.40}$$

Thus, when we are given an $e^{j\omega_0 kT}$, we can always reduce ω_0 to ω with ω lying inside the range in (1.37). Then the frequency of $e^{j\omega_0 kT}$, whether it is periodic or not, is ω. Complex exponential sequences are fundamental in signal analysis and will be used extensively in Chapter 7.

1.7   SUMMARY

1. Continuous-time signals are defined for all t in $(-\infty, \infty)$. A continuous-time signal $f(t)$ is a positive-time signal if $f(t) \equiv 0$ for $t < 0$ and a negative-time signal if $f(t) \equiv 0$ for $t > 0$. A continuous-time signal can be discontinuous; that is, it need not be a continuous function of t.

2. Discrete-time signals studied in this text are defined only at equally spaced discrete instants of time and can be represented by $f[k] = f(kT)$, where T is the sampling period and k is an integer ranging from $-\infty$ to ∞. The variable k is reserved in this text to denote discrete instants of time. $f[k]$ is positive time if $f[k] \equiv 0$ for all $k < 0$ and negative time if $f[k] \equiv 0$ for all $k > 0$.

3. The value of a continuous-time signal at discrete instants of time (with zero time duration) does not carry any information and has no effect on continuous-time systems. Thus, nonzero time duration is crucial in continuous-time signals. For this reason, the value at $t = 0$ of the unit step function $q(t)$ in (1.2) can be defined as 1, 0, or any number and we can write

$$f(t) = f(t)q(t) + f(t)q(-t) =: f_+(t) + f_-(t) \tag{1.41}$$

even the equality does not hold at $t = 0$.

4. Discrete-time signals are defined only at discrete instants of time, and the concept of time duration does not arise. Thus, the value of $f[k]$ at every k is important. The discrete counterpart of (1.41) is

$$f[k] = f[k]q[k] + f[k]q[-k] - f[0]\delta[k] \qquad (1.42)$$

where $q[k]$ is the unit step sequence and $\delta[k]$ the impulse sequence. The impulse sequence is very useful in representing a sequence. For example, $f[k]$ can be written in terms of $f[0]$, $f[\pm 1]$, $f[\pm 2]$, ..., as

$$f[k] = \sum_{i=-\infty}^{\infty} f[i]\delta[k-i]$$

5. A discrete-time sinusoidal sequence $f[k] = \sin \omega_0 kT$ is periodic if there exists an integer N such that $f[k+N] = f[k]$ for all integer k. The sequence is periodic if and only if $\omega_0 T$ can be expressed as $r \times \pi$, where r is a rational number. However, a continuous-time sinusoidal function $f(t) = \sin \omega_0 t$ is periodic for every ω_0.

6. For sinusoidal sequences, we have

$$\sin \omega_1 kT = \sin \omega_2 kT \qquad \text{if} \qquad \omega_1 = \omega_2 \ (\text{modulo} \ \frac{2\pi}{T})$$

Therefore, every $\sin \omega_0 kT$ can be reduced, by subtracting $2\pi / T$ or its multiple from ω_0, to

$$\sin \omega_0 kT = \sin \omega kT$$

with ω in $(-\pi/T, \pi/T]$. The frequency of $\sin \omega_0 kT$ is then defined as ω, which is, in fact, the frequency of the *analog sinusoid* which has the smallest frequency in magnitude and whose sample yields $\sin \omega_0 kT$. This definition applies to every $\sin \omega_0 kT$, whether it is periodic or not.

7. The frequency of $\sin \omega t$ can be determined from the frequency of its sampled sequence only if the sampling period is less than half of the fundamental period of $\sin \omega t$ or more than two points are sampled from one period of $\sin \omega t$. This is the simplest version of the sampling theorem.

PROBLEMS

1.1 Consider the functions $f_1(t)$ and $f_2(t)$ shown in Figure P1.1.

(a) Plot $f_1(t+1)$, $f_1(t-2)$, $f_1(-t-3)$ and $f_1(-t+4)$.

(b) Establish the relationship between $f_1(t)$ and $f_2(t)$.

1.2 Express $f_1(t)$ and $f_2(t)$ in Problem 1.1 in terms of the unit step and ramp functions defined in (1.2) and (1.3).

1.3 Express the functions in Figure P1.3 in terms of step and ramp functions.

Figure P1.1

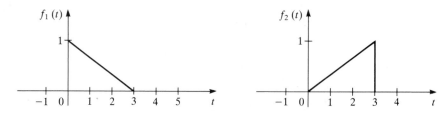

1.3 Express the functions in Figure P1.3 in terms of step and ramp functions.

1.4 A function $f(t)$ is called *even* if $f(t) = f(-t)$, *odd* if $f(t) = -f(-t)$. Which functions in Figure P1.3 are even? Which are odd?

1.5 Consider a function $f(t)$. Define

$$f_e(t) := \frac{f(t) + f(-t)}{2} \qquad f_o(t) := \frac{f(t) - f(-t)}{2}$$

Show that $f_e(t)$ is even and $f_o(t)$ is odd. Show

$$f(t) = f_e(t) + f_o(t)$$

Thus, every function can be expressed as the sum of an even function and an odd function.

1.6 Is $\sin \omega_0 t$ odd or even? How about $|\sin \omega_0 t|$? Are they periodic? What are their fundamental periods?

Figure P1.3

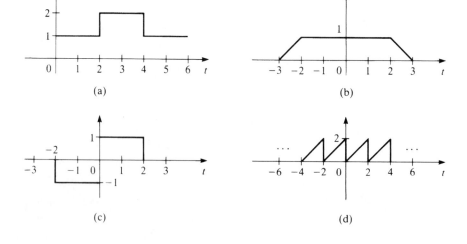

1.7 Roughly sketch the following functions:

(a) $f_1(t) = e^{-0.2t}$

(b) $f_2(t) = \sin 4t$

(c) $f_1(t)f_2(t)$

1.8 Roughly sketch $f_1(t) = 2e^{-0.01t}$ and $f_2(t) = -e^{-3t}$ for $t \geq 0$. Do they approach zero as $t \to \infty$? If a signal is considered to have reached 0 if its magnitude is less than 1% of its largest magnitude and that instant is considered to be ∞, what are the infinite times for $f_1(t)$ and $f_2(t)$?

1.9 Roughly sketch the functions in Figure P1.3 parts (a) and (c) modulated by $\cos 2\pi t$.

1.10 Consider the sequence $f_1[k]$ and $f_2[k]$ shown in Figure P1.10.

(a) Plot $f_1[k + 1]$, $f_1[k - 2]$, $f_1[-k - 3]$, and $f_1[-k + 4]$.

(b) Establish the relationship between $f_1[k]$ and $f_2[k]$.

1.11 Express $f_1[k]$ and $f_2[k]$ in Problem 1.10 in terms of the unit step and ramp sequences defined in (1.21) and (1.25).

1.12 Consider a discrete-time sequence $f[k]$. Show that

$$f_e[k] := \frac{f[k] + f[-k]}{2} \quad \text{and} \quad f_o[k] := \frac{f[k] - f[-k]}{2}$$

are even and odd, respectively. Show also

$$f[k] = f_e[k] + f_o[k]$$

1.13 Is the following true?

$$f[k] = f[k]q[k] + f[k]q[-k]$$

If not, modify it to make it correct.

1.14 Show

$$f[k]\delta[k - m] = f[m]\delta[k - m] \tag{1.43}$$

Figure P1.10

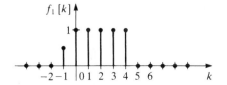
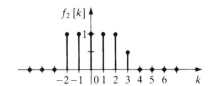

where *m* is a fixed integer. Show also

$$\sum_{k=-\infty}^{\infty} f[k]\delta[k-m] = f[m] \qquad (1.44)$$

This is often referred to as the *sifting property* because the multiplication of the impulse sequence $\delta[k-m]$ and the summation over all k pick up or sift out the value of $f[k]$ at $k = m$. Compare with (1.20).

1.15 Consider the sequence $p_4[k]$ in Figure 1.20. Compute

$$g[k] = \sum_{i=-\infty}^{k} p_4[i]$$

for *k* from −5 to 10. Note that in the summation, *i* is a dummy variable and can be chosen as *m*, *n*, or any variable except *k*. If we use *k*, then confusion will arise. Compute also

$$h[k] = \sum_{i=-\infty}^{\infty} p_4[i]$$

for *k* from −5 to 10.

1.16 Which of $\sin 6.8\pi k$, $\cos 0.2k$, $\sin 4.9k$, $\sin 1.6\pi k$, and $-\sin 1.1k$ are periodic sequences? If any are, find their fundamental periods.

1.17 Are $\sin 6.9\pi k$, $\sin 4.9\pi k$, $\sin 0.9\pi k$, and $-\sin 1.1\pi k$ the same periodic sequence? What is its frequency?

1.18 Find the frequencies of the sequences in Problem 1.16. Assume that the sampling period equals 1.

1.19 Consider $f[k] = (-1)^k$ for all integer *k*. Show that samples of $\cos \pi t$, $\cos(-\pi t)$, and $\cos 3\pi t$, with sampling period $T = 1$, equal $f[k]$ for all *k*. Thus, they are all envelopes of $f[k]$. What is the frequency of $f[k]$?

1.20 Find the frequencies of $\sin 2k$ if $T = 0.01, 0.1, 1, 2$, and 10. Is $\sin 2k$ periodic? If not, how is the frequency defined?

1.21 Consider analog sinusoid $\sin 2.2t$. What are the frequencies of its sampled sequences if $T = 0.01, 0.1, 1, 2$, and 10.

1.22 Consider the complex exponential sequence $f[k] = e^{j\omega_0 k}$ with $T = 1$. Is the sequence periodic if

$$\omega_0 = \frac{2\pi}{N}$$

for $N = 2, 3, 4, 7$, and 10? If yes, what is its period?

1.23 Plot the sequences in Problem 1.22 on the complex plane for $k = 0, 1, \ldots, 10$.

1.24 Write explicitly the values of $f[k]$ in Problem 1.23 for $N = 2, 3,$ and 4. Are they all complex-valued sequences?

1.25 Consider $\sin \omega_0 k$ with $T = 1$. Show that if

$$\omega_0 = \frac{2\pi m}{N}$$

where m and N are integers and are coprime (have no common factors), then $\sin \omega_0 k$ is periodic with fundamental period N. What is the frequency of $\sin \omega_0 k$? Is it equal to $\omega_0 / m = 2\pi / N$?

1.26 Find the largest sampling period of $\sin 10t$ so that the frequency of its sampled sequence equals the frequency of $\sin 10t$. Repeat the problem for $\sin 10\pi t$, $\cos 3t$, and $\sin 1000t$.

1.22:

$$\frac{\omega_0 T}{\pi} = \frac{2n}{N}$$

$$\frac{(2\pi/N)}{\pi} = \frac{2n}{N}$$

$$\frac{2\pi 1}{N\pi} = \frac{2n}{N}$$

$$\frac{2}{N} = \frac{2n}{N}$$

$$2$$

for $N=2$: $\quad 1 = n$

2

Systems

2A

2.1   INTRODUCTION

2

Just like signals, systems exist everywhere. There are many types of systems. A system that transforms a signal from one form to another is called a *transducer*. There are two transducers in every telephone set: a microphone that transforms the voice into an electrical signal and a loudspeaker that transforms the electrical signal into the voice. The temperature control knob in an oven is a transducer that transforms the desired temperature into an electrical voltage. Other examples of transducers are strain gauges, flow-meters, thermocouples, accelerometers, and seismometers. Signals acquired by transducers often are corrupted by noise. The noise must be suppressed or, if possible, completely eliminated. This can be achieved by designing a system called *filter*. If the signal level is too small or too large to be processed by the next system, the signal can be amplified or attenuated by designing an *amplifier*. Motors are systems that are used to drive compact discs, audio tapes, or huge satellite dishes. Generators are also systems; commonly used to generate electricity, they can also be used as power amplifiers in control systems.

In this text, a system will be modeled as a *black box* with at least one input terminal and one output terminal, as shown in Figure 2.1. Note that a terminal does not necessarily mean a physical terminal, such as a wire sticking out of the box. It merely indicates that a signal may be applied or measured from the terminal. We may or may not know the content or internal structure of the box; therefore, we call it a black box. Even if we know the internal structure, the information will not be used in our discussion. Thus, our only access to the system is its input and output terminals. We assume that if an excitation or input signal is applied to the input terminal, a *unique* response or output signal will be measurable or observable at the output terminal. This unique relationship

Figure 2.1 *A black box*

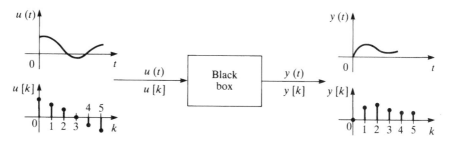

between the excitation and response, input and output, or cause and effect is essential in defining a system. A system is called a *continuous-time system* if it accepts continuous-time signals as its input and generates continuous-time signals as its output. Similarly, a system is a *discrete-time system* if it accepts discrete-time signals as its input and generates discrete-time signals as its output. We use $u(t)$ or $u[k]$ to denote the input and $y(t)$ or $y[k]$ as the output.

A system with one input terminal and one output terminal is called a single-input single-output (SISO) system. A system with two or more input terminals and/or two or more output terminals is called a multiple-input multiple-output (MIMO) system. We study only SISO systems in this text. All results in the SISO case can, however, be extended to the MIMO case. See Reference [4].

The concept of systems is very general and is applicable to the fields of electrical, mechanical, chemical, and aeronautical engineering, bioengineering, economics, and others. For example, the U.S. economy can be considered as a system. Its inputs are government spending, interest rates, and monetary and tax policies; its outputs are the gross national product (GNP), the rate of unemployment, average hourly wage, and consumer price index. The cause and effect relationships of this system are very complex and are not completely known. Furthermore, there are many uncertainties, such as consumer psychology, labor disputes, or international crises. Therefore, the economic system is a very complicated system.

A system can be classified dichotomously as causal or noncausal, linear or nonlinear, time invariant or time varying, lumped or distributed. In this chapter, we will introduce these concepts and discuss their implications. We will study continuous-time systems first, and then we will move on to discrete-time systems.

2.2   SYSTEMS WITH AND WITHOUT MEMORY

A system is called a *memoryless system* if its output $y(t_0)$ depends only on the input applied at time t_0; it is *independent* of the input applied before t_0 or after t_0. This can be stated succinctly as follows: Present output[1] of a memoryless

---

[1]Present time is a relative one; it is the instant at which the output will be computed. Thus, if we want to compute $y(t_1)$, then present time is t_1, past time is $t < t_1$, and future time is $t > t_1$.

system depends only on present input; it is independent of past input and future input. If we apply an input $u(t)$ to an amplifier with gain a, then the output $y(t)$ is

$$y(t) = au(t)$$

This output depends only on the input applied at the same instant, thus the amplifier is a memoryless system. The voltage divider shown in Figure 2.2 is also a memoryless system because its input $u(t)$ and output $y(t)$ are related by

$$y(t) = \frac{R_2}{R_1 + R_2} u(t)$$

Most systems, however, have memory. By this, we mean that the output $y(t_0)$ of a system with memory depends on $u(t)$ for $t < t_0$, $t = t_0$, and $t > t_0$. That is, present output depends on past, present, and future inputs. The system with the following input-output relation

$$y(t) = u(t-1) + 2u(t) - 3u(t+2) \tag{2.1}$$

is such a system. For convenience, every system will be assumed to have memory unless it is explicitly stated as a memoryless system.

 A system is called a *causal* or *nonanticipatory* system if its present output depends on past and present inputs but not on future input. An *integrator* has the following input-output relation

$$y(t_0) = \int_{-\infty}^{t_0} u(t)dt \tag{2.2}$$

Its output does not depend on $u(t)$ for $t > t_0$, thus it is a causal system. A *unit-time delay system* is defined by

$$y(t) = u(t-1) \tag{2.3}$$

Figure 2.2 *Voltage divider*

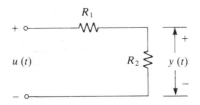

Its output equals the input delayed by one second or shifted to the right by one second, as shown in Figure 2.3(a). Clearly, the unit-time delay system is a system with memory and is causal.

If a system is noncausal or anticipatory, then its present output will depend on the future input. For example, consider the *unit-time advance system* defined by

$$y(t) = u(t + 1) \tag{2.4}$$

Its output equals the input advanced by one second or shifted to the left by one second, as shown in Figure 2.3(b). The output at t_0 depends on the input at $t = t_0 + 1$, a future input. Thus, the unit-time advance system is a noncausal system. Because the output of a noncausal system depends on the input that is not yet applied, a noncausal system can *predict* what will be applied in the future. No physical system has such capability. Therefore, every physical system is causal and causality is a necessary condition for a system to be built or implemented in the real world.

The concept of causality is important in the design of filters. The best or *ideal filter* is noncausal and cannot be implemented using physical elements. Therefore, a major task in filter design is to design a real physical system that approximates as closely as possible a noncausal ideal filter. The concept of causality is implicitly associated with time (past, present, and future); thus, it is an important concept for systems, such as communication and control systems, that operate in real time. There are applications that do not require real-time processing. For example, in oil exploration, seismic data collected from a man-made explosion can be stored in tapes and then processed later. In picture processing, we can utilize all information surrounding a point to make it sharper. This is, in some sense, a noncausal operation. Therefore, noncausal systems can be used in non-real-time processing. In this text, we study mostly causal systems.

Figure 2.3 *(a) Causal system (b) Noncausal system*

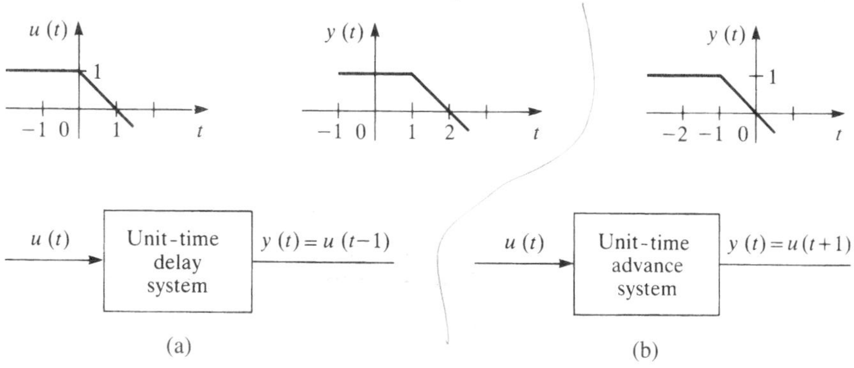

EXERCISE 2.2.1

Are the following systems with or without memory, causal or noncausal?

(a) $y(t) = 5u(t)$

(b) $y(t) = \sin u(t)$

(c) $y(t) = \sin u(t) + \sin u(t - 1)$.

(d) $y(t) = \sin u(t) + \sin u(t + 1)$ ·

(e) $y(t) = (\sin t)u(t)$

(f) $y(t) = (\sin t)u(t - 1)$ ·

[**ANSWERS:** (a) without, causal; (b) without, causal; (c) with, causal; (d) with, noncausal; (e) without, causal; (f) with, causal.]   ■

2.3

4.5

THE CONCEPT OF STATE—SET OF INITIAL CONDITIONS

Present output of a causal system is affected by past input. How far back in time will the past input affect the present output? Generally, the time should go all the way back to minus infinity. In other words, the input from $-\infty$ to time t has an effect on $y(t)$. The tracking of $u(t)$ from $-\infty$ is, if not impossible, very inconvenient. The concept of state can deal with this problem.

Consider the block with mass m sliding on a frictionless floor, shown in Figure 2.4(a). The applied force $u(t)$ is considered as the input and the displacement $y(t)$, the output. Using Newton's law, we have

$$m\frac{d^2y(t)}{dt^2} = u(t) \tag{2.5}$$

Thus, the velocity $v(t)$ of the mass is

$$v(t) := \frac{dy(t)}{dt} = \int_{t_0}^{t} \frac{u(\tau)}{m}d\tau + v(t_0) \tag{2.6}$$

and the displacement is

$$y(t) = \int_{t_0}^{t} v(\tau)d\tau + y(t_0) \tag{2.7}$$

We see that $y(t)$ depends on $y(t_0)$ and $v(t)$ for t in $[t_0, t]$, which, in turn, depends on $v(t_0)$ and $u(t)$ for t in $[t_0, t]$. In other words, in order to determine $y(t)$, for $t \geq t_0$, from $u(t)$, for $t \geq t_0$, we also need the knowledge of the

Figure 2.4 *(a) Mechanical system (b) Electrical system*

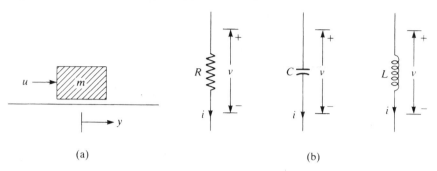

(a)   (b)

displacement $y(t_0)$ and velocity $v(t_0)$. The input $u(t)$ applied before $t = t_0$ is not needed. Therefore, in some sense, the *effect* of $u(t)$ for t in $(-\infty, t_0)$ on $y(t)$, for $t \geq t_0$, is summarized in the two numbers $v(t_0)$ and $y(t_0)$. The set of the two numbers is called the state at time t_0. To state succinctly, *the state summarizes the effect of past input on future output.* We use a column vector $\mathbf{x}(t)$ defined as

$$\mathbf{x}(t) := \begin{bmatrix} y(t) \\ v(t) \end{bmatrix}$$

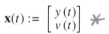

to denote the state. The components of $\mathbf{x}(t)$ are called *state variables*. (A word of notation: scalars such as u and y are denoted by regular-face letters; vectors, by boldface letters.) For easy reference, here is a definition of state.

Definition 2.1 The state $\mathbf{x}(t_0)$ of a system at time t_0 is the information at t_0 that, together with the input $u(t)$, $t \geq t_0$, determines uniquely the output $y(t)$ for all $t \geq t_0$.

Roughly speaking, the state at time t_0 is the set of initial conditions at t_0. In order to determine the output uniquely, we must know the input as well as the initial conditions or initial state. For the same input, different initial states will yield different outputs. Note that how the system reaches the state $\mathbf{x}(t_0)$ is immaterial in determining the future behavior of the system. That is, if several different past inputs $u_1(t)$, $u_2(t)$, $\ldots$, for $t < t_0$, yield the same $\mathbf{x}(t_0)$, the behavior of the system after t_0 will be the same if the same input $u(t)$, $t \geq t_0$, is applied. Using the state, the input-output pair of a system can now be expressed as

$$\left. \begin{array}{l} \mathbf{x}(t_0) \\ u(t),\ t \geq t_0 \end{array} \right\} \rightarrow y(t),\ t \geq t_0 \tag{2.8}$$

We call this the state-input-output pair. The pair means that the output depends on the input and the initial state. For a memoryless system, there is no state, and the state-input-output pair reduces to $\{u(t)\} \rightarrow \{y(t)\}$ or, simply, $\{u(t), y(t)\}$.

We turn now to the state of electrical networks. Consider the electrical components shown in Figure 2.4(b). The voltage $v(t)$ across the resistor and the current $i(t)$ through it are related by $v(t) = Ri(t)$. It is a memoryless element. The relationship between the current and voltage of the capacitor is, if the capacitance C is a constant,

$$i(t) = \frac{d}{dt}[Cv(t)] = C\frac{dv(t)}{dt}$$

Its integration yields

$$v(t) = \int_{t_0}^{t} \frac{i(\tau)}{C} d\tau + v(t_0) \tag{2.9}$$

In order to determine $v(t)$ for $t \geq t_0$ from $i(t)$ for $t \geq t_0$, we need the information of $v(t)$ at $t = t_0$. Thus, $v(t_0)$ qualifies as the initial state of the capacitor. Similarly, the voltage and current of an inductor with inductance L are related by

$$v(t) = \frac{d}{dt}[Li(t)] = L\frac{di(t)}{dt}$$

or

$$i(t) = \int_{t_0}^{t} \frac{v(\tau)}{L} d\tau + i(t_0) \tag{2.10}$$

Thus, the inductor current $i(t)$ qualifies as the state of the inductor. In general, for RLC networks, all capacitor voltages and all inductor currents can be assigned as state variables. For example, for the network in Figure 2.5, the voltages v_1 and v_2 across the two capacitors and the current i_3 through the inductor can be assigned as state variables. Hence, the state of the network is

$$\mathbf{x}(t) = \begin{bmatrix} v_1(t) \\ v_2(t) \\ i_3(t) \end{bmatrix} \tag{2.11}$$

This state $\mathbf{x}(t)$ at $t = t_0$ summerizes the effect of the input applied before t_0 on the output after t_0. Thus, the use of state greatly simplifies the analysis of systems.

The number of state variables in a system may be finite or infinite. A system is a *lumped* system if its state has a finite number of state variables; it is a *distributed* system if its state has infinitely many state variables. The mechanical system in Figure 2.4(a) has two state variables and the network in Figure 2.5 has three state variables; thus, they are lumped systems. Every memoryless system is a lumped system because its number of state variables is zero.

Figure 2.5 *A network with three state variables*

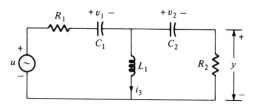

EXAMPLE 2.3.1

Consider the unit-time delay system shown in Figure 2.3(a). In order to determine $y(t)$, for $t \geq t_0$ from $u(t)$, for $t \geq t_0$, we need the information $u(t)$ for *all* t in $[t_0 - 1, t_0)$. Therefore, the state of the system at time t_0 is $u(t)$ for t in $[t_0 - 1, t_0)$. There are infinitely many points between $(t_0 - 1)$ and t_0; therefore, the state consists of infinitely many state variables, and the system is a distributed system.

EXAMPLE 2.3.2

Consider the unit-time advance system shown in Figure 2.3(b). The output $y(t)$, for $t \geq t_0$, can be determined uniquely from $u(t)$, for $t \geq t_0$, without any other information. Thus, the system has no state variable and is lumped.

In conclusion, a system is lumped if the effect of past input on future output can be summarized in a finite number of initial conditions. Otherwise, it is distributed. Continuous-time systems that consist of transmission lines or time-delay elements are distributed systems.

EXERCISE 2.3.1

Which of the systems in Exercise 2.2.1 are lumped.

[**ANSWERS:** (a), (b), (d), and (e).] ■

2.3.1 Zero-input response and zero-state response

The output of a system (with memory) depends on the input and initial state. If the input is identically zero for all $t \geq t_0$, then the output $y(t)$ for $t \geq t_0$ is excited *exclusively* by the initial state $\mathbf{x}(t_0)$. This output is called the *zero-input response* of the system. Conversely, if initial state at t_0 is zero, then the output $y(t)$ for $t \geq t_0$ is excited *exclusively* by the input $u(t)$, for $t \geq t_0$. This output is called the *zero-state response.* Using the state-input-output pair in (2.8), we can express these two responses as

$$\left.\begin{array}{l} \mathbf{x}(t_0) \neq \mathbf{0} \\ u(t) \equiv 0, \text{ for } t \geq t_0 \end{array}\right\} \rightarrow y_{zi}(t), \text{ for } t \geq t_0 \quad \text{(zero-input response)} \qquad \textbf{(2.12a)}$$

and

$$\left.\begin{array}{l} \mathbf{x}(t_0) = \mathbf{0} \\ u(t) \not\equiv 0, \text{ for } t \geq t_0 \end{array}\right\} \rightarrow y_{zs}(t), \text{ for } t \geq t_0 \quad \text{(zero-state response)} \qquad \textbf{(2.12b)}$$

where the subscripts *zi* and *zs* stand for zero input and zero state, respectively.

A system is said to be *initially relaxed* at t_0 if the initial state at t_0 is zero. In other words, if the input applied before t_0 has no more effect on the output after t_0, then the system is relaxed at t_0. For example, consider the unit-time delay system in Figure 2.3(a). For the inputs that end at time t_1, as shown in Figure 2.6, the system is relaxed at any t_0 that is larger than $t_1 + 1$. If a system is initially relaxed at t_0, then the output after t_0 is excited exclusively by the input applied after t_0 and the output is the zero-state response. Note that, if a system is initially relaxed and is causal, no output can appear before the application of an input.

2.4   LINEARITY AND ITS IMPLICATIONS

This section introduces the concept of linearity. Roughly speaking, a system is linear if the property of superposition holds; that is, the output of the system excited by any linear combination of two inputs equals the same linear combination of the outputs excited by individual inputs. First memoryless systems and then systems with memory will be discussed.

For memoryless systems, the output $y(t)$ depends only on $u(t)$. We call $\{u(t), y(t)\}$ or $\{u(t)\} \rightarrow \{y(t)\}$ a permissible input-output pair of a system if the output excited by the input $u(t)$ equals $y(t)$. For example, consider the system with input and output related by the curve shown in Figure 2.7(a). If we apply the input $u = 1$, then the output can be read from the graph as $y = 0.6$. Thus, $\{1, 0.6\}$ is permissible, but $\{1, -1\}$, $\{1, 2\}$, and $\{1, a\}$ for any $a \neq 0.6$ are not. A memoryless system is said to be linear if, for any permissible pairs $\{u_1(t)\} \rightarrow \{y_1(t)\}$ and $\{u_2(t)\} \rightarrow \{y_2(t)\}$, the following pairs

Figure 2.6   *Initially relaxed after* $t_1 + 1$

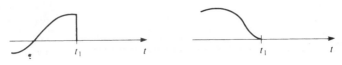

Figure 2.7 *(a) Nonlinear system (b) Nonlinear system (c) Linear system*

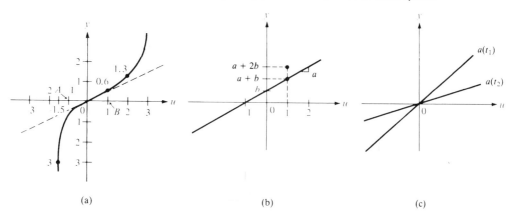

(a) (b) (c)

$$\{u_1(t) + u_2(t)\} \rightarrow \{y_1(t) + y_2(t)\} \quad \text{(additivity)} \quad (2.13)$$

and

$$\{\alpha u_1(t)\} \rightarrow \{\alpha y_1(t)\} \quad \text{(homogeneity)} \quad (2.14)$$

for any α are also permissible. Otherwise, the system is said to be nonlinear. The relationships in (2.13) and (2.14) are called, respectively, the property of additivity and the property of homogeneity. Jointly, they are called the *principle of superposition*. The additivity property *almost* implies the homogeneity property but not conversely. See Problems 2.3 and 2.4. The two relationships in (2.13) and (2.14) can be combined and written as, for any α_1 and α_2,

$$\{\alpha_1 u_1(t) + \alpha_2 u_2(t)\} \rightarrow \{\alpha_1 y_1(t) + \alpha_2 y_2(t)\} \quad (2.15)$$

EXAMPLE 2.4.1

Consider two memoryless systems with input u and output y related by $y(t) = \sin u(t)$ and $y(t) = (\sin t)u(t)$. Because $\sin(u_1 + u_2) \neq \sin u_1 + \sin u_2$, the first system is not linear. Because $(\sin t)(\alpha_1 u_1(t) + \alpha_2 u_2(t)) = \alpha_1(\sin t)u_1(t) + \alpha_2(\sin t)u_2(t)$, the second system is linear.

EXERCISE 2.4.1

Show that $y(t) = u^2(t)$ is not linear and that $y(t) = (\sin t^2)u(t)$ is linear. ∎

A memoryless system with input and output related by $y = au + b$ for some constants a and b, as shown in Figure 2.7(b), is not linear. For example, $\{0, b\}$ and $\{1, a + b\}$ are permissible but $\{0 + 1, b + a + b\} = \{1, a + 2b\}$ is not. A memoryless system is linear if and only if its input and output can be related by

$$y(t) = a(t)u(t) \tag{2.16}$$

for some time function $a(t)$, as shown in Figure 2.7(c).

The definition of linearity for systems with memory is slightly more complicated. The response of a system with memory depends not only on the input but also on the initial state; therefore, the additivity and homogeneity properties must also apply to the initial state. Let $y_i(t)$, for $t \geq t_0$ denote the output excited by the initial state $\mathbf{x}_i(t_0)$ and the input $u_i(t)$ for $t \geq t_0$. Then the system is defined to be *linear* if for any two permissible state-input-output pairs, with $i = 1,2$,

$$\left. \begin{array}{c} \mathbf{x}_i(t_0) \\ u_i(t),\ t \geq t_0 \end{array} \right\} \to y_i(t),\ t \geq t_0$$

the following pairs

$$\left. \begin{array}{c} \mathbf{x}_1(t_0) + \mathbf{x}_2(t_0) \\ u_1(t) + u_2(t),\ t \geq t_0 \end{array} \right\} \to y_1(t) + y_2(t),\ t \geq t_0 \qquad \text{(additivity)} \tag{2.17}$$

and

$$\left. \begin{array}{c} \alpha \mathbf{x}_1(t_0) \\ \alpha u_1(t),\ t \geq t_0 \end{array} \right\} \to \alpha y_1(t),\ t \geq t_0 \qquad \text{(homogeneity)} \tag{2.18}$$

for any α, are also permissible. Otherwise, the system is said to be *nonlinear*. As for memoryless systems, the first property is called the property of *additivity* and the second, the property of *homogeneity*. The property of superposition includes both properties. The two conditions in (2.17) and (2.18) can also be combined into one, as was done in (2.15). Note that the input and output of a linear system with memory cannot be related by a straight line, as in Figure 2.7(c).

EXAMPLE 2.4.2

Consider the network shown in Figure 2.8(a). The input is a current source $i(t)$ and the output is the voltage across the resistor with resistance R and the

Figure 2.8 *Linear system*

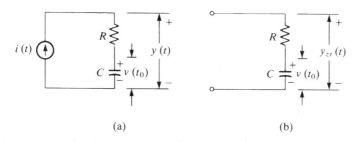

capacitor with capacitance C as shown. The voltage drop across the resistor is $Ri(t)$; the voltage drop across the capacitor is computed in (2.9). Thus, the output is

$$y(t) = Ri(t) + \frac{1}{C}\int_{t_0}^{t} i(\tau)d\tau + v(t_0) \tag{2.19}$$

where $v(t_0)$ is the initial state. Let $y_1(t)$ be the output excited by $\alpha i(t)$ and $\alpha v(t_0)$. Then we have

$$y_1(t) = R\alpha i(t) + \frac{1}{C}\int_{t_0}^{t} \alpha i(\tau)d\tau + \alpha v(t_0) = \alpha y(t)$$

Thus, the property of homogeneity holds. Similarly, the system can be shown to satisfy the additivity property and is, therefore, linear.

EXERCISE 2.4.2

Which of the systems in Exercise 2.2.1 are linear?

[**ANSWERS:** (a), (e), and (f).]

Consider now some important implications of linearity. If $\alpha = 0$ in (2.18), then

$$\left.\begin{array}{c} \mathbf{0} \\ 0, t \geq t_0 \end{array}\right\} \to 0, t \geq t_0$$

It means that if $\mathbf{x}(t_0) = \mathbf{0}$ and if $u(t) = 0$, for $t \geq t_0$, then $y(t) = 0$ for all $t \geq t_0$. Therefore, a necessary condition for a system to be linear is that when the initial state is zero, the output must be identically zero if no input is applied. The output of the system in Figure 2.7(b) is b due to $u = 0$; thus, the system violates the necessary condition and is, therefore, not linear. There is no need to check the superposition property.

If u_1 and u_2 in (2.17) and (2.18) are chosen as zero, then the additivity and homogeneity properties apply only to the zero-input response. If initial conditions in (2.17) and (2.18) are zero, then the two properties apply only to the zero-state response. In conclusion, if a system is linear, then the additivity and homogeneity properties apply to the zero-input response, the zero-state response, and the response excited jointly by nonzero initial state and nonzero input. For example, if the zero-input response of a linear system excited by $\mathbf{x}(t_0)$ is $y(t)$, as shown by the solid line in Figure 2.9(a), then the response excited by $2\mathbf{x}(t_0)$ equals the one denoted by the dashed line, twice the value of $y(t)$. If the system is nonlinear, then the zero-input response excited by $2\mathbf{x}(t_0)$ could be as shown by the dashed line in Figure 2.9(b); it does not resemble $y(t)$ in any way. If the zero-state response of a linear system excited by a step function with magnitude 0.6 is as shown by the solid line in Figure 2.10(a), then the response excited by a step function with magnitude 1.2 equals the one denoted by the dashed line, twice the value of the solid line. If the system is nonlinear, then the response can be as shown by the dashed line in Figure 2.10(b), which does not bear any relation with the solid line (see Example 6.7.1 of Reference [8]).

Another important property of every linear system is that its response can always be decomposed as

Figure 2.9 *Zero-input response (a) Linear system (b) Nonlinear system*

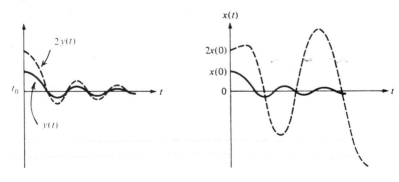

Figure 2.10 *Zero-state response (a) Linear system (b) Nonlinear system*

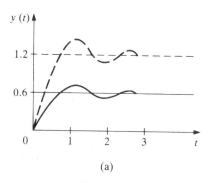
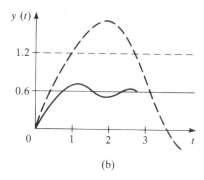

(a) (b)

$$\text{response due to } \begin{cases} \mathbf{x}(t_0) \\ u(t),\, t \geq t_0 \end{cases} = \text{response due to } \begin{cases} \mathbf{x}(t_0) \\ u(t) \equiv 0,\, t \geq t_0 \end{cases}$$

$$+ \text{ response due to } \begin{cases} \mathbf{x}(t_0) = \mathbf{0} \\ u(t),\, t \geq t_0 \end{cases} \quad \text{(2.20a)}$$

or

$$\boxed{\text{total response = zero-input response + zero-state response}} \quad \text{(2.20b)}$$

This follows directly from (2.17) by choosing $\mathbf{x}_1(t_0) = \mathbf{x}(t_0)$, $u_1(t) = 0$ and $\mathbf{x}_2(t_0) = \mathbf{0}$, $u_2(t) = u(t)$. Because of this property, the zero-state response and the zero-input response of linear systems are often studied separately. Their sum then yields the total response. If we are interested in the total response for a nonlinear system, however, it is useless to study the zero-state and zero-input responses, because the total response can be very different from their sum.

EXAMPLE 2.4.3

Consider the linear system illustrated in Figure 2.8(a). If $v(t_0) = 0$, the zero-state response $y_{zs}(t)$ is

$$y_{zs}(t) = Ri(t) + \frac{1}{C}\int_{t_0}^{t} i(\tau)d\tau$$

Now, if $i(t) = 0$, the system reduces to the one shown in Figure 2.8(b). Because of the open circuit, the current is zero, and the zero-input response $y_{zi}(t)$ is

$$y_{zi}(t) = v(t_0)$$

Thus, the total response $y(t)$ is

$$y(t) = y_{zs}(t) + y_{zi}(t) = Ri(t) + \frac{1}{C}\int_{t_0}^{t} i(\tau)d\tau + v(t_0)$$

for $t \geq t_0$, which is indeed the same as (2.19).

2.5   TIME INVARIANCE AND ITS IMPLICATIONS

If the characteristics or properties of a system do not change with time, then the system is said to be time invariant. Otherwise, it is time varying. For a time-invariant memoryless system, no matter at what time an input is applied, the output is always the same. For example, consider $y(t) = \sin u(t)$ studied in Example 2.4.1. If $u(0) = 1$, then $y(0) = \sin 1$. If $u(2) = 1$, then $y(2) = \sin 1$. In general, if $u(t) = 1$, the output is always sin 1 no matter what t is. Thus, the system is time invariant. Consider $y(t) = (\sin t)u(t)$. If we apply $u(t) = 1$ at $t = 0$, then the output is $y(0) = (\sin 0) \times 1 = 0$. If the same input is applied at $t = 1$, then $y(1) = (\sin 1) \times 1 = 0.84 \times 1 = 0.84$. We see that the output is different if the same input is applied at a different time. Thus, the system described by $y(t) = (\sin t)u(t)$ is time varying. A memoryless system is linear if and only if its input and output can be described by $y(t) = a(t)u(t)$. It is time invariant if and only if $a(t)$ is constant, independent of time. For example, the voltage divider in Figure 2.2, with constant R_1 and R_2, is a time-invariant memoryless system. If R_1 and/or R_2 change with time, then the divider is a time-varying system.

A system with memory is time invariant if for any permissible state-input-output pair

$$\left.\begin{array}{r} \mathbf{x}(t_0) = \mathbf{x}_0 \\ u(t),\ t \geq t_0 \end{array}\right\} \rightarrow y(t),\ t \geq t_0$$

and any T, the pair

$$\left.\begin{array}{r} \mathbf{x}(t_0 + T) = \mathbf{x}_0 \\ u(t - T),\ t \geq t_0 + T \end{array}\right\} \rightarrow y(t - T),\ t \geq t_0 + T \quad \text{(time invariance)} \quad \textbf{(2.21)}$$

is also permissible. Otherwise, it is time varying. This means that, for time-invariant systems, if the initial state and the input are the same, no matter at what time they are applied, the output waveform will always be the same. Or, equivalently, if the initial state and the input are shifted by T, the output waveform remains the same except it is shifted by T, as shown in Figure 2.11.

Figure 2.11 *Responses of a time-invariant system*

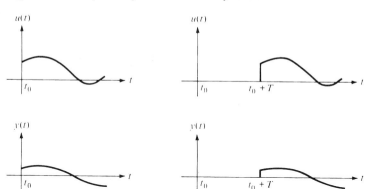

Note that time invariance is defined for systems, not for signals. Signals are mostly time varying; if a signal is time invariant, such as $f(t) = 1$ or $q(t)$ (unit step function), then it is a very simple or a trivial signal. The characteristics or properties of time-invariant systems must be independent of time. For example, the network in Figure 2.5 is time invariant if R_i, C_i, and L_1 are all constants. For time-invariant systems, the initial time t_0 is not critical; therefore, we may assume, without loss of generality, that $t_0 = 0$. The initial time $t_0 = 0$ is a relative one; it is the instant we start to study or to apply an input.

Some physical systems are time-varying systems. For example, a burning rocket is a time-varying system, because its mass decreases rapidly with time. Although the performance of an automobile or a TV set may deteriorate over a long period of time, say five years, its performance and characteristics do not change appreciably over a short period of time. Thus, an automobile can be considered as a time-invariant system in the first year. In fact, a large number of physical systems can be considered as time invariant over a limited period of time.

EXAMPLE 2.5.1

Consider an initially relaxed linear system with two input-output pairs, as shown in Figure 2.12(a) and (b). Is the system time invariant? Is it possible to find the outputs excited by the inputs shown in Figure 2.12(c) and (d)?

The system is linear; therefore, the properties of homogeneity and additivity can be employed. If we double the amplitude of u_1, then the amplitude of the output y_1 will be doubled. Because u_2 is a shifting of $2u_1$ or $u_2(t) = 2u_1(t-1)$, if the system were time invariant, $y_2(t)$ would equal $2y_1(t-1)$. This is not the case; therefore, the system is a time-varying system.

Figure 2.12   *Responses of a time-varying system*

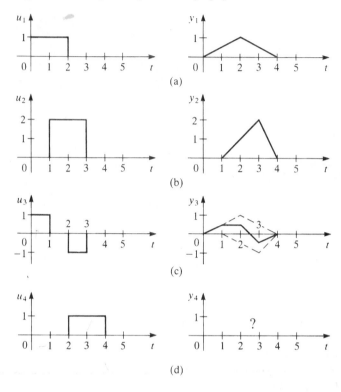

(a)

(b)

(c)

(d)

The input u_3 in Figure 2.12(c) equals $u_1 - 0.5u_2$. Therefore, the output y_3 equals $y_1 - 0.5y_2$, as shown in Figure 2.12(c). Note that the summation is carried out point by point (check the values at $t = 0, 1, 2, 3,$ and 4). In computing the response in Figure 2.12(c), we use only the superposition property; time shifting is not used. The input u_4 in Figure 2.12(d) is the shifting of u_1 to the right by 2 units of time or $u_4(t) = u_1(t - 2)$; however, the output $y_4(t)$ may not equal $y_1(t - 2)$ because the system is time varying. Therefore, it is not possible to find y_4 from the two input-output pairs given.

EXERCISE 2.5.1

Which of the systems in Exercise 2.2.1 are time invariant?

[**ANSWERS:** (a), (b), (c), and (d).]

2.6   LINEAR TIME-INVARIANT LUMPED SYSTEMS

The response of a linear time-invariant lumped (LTIL) system can always be decomposed into the zero-state response and zero-input response. In this section, we will discuss some general properties of these two responses. Note that we may choose $t = 0$ as the initial time for LTIL systems.

Consider the LTIL network shown in Figure 2.5, its state is defined in (2.11). Let y_i, for $i = 1$, 2, and 3, respectively, be the zero-input response excited by the following three initial states

$$\mathbf{x}_1 = \begin{bmatrix} 1 \\ 0 \\ 0 \end{bmatrix} \qquad \mathbf{x}_2 = \begin{bmatrix} 0 \\ 1 \\ 0 \end{bmatrix} \qquad \mathbf{x}_3 = \begin{bmatrix} 0 \\ 0 \\ 1 \end{bmatrix} \tag{2.22}$$

Then, the zero-input response excited by any initial state can be expressed as a linear combination of y_i. Indeed, we can express any initial state as

$$\mathbf{x} = \begin{bmatrix} \alpha_1 \\ \alpha_2 \\ \alpha_3 \end{bmatrix} = \alpha_1 \mathbf{x}_1 + \alpha_2 \mathbf{x}_2 + \alpha_3 \mathbf{x}_3 \tag{2.23}$$

Then, the zero-input response excited by $\mathbf{x}$ is

$$y(t) = \alpha_1 y_1(t) + \alpha_2 y_2(t) + \alpha_3 y_3(t) \tag{2.24}$$

for all $t \geq 0$. This fact can be extended to the general case as follows. If the LTIL system has a set of n initial conditions, then every zero-input response can be expressed as a linear combination of n zero-input responses, as in (2.24), and each of the n zero-input responses is excited by setting one initial condition to 1 and the remaining initial conditions to 0, as in (2.22). Note that a system is lumped if it has a finite set of initial conditions, it is distributed if it has an infinite set of initial conditions. Thus, the preceding discussion applies only to linear time-invariant lumped (LTIL) systems. For LTI distributed systems, the situation is more complicated.

We will now discuss the zero-state response of LTIL systems, considering first memoryless systems and then systems with memory. Every LTIL memoryless system is completely determinable from a single input-output pair. That is, if we know the output excited by, say, $u = 1$ is 2, then we can *predict* the output excited by any other input. Indeed, every LTIL memoryless system is describable by $y = au$. If $u = 1$ excites $y = 2$, then $a = 2$. Thus, the output excited by any u is simply $y = 2u$. Surprisingly, we have a similar situation for systems with memory. That is, if we know the zero-state response of an LTIL system excited by a single arbitrary input, then we can predict the zero-state response of the system excited by any input. An example can illustrate the basic idea: Consider an LTIL system with the input-output pair shown in Figure 2.12(a) and

Figure 2.13 *Responses of an LTIL system*

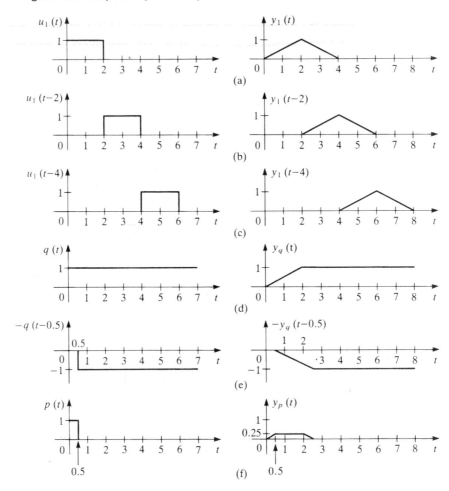

repeated in Figure 2.13(a). As was discussed in Figure 2.12, if the system is
time varying, then there is no way to know the output y_4 excited by the input u_4
shown in Figure 2.12(d). However, if the system is time invariant, then the output y_4 is simply the shifting of y_1 to the right two units of time, as shown in
Figure 2.13(b). If we shift u_1 two more units of time, then the output is
correspondingly shifted, as shown in Figure 2.13(c). We repeat the process to
infinite time.[2] The sum of all inputs yields the unit step function $q(t)$ shown in

---

[2]The process must be extended all the way to infinity. If we stop the process at a very large but finite time, then the subsequent discussion will not hold. Here we use the ambiguity of infinity to
achieve our purpose.

Figure 2.13(d), where the corresponding output $y_q(t)$ is also computed by adding up all preceding outputs. Now we shift $q(t)$ to the right 0.5 second and reverse the magnitude, as shown in Figure 2.13(e), to yield $-q(t - 0.5)$. The corresponding output is $-y_q(t - 0.5)$. The summation of $q(t)$ and $-q(t - 0.5)$ yields the pulse $p(t)$ with width 0.5, as shown in Figure 2.13(f). The corresponding output $y_p(t)$ is also plotted; it is obtained by adding $y_q(t)$ and $-y_q(t - 0.5)$ point by point. In conclusion, from the input-output pair shown in Figure 2.13(a), we can compute the zero-state response excited by a pulse; the width of the pulse can be any value, for example, 0.5 or 0.01. Note that it is difficult to obtain graphically the response of the pulse $p(t)$ from that of $u_1(t)$ without going through the unit step function.

Now consider the arbitrary input $u(t)$ shown in Figure 2.14. The input signal can be approximated by a sequence of pulses, as shown. Thus, using the additivity, homogeneity, and shifting properties, we can obtain the zero-state response of the system excited by $u(t)$. This justifies the assertion that the zero-state response of an LTIL system excited by any input can be computed or predicted from the zero-state response of a single arbitrary input. In Chapter 3, we shall use the preceding idea to develop mathematical equations to describe LTIL systems, and then we will formally establish the assertion in Chapter 4.

In conclusion, because of all the nice properties discussed so far, it is possible to develop a complete theory for LTIL systems. The theory for LTI distributed systems is not as simple. This text studies LTIL systems almost exclusively.

EXERCISE 2.6.1

Consider a linear time-invariant (LTI) system. The zero-state response of the system excited by the unit step function is shown in Figure 2.15(a). This output is called the *unit step response*.  Verify the input-output pairs shown in Figure 2.15(b) and (c). ■

Figure 2.14 *Input signal*

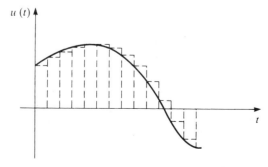

Figure 2.15   *Some input-output pairs*

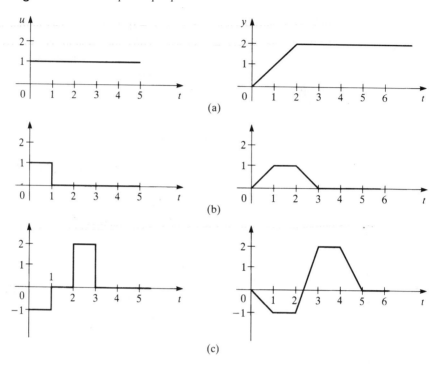

(a)

(b)

(c)

2.7   LINEAR TIME-INVARIANT LUMPED DISCRETE-TIME SYSTEMS

A.5

In this section, we will develop the discrete counterpart of LTIL analog systems. Because most concepts can be applied directly to discrete-time systems, the discussion will be brief.

A system is called a discrete-time system if an input sequence $u[k]$ will produce or excite a unique output sequence $y[k]$. In other words, if the input and output of a system are discrete-time signals, then the system is a discrete-time system. Let us look at how a discrete-time system can arise in practice.

EXAMPLE **2.7.1**

Consider a savings account in a bank. For convenience of discussion, we assume that the monthly interest rate is $r\%$ and the interest is added to the principal at the first day of each month (that is, the interest is *compounded* monthly). We assume that money is deposited into the account only on the first day of each month. Let $u[k] := u(kT)$, $k = 0, 1, 2, \ldots$, with T = one month, be the amount of money deposited at the kth month, and $y(t)$ be the total amount of money in the account at time t. The relationship between the deposit (input) and the money in the account (output) is shown in Figure 2.16(a) and (b). The input $u(t)$ is zero except at $t = kT$, $k = 0, 1, 2, \ldots$. The output is defined and nonzero at all times, as shown. Thus the system is, strictly speaking, a continuous-time system. However, because the output changes value only at $t = kT$, there is no loss of information to consider the sequence $y[k] := y(kT)$, shown in Figure 2.16(c), as the output. Thus, the savings account can be considered as a discrete-time system with $u[k]$ as the input and $y[k]$ as the output.

A discrete-time system is called a memoryless system if its output $y[k]$ depends only on $u[k]$. Otherwise, it is a system with memory. Therefore, the output of a discrete-time system with memory may depend on past, present, and future inputs. A discrete-time system is said to be causal or nonanticipatory if its output at time instant k does not depend on the input applied after k. In other words, for a causal discrete-time system, we have

$$y[k] = f(u[j], \text{ for } j \leq k) \tag{2.25}$$

where f is some function and k and j are integers. For discrete-time systems with memory, we define the state $\mathbf{x}(k_0)$ as the information that, together with $u[k]$, $k \geq k_0$, uniquely determines $y[k]$ for $k \geq k_0$. This is denoted as

$$\left.\begin{array}{l} \mathbf{x}[k_0] \\ u[k], k \geq k_0 \end{array}\right\} \rightarrow y[k], k \geq k_0 \tag{2.26}$$

If the number of components or the state variables in $\mathbf{x}[k]$ is finite, the system is a lumped system. Otherwise, it is a distributed system.

A discrete-time system is linear if for any permissible pairs

$$\left.\begin{array}{l} \mathbf{x}_i[k_0] \\ u_i[k], k \geq k_0 \end{array}\right\} \rightarrow y_i[k], k \geq k_0 \tag{2.27}$$

Figure 2.16 *A savings account*

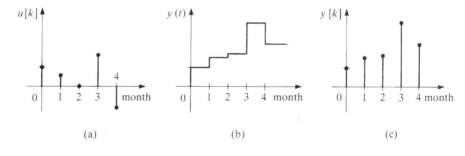

(a)   (b)   (c)

with $i = 1, 2$, the following pairs

$$\left.\begin{array}{l} \mathbf{x}_1[k_0] + \mathbf{x}_2[k_0] \\ u_1[k] + u_2[k],\ k \geq k_0 \end{array}\right\} \rightarrow y_1[k] + y_2[k],\ k \geq k_0 \quad \text{(additivity)} \quad \textbf{(2.28)}$$

and with any constant α,

$$\left.\begin{array}{l} \alpha\mathbf{x}_1[k_0] \\ \alpha u_1[k],\ k \geq k_0 \end{array}\right\} \rightarrow \alpha y_1[k],\ k \geq k_0 \quad \text{(homogeneity)} \quad \textbf{(2.29)}$$

are also permissible. These two conditions are similar to (2.17) and (2.18). Thus, the concept of linearity in the discrete-time case is identical to the concept in the continuous-time case. As in the continuous-time case, the response of every linear discrete-time system can be decomposed as

$$\boxed{\text{total response} = \text{zero-state response} + \text{zero-input response}} \quad \textbf{(2.30)}$$

A discrete-time system is time invariant if for any permissible state-input-output pair

$$\left.\begin{array}{l} \mathbf{x}[k_0] = \mathbf{x}_0 \\ u[k],\ k \geq k_0 \end{array}\right\} \rightarrow y[k],\ k \geq k_0 \quad \textbf{(2.31)}$$

Figure 2.17 *Two zero-state input-output pairs*

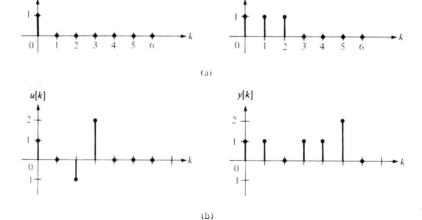

(a)

(b)

Figure 2.18 *A zero-state input-output pair*

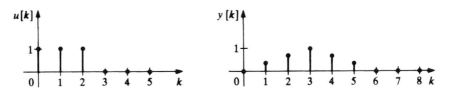

and any integer k_1, the pair

$$\left.\begin{array}{r} \mathbf{x}[k_0 + k_1] = \mathbf{x}_0 \\ u[k - k_1], \ k \ge k_0 + k_1 \end{array}\right\} \rightarrow y[k - k_1], \ k \ge k_0 + k_1 \quad \text{(time invariance)}$$

is also permissible. Otherwise, it is time varying. Thus, for a time-invariant system, the response will be the same no matter when the input and the initial state are applied. Consequently, we may assume without loss of generality that $k_0 = 0$.

The savings account problem discussed in Example 2.7.1 is linear if the interest rate r is independent of the amount of money in the account. If the interest rate is 0% if the account is below, say, \$500.00; 5¼% if between \$500.00 and \$5,000.00; and 7% if above \$5,000.00, then the account is a nonlinear system. If the interest rate r is fixed, then it is a time-invariant system. The interest rate of a money-market account is floating, changing with the prime rate. Therefore, a money-market account is a time-varying system.

In the savings account, the knowledge of $y[k_0]$ together with $u[k]$, $k \ge k_0$ determines uniquely $y[k]$ for $k \ge k_0$. Thus, the state of the system consists of only $y[k_0]$. Therefore, the system has one state variable and is a lumped system. In conclusion, then, the savings account with a fixed interest rate is a linear time-invariant lumped system.

As in the continuous-time case, the zero-state response of any LTIL discrete-time system excited by any input sequence can be obtained from any single input-output pair. For example, if the zero-state response of an LTIL system excited by the impulse sequence $\delta[k]$ is as shown in Figure 2.17(a), then the zero-state response of the system excited by the input sequence shown in Figure 2.17(b) can be obtained as shown.

EXERCISE 2.7.I

Verify the output shown in Figure 2.17(b). If the system is linear but not time invariant, can you determine the output?

[**ANSWER:** No.] ■

EXERCISE 2.7.2

Consider an LTI system with the input-output pair shown in Figure 2.18. What is the output of the system excited by the impulse sequence $\delta[k]$?

[**ANSWER:** $y(0) = 0$, $y[k] = 1/3$, $k = 1, 2, 3$, and $y[k] = 0$, for $k > 3$.]   ■

2.8   ## SUMMARY

1. A system is defined as a black box whose output $y(t)$ is uniquely determined by the applied input $u(t)$. It is memoryless if $y(t_0)$ depends only on $u(t_0)$; it is a system with memory if $y(t_0)$ depends on $u(t)$ for all t. The system is causal if $y(t_0)$ depends on $u(t)$ for $t \leq t_0$.

2. The state of a system at time t_0 summarizes the effect of the input $u(t)$, for $t < t_0$, on the output $y(t)$, for $t \geq t_0$. Thus, $y(t)$ for $t \geq t_0$ can be determined uniquely from the input applied on and after t_0 and the inital state at t_0; the knowledge of the input applied before t_0 is not needed. The output excited exclusively by the initial state is called the zero-input response, and the output excited exclusively by the input is called the zero-state response.

3. A system is called a lumped system if its state consists of a finite number of state variables or initial conditions. It is a distributed system if its state consists of an infinite number of state variables.

4. A system is linear if additivity and homogeneity properties hold. These properties apply to the zero-state response, zero-input response, and the response excited jointly by the input and initial conditions. If a system is linear, then

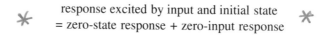

$$\text{response excited by input and initial state} = \text{zero-state response} + \text{zero-input response}$$

This equality does not hold if the system is nonlinear.

5. A system is time invariant if the time-shifting property holds; that is, if the excitation is shifted by T, the output will be shifted by T.

6. Roughly speaking, if the zero-state response of a linear time-invariant lumped (LTIL) system excited by an arbitrary input is known, then the zero-state response of the system excited by any other input can be computed from the known input-output pair. Because of this and other nice properties, it is possible to develop a complete theory for LTIL systems. This text focuses on LTIL systems and only touches on LTI distributed systems.

PROBLEMS

2.1 Consider a linear time-invariant lumped system. Let y_i be the output excited by the input u_i and initial state $x_i(0)$, $i = 1,2,3$. If $x_1(0) = x_2(0) = x_3(0) = a \neq 0$, which of the following statements are correct?

(a) If $u_3 = u_1 + u_2$, then $y_3 = y_1 + y_2$.

(b) If $u_3 = 0.5 \times (u_1 + u_2)$, then $y_3 = 0.5 \times (y_1 + y_2)$.

(c) If $u_3 = u_1 - u_2$, then $y_3 = y_1 - y_2$.

[**ANSWERS:** (a) incorrect, (b) correct, (c) incorrect. Note that initial states must also meet the additivity and homogeneity properties.]

2.2 In Problem 2.1, if $x_1(0) = x_2(0) = x_3(0) = \mathbf{0}$, which of the three statements are correct? Give your reasons.

[**ANSWERS:** All are correct.]

2.3 Consider a system whose input and output are related by

$$y(t) = \begin{cases} u^2(t)/u(t-1) & \text{if } u(t-1) \neq 0 \\ 0 & \text{if } u(t-1) = 0 \end{cases}$$

Show that the system satisfies the homogeneity property but not the additivity property. This supports the assertion that the homogeneity property does not imply the additivity property.

2.4 Show that if $\{u_1 + u_2\} \to \{y_1 + y_2\}$, then

$$\{nu\} \to \{ny\}$$

for any integer n, and that

$$\left\{\frac{n}{m}u\right\} \to \left\{\frac{n}{m}y\right\}$$

for any integer n and any nonzero integer m. Thus, the additivity property implies $\{\alpha u\} \to \{\alpha y\}$ for any rational number α. This supports the assertion that the additivity property almost implies the homogeneity property.

2.5 (a) Show the equivalenc of (2.15) and the set of (2.13) and (2.14).

(b) Combine Equations (2.17) and (2.18) into one equation.

2.6 Discuss whether or not each of the following equations is memoryless, linear, time-invariant, and causal:

(a) $y(t) = -2 + 3u(t)$ *w/o mem ; causal ; not lin ; t. inv.*

(b) $y(t) = u^2(t)$ *w/o mem ; causal ; not lin , t. inv.*

(c) $y(t) = u(t)u(t-1)$ *w/mem; causal ; not lin ; t. van.*

(d) $y(t) = tu(t)$ *w/o mem ; causal ; lin ; t.invariant;*

(e) $y(t) = u(-t)$ *w/ mem ; partial non- causal ; lin ; t. inv.*

(f) $y(t) = u(t/2)$ *w/ mem , partial non causal . lin ; t. inv*

(g) $y(t) = u(2t)$ *w/o mem; partial causal ; lin ; t. inv.*

(h) $y(t) = \int_{t_0}^{t} u(\tau)d\tau + y(t_0)$ *w/mem ; nonlin ; t.invar; noncausal*

(i) $y(t) = \int_{t_0}^{t} \tau u(\tau)d\tau + y(t_0)$ *w/mem ; lin ; t. var ; noncausal*

2.7 Consider the system shown in Figure P2.7, in which the diodes are assumed to be ideal. Are the input u and output y linearly related? Is it true that a linear system must consist of only linear components?

2.8 Consider the network shown in Figure P2.8. If the element T is a 10 ohm resistor, is the network linear? time invariant? lumped? How many state variables are there in the system?

2.9 Consider again the network in Figure P2.8. If the element T is a tunnel diode with the characteristic shown in Figure P2.9, is the system linear? time invariant? lumped?

2.10 Consider a system whose input $u(t)$ and output $y(t)$ are related by

$$y(t) = \begin{cases} u(t) & \text{for } t \le b \\ 0 & \text{for } t > b \end{cases}$$

for a fixed b. This is called a *truncation* operator because it truncates the input after time b. Is it memoryless? linear? time invariant? lumped?

[**ANSWERS:** Yes, yes, no, yes.]

$x(t) dt$

x^2

$\dfrac{}{2}$

Figure P2.7

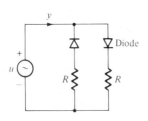

$$(u_1 + u_2)(v_1 + u_2)$$
$$u_1 u_1 + u_1 u_2 + u_1 u_2 + u_2 u_2^g$$
$$= u_1^2 + 2u_1 u_2 + u_2^2$$

Figure P2.8

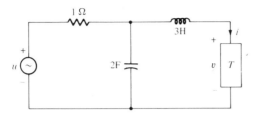

2.11 Consider a system whose input and output are related by

$$y(t) = \operatorname{sgn} u(t) := \begin{cases} 1 & \text{for } u(t) > 0 \\ 0 & \text{for } u(t) = 0 \\ -1 & \text{for } u(t) < 0 \end{cases}$$

Is it memoryless? linear? time invariant? lumped?

2.12 Consider a linear system. Its zero-state responses due to u_1 and u_2 are shown in Figure P2.12(a). Is the system time invariant? Can you find the zero-state responses due to the inputs u_3, u_4, and u_5 shown in Figure P2.12(b)?

2.13 A thermostat is designed to operate as follows: If the temperature falls below the desired temperature by, say, 3 degrees, the thermostat will turn on a burner. When the temperature reaches the desired temperature, it will turn off the burner. Show that its characteristic can be modeled as shown in Figure P2.13. Is it linear? Is it time invariant?

2.14 Consider the system shown in Figure P2.14. Is it linear? time invariant? lumped?

2.15 Consider an LTI continuous-time system. Suppose its unit step response (the zero-state response excited by the unit step function) is as shown in Figure P2.15(a). What are the zero-state responses excited by the inputs shown in Figure P2.15(b), (c), and (d)?

Figure P2.9

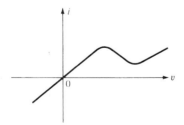

Figure P2.12

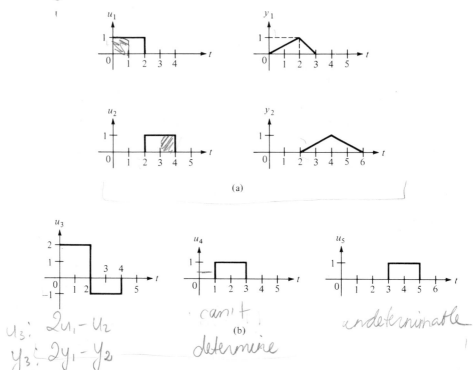

(a)

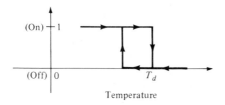

$u_3: \ 2u_1 - u_2$

$y_3: \ 2y_1 - y_2$

can't
determine

(b)

undeterminable

Figure P2.13

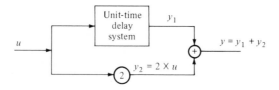

Figure P2.14

Figure P2.15

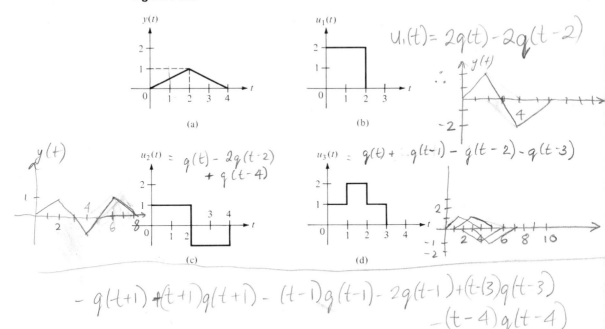

(a)   (b)   (c)   (d)

$$u_1(t) = 2q(t) - 2q(t-2)$$

$$u_2(t) = q(t) - 2q(t-2) + q(t-4)$$

$$u_3(t) = q(t) + q(t-1) - q(t-2) - q(t-3)$$

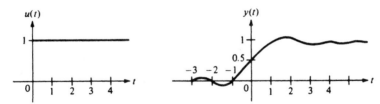

$$-q(t+1) + t(t+1)q(t+1) - (t-1)q(t-1) - 2q(t-1) + (t-3)q(t-3) - (t-4)q(t-4)$$

2.16 The input-output pair shown in Figure P2.16 is the unit step response of an ideal low-pass filter. The filter is known to be linear and time invariant. Is it causal?

2.17 (a) Consider a system with its input and output related by the characteristic shown in Figure P2.17(a). If $u(t) = \sin t$, what is its output? Is the system memoryless? lumped? time invariant? (This system is called a *half-wave rectifier*.)

(b) Consider a system with its input and output related by the characteristic shown in Figure P2.17(b). If $u(t) = \sin t$, what is its output? Is the system memoryless? lumped? time invariant? (This system is called a *full-wave rectifier*.)

Figure P2.16

Figure P2.17

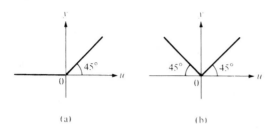

(a)   (b)

2.18 Consider an LTI discrete-time system. Suppose the zero-state response excited by the impulse sequence $\delta[k]$ is $h[k] = 0.5k$, $k = 0, 1, 2, \dots$. What are the zero-state responses excited by the input sequences shown in Figure P2.18?

Figure P2.18

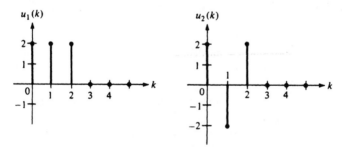

3

Convolution, Difference and Differential Equations

57

3.1   LINEAR TIME-INVARIANT SYSTEMS WITH MEMORY

In this chapter, we will develop mathematical equations to describe linear time-invariant (LTI) systems. The response of such systems can always be decomposed as

> total response = zero-state response + zero-input response

Furthermore, the zero-state response and the zero-input response can be studied separately and independently. We will develop an equation to describe the zero-state response and another to describe the total response.

Every LTI system without memory can be described by $y(t) = au(t)$. This equation is applicable at every instant of time; thus, the equation is often written simply as $y = au$ without specifying explicitly the time. The output of an LTI system with memory depends on present input, past input, and initial conditions. For different t, the segment of the past input will be different. Therefore, the time must appear explicitly in mathematical equations that describe LTI systems with memory.

In this chapter, both the discrete-time and continuous-time systems will be studied. Because the former is considerably simpler than the latter, we will begin with the former. Because we will study only time-invariant systems, we may assume without loss of generality that the initial time is zero, that is, $k_0 = 0$ in the discrete-time case and $t_0 = 0$ in the continuous-time case.

3.2   LTI DISCRETE-TIME SYSTEMS— IMPULSE RESPONSE SEQUENCES

13

In this section, we will develop a mathematical equation, called the discrete convolution, to describe LTI discrete-time systems. The description will be developed from a special input-output pair. No knowledge of the internal structure of systems is required. However, the description describes only the zero-state response. Therefore, before the application of the special input, the system must be assumed to be initially relaxed.

Consider an LTI discrete-time system with input $u[k]$ and output $y[k]$. Let us first look at a special input-output pair. Let the input be the impulse sequence defined in (1.19), or

$$u[k] = \delta[k] = \begin{cases} 1 & \text{for } k = 0 \\ 0 & \text{for } k \neq 0 \end{cases}$$

Then the excited output is called the *impulse response sequence* or *discrete-time impulse response* or, simply, *impulse response*.[1] We use $h[k]$ to denote the impulse response. Therefore, we have

$$\boxed{h[k] = \text{impulse response} := \text{output excited by } \delta[k]} \qquad (3.1)$$

Certainly, different LTI discrete-time systems have different impulse responses. If a system is causal, then the output will not appear before the application of an input. Thus, a necessary and sufficient condition for a system to be causal is

$$\boxed{h[k] = 0 \qquad \text{for } k < 0} \qquad (3.2)$$

Here are some examples of impulse responses.

EXAMPLE 3.2.1

Consider the savings account studied in Example 2.7.1. If the interest rate is 0.5% per month and is compounded monthly, what is its impulse response?

If we deposit one dollar at the beginning of a month (that is, $u[0] = 1$) and nothing thereafter ($u[k] = 0$, $k = 1, 2, \ldots$), then $y[0] = u[0] = 1$ and $y[1] = 1 + 0.005 \times 1 = 1.005$. Because the money is compounded monthly, we have

$$y[2] = y[1] + 0.005y[1] = y[1] \times (1 + 0.005) = (1.005)^2$$

and, in general,

$$y[k] = (1.005)^k$$

Because the input $u[k]$ is an impulse sequence, the output is, by definition, the impulse response. Thus, we have

---

[1]The impulse response will also be defined for continuous-time systems. The heading, discrete-time, will be dropped when no confusion will arise.

$$h[k] = (1.005)^k \qquad \text{for } k = 0, 1, 2, \ldots \tag{3.3}$$

This is the impulse response of the system. Because the system is causal, we have $h[k] = 0$, for $k < 0$.

EXAMPLE 3.2.2

Consider the systems shown in Figure 3.1. Each system consists of a nonzero gain and a unit-time delay element, that is, $f[k] = ae[k]$ and $y[k] = f[k-1]$. The output in Figure 3.1(a) is fed positively back into the input to yield $e[k] = u[k] + y[k]$; therefore, the system is called a <u>positive feedback system</u>. The system in Figure 3.1(b) is called a <u>negative feedback system</u> because the output is fed negatively back into the input to yield $e[k] = u[k] - y[k]$.

Now let us compute the impulse response of the positive feedback system. The impulse response describes only the <u>zero-state response; therefore, in its computation, the system is implicitly assumed to be initially relaxed or</u> $u[k] = e[k] = f[k] = y[k] = 0$ for $k < 0$. Now, if $u[k] = \delta[k]$, then the response of the system is

k	0	1	2	3	4	5
$u[k]$	1	0	0	0	0	0
$y[k] = f[k-1]$	0	a	a^2	a^3	a^4	a^5
$e[k] = u[k] + y[k]$	1	a	a^2	a^3	a^4	a^5
$f[k] = ae[k]$	a	a^2	a^3	a^4	a^5	a^6

It is obtained as follows: The input $u[k]$ is an impulse sequence; therefore, its values at $k = 0, 1, 2, \ldots$ are $1, 0, 0, \ldots$. Because the system is initially relaxed, we have $f[-1] = 0$ and, consequently, $y[0] = f[-1] = 0$. The sum of $u[0]$ and $y[0]$ yields $e[0] = 1$. Because $f[k] = ae[k]$, we have $f[0] = a \times 1 = a$. Once $f[0]$ is computed, we have $y[1] = f[0] = a$. We then compute $e[1], f[1]$, and then $y[2]$. Proceeding forward, the table can be

Figure 3.1 *(a) Positive feedback system (b) Negative feedback system*

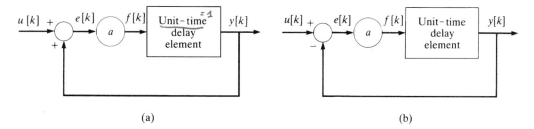

(a)   (b)

completed. The output excited by $u[k] = \delta[k]$ is, by definition, the impulse response $h[k]$ of the system. Thus, we have

$$h[0] = 0 \quad h[1] = a \quad h[2] = a^2 \tag{3.4a}$$
$$h[3] = a^3 \quad \cdots \quad h[i] = a^i \quad \cdots$$

This is the impulse response of the positive feedback system in Figure 3.1(a). Using the notation of (1.20), we can express the impulse response as

$$h[k] = \sum_{i=1}^{\infty} a^i \delta[k - i] \tag{3.4b}$$

EXERCISE 3.2.1

Show that the impulse response of the negative feedback system in Figure 3.1(b) is

$$h[0] = 0 \quad h[1] = a \quad h[2] = -a^2 \quad h[3] = a^3 \quad \cdots \quad h[i] = (-1)^{i-1}a^i$$

or

$$h[k] = \sum_{i=1}^{\infty} (-1)^{i-1} a^i \delta[k - i] \qquad \blacksquare$$

3.2.1   Discrete Convolutions

Consider an LTI discrete-time system with impulse response $h[k]$, $k = 0, 1, 2, \ldots$. We show that the zero-state response of the system excited by any input sequence is determinable from $h[k]$. Let $u[k]$, $k = 0, 1, 2, \ldots$ be an arbitrary input sequence, as shown in Figure 3.2(a). The input can be decomposed into a sequence of inputs, as shown in Figure 3.2(b), (c), and (d), where each input consists of only one nonzero entry at all k. Using (1.19) the inputs in Figure 3.2(b), (c), and (d) can be represented as $u[0]\delta[k - 0]$, $u[1]\delta[k - 1]$, and $u[2]\delta[k - 2]$. Note that $u[i]\delta[k - i]$ is zero at all integer k except at $k = i$, where its magnitude is $u[i] \times 1 = u[i]$. Thus, the input $u[k]$ can be written as

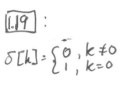

$$u[k] = u[0]\delta[k] + u[1]\delta[k - 1] + u[2]\delta[k - 2] + \cdots$$
$$= \sum_{i=0}^{\infty} u[i]\delta[k - i] \tag{3.5}$$

as in (1.20).

If a system is linear and time invariant, then its output excited by $u[k]$ equals the sum of the outputs excited by $u[i]\delta[k - i]$, for all integer $i \geq 0$. Let

Figure 3.2 *Decomposition of an input sequence*

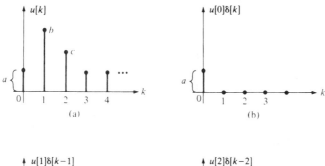

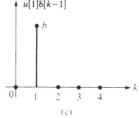

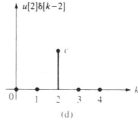

$$\delta[k] \rightarrow h[k]$$

denote that the input $\delta[k]$ excites the output $h[k]$. Then we have

$$\delta[k - i] \rightarrow h[k - i] \quad \text{(time invariance)} \quad ✳$$
$$u[i]\delta[k - i] \rightarrow u[i]h[k - i] \quad \text{(homogeneity)} \quad ✳$$

and[2]

$$\sum_{i=0}^{\infty} u[i]\delta[k - i] \rightarrow \sum_{i=0}^{\infty} u[i]h[k - i] \quad \text{(additivity)} \quad ✳$$

Note that $u[i]$ in the preceding equations is considered as a constant because it is independent of k. Thus, if the system is linear and time invariant, the zero-state output $y[k]$ excited by the input $u[k]$ equals

$$y[k] = \sum_{i=0}^{\infty} h[k - i]u[i] \qquad \textbf{(3.6)}$$

This summation contains $u[i]$ with i ranging from zero to infinity; therefore, $y[k]$ depends on past input $u[i]$, $i < k$, present input $u[k]$, and future input

---

[2]Wherever the sign of infinity appears in an equation, the questions of convergence and the validity of its operations may arise. See Reference [18]. In this text, we will not be concerned with these mathematical subtleties and will proceed intuitively.

$u[i], i > k.$ Although this appears to contradict the causality assumption of the system, it does not. Recall that $h[k] = 0$ for $k < 0$. This implies $h[k - i] = 0$ for $i > k$. Therefore, the upper limit of the summation in (3.6) can be replaced by k and (3.6) reduces to

① *discrete convolution*

$$y[k] = \sum_{i=0}^{k} h[k - i]u[i] \qquad (3.7)$$

for $k = 0, 1, 2, \ldots$. This is called the *discrete convolution*. It gives the zero-state response of the system excited by the input $u[k], k = 0, 1, 2, \ldots$. Because it relates the input and output, it is also called an *input-output description* of the system. This description is developed without using any information of the system other than linearity, time invariance, and causality. Therefore, it is a general formula applicable to *any* system so long as it is linear, time invariant, causal, and initially relaxed at $k = 0$. It is applicable to lumped as well as distributed systems.

An alternative form of (3.7) can be developed. Define $i' := k - i$. Then $i = k - i'$ and (3.7) can be written as

$$y[k] = \sum_{i=0}^{k} h[k - i]u[i] = \sum_{i'=k}^{0} h[i']u[k - i'] = \sum_{i'=0}^{k} h[i']u[k - i']$$

Because i' is a dummy variable, it can be replaced by any symbol. If i' is replaced by i, then the equation becomes

②

$$y[k] = \sum_{i=0}^{k} h[k - i]u[i] = \sum_{i=0}^{k} h[i]u[k - i] \qquad (3.8)$$

for $k = 0, 1, 2, \ldots$. Thus, the discrete convolution has two equivalent forms. Either form can be used.

Equation (3.8) is an algebraic equation and its computation is simple and straightforward. By direct substitution, we have

$$\begin{aligned}
y[0] &= h[0]u[0] \\
y[1] &= h[1]u[0] + h[0]u[1] \\
y[2] &= h[2]u[0] + h[1]u[1] + h[0]u[2]
\end{aligned} \qquad (3.9)$$
$$\vdots$$

Thus, if the impulse response of a system is known, its zero-state output excited by any input can be computed from (3.8) or (3.9).

Discrete convolution is often written more generally as

$$y[k] = \sum_{i=-\infty}^{\infty} h[k-i]u[i] =: h[k] * u[k] \qquad (3.10a)$$

$$= \sum_{i=-\infty}^{\infty} u[k-i]h[i] =: u[k] * h[k] \qquad (3.10b)$$

for all integer k in $(-\infty, \infty)$. If $u[k] = 0$ for $k < 0$ and $h[k] = 0$ for $k < 0$, then (3.10) reduces to (3.8). If the input is applied from $k = -\infty$, then we must use (3.10). Note that every system is assumed to be initially relaxed at $k = -\infty$. Convolution is often denoted by an asterisk; it has the commutative property, that is, $h[k] * u[k] = u[k] * h[k]$. ✴

EXAMPLE 3.2.3

Consider the savings account discussed in Example 3.2.1. If we deposit $u[0] = \$100.00$, $u[1] = -\$50.00$ (withdraw), $u[2] = \$200.00$, $u[3] = \$0$, and $u[4] = \$50.00$, what is the total amount of money in the account on the first day of the fifth month? Note that $k = 0$ is the first month. Thus the fifth month is $k = 4$.

The impulse response of the system is computed in Example 3.2.1 as

$$h[k] = (1.005)^k \qquad \text{for } k = 0, 1, 2, \ldots$$

Therefore, if the system is initially relaxed, or $y[0] = 0$, the output is given by

$$y[k] = \sum_{i=0}^{k} h[k-i]u[i] = \sum_{i=0}^{k}(1.005)^{k-i}u[i]$$

This equation becomes, at $k = 4$,

$$y[4] = \sum_{i=0}^{4}(1.005)^{4-i}u[i]$$

$$= (1.005)^4 u[0] + (1.005)^3 u[1] + (1.005)^2 u[2] + (1.005)u(3) + u(4)$$

which becomes, after substituting $u[i]$,

$$y[4] = (1.005)^4 \times 100 - (1.005)^3 \times 50 + (1.005)^2 \times 200 + (1.005) \times 0 + 50$$
$$= 102.02 - 50.75 + 202.01 + 50 = 303.28$$

Thus, the total amount of money is $303.28.

EXAMPLE 3.2.4

Consider a memoryless LTI system with its input $u[k]$ and output $y[k]$ related by $y[k] = au[k]$. If the input is $\delta[k]$, then the output is $a\delta[k]$. Thus, the impulse response is $h[k] = a\delta[k]$ or

$$h[0] = a \qquad h[k] = 0 \qquad \text{for } k = 1, 2, 3, \dots$$

In this case, the convolution

$$y[k] = \sum_{i=0}^{k} h[k-i]u[i] = \sum_{i=0}^{k} h[i]u[k-i]$$

reduces to

$$y[k] = h[0]u[k] + h[1]u[k-1] + h[2]u[k-2] + \dots + h[k]u[0] = au[k]$$

Thus, $y[k] = au[k]$ is a special case of convolutions.

In addition to direct substitution, as in (3.9), discrete convolution can also be carried out using the table shown in Figure 3.3. The table lists $h[k]$ and $u[k]$ on the top row and left-most column. The products of these elements yield the entries of the table. The sums of those along the dotted lines yield the convolution. Note that the positions of $h[k]$ and $u[k]$ can be interchanged and the result will still be the same. To illustrate the application of the table, we compute the convolution of $\{1, 2, 3, 4, -1, \dots\}$ and $\{-1, 1, -3, 2, \dots\}$. We list the two sequences on the top row and left-most column in Figure 3.3. The products of

Figure 3.3 *Computation of discrete convolutions*

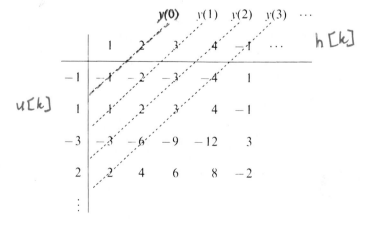

these elements yield the entries of the table. From the entries along the dotted lines, we have

$$y[0] = -1 \qquad y[1] = -2 + 1 = -1 \qquad y[2] = -3 + 2 - 3 = -4, \ldots$$

This is a convenient method of computing discrete convolutions by hand.

EXERCISE 3.2.2

Discrete convolution also arises in the product of two polynomials. Let

$$a(x) = a_0 + a_1 x + a_2 x^2$$

and

$$b(x) = b_0 + b_1 x + b_2 x^2 + b_3 x^3$$

Show that the coefficients of $c(x) = a(x)b(x) = \sum_{k=0}^{5} c_k x^k$ are the convolution of the coefficients of $a(x)$ and $b(x)$, this is,

$$c_k = \sum_{i=0}^{k} a_i b_{k-i}$$

where $a_3 = a_4 = a_5 = 0$ and $b_4 = b_5 = 0$. ■

3.2.2   Graphical Computation of Discrete Convolutions

In this section, we'll discuss how to compute discrete convolutions graphically. Consider

$$y[k] = \sum_{i=0}^{k} h[k-i]u[i]$$

with $h[k] = k - 2$, for $k = 0, 1, 2, \ldots$, and $u[k] = 2$, for $k = 0, 1, 2, \ldots$. Although the equation holds for every integer k, it is not possible to compute $y[k]$, for all k, at the same time. To compute the equation, we must choose a k, for example, $k = 3$. Then the equation becomes

$$y[3] = \sum_{i=0}^{3} h[3-i]u[i] \qquad (3.11)$$

In this equation, integer i is the variable and we must plot u and h as functions of i. We plot $h[i] = i - 2$, for $i = 0, 1, 2, \ldots$, and $u[i] = 2$, for $i = 0, 1, 2, \ldots$

in Figure 3.4(a) and (c). Note that we have changed the independent variable of u and h from k to i. As was discussed in Chapter 1, $h[-i+3]$ is the flipping of $h[i]$ and then the shifting to the right of three sampling periods, as shown in Figure 3.4(b). The product of parts (b) and (c) in the figure point by point yields the plot in Figure 3.4(d). Note that $u[i] = 0$ for $i < 0$ and $h[3-i] = 0$ for $i > 3$; thus, the plot in Figure 3.4(d) has nonzero values only in $0 \le i \le 3$. The summation of these values is $2 + 0 - 2 - 4 = -4$, which yields $y[3]$ as plotted with a solid dot in Figure 3.4(e). Thus, to compute graphically $y[k]$ for a given k, we must carry out the following four steps:

1. Flipping $h[i]$ to yield $h[-i]$.
2. Shifting $h[-i]$ to yield $h[k-i]$.
3. Multiplication of $h[k-i]$ and $u[i]$.
4. Summation of $h[k-i]u[i]$ from $i = 0, 1, \ldots, k$.

For example, to compute $y[0]$, we flip $h[i]$ to yield $h[-i]$. No shifting is needed for $k = 0$. The multiplication of $h[-i]$ and $u[i]$ yields only one nonzero value -4. Its summation is -4. Thus $y[0] = -4$ and is plotted in Figure 3.4(e). To compute $y[1]$, we shift $h[-i]$ one unit to the right to yield $h[1-i]$. The multiplication of $h[1-i]$ and $u[i]$ yields -4 and -2. Their sum yields $y[1] = -4 - 2 = -6$. To compute $y[2]$, we shift $h[-i]$ two units to the right to yield $h[2-i]$. The multiplication of $h[2-i]$ and $u[i]$ yields 0, -2, and -4. Their sum yields $y[2] = 0 - 2 - 4 = -6$. To compute $y[3]$, we shift $h[-i]$ three units to yield $h[3-i]$, as shown in Figure 3.4(b). The multiplication of $h[3-i]$ and $u[i]$ yields 2, 0, -2, and -4, as shown in Figure 3.4(d). Their sum yields $y[3] = 2 + 0 - 2 - 4 = -4$, as shown in Figure 3.4(e). Using the same procedure, we can compute $y[k]$ for any k. We see that, graphically, (3.8) is rolling or convolving two sequences; therefore, (3.8) is called a convolution.

In the preceding computation, we computed the first equality in (3.8). If we used the second equality, then we must flip $u[i]$ rather than $h[i]$; otherwise, the procedure is identical. For the example in Figure 3.4, it is actually simpler to flip $u[i]$.

EXERCISE 3.2.3

Repeat the convolution in Figure 3.4 by flipping $u[i]$ rather than $h[i]$. Compute $y[k]$ from $k = 0$ to 5. Are the results the same? ∎

EXAMPLE 3.2.5

Consider the two sequences $h[k]$ and $u[k]$ shown in Figure 3.4(a') and (b'). Their lengths are, respectively, 3 and 4. To compute $y[-1]$, we flip $h[i]$ and shift it to -1 to yield $h[-1-i]$, as shown in Figure 3.4(c'). Because nonzero entries of $u[i]$ and $h[-1-i]$ do not overlap, their products are zero for all i.

Figure 3.4 *Graphical computation of convolutions*

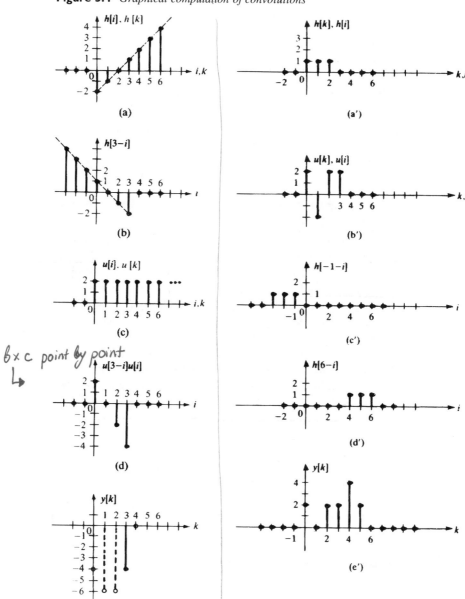

Thus, we have $y[-1] = 0$. To compute $y[6]$, we shift $h[-i]$ to 6 to yield $h[6-i]$. Again, nonzero entries of $u[i]$ and $h[6-i]$ do not overlap, and their products are zero for all i. Thus, $y[6] = 0$. In fact, $y[k] = 0$ for all $k < 0$ and $k > 5$. Thus, $y[k]$ is possibly nonzero only for $k = 0, 1, \ldots, 5$, or $y[k]$ has at most six nonzero entries. The values of $y[k]$ for $k = 0, 1, \ldots, 5$, are computed

and plotted in Figure 3.4(*e′*) (Verify). From this example, we can induce that
the discrete convolution of two sequences of lengths P and Q has at most
$P + Q - 1$ nonzero entries.

EXERCISE 3.2.4

Use the graphical method to compute and plot the discrete convolution of
$u[k] = h[k] = 1$ for $k = 0, 1, 2, 3$, and $u[k] = h[k] = 0$ otherwise.

[**ANSWER:** $y[0] = 1, y[1] = 2, y[2] = 3, y[3] = 4, y[4] = 3, y[5] = 2, y[6] = 1$,
and 0 otherwise.] ∎

3.2.3   Finite Impulse Response and Infinite Impulse Response Discrete-Time Systems

Every LTI discrete-time system has an impulse response $h[k], k = 0, 1, 2, \ldots$.
Now we will use $h[k]$ to classify discrete-time systems. A discrete-time system
is said to be a finite impulse response (FIR) system if its impulse response has
only a finite number of nonzero elements. Otherwise it is said to be an infinite
impulse response (IIR) system. Thus, an FIR system has a finite integer N such
that $h[k] = 0$ for $k \geq N$. In this case, the FIR system has at most N nonzero elements $\{h[0], h[1], \ldots, h[N-1]\}$ and is said to have length N. If no such N
exists, the system is an IIR system. The systems in Examples 3.2.1 and 3.2.2
have infinitely many nonzero $h[k]$; thus, they are IIR systems. For an FIR
system of length N, because $h[k] = 0$ for $k < 0$ and $k \geq N$, the convolution reduces to

$$y[k] = \sum_{i=0}^{k} h[i]u[k-i] = \sum_{i=0}^{k} h[k-i]u[i] \qquad \text{for } k < N \qquad \text{(3.12a)}$$

$$y[k] = \sum_{i=0}^{N-1} h[i]u[k-i] = \sum_{i=k-N+1}^{k} h[k-i]u[i] \qquad \text{for all } k \geq N \qquad \text{(3.12b)}$$

Note that the upper limit of the first summation in (3.12b) is independent of k.

EXAMPLE 3.2.6

Consider a discrete-time system with impulse response

$$h[k] = \begin{cases} 1/3 & \text{for } k = 0,1,2 \\ 0 & \text{for } k \geq 3 \end{cases}$$

It is an FIR system of length 3. The output $y[k]$ of the system excited by any
$u[k]$ is, for $k \geq 3$,

$$N - 1 = 3 - 1$$

$$y[k] = \sum_{i=0}^{2} h[i]u[k-i] = h[0]u[k] + h[1]u[k-1] + h[2]u[k-2]$$

$$= \frac{1}{3}(u[k] + u[k-1] + u[k-2])$$

We see that $y[k]$ is the average of its preceding three inputs. Therefore, the system is called a *moving-average* filter.

The moving-average filter has many applications in practice. It can be used to detect the trend of a fast-changing movement. For example, the daily temperature at noon may change rapidly, as shown in Figure 3.5(a). In order to see the trend better, we may use a five-day moving-average filter to process the data. The result (crosses) is plotted in Figure 3.5(a) with the coordinate shown on the right-hand side. The trend is indeed better shown. The same technique is widely used in stocks markets. If we want to see the trend over one month, we may use the five-day moving average. The trend over ten years, however, can be better seen using a thirty-day moving average. Figure 3.5(b) shows the thirty-day moving average of the U.S.-Japanese currency exchange rate.

EXERCISE 3.2.5

What is the impulse response of a five-day moving-average system?

[**ANSWER:** $h[k] = 1/5$, for $k = 0, 1, 2, 3, 4$; $h[k] = 0$, for $k > 4$.]

Figure 3.5 *The application of a moving-average filter*

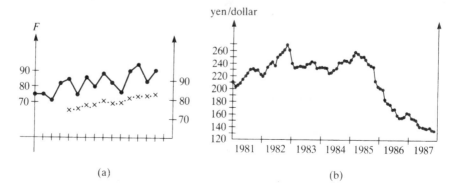

(a)   (b)

3.3   LTIL DISCRETE-TIME SYSTEMS— DIFFERENCE EQUATIONS

Every linear time-invariant discrete-time system, distributed or lumped, can be described by a discrete-time convolution as shown in (3.8). In this section, we will show that some convolutions can be transformed into difference equations.

EXAMPLE 3.3.1

Consider the savings account studied in Example 3.2.3. Its output $y[k]$ (total amount of money in the account) and input $u[k]$ (the amount of money deposited each month) are related by

$$y[k] = \sum_{i=0}^{k}(1.005)^{k-i}u[i] \tag{3.13}$$

where we have assumed $y[0] = 0$ (the initial condition is zero). Equation (3.13) holds for every positive integer k; therefore, we have

$$y[k+1] = \sum_{i=0}^{k+1}(1.005)^{k+1-i}u[i]$$

which can be expanded as

$$
\begin{aligned}
y[k+1] &= (1.005)^{k+1-(k+1)}u[k+1] + \sum_{i=0}^{k}(1.005)^{k+1-i}u[i] \\
&= u[k+1] + (1.005)\sum_{i=0}^{k}(1.005)^{k-i}u[i]
\end{aligned} \tag{3.14}
$$

Substituting (3.13) into (3.14) yields

$$y[k+1] = u[k+1] + 1.005y[k]$$

or

$$y[k+1] - 1.005y[k] = u[k+1] \qquad \text{for } k = 0,1,2,\ldots \tag{3.15}$$

This is called a first-order linear difference equation with constant coefficients, or an LTIL difference equation.

If the discrete convolution of a system can be transformed into a difference equation as in the preceding example, then the system is said to be lumped.

According to this definition, the savings account is a lumped system. This is consistent with the definition given in Chapter 2, where a lumped system is defined to have a finite number of state variables. The savings account has one state variable and is, therefore, lumped.

It is important to mention that *not* every discrete convolution can be transformed into an LTIL difference equation. For example, the system with the following impulse response

$$h[k] = \begin{cases} 0 & k = 0 \\ 1/k & k = 1, 2, 3, \ldots \end{cases} \qquad \text{or} \qquad h[k] = \begin{cases} 0 & k = 0, 2, 4, \ldots \\ 1/k! & k = 1, 3, 5, \ldots \end{cases}$$

cannot be described by an LTIL difference equation. Such a system is called a distributed system.

3.3.1   From Difference Equation to Impulse Response

Not every convolution can be transformed into a difference equation. However, if the difference-equation description of a system is known, then the impulse response of the system can be readily obtained. For example, suppose a savings account is known to be described by (3.15), but its impulse response is not known. Then the impulse response can be computed as follows: Let us rewrite (3.15) as

$$y[k + 1] = 1.005y[k] + u[k + 1]$$

or

$$y[k] = 1.005y[k - 1] + u[k] \qquad (3.16)$$

Now, if the initial condition is zero (that is, $y[-1] = 0$) and $u[k] = \delta[k]$, then the output is, by definition, the impulse response. Substituting $u[0] = 1$ and $u[k] = 0$, for $k = 1, 2, \ldots$ and $y[-1] = 0$ into (3.16) yields, for $k = 0, 1, 2, \ldots$,

$$h[0] = y[0] = 1.005y[-1] + u[0] = 1.005 \times 0 + 1 = 1$$
$$h[1] = y[1] = 1.005y[0] + u[1] = 1.005 \times 1 + 0 = 1.005$$
$$h[2] = y[2] = 1.005y[1] + u[2] = 1.005 \times 1.005 + 0 = (1.005)^2$$

and, in general,

$$h[k] = y[k] = (1.005)^k$$

This is the impulse response of the savings account. Thus, the impulse response can be readily obtained from a difference equation. In conclusion, every LTIL system can be described by a convolution and a difference equation. However, even though every LTI distributed system can be described by a convolution, no simple difference equation can be developed to describe it.

3.3.2   **Comparison of Discrete Convolutions and Difference Equations**

Because every LTIL system can be described by either a discrete convolution or a difference equation, it is natural to ask which description is preferable in analysis and design. To answer this question, let us compare the two descriptions.

1. The convolution generally requires an infinite number of coefficients $h[k]$, $k = 0, 1, 2, \ldots$, to describe a system. A difference equation, however, consists of only a finite number of terms and, consequently, requires only a finite number of coefficients. For example, (3.15) requires only the three coefficients $\{1, -1.005, 1\}$, whereas, (3.13) requires $h[k] = (1.005)^k$ for all positive integer k. Therefore, difference equations generally require less memory than convolutions.

2. The convolution describes only the zero-state response of a system. If the system is not initially relaxed at $k_0 = 0$, then the response of the system must be written as

$$y[k] = \sum_{i=0}^{k} h[k-i]u[i] + g[k] \qquad (3.17)$$

where $g[k]$ denotes the zero-input response. For some systems, $g[k]$ is easily computable; for others, it is not. Therefore, if a system is not initially relaxed, (3.17) is rarely used. Although the difference equation is developed from the convolution, it describes not only the zero-state response but also the zero-input response. For example, the equation in (3.15) is applicable whether $y[0]$ is zero or not. Therefore, the difference equation is more general than the convolution in the sense that it also describes the zero-input response.

3. The number of operations in difference equations is smaller than the one in convolutions. For example, the number of multiplications in the *direct* computation of (3.13) equals k.[3] Thus, the computation of $y[k]$, $k = 1, 2, \ldots, 10$, requires

$$1 + 2 + \ldots + 10 = 55$$

multiplications. On the other hand, the difference equation (3.15) requires only one multiplication for each k; thus, the computation of $y[k]$, $k = 1, 2, \ldots, 10$, requires only 10 multiplications. Therefore, the difference equation is simpler in computation then is the convolution.

---

[3] An alternative method of computing (3.13) is to use the discrete Fourier transform and then employ the fast Fourier transform. This is a more efficient way of computing (3.13) than the direct computation. See Chapter 7 and Reference [3].

Based on these reasons, we can conclude that the difference equation is preferable to the convolution in describing LTIL systems.

3.4 SETTING UP DIFFERENCE EQUATIONS

12

In this section, we will discuss two different approaches to develop difference equations. The first approach requires no knowledge of the internal structure of LTIL systems; the description is obtained from measurement. An input sequence is applied, and the corresponding output sequence is measured. From such input and output sequences, in particular, the impulse response sequence, we can then develop a difference equation, as demonstrated in Example 3.3.1. In the second approach, we must know the internal structure of systems; we can then apply physical laws to develop equations to describe the systems. In this approach, no experimentation or measurement is required. These two approaches are best illustrated by the problem of the savings account. Suppose we ask the bank: What will be the amount of money $y[k]$ at the beginning of the $(k + 1)$th month if we deposit $u[0] = 1.00$ (an impulse sequence) in the first month? If we are told that $y[k] = (1.005)^k$, $k = 0, 1, 2, \ldots$ (the impulse response sequence), then we can develop the difference equation (3.15) to describe the system. From the difference equation, we can deduce that the monthly interest rate is 0.5% and is compounded monthly. It is important to mention that these two pieces of information (interest rate and compounded monthly) are not needed in developing the difference equation. They become known only after the equation is developed.

If we are told that the monthly interest rate is 0.5% and is compounded monthly, we can use the second approach to develop a difference equation to describe the system. Let $y[k]$ be the amount of money in the account at the $(k + 1)$th month and $u[k + 1]$ be the amount of money deposited into the account at the $(k + 2)$th month. Clearly, we have

$$y[k + 1] = y[k] + 0.005y[k] + u[k + 1] = 1.005y[k] + u[k + 1]$$

This is the same as (3.15). This equation is developed without doing any experiment. Therefore, the two approaches are quite different.

If the internal structure of an LTIL system is not known, there is no choice but to use the first approach to develop an equation. If the internal structure is known, then either approach can be used. In practice, measurements are often corrupted by noise; therefore, the first approach may encounter practical difficulties. Even if an impulse response sequence can be measured exactly, we still need some manipulation to develop a difference equation from the impulse response. This problem is called the realization problem or, more generally, the identification problem. See References [4] and [5]. The second approach does not require any measurements and can be carried out analytically on paper. We can use this approach to develop a number of difference equations.

EXAMPLE 3.4.1

(**Population model**) Let $y[k]$ be the U.S. population in year k and let $b[k]$ and $d[k]$ be, respectively, the birth rate and death rate. If $u[k]$ is the net number of immigrants entering the United States, then we have

$$y[k+1] = y[k] + b[k]y[k] - d[k]y[k] + u[k] = (1 + b[k] - d[k])y[k] + u[k]$$

This is a first-order difference equation. Its coefficient may change year by year, however; therefore, it is a linear time-varying difference equation. If $b[k]$ and $d[k]$ are constants, then the equation becomes

$$y[k+1] = (1 + b - d)y[k] + u[k] \tag{3.18}$$

This is an LTIL difference equation.

EXAMPLE 3.4.2

(**Inventory**) The inventory of a product can be described by a difference equation. Let orders and deliveries be carried out weekly and $y[k]$ be the inventory at week k. Let $u[k]$ be the order to the manufacturer at week k. It is assumed that it will take three weeks for the product to come into the inventory. If $v[k]$ is the total delivery to the customers in week k, then we have

$$y[k+1] = y[k] - v[k] + u[k-2]$$

or

$$y[k+1] - y[k] = u[k-2] - v[k] \tag{3.19}$$

This is an LTIL difference equation. Note that this equation has two inputs, u and v, and one output, y.

*EXAMPLE 3.4.3[4]

(**Amortization**) In purchasing a car or house, we may assume some debt and then pay it off by monthly installment. This is known as amortization. Let $y[0]$ be the initial total amount of money borrowed. We assume that we pay back

[4]Examples with asterisks may be omitted without loss of continuity.

$u[k]$ at month k. The unpaid debt will carry a monthly charge of $100r\%$. If $y[k]$ is the amount of debt at month k, then we have

$$y[k+1] = y[k] + ry[k] - u[k+1] = (1+r)y[k] - u[k+1] \qquad \textbf{(3.20)}$$

This is an LTIL difference equation. This is an algebraic equation and can be solved using addition, subtraction, multiplication, and division. For example, if we want to pay off the debt in N installments and in equal monthly payments, then the payment can be computed as follows. Let $u[k] = p$, for $k = 1$, $2, \ldots, N$. The substitution of $k = 0$ and 1 into (3.20) yields

$$y[1] = (1+r)y[0] - p$$

and

$$y[2] = (1+r)y[1] - p$$

which becomes, after substituting y[1],

$$y[2] = (1+r)^2 y[0] - (1+r)p - p$$

The substitution of $k = 2$ and $y[2]$ into (3.20) yields

$$y[3] = (1+r)y[2] - p = (1+r)^3 y[0] - p[(1+r)^2 + (1+r) + 1]$$

Proceeding forward, we can show

$$y[N] = (1+r)^N y[0] - p[(1+r)^{N-1} + (1+r)^{N-2} + \cdots + (1+r) + 1] \quad \text{✳}$$
$$\textbf{(3.21)}$$

Using the formula

$$1 + a + a^2 + \cdots + a^n = \frac{1 - a^{n+1}}{1-a} \qquad \text{✳} \qquad \textbf{(3.22)}$$

(see Problem 3.9), we can write (3.21) as

$$y[N] = (1+r)^N y[0] - p\frac{1-(1+r)^N}{1-(1+r)} \qquad \textbf{(3.23)}$$

If we pay off the debt completely after N installments, then $y(N) = 0$ and (3.23) becomes

$$p = \frac{r(1+r)^N}{(1+r)^N - 1} y[0] \qquad \textbf{(3.24)}$$

This is the amount we must pay each month in order to pay off the original debt $y[0]$ in N equal installments.

EXAMPLE 3.4.4

(**Block diagram**) Consider the three elements shown in Figure 3.6. The element denoted by a box is a unit-time delay element; its output $y[k]$ is equal to the input $u[k]$ delayed by one sampling period, that is,

$$y[k] = u[k-1] \quad \text{(unit-delay element)}$$

The element denoted by a big circle with a real number α is called a multiplier with gain α; its input $u[k]$ and output $y[k]$ are related by

$$y[k] = \alpha u[k] \quad \text{(multiplier)}$$

It is also called an amplifier if $|\alpha| > 1$ or an attenuator if $|\alpha| < 1$. The element denoted by a small circle with a plus sign is called an adder or summer. Every adder has two or more inputs $u_i[k]$, $i = 1, 2, \ldots, m$, but has one and only one output $y[k]$. They are related by

$$y[k] = u_1[k] + u_2[k] + \cdots + u_m[k] \quad \text{(adder)}$$

A diagram built by interconnecting these three types of elements, such as the one in Figure 3.7, is called a discrete-time *basic block diagram*. Every discrete-time basic block diagram is an LTIL system and can be described by a difference equation. To develop a difference equation to describe the diagram in Figure 3.7, we assign the input of the left-most unit-time delay element as $y[k]$. Then the outputs of the three unit-time delay elements are, respectively, $y[k-1]$, $y[k-2]$ and $y[k-3]$. Equating the inputs and output of the summer in Figure 3.7 yields

Figure 3.6 *Three basic elements*

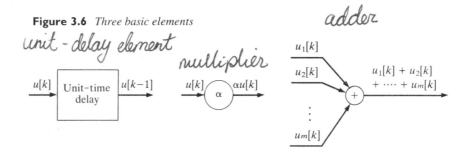

Figure 3.7 *Discrete-time basic block diagram*

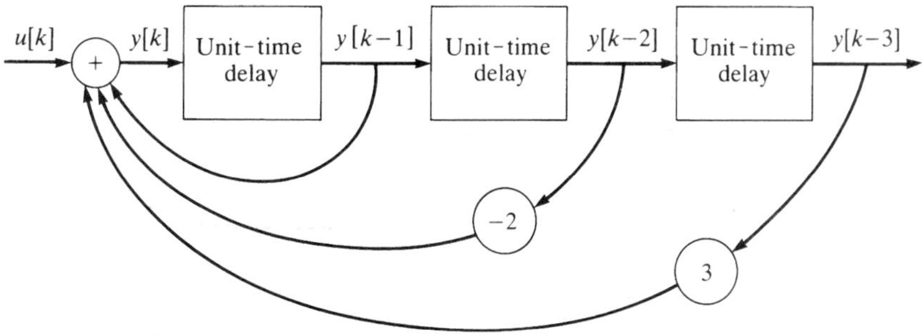

$$y[k] = u[k] + y[k-1] - 2y[k-2] + 3y[k-3] \qquad (3.25)$$

or

$$y[k+3] - y[k+2] + 2y[k+1] - 3y[k] = u[k+3] \qquad (3.26)$$

This third-order LTIL difference equation describes the block diagram in Figure 3.7.

EXERCISE 3.4.1

Show that the feedback systems in Figure 3.1 can be described by the first-order difference equations

$$y[k+1] - ay[k] = au[k] \qquad y[k+1] + ay[k] = au[k] \qquad ∎$$

---

*EXAMPLE 3.4.5

To find difference equations to describe the basic block diagrams shown in Figure 3.8, we assign the output of the right-hand side unit-time delay element as $y[k]$. Then the inputs of the delay elements are $y[k+1]$ and $y[k+2]$, as shown. Equating all signals at the summer yields

$$y[k+2] + 2y[k+1] + 3y[k] = 10u[k] \qquad (3.27)$$

This second-order difference equation describes the block diagram in Figure 3.8(a).

Consider the basic block diagram in Figure 3.8(b), which is obtained from the one in Figure 3.8(a) by adding two branches with gains 1 and 4. In this

Figure 3.8 *Basic block diagrams*

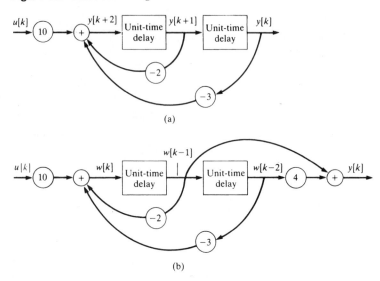

(a)

(b)

block diagram, an intermediate variable $w[k]$ is introduced. Then the outputs of the delay elements are $w[k-1]$ and $w[k-2]$. From the summer on the left-hand side, we can readily obtain

$$w[k] + 2w[k-1] + 3w[k-2] = 10u[k] \tag{3.28}$$

From the summer on the right-hand side, we have

$$y[k] = w[k-1] + 4w[k-2] \tag{3.29}$$

In order to obtain a difference equation to relate the input $u[k]$ and output $y[k]$, we must eliminate $w[k]$ from (3.28) and (3.29). From (3.28) and (3.29), we have

$$w[k+1] + 2w[k] + 3w[k-1] = 10u[k+1]$$
$$y[k+1] = w[k] + 4w[k-1]$$

and

$$y[k+2] = w[k+1] + 4w[k]$$

Using these equations, it is straightforward to verify

$$y[k+2] + 2y[k+1] + 3y[k] = 10u[k+1] + 40u[k] \tag{3.30}$$

This second-order difference equation describes the basic block diagram in Figure 3.8(b).

⌈Although every basic block diagram can be described by an LTIL difference equation, it is not necessarily simple to develop such an equation, as the preceding example showed. However, it is simple and straightforward to develop a state-variable equation, which will be introduced in Chapter 10, to describe any basic block diagram. From the state-variable equation, we can then develop a difference equation to describe the diagram. ⌋

The independent variable k in the preceding examples was time. Here is an example whose independent variable k is not time.

*EXAMPLE 3.4.6

Consider the resistive network shown in Figure 3.9. It is called a ladder network. Let V_k, $k = 0, 1, \ldots, m$, denote the node voltages, as shown. Clearly, we have $V_0 = E$ and $V_m = 0$. At the kth node, we have $i_1 = i_2 + i_3$ which implies, using Ohm's law,

$$\frac{V_{k-1} - V_k}{R} = \frac{V_k}{R} + \frac{V_k - V_{k+1}}{R}$$

This equation can be simplified as

$$V_{k-1} - 3V_k + V_{k+1} = 0 \qquad (3.31)$$

This equation holds for $k = 1, 2, \ldots, m - 1$, and is an LTIL difference equation. Note that the independent variable k is position or location; it is not time.

If $m = 4$, then V_k, $k = 1, 2, 3$, can be solved as follows. Equation (3.31) implies, for $k = 1$,

Figure 3.9 *Ladder network*

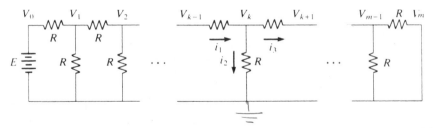

$$V_0 - 3V_1 + V_2 = 0$$

which implies, using $V_0 = E$,

$$V_2 = 3V_1 - E \tag{3.32}$$

If $k = 2$, then (3.31) becomes

$$V_1 - 3V_2 + V_3 = 0$$

which, together with (3.32), implies

$$V_3 = 3V_2 - V_1 = 8V_1 - 3E \tag{3.33}$$

Similarly, from (3.31) with $k = 3$ and (3.32) and (3.33), we can obtain

$$V_4 = 3V_3 - V_2 = 21V_1 - 8E \tag{3.34}$$

Because $V_4 = 0$, V_1 can be solved from (3.34) as

$$V_1 = \frac{8E}{21}$$

The substitution of V_1 into (3.33) and (3.32) yields

$$V_2 = \frac{E}{7} \quad \text{and} \quad V_3 = \frac{E}{21}$$

This solves the problem. Note that the network is a memoryless system and can also be formulated as a problem of linear algebraic equations. See Problem A.32.

3.4.1 General Forms of Difference Equations

A number of difference equations for physical systems were developed in the preceding section. Now we will discuss the general case. Consider the equation

advanced form

$$\boxed{\begin{aligned} &y[k + n] + a_{n-1}y[k + n - 1] + \cdots + a_1 y[k + 1] + a_0 y[k] \\ &= b_m u[k + m] + b_{m-1} u[k + m - 1] + \cdots + b_1 u[k + 1] + b_0 u[k] \end{aligned}} \tag{3.35}$$

where n and m are fixed integers and k is a variable ranging from $k = -n, -n + 1, \ldots, 0, 1, 2, \ldots$. The coefficients a_i and b_i are real constants, not necessarily nonzero. The coefficient a_n of $y[k + n]$ has been normalized, without loss of generality, to 1. If $m > n$, then $y[k + n]$ depends on future input

$u[k + m]$, and the system is not causal. Thus, for causal systems, we require
$n \geq m$ in (3.35). Equation (3.35) is called an nth order linear ordinary difference
equation with constant coefficients or an nth order LTIL difference equation.

If we define $\bar{k} = k + n$, then (3.35) can be written as

$$y[\bar{k}] + a_{n-1}y[\bar{k} - 1] + \cdots + a_1 y[\bar{k} - n + 1] + a_0 y[\bar{k} - n]$$
$$= b_m u[\bar{k} - n + m] + b_{m-1}u[\bar{k} - n + m - 1] \qquad (3.36)$$
$$+ \cdots + b_1 u[\bar{k} - n + 1] + b_0 u[\bar{k} - n]$$

which, by renaming $\bar{k}$ as k and renaming the coefficients, can be written as

delayed
form

$$y[k] + a_{-1}y[k - 1] + \cdots + a_{-n+1}y[k - n + 1] + a_{-n}y[k - n]$$
$$= b_{-n+m}u[k - n + m] + b_{-n+m-1}u[k - n + m - 1] + \cdots \qquad (3.37)$$
$$+ b_{-n+1}u[k - n + 1] + b_{-n}u[k - n]$$

This is an alternative form of the nth order LTIL difference equation in (3.35).
To differentiate them, we call (3.35) the *advanced form* and (3.37) the *delayed
form*. Both forms are widely used in application. To find the response $y[k]$,
$k = 0, 1, 2, \ldots$ of (3.35) or (3.37) excited by the input sequence $u[k]$,
$k = 0, 1, 2, \ldots$, we need $2n$ initial conditions $\{y[k], u[k], k = -1, -2,
\ldots, -n\}$. If $u[k] = 0$ for $k < 0$, as is often assumed, then we need n initial
conditions $\{y[k], k = -1, -2, \ldots, -n\}$ to find the response of (3.35) or (3.37).
Once the initial conditions are given, the response can be obtained by direct
substitution. This can be illustrated by an example.

EXAMPLE 3.4.7

Compute the response $y[k]$, for $k \geq 0$, of the LTIL advanced-form difference
equation

$$-2y[k + 2] + y[k] = 3u[k + 1] - 2u[k] \qquad (3.38)$$

excited by the initial conditions $y[-1] = 1$, $y[-2] = 2$, and the input $u[k] = 1$,
for $k = 0, 1, 2, \ldots$. We write (3.38) as

$$y[k + 2] = \frac{1}{-2}(-y[k] + 3u[k + 1] - 2u[k]) \qquad (3.39)$$

If $k = -2$, then (3.39) becomes, using $y[-2] = 2$, $u[-2] = u[-1] = 0$,

$$y[0] = -0.5(-y[-2] + 3u[-1] - 2u[-2]) = -0.5(-2 + 0 - 0) = 1$$

If $k = -1$, then (3.39) becomes

$$y[1] = -0.5(-y[-1] + 3u[0] - 2u[-1]) = -0.5(-1 + 3 - 0) = -1$$

If $k = 0$, then

$$y[2] = -0.5(-y[0] + 3u[1] - 2u[0]) = -0.5(-1 + 3 - 2) = 0$$

Proceeding forward, the output $y[k]$ for all positive k can be computed.

EXERCISE 3.4.2

Repeat Example 3.4.7 using the following delayed form of (3.39).

$$-2y[k] + y[k-2] = 3u[k-1] - 2u[k-2] \tag{3.40}$$

■

3.4.2   Recursive and Nonrecursive Difference Equations

The computation of $y[k]$ in (3.35) or (3.37) requires n previous values of $y[k]$; therefore, its computation must be carried out in the order of $y[0]$, $y[1]$, $y[2]$, ... and so forth, as in Example 3.4.7. For this reason, (3.35) and (3.37) are called *recursive* difference equations. If $a_i = 0$, $i = 0, 1, \ldots, n-1$, then (3.35) reduces to

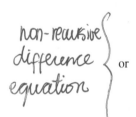

non-recursive difference equation

$$y[k+n] = b_m u[k+m] + b_{m-1}u[k+m-1] + \cdots$$
$$+ b_1 u[k+1] + b_0 u[k] \tag{3.41a}$$

or

$$y[k] = b_m u[k-n+m] + b_{m-1}u[k-n+m-1] + \cdots$$
$$+ b_1 u[k-n+1] + b_0 u[k-n] \tag{3.41b}$$

In this case, the computation of $y[k]$ does not require any $y[k-i]$ with $i > 0$, and we can compute, say, $y[10]$ without computing first $y[k]$, $k = 0$, $1, 2, \ldots, 9$. Thus, (3.41) is called a *nonrecursive* difference equation of order n.

Finite Impulse Response

Every FIR filter can be described by a nonrecursive difference equation. For example, as was discussed regarding (3.12), the output $y[k]$ of an FIR filter of length 4 can be described by the convolution

$$y[k] = \sum_{i=0}^{3} h[i]u[k-i] \tag{3.42a}$$

for $k \geq 4$. It can be written explicitly as

$$y[k] = h[0]u[k] + h[1]u[k-1] + h[2]u[k-2]$$
$$+ h[3]u[k-3] \tag{3.42b}$$

This equation becomes (3.41b) if we set $n = m = 3$ and define $h[i] = b_{m-i}$. Thus, the convolution in (3.42) is also a nonrecursive difference equation of order 3. In conclusion, every FIR filter of length N can be described by a nonrecursive difference equation of order $N - 1$.

Every nonrecursive difference equation can be written as a recursive equation. For example, (3.42b) can be written as

$$y[k + 1] - y[k] = h[0]u[k + 1] + h[1]u[k] + h[2]u[k - 1] + h[3]u[k - 2]$$
$$- (h[0]u[k] + h[1]u[k - 1] + h[2]u[k - 2] + h[3]u[k - 3])$$
$$= h[0]u[k + 1] + (h[1] - h[0])u[k] + (h[2] - h[1])u[k - 1]$$
$$+ (h[3] - h[2])u[k - 2] - h[3]u[k - 3]$$

This is a recursive equation. This recursive equation does not offer any advantage over (3.42) and is, therefore, not used. However, in some special cases, it is preferable to use a recursive equation over a nonrecursive one. This can be illustrated by an example.

EXAMPLE 3.4.8

Consider a moving-average filter of length 20. Its impulse response is

$$h[k] = \frac{1}{20} \qquad \text{for } 0 \le k \le 19$$

and $h[k] = 0$ otherwise. The output of this filter is

$$y[k] = \frac{1}{20}(u[k] + u[k - 1] + u[k - 2] + \cdots + u[k - 19]) \qquad (3.43)$$

It is the average of its previous twenty values of $u[k]$. Equation (3.43) is clearly a nonrecursive difference equation.

Let k be replaced by $k + 1$. Then (3.43) becomes

$$y[k + 1] = 0.05(u[k + 1] + u[k] + u[k - 1] + \cdots$$
$$+ u[k - 18] + u[k - 19] - u[k - 19]) \qquad (3.44)$$

where we have added and subtracted the term $u[k - 19]$. Substituting (3.43) into (3.44) yields

$$y[k + 1] = 0.05u[k + 1] + y[k] - 0.05u[k - 19] \qquad (3.45a)$$

or

$$y[k + 1] - y[k] = 0.05(u[k + 1] - u[k - 19]) \qquad (3.45b)$$

This is a recursive difference equation.

The computation of $y[k]$ requires 19 additions and one multiplication in using (3.43); it requires 2 additions and one multiplication in using (3.45). Thus, the recursive equation (3.45) is preferable to the nonrecursive equation (3.43) in describing the twenty-point moving-average filter.

A discrete-time system with an infinite impulse response is not always describable by a difference equation. It can be so described only if the system is lumped. In this case, the difference equation must be a recursive difference equation, as shown in Example 3.4.1. Therefore, difference equations that describe IIR systems must be recursive equations.

EXERCISE 3.4.3

Classify the following difference equations as recursive or nonrecursive and compute their impulse responses. Classify the systems they describe as IIR or FIR.

(a) $y[k + 2] + 2y[k + 1] - 3y[k] = u[k + 1]$

(b) $y[k] = 4u[k] - 5u[k - 1]$

(c) $y[k + 1] + y[k] = u[k + 1] + 2u[k] + u[k - 1]$

[**ANSWERS:** (a) Recursive; $h[0] = 0$, $h[1] = 1$, $h[2] = -2$, $h[3] = -7$, . . . ; an IIR system. (b) Nonrecursive; $h[0] = 4$, $h[1] = -5$, $h[k] = 0$ for $k \geq 2$; an FIR system. (c) Recursive; $h[0] = h[1] = 1$, $h[k] = 0$ for $k \geq 2$; an FIR system. The recursive difference equation is equivalent to the nonrecursive equation $y[k] = u[k] + u[k - 1]$.]

3.5 LTI CONTINUOUS-TIME SYSTEMS—CONVOLUTIONS

In this section, we will develop a mathematical description for continuous-time systems. This description is the counterpart of the discrete convolution discussed in Section 3.2.1. The description will be obtained from a special input-output pair and describes only the zero-state response of systems. Therefore, we assume in this section that every system is initially relaxed before the application of an input.

The discrete convolution for discrete-time systems is obtained by decomposing an input sequence into the summation of a number of impulse sequences. We shall use the same procedure for continuous-time systems. Before proceeding, we need the concept of impulse that is equivalent to the impulse sequence in the discrete-time case.

3.5.1   The Impulse

We first need the counterpart of the impulse sequence for the continuous-time case. Consider the five *different* time functions shown in Figure 3.10. They are defined over the entire time interval $(-\infty, \infty)$ and assume nonzero value only in the neighborhood of $t = 0$. The areas of these functions all equal 1. Let $\delta_\varepsilon(t)$ denote any one of them. We define

$$\delta(t) := \lim_{\varepsilon \to 0} \delta_\varepsilon(t) \qquad\qquad (3.46)$$

It is called the Dirac delta function, δ-function, impulse function or, simply, *the impulse*.[5] Clearly, we have

$$\delta(t) = 0 \qquad \text{for } t \neq 0$$

The value of $\delta(t)$ at $t = 0$, however, is not uniquely defined. It may assume 0, as in Figure 3.10(c) and (e), or ∞ as in Figure 3.10 (a), (b), and (d). Every function in Figure 3.10 has area 1 for any $\varepsilon > 0$; therefore, we have

$$\int_{-\infty}^{\infty} \delta(t)dt = \int_{-\Delta}^{\Delta} \delta(t)dt = 1 \qquad\qquad (3.47a)$$

for every $\Delta > \varepsilon > 0$. This also holds as $\Delta \to 0$; therefore, (3.47a) is often written simply as

$$\int_{-\infty}^{\infty} \delta(t)dt = \int_{0-}^{0+} \delta(t)dt = 1 \qquad\qquad (3.47b)$$

[5]Note that $\delta[k]$, where k is an integer, is used to denote the impulse sequence and $\delta(t)$, where t is a continuous variable, is used to denote the impulse. This convention will be used throughout this text. For example, $f[k]$ is a discrete-time sequence, and $f(t)$ is a continuous-time function.

Figure 3.10 *Various forms of pulse*

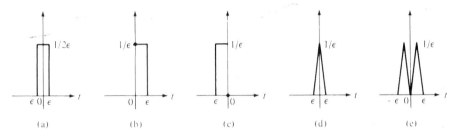

where 0– is zero approaching from the negative and 0+ is zero approaching from the positive. It is implicitly assumed that the impulse is included *wholely* inside the integration interval from 0– to 0+.

In order for the integration in (3.47) to equal 1, the impulse must be included *wholely* inside the integration interval. If it is not included, then the integration is 0, as in the following example.

$$\int_{-\infty}^{-0.5} \delta(t)dt = \int_{0.1}^{10} \delta(t)dt = \int_{0+}^{\infty} \delta(t)dt = 0$$

If the impulse is included only partially inside the integration interval, then the integration is not defined. For example, we have

$$\int_{0}^{\infty} \delta(\tau)d\tau = \int_{0}^{0+} \delta(\tau)d\tau = \begin{cases} 0 & \text{using the pulse in Figure 3.10(c)} \\ 1/2 & \text{using the pulse in Figure 3.10(a), (d), or (e)} \\ 1 & \text{using the pulse in Figure 3.10(b)} \end{cases}$$

Therefore, whenever we encounter an impulse, if the integration does not wholely include the impulse, then ambiguity will arise.

The impulse is a peculiar "function." The value of $\delta(t)$ is zero everywhere except at $t = 0$. Strictly speaking, $\delta(0)$ is not defined; however, we often write $\delta(0) = \infty$. Even though the width of the integration interval [0–, 0+] is zero, we have

$$\int_{0-}^{0+} \delta(t)dt = 1$$

If $f(t)$ is an ordinary function, then

$$\int_{0-}^{0+} f(t)dt = 0 \tag{3.48a}$$

and, for any $a > 0$,

$$\int_{0}^{a} f(t)dt = \int_{0-}^{a} f(t)dt = \int_{0+}^{a} f(t)dt \tag{3.48b}$$

Therefore, if an integrand does not contain impulses, there is no difference between using 0, 0–, or 0+.

Customarily, the impulse with area or weight 1 is denoted by an arrow with height 1, as shown in Figure 3.11(a). The impulse $\delta(t) = \delta(t - 0)$ appears at $t = 0$. If it appears at t_0 and with area 3, then it is written as $3\delta(t - t_0)$ and is plotted as shown in Figure 3.11(b). In terms of $\delta(t - t_0)$, (3.47) must be generalized as, for any $\Delta > 0$,

$$\int_{-\infty}^{\infty} \delta(t - t_0)dt = \int_{t_0-\Delta}^{t_0+\Delta} \delta(t - t_0)dt = \int_{t_0-}^{t_0+} \delta(t - t_0)dt = 1$$

Figure 3.11 *The impulses*

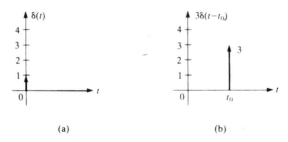

(a)   (b)

Now we shall establish some important properties of the impulse. Let $f(t)$ be an arbitrary time function. If $f(t)$ is continuous at t_0, then

$$f(t)\delta(t - t_0) = f(t_0)\delta(t - t_0) \qquad (3.49)$$

and

$$\int_{-\infty}^{\infty} f(t)\delta(t - t_0)dt = f(t_0)\int_{t_0-}^{t_0+} \delta(t - t_0)dt = f(t_0) = f(t)\Big|_{t=t_0} \qquad (3.50)$$

sifting property of the impulse

Equation (3.49) follows from the property that $\delta(t - t_0) = 0$ everywhere except at $t = t_0$ and (3.50) follows immediately from (3.49). The integral in (3.50) is often referred to as the *sifting* property of the impulse, because it picks or sifts out the value of $f(t)$ at time t_0.[6] (Compare with the discrete sifting property in Problem 1.14.) It is important that the integration interval must wholly include the impulse; otherwise, the integration will be zero or undefined. Note that (3.50) holds no matter which function in Figure 3.10 is used to define the impulse. Strictly speaking, the impulse should be defined through the use of (3.50).

EXAMPLE 3.5.1

Because the impulse $\delta(t - 1)$ appears at $t = 1$, we have

$$\int_{-\infty}^{\infty} \sin t\,\delta(t - 1)dt = \int_{0}^{2} \sin t\,\delta(t - 1)dt = \int_{1-}^{1+} \sin t\,\delta(t - 1)dt = \sin t\Big|_{t=1}$$

$$= \sin 1 = \sin 57.3^{\circ} = 0.841$$

[6]The impulse can also be used to achieve *shifting*. See Problem 3.23. Sifting and shifting are both important properties of the impulse.

because each integration interval includes the impulse wholely. However, we have

$$\int_{1+}^{\infty} \sin t\delta(t-1)dt = \int_{-\infty}^{0} \sin t\delta(t-1)dt = 0$$

because the impulse appears outside the integration intervals. The following

$$\int_{1}^{\infty} \sin t\delta(t-1)dt \quad \text{and} \quad \int_{-\infty}^{1} \sin t\delta(t-1)dt$$

are not well defined because the integration intervals cover only part of the impulse.

EXERCISE 3.5.1

Compute the following:

(a) $\displaystyle\int_{-\infty}^{\infty} \sin t\delta(t - \frac{\pi}{2})dt$

(b) $\displaystyle\int_{-1}^{1} \sin t\delta(t)dt$

(c) $\displaystyle\int_{0}^{1} \cos t\delta(t)dt$

(d) $\displaystyle\int_{0-}^{1} \cos t\delta(t)dt$

(e) $\displaystyle\int_{0+}^{1} \cos t\delta(t)dt$

(f) $\displaystyle\int_{0}^{3+} \cos t\delta(t-3)dt$

(g) $\displaystyle\int_{3-}^{3} \cos t\delta(t-3)dt$

[**ANSWERS:** (a) 1, (b) 0, (c) not defined, (d) 1, (e) 0, (f) cos 3 = −0.99, (g) not defined.] ▉

3.5.2   ## Convolution Integrals

Consider an LTI continuous-time system. The system is assumed to be initially relaxed at $t = 0$. We apply the input $\delta(t)$ and measure the output. This particular output will be denoted by $h(t)$ and will be called the (continuous-time) *impulse response* of the system, that is,

$$\text{impulse response} = h(t) := \text{output excited by } \delta(t)$$

If the system is causal, no output will appear before the application of the impulse $\delta(t)$. Therefore, we have $h(t) = 0$, for $t < 0$. In fact, the condition

$$h(t) = 0 \quad \text{for } t < 0$$

is necessary and sufficient for any LTI system to be causal. Using the impulse response, we can develop a mathematical equation to relate the output and input of the system. Let $u(t)$ be an arbitrary input, as shown in Figure 3.12(a). The input can be approximated by a sequence of pulses, as shown in Figure 3.12(b). The heights of the pulses are dictated by $u(t)$ at $t = k\varepsilon$ for $k = 0, 1, 2, \ldots$. Clearly, the smaller the ε, the better the approximation. We used $\delta_\varepsilon(t)$ to denote the pulse in Figure 3.10(b). Because the height of the pulse $\delta_\varepsilon(t)$ is $1/\varepsilon$, the height of $\delta_\varepsilon(t) \cdot \varepsilon$ is 1. Therefore $u_i(t)$ in Figure 3.12(b) can be expressed as

$$u_0(t) = \delta_\varepsilon(t)u(0)\varepsilon$$
$$u_1(t) = \delta_\varepsilon(t - \varepsilon)u(\varepsilon)\varepsilon$$
$$u_2(t) = \delta_\varepsilon(t - 2\varepsilon)u(2\varepsilon)\varepsilon$$
$$\vdots$$

and

$$u(t) \approx \sum_{i=0}^{\infty} \delta_\varepsilon(t - i\varepsilon)u(i\varepsilon)\varepsilon \tag{3.51}$$

$\varepsilon = $ sampling period

Figure 3.12 *Decomposition of an input function*

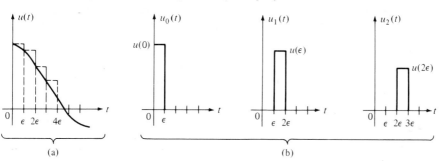

(a)

(b)

Now we use

$$\delta_\varepsilon(t) \to h_\varepsilon(t) \tag{3.52}$$

to denote that the input $\delta_\varepsilon(t)$ excites the output $h_\varepsilon(t)$. Then we have

$$\delta_\varepsilon(t - i\varepsilon) \to h_\varepsilon(t - i\varepsilon) \quad \text{(time invariance)}$$

$$\delta_\varepsilon(t - i\varepsilon)u\,(i\varepsilon)\varepsilon \to h_\varepsilon(t - i\varepsilon)u\,(i\varepsilon)\varepsilon \quad \text{(homogeneity)}$$

and

$$\sum_{i=0}^{\infty} \delta_\varepsilon(t - i\varepsilon)u\,(i\varepsilon)\varepsilon \to \sum_{i=0}^{\infty} h_\varepsilon(t - i\varepsilon)u\,(i\varepsilon)\varepsilon \quad \text{(additivity)}$$

Therefore, if the system is linear and time invariant, the zero-state response $y\,(t)$ is given by

$$y\,(t) \approx \sum_{i=0}^{\infty} h_\varepsilon(t - i\varepsilon)u\,(i\varepsilon)\varepsilon \tag{3.53}$$

We see that the derivation of (3.53) is identical to the discrete-time case in (3.6).

Now consider the case as $\varepsilon \to 0$. The input in (3.52) then becomes the impulse and the output becomes the impulse response, that is,

$$h\,(t) := \lim_{\varepsilon \to 0} h_\varepsilon(t) = \text{output excited by } \delta(t)$$

Let $\tau := i\varepsilon$, $i = 0, 1, 2, \ldots$. Then as $\varepsilon \to 0$, τ becomes a continuous variable and, therefore, we may write ε as $d\tau$. Furthermore, the summation in (3.53) becomes an integration, and the approximation becomes an equality. Thus, as $\varepsilon \to 0$, (3.53) becomes

$$y\,(t) = \int_0^{\infty} h\,(t - \tau)u\,(\tau)d\tau \tag{3.54}$$

This is called the *convolution integral* or, simply, the *convolution*. If a system is causal, then $h\,(t) = 0$, for $t < 0$, which implies $h\,(t - \tau) = 0$ for $t < \tau$. Thus, (3.54) can be reduced to, under the causality condition,

$$y\,(t) = \int_0^{t} h\,(t - \tau)u\,(\tau)d\tau \tag{3.55}$$

This is a basic equation in system analysis and is also called the input-output description of the system. There is an alternative form of (3.55). Let $\tau' := t - \tau$.

Then $\tau = t - \tau'$ and $d\tau = -d\tau'$. Substituting these into (3.55) yields

$$y(t) = -\int_{\tau'=t}^{\tau'=0} h(\tau')u(t-\tau')d\tau' = \int_0^t h(\tau')u(t-\tau')d\tau'$$

Thus, by dropping the prime of τ, we have

se if no - impulses ✳

$$\boxed{y(t) = \int_0^t h(t-\tau)u(\tau)d\tau = \int_0^t h(\tau)u(t-\tau)d\tau}$$ ✳ (3.56)

for $t \geq 0$. This is the analog counterpart of (3.8). Discrete-time systems are often classified as FIR or IIR. This is not done in the continuous-time case, because almost all practical continuous-time systems are IIR systems.

Different systems have different impulse responses. Some impulse responses may contain impulses; some may not. If $h(t)$ contains impulses, then the convolution in (3.56) must be modified as

use if impulses exist

$$\boxed{y(t) = \int_{0-}^{t+} h(t-\tau)u(\tau)d\tau = \int_{0-}^{t+} h(\tau)u(t-\tau)d\tau \qquad \text{for } t \geq 0}$$ (3.57)

so that the integration interval will wholly include the impulses; otherwise, the integration will not be well defined. If $h(t)$ does not contain impulses, there is no difference between using (3.56) or (3.57).

Convolution integral is often written more generally as

$$\boxed{\begin{aligned} y(t) &= \int_{-\infty}^{\infty} h(t-\tau)u(\tau)d\tau =: h(t)*u(t) \\ &= \int_{-\infty}^{\infty} u(t-\tau)h(\tau)d\tau =: u(t)*h(t) \end{aligned}}$$

(3.58a)

(3.58b)

for all t in $(-\infty, \infty)$. This equation reduces to (3.56) if $h(t) = u(t) = 0$ for $t < 0$, and must be employed if the input is applied from $t = -\infty$. Note that every ✳ system is assumed to be initially relaxed at $t = -\infty$. Equation (3.58) is the analog counterpart of (3.10). The convolution integral describes lumped as well as distributed linear time-invariant systems.

EXAMPLE 3.5.2

Consider a memoryless system with gain a, that is, its output $y(t)$ and input $u(t)$ are related by $y(t) = au(t)$. If we apply $u(t) = \delta(t)$, then the output is $a\delta(t)$. Thus, the impulse response $h(t)$ of the system is

$$h(t) = a\delta(t)$$

In this case, the convolution in (3.57) becomes, using the sifting property,

$$y(t) = \int_{0-}^{t+} h(t-\tau)u(\tau)d\tau = \int_{0-}^{t+} a\delta(t-\tau)u(\tau)d\tau = au(\tau)\Big|_{\tau=t} = au(t)$$

Thus, the convolution reduces to $y(t) = au(t)$ for memoryless systems.

EXERCISE 3.5.2

Consider an LTI system that delays its input by two units of time. What is its impulse response? What is its convolution integral?

[**ANSWERS:** $\delta(t-2)$, $y(t) = u(t-2)$.]

EXAMPLE 3.5.3

Consider the feedback systems shown in Figure 3.1. As in Example 3.2.2, the impulse response of the positive feedback system in Figure 3.1(a) is

$$h(t) = a\delta(t-1) + a^2\delta(t-2) + a^3\delta(t-3) + \cdots = \sum_{i=1}^{\infty} a^i\delta(t-i)$$

This is the analog counterpart of (3.4b). The impulse response of the negative feedback system in Figure 3.1(b) is

$$h(t) = a\delta(t-1) - a^2\delta(t-2) + a^3\delta(t-3) - \cdots = \sum_{i=1}^{\infty} (-1)^{i-1}a^i\delta(t-i)$$

EXAMPLE 3.5.4

Consider the network shown in Figure 3.13(b). The input $u(t)$ is a voltage source; the output $y(t)$ is the voltage across the capacitor. If we apply $u(t) = \delta(t)$, then the voltage across the capacitor can be measured, using a voltmeter, as[7]

$$h(t) = 0.5e^{-0.5t} \qquad \text{for } t \geq 0 \tag{3.59}$$

---

[7]In practice, it is not possible to generate an impulse. Actual measurement generally contains noise. Therefore, it is difficult, if not impossible, to obtain (3.59) by measurement. It is introduced here to illustrate the use of convolution. In the next chapter, a method will be introduced to develop (3.59) analytically.

Figure 3.13 *A network and its impulse response*

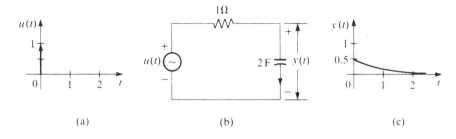

(a)   (b)   (c)

as shown in Figure 3.13(c). Once $h(t)$ is known, the zero-state response of the network excited by any input $u(t)$ can be described by

$$y(t) = \int_0^t [0.5e^{-0.5(t-\tau)}]u(\tau)d\tau = \int_0^t 0.5e^{-0.5\tau}u(t-\tau)d\tau \qquad \textbf{(3.60)}$$

Because $h(t)$ in (3.59) does not contain any impulse, there is no need to use $0-$ or $t+$ as the limits of the integration, as in (3.57). Equation (3.60) is the convolution description of the network.

3.5.3 Graphical Computation of Convolutions

This subsection discusses the graphical computation of the convolution in (3.55). As in the discrete-time case, the computation consists of four steps: flipping, shifting, multiplication, and integration (instead of the summation used in the discrete-time case). This can be illustrated by an example. Consider the impulse response $h(t)$ shown in Figure 3.14. Let $u(t) = 2$ for $t \geq 0$. Plotted in Figure 3.14 is the computation of

$$y(0.5) = \int_0^{0.5} h(0.5 - \tau)u(\tau)d\tau$$

and

$$y(2) = \int_0^2 h(2 - \tau)u(\tau)d\tau$$

First $h(\tau)$ is flipped to obtain $h(-\tau)$, and then it is shifted to 0.5 and 2 to yield Figure 3.14(b). Multiplying $h(t-\tau)$ and $u(\tau)$ in Figure 3.14(b) and (c) point by point yields Figure 3.14(d). The *net* area in the graph on the right-hand side in Figure 3.14(d) is clearly 0; thus, $y(2) = 0$. The area in the graph on the left-hand side is

Figure 3.14 *Graphical computation of convolution integral*

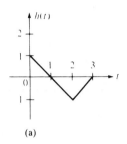

(a)

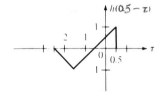

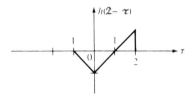

(b)

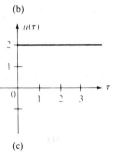

(c)

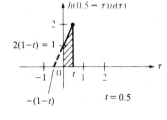

(c)

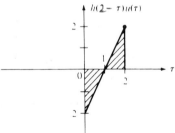

(d)

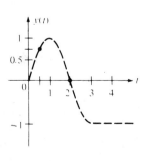

(e)

$$\frac{1}{2}(2 \times 1) - \frac{1}{2}(1 \times 0.5) = \frac{2 - 0.5}{2} = 0.75$$

Thus, $y(0.5) = 0.75$. In fact, for $0 < t < 1$, the shaded area can be computed analytically as

$$\frac{1}{2}(2 \times 1) - \frac{1}{2}((1 - t) \times 2(1 - t)) = 1 - (1 - t)^2 = 2t - t^2$$

thus, $y(t) = 2t - t^2$ for $0 < t < 1$. For $t > 3$, the flipped and shifted $h(t - \tau)$ overlaps completely with $u(\tau) = 2$; thus, the net area is $2(0.5 - 1) = -1$, which implies $y(t) = -1$ for $t > 3$, as plotted in Figure 3.14(e). This completes the graphical computation of the convolution.

EXERCISE 3.5.3

Verify that $y(1) = 1$ and $y(4) = -1$ in Figure 3.14. ■

EXERCISE 3.5.4

Depending on which form of (3.56) is used, we can flip either $h(t)$ or $u(t)$. Recompute the convolution in Figure 3.14 by flipping $u(t)$. ■

*3.5.4 **Impulse Response and Unit Step Response**

In the discrete-time case, the impulse sequence $\{1, 0, 0, 0, \ldots\}$ can be easily generated. In the continuous-time case, however, the impulse cannot be generated in practice because it has a zero width and yet a finite area. If no impulse can be generated, there is no impulse response to be measured. In this section, we will discuss a method of obtaining the impulse response without generating an impulse. Let $q(t - t_0)$ be the unit step function defined by

$$q(t - t_0) := \begin{cases} 1 & \text{for } t \geq t_0 \\ 0 & \text{for } t < t_0 \end{cases} \tag{3.61}$$

as shown in Figure 3.15(a). We show that the impulse and unit step function are related by

$$\boxed{\delta(t - t_0) = \frac{d}{dt}[q(t - t_0)]} \tag{3.62}$$

Indeed, if we approximate $q(t - t_0)$ by the function $q_\varepsilon(t - t_0)$ shown in Figure 3.15(b), then the derivative of $q_\varepsilon(t - t_0)$ equals the pulse defined in Figure 3.10(c), that is,

Figure 3.15 *Unit step function*

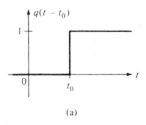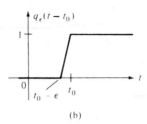

(a)   (b)

$$\delta_\varepsilon(t - t_0) = \frac{d}{dt} q_\varepsilon(t - t_0)$$

 which implies (3.62) as ε approaches zero. Thus, the impulse can be generated by differentiating a unit step function.

EXAMPLE 3.5.5

Consider the discontinuous function shown in Figure 3.16(a). It can be expressed, using (3.61), as

$$f(t) = 2q(t) + q(t-1) - 4q(t-2) + 2q(t-3) - q(t-5) \qquad \text{for } t \geq 0$$

Its derivative is

$$\dot{f}(t) := \frac{df(t)}{dt} = 2\delta(t) + \delta(t-1) - 4\delta(t-2) + 2\delta(t-3) - \delta(t-5)$$

Figure 3.16 *A time function and its derivative*

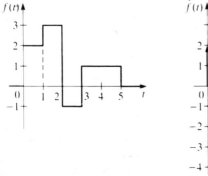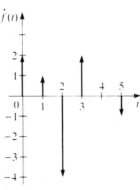

(a)   (b)

and is shown in Figure 3.16(b). The magnitude or, more precisely, the area of each impulse is indicated by the height of the arrow.

EXAMPLE 3.5.6

Let $f(t) = e^{-0.5t}$ for $t \geq 0$ and $f(t) = 0$ for $t < 0$. This is an exponentially decreasing function with a discontinuity at $t = 0$. The differentiation of $f(t)$ yields

$$\dot{f}(t) = \delta(t) - 0.5e^{-0.5t} \tag{3.63}$$

Because of the discontinuity at $t = 0$, $\dot{f}(t)$ contains an impulse with area 1 at $t = 0$, as shown in Figure 3.17. Equation (3.63) can also be obtained as follows. We use (3.61) to write $f(t)$ as $f(t) = e^{-0.5t}q(t)$. Then we have

$$\dot{f}(t) = e^{-0.5t}\dot{q}(t) + (-0.5)e^{-0.5t}q(t) = e^{-0.5t}\delta(t) - 0.5e^{-0.5t}q(t)$$
$$= \delta(t) - 0.5e^{-0.5t}q(t)$$

where we have used (3.49). This is the same as (3.63).

Now we use (3.62) to develop the impulse response of LTI systems. Consider an LTI system with impulse response $h(t)$, $t \geq 0$. Let $y_q(t)$ be the unit step response, or the zero-state response of the system excited by the unit step input $q(t)$. Because $q(t - \tau) = 1$ for $0 \leq \tau \leq t$, (3.56) becomes

$$y_q(t) = \int_0^t h(\tau)q(t - \tau)d\tau = \int_0^t h(\tau)d\tau \qquad \text{for } t \geq 0 \tag{3.64}$$

$$\text{cause } q = 1$$

Figure 3.17 *Differentiation of* $e^{-0.5t}$

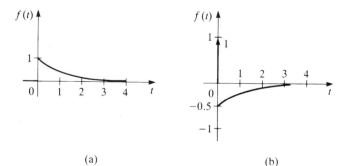

(a)

(b)

The differentiation of (3.64) yields

$$h(t) = \frac{d}{dt}y_q(t)$$ (3.65)

It means that the impulse response equals the differentiation of the unit step response. The unit step function, unlike the impulse, can be easily generated, and the corresponding unit step response can then be measured. Once the step response is obtained, the impulse response can be obtained by differentiation.

The use of (3.65) to compute the impulse response does not require the generating of an impulse and appears to be feasible in practice. Unfortunately, this is not necessarily so. In practice, measurement is often corrupted by noise. If $y_q(t)$ is corrupted by noise, then (3.65) cannot be used, as is demonstrated in the following example.

EXAMPLE 3.5.7

Consider the network shown in Figure 3.13. The voltage source is replaced by a 1 volt battery and a switch, as shown in Figure 3.18. The closure of the switch at $t = 0$ is the same as applying a unit step function. Suppose the response is measured as

$$y_q(t) = 1 - e^{-0.5t} \qquad \text{for } t \geq 0$$ (3.66)

with $y_q(t) = 0$, $t < 0$ as shown in Figure 3.18(b). The differentiation of $y_q(t)$ yields

$$h(t) = \dot{y}_q(t) = 0.5e^{-0.5t}$$

This is the same as (3.59) and is plotted in Figure 3.18(c).

Suppose now the measured $y_q(t)$ is corrupted by high frequency noise. To simplify discussion, we will assume that the noise can be represented by $0.01\sin 1000t$ and the measured step response becomes

Figure 3.18 *Impulse response from step response*

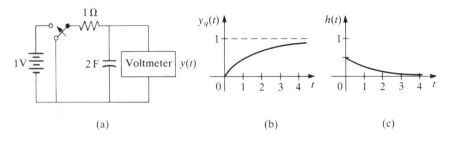

(a)   (b)   (c)

$$\overline{y}_q(t) = 1 - e^{-0.5t} + 0.01\sin 1000t \tag{3.67}$$

Because the noise has a very small amplitude, its presence can hardly be detected in $\overline{y}_q$ and $y_q \approx \overline{y}_q$. The differentiation of (3.67) yields

$$\overline{h}(t) := \frac{d}{dt}\overline{y}_q(t) = 0.5e^{-0.5t} + 0.01 \times 1000 \cos 1000t \tag{3.68}$$

$$= 0.5e^{-0.5t} + 10 \cos 1000t$$

Although the noise is negligible in $\overline{y}_q$, it becomes the dominant term in $\overline{h}$ after differentiation. We plot $\overline{h}(t)$ in Figure 3.19; it is quite different from $h(t)$ plotted in Figure 3.18(c). Therefore, although the impulse response can be obtained by measurement and differentiation in theory, we may encounter difficulties in practice.

In addition to measurement and differentiation, the impulse response can also be obtained by analysis, as will be discussed in Chapter 4. Although the impulse is a mathematical entity, it is a very useful concept in engineering. It leads to the concept of convolution, which is basic in system and signal analysis. In deriving the convolution, the concepts of linearity, time invariance, and causality are employed.

3.6   LTIL CONTINUOUS-TIME SYSTEMS— DIFFERENTIAL EQUATIONS

In the preceding section, we showed that the zero-state response of every linear time-invariant continuous-time system can be described by

Figure 3.19 *Impulse response obtained by differentiation of* $\overline{y}_q(t)$

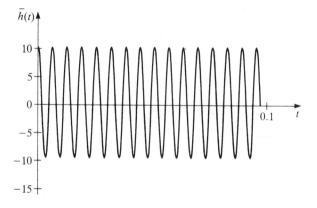

$$y(t) = \int_0^t h(t - \tau)u(\tau)d\tau$$

It is developed using only the conditions of linearity, time invariance, and causality. The description, although very general, is not convenient for analysis and design. In this section, we shall develop a different mathematical description by requiring systems to be lumped. The development follows Section 3.3 closely; therefore, the discussion will be brief.

EXAMPLE 3.6.1

Consider the network in Figure 3.13(b). Its impulse response is, as was discussed in (3.59),

$$h(t) = 0.5e^{-0.5t} \qquad \text{for } t \geq 0 \tag{3.69}$$

and the zero-state response of the network is

$$y(t) = \int_0^t h(t - \tau)u(\tau)d\tau = \int_0^t 0.5e^{-0.5(t-\tau)}u(\tau)d\tau \tag{3.70}$$

Using the formula

$$\frac{d}{dt}\int_0^t g(t,\tau)d\tau = \int_0^t \frac{d}{dt}[g(t,\tau)]d\tau + g(t,\tau)\Big|_{\tau=t} \tag{3.71}$$

we differentiate (3.70) as

$$\dot{y}(t) := \frac{dy(t)}{dt} = \frac{d}{dt}\left[\int_0^t 0.5e^{-0.5(t-\tau)}u(\tau)d\tau\right]$$

$$= \int_0^t 0.5(-0.5)e^{-0.5(t-\tau)}u(\tau)d\tau + 0.5e^{-0.5(t-\tau)}u(\tau)\Big|_{\tau=t}$$

$$= -0.5\int_0^t 0.5e^{-0.5(t-\tau)}u(\tau)d\tau + 0.5u(t)$$

This equation becomes, after the substitution of (3.70),

$$\dot{y}(t) = -0.5y(t) + 0.5u(t)$$

or

$$\dot{y}(t) + 0.5y(t) = 0.5u(t) \tag{3.72}$$

This is called a first-order linear ordinary differential equation with constant coefficients or a first-order LTIL differential equation.

___

As in the discrete-time case, the LTIL differential equation is simpler than the convolution in representation and computation. The differential equation is also more general in the sense that it describes both the zero-state response and zero-input response. Therefore, if a system is describable by both a convolution and an ordinary differential equation, we prefer to use the latter. However, not every LTI continuous-time system can be described by a differential equation. Only if the system is also lumped can it be so described.

3.7 SETTING UP DIFFERENTIAL EQUATIONS

As in the discrete-time case, there are two approaches to develop differential equations for LTIL continuous-time systems. In the first approach, we do not have to know the internal structure of a system. We apply an input, in particular, an impulse, and measure its zero-state response or impulse response. We then try to develop a differential equation from the measured data. In the second approach, we must know the internal structure of a system. We then apply physical laws, such as Newton's and Kirchhoff's laws, to develop differential equations. This approach can be illustrated by examples.

EXAMPLE 3.7.1

(**Mechanical system**) Consider the mechanical system shown in Figure 3.20(a). It consists of a block with mass m connected to a wall by a spring. The input $u(t)$ is the force applied to the block, and the output $y(t)$ is the distance measured from the equilibrium position. Strictly speaking, the spring is a nonlinear

Figure 3.20 *(a) Mechanical system (b) Friction*

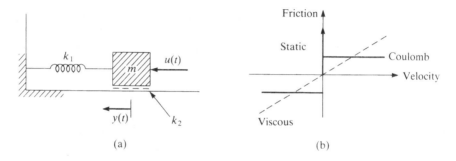

(a)

(b)

element with the characteristic shown in Figure 2.7(a). However, if the displacement is limited to the range $[A, B]$, then it can be modeled as a linear element with spring constant k_1. This means that the force generated by the spring equals $k_1 y(t)$. The friction between the block and floor is very complex and generally consists of three parts: static, Coulomb, and viscous frictions, as shown in Figure 3.20(b). Note that the coordinates are friction versus velocity. When the block is stationary (i.e., the velocity is zero), we need a certain amount of force to overcome the static friction to start its movement. Once the block is moving, there is a constant friction, called the Coulomb friction, which is independent of velocity. The viscous friction is modeled as

$$\text{viscous friction} = k_2 \times \text{velocity}$$

where k_2 is called the viscous friction coefficient. It is a linear relationship. Most texts on general physics discuss only static and Coulomb frictions. We consider, however, only viscous friction; static and Coulomb frictions will be disregarded. By so doing, we can consider the friction as a linear phenomenon.

Consider now the mechanical system in Figure 3.20(a) with spring modeled as $k_1 y$ and friction modeled as $k_2 \dot{y}(t) := k_2 dy(t)/dt$. Newton's law states that the applied force must overcome the spring and friction forces, and the remainder is used to accelerate the block. Therefore, we have

$$u(t) - k_1 y(t) - k_2 \dot{y}(t) = m\ddot{y}(t)$$

where $\ddot{y}(t) := d^2 y(t)/dt^2$, or

$$m\ddot{y}(t) + k_2 \dot{y}(t) + k_1 y(t) = u(t) \tag{3.73}$$

This second-order LTIL differential equation describes the mechanical system.

EXAMPLE 3.7.2

(**Suspension system**) The suspension system of an automobile, shown in Figure 3.21(a), can be modeled as shown in Figure 3.21(b). The model consists of a block with mass m, which denotes the weight of the automobile. When the automobile hits a pothole, a vertical force $u(t)$ is applied to the mass and the vertical displacement $y(t)$ of the block will vary with time. The displacement is to be measured from equilibrium, that is, if $u(t) = 0$ then $y(t) = 0$. The suspension system consists of a spring with spring constant k_1 and a dashpot which represents the shock absorber. The dashpot generates a viscous friction proportional to $\dot{y}(t)$. If the viscous friction coefficient is k_2, then we have

Figure 3.21 *Suspension system*

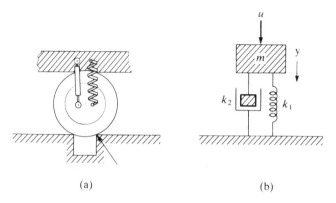

(a)   (b)

$$u(t) - k_1 y(t) - k_2 \dot{y}(t) = m\ddot{y}(t)$$

or

$$m\ddot{y}(t) + k_2 \dot{y}(t) + k_1 y(t) = u(t)$$

which is the same as (3.73) and describes the suspension system in Figure 3.21.

EXAMPLE 3.7.3

(**Electrical system**) Consider the network shown in Figure 3.13(b). Its impulse response is $h(t) = 0.5e^{-0.5t}$. From the impulse response, we have developed the differential equation (3.72) to describe the network. Now we shall develop the differential equation from the structure of the network by using Kirchhoff's current and voltage laws. Let $y(t)$ be the voltage across the capacitor with capacitance $C = 2$ farads. Then its current is $C dy(t)/dt = 2\dot{y}(t)$. This current passes through the 1 Ω resistor and causes a voltage drop of $1 \times 2\dot{y}(t) = 2\dot{y}(t)$. The application of Kirchhoff's voltage law around the loop yields

$$u(t) - 2\dot{y}(t) - y(t) = 0$$

or

$$2\dot{y}(t) + y(t) = u(t) \tag{3.74}$$

This is the same as (3.72).

*EXAMPLE 3.7.4

(**Electromechanical system**) The torque required to drive antennas, telescopes, or video tapes in a VCR can be generated by motors. Consider the DC motor shown in Figure 3.22. It is called a field-controlled DC motor if the input signal $u(t)$ is applied to the field circuit, as shown, and the armature current i_a is kept constant. The generated torque $T(t)$ is proportional to the field current i_f, written as $T(t) = k_t i_f(t)$. The torque is used to drive a load with moment of inertia J. The viscous friction coefficient between the motor shaft and bearing is f. The angular displacement of the load is denoted by $\theta(t)$. If we identify the following between translational and angular movements,

Translational movement		Angular movement
Force	$\longleftrightarrow$	Torque
Mass	$\longleftrightarrow$	Moment of inertia
Displacement	$\longleftrightarrow$	Displacement

then we have, as in (3.73) with $k_1 = 0$,

$$J\ddot{\theta}(t) + f\dot{\theta}(t) = T(t) \tag{3.75}$$

where $\dot{\theta} = d\theta / dt$ and $\ddot{\theta} = d^2\theta / dt^2$. That is, the torque is used to overcome the viscous friction and to accelerate the load. In the equation, we have disregarded the Coulomb and static frictions and considered only the viscous friction, as in

Figure 3.22 *Field-controlled DC motor*

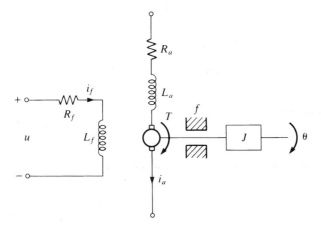

the translational case. Because of the absence of a spring, (3.75) has one term less than (3.73). From the field circuit, we can readily obtain

$$u(t) = R_f i_f(t) + L_f \dot{i}_f(t) \tag{3.76}$$

where R_f and L_f are, respectively, the resistance and inductance of the field circuit. Now we shall develop a differential equation from (3.75), (3.76), and $T(t) = k_t i_f$ to relate the input u and output θ of the motor. The differentiation of (3.75) yields

$$J\theta^{(3)}(t) + f\ddot{\theta}(t) = \dot{T}(t)$$

where $\theta^{(3)} = d^3\theta / dt^3$. The equation becomes, after the substitution of $\dot{T} = k_t \dot{i}_f$ and (3.76),

$$J\theta^{(3)}(t) + f\ddot{\theta}(t) = k_t \dot{i}_f(t) = \frac{k_t}{L_f}[u(t) - R_f i_f(t)] \tag{3.77}$$

Substituting $i_f = T / k_t$ and (3.75) into (3.77) yields

$$J\theta^{(3)}(t) + f\ddot{\theta}(t) = \frac{k_t}{L_f}u(t) - \frac{R_f}{L_f}[f\dot{\theta}(t) + J\ddot{\theta}(t)]$$

which can be simplified as

$$J\theta^{(3)}(t) + (f + \frac{JR_f}{L_f})\ddot{\theta}(t) + \frac{fR_f}{L_f}\dot{\theta}(t) = \frac{k_t}{L_f}u(t) \tag{3.78}$$

This third-order LTIL differential equation describes the DC motor.

---

*EXAMPLE 3.7.5

(**Hydraulic tanks**) In chemical plants, it is often necessary to maintain the levels of liquids. A simplified model is shown in Figure 3.23 in which

  q_i, q_1, q_2 = rates of the flow of liquid
  A_1, A_2 = areas of the cross-section of tanks
  h_1, h_2 = liquid levels
  R_1, R_2 = flow resistance, controlled by valves

It is assumed that q_1 and q_2 are governed by

Figure 3.23   *Control of liquid levels*

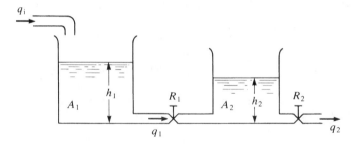

$$q_1 = \frac{h_1 - h_2}{R_1} \quad \text{and} \quad q_2 = \frac{h_2}{R_2} \tag{3.79}$$

They are proportional to relative liquid levels and inversely proportional to flow resistances. The changes of liquid levels are governed by

$$A_1 dh_1 = (q_i - q_1)dt$$

and

$$A_2 dh_2 = (q_1 - q_2)dt$$

or

$$A_1 \dot{h}_1(t) = q_i - q_1 \tag{3.80a}$$

and

$$A_2 \dot{h}_2(t) = q_1 - q_2 \tag{3.80b}$$

Let q_i and q_2 be the input and output of the system. Now we shall develop a differential equation to describe them. The differentiation of (3.79) yields

$$\dot{h}_2 = R_2 \dot{q}_2 \quad \text{and} \quad \dot{h}_1 = R_1 \dot{q}_1 + \dot{h}_2 = R_1 \dot{q}_1 + R_2 \dot{q}_2$$

The substitution of these into (3.80) yields

$$q_i - q_1 = A_1(R_1 \dot{q}_1 + R_2 \dot{q}_2) \tag{3.81a}$$

$$q_1 - q_2 = A_2 R_2 \dot{q}_2 \tag{3.81b}$$

Now we eliminate q_1 from (3.81a) to yield, using (3.81b) and its derivative,

$$q_i - (q_2 + A_2 R_2 \dot{q}_2) = A_1 R_1 (\dot{q}_2 + A_2 R_2 \ddot{q}_2) + A_1 R_2 \dot{q}_2$$

which can be simplified as

$$A_1 A_2 R_1 R_2 \ddot{q}_2(t) + (A_1 R_1 + A_2 R_2 + A_1 R_2) \dot{q}_2(t) + q_2(t) = q_i(t) \quad \textbf{(3.82)}$$

This second-order LTIL differential equation describes the input q_i and output q_2 of the system in Figure 3.23.

---

*EXAMPLE 3.7.6

(**RC network**) Consider the network shown in Figure 3.24. The input u is a current source. The output is chosen as the voltage across the branch AC, as shown. We assume that the capacitor in Figure 3.24 has capacitance -1 farad; such a negative capacitance can be generated using an operational amplifier circuit. Let the voltage across the capacitor with the chosen polarity be x. Then its current is equal to $-\dot{x}(t)$. The current in the branch AC is clearly equal to $y/2 = 0.5y$, and the current in the branch AB is $u - 0.5y$. The currents in the branches CD and BD are, respectively, $0.5y - \dot{x}$ and $u - 0.5y + \dot{x}$.

$\iota = C\dfrac{dx}{dt}$ ⟶

Now we shall use Kirchhoff's voltage law to develop a differential equation to describe the network. From the loop ABC, we have

$$2 \times (u - 0.5y) + x = y$$

which can be simplified as

$$\therefore \ y = 0.5x + u \quad \textbf{(3.83)}$$

(handwritten annotations:)
$2 = R_1, \quad y = 2(u-0.5y)+x$
$\Rightarrow 2u - 0.5(2)y + x = y$
$2u - y + x = y$
$2u + x = 2y$

Figure 3.24 *Network*

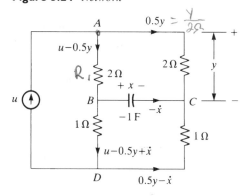

(figure labels:) A; $0.5y = \frac{y}{2\Omega}$; $u-0.5y$; $R_1 \ 2\Omega$; $+x-$; 2Ω; y; u; B; $-\dot{x}$; $-1\,\mathrm{F}$; C; 1Ω; 1Ω; $u-0.5y+\dot{x}$; D; $0.5y-\dot{x}$

From the loop *BCD*, we can readily obtain

$$x + 0.5y - \dot{x} = u - 0.5y + \dot{x}$$

$$x + 0.5y + 0.5y - u = \dot{x} + \dot{x}$$

or

$$x + y - u = 2\dot{x} \tag{3.84}$$

The substitution of (3.83) into (3.84) yields

$$x + 0.5x + u - u = 2\dot{x}$$

which implies

$$\dot{x}(t) = 0.75x(t) \tag{3.85}$$

The differentiation of (3.83) and substitution of (3.85) and (3.83) yield

$$\dot{y} = 0.5\dot{x} + \dot{u} = 0.5 \times 0.75x + \dot{u} = 0.375 \times 2(y - u) + \dot{u}$$

which can be simplied as

$$\dot{y}(t) - 0.75y(t) = \dot{u}(t) - 0.75u(t) \tag{3.86}$$

This first-order LTIL differential equation describes the network in Figure 3.24.

EXAMPLE 3.7.7

(**Block diagram**) Consider the system shown in Figure 3.25(a). It is built by interconnecting the three types of elements shown in Figure 3.25(b). The element denoted by a box is called an integrator. The input $u(t)$ and output $y(t)$ of an integrator is related by

$$y(t) = \int_{t_0}^{t} u(\tau)d\tau + u(t_0) \quad \text{(integrator)}$$

It is, however, much more convenient to write its input and output as $\dot{u}(t)$ and $u(t)$. Note that an initial condition must be assigned to every integrator. The element denoted by a big circle with real number α is called a multiplier with gain α. Its input $u(t)$ and output $y(t)$ are related by

$$y(t) = \alpha u(t) \quad \text{(multiplier)}$$

Figure 3.25 *Continuous-time basic block diagram*

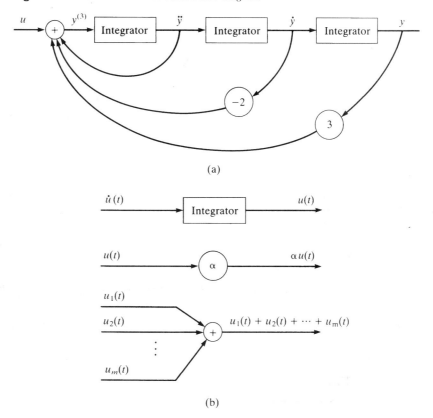

(a)

(b)

It is also called an amplifier if $|a| > 1$ and an attenuator if $|a| < 1$. The element denoted by a small circle with a plus sign is called an adder. Every adder has two or more inputs $u_i(t)$, $i = 1, 2, \ldots$, but has one and only one output $y(t)$. They are related by

$$y(t) = u_1(t) + u_2(t) + \cdots + u_m(t) \quad \text{(adder)}$$

These three elements can be easily built using operational amplifier circuits. Note that differentiators are not used because they will amplify high frequency noise and are difficult to build in practice (see Chapter 9). A diagram built from these elements is called a continuous-time basic block diagram. To develop a differential equation to describe the basic block diagram in Figure 3.25(a), we assign the output of the right-most integrator as y. Then the inputs of the three integrators are, respectively, $\dot{y}$, $\ddot{y}$, and $y^{(3)}$, as shown. Equating the inputs and output of the summer in Figure 3.25(a) yields

$$y^{(3)}(t) = \ddot{y}(t) - 2\dot{y}(t) + 3y(t) + u(t)$$

or

$$y^{(3)}(t) - \ddot{y}(t) + 2\dot{y}(t) - 3y(t) = u(t) \tag{3.87}$$

This third-order LTIL differential equation describes the block diagram in Figure 3.25(a). Note that the block diagram in Figure 3.25(a) is identical to the one in Figure 3.7 if every integrator is replaced by a unit-time delay element. Compare also Equations (3.87) and (3.26).

---

*EXAMPLE 3.7.8

(**Block diagram**) Consider the basic block diagram shown in Figure 3.26. It is obtained from Figure 3.8(b) by replacing every unit-time delay element by an integrator. The input $u(t)$ and output $y(t)$ are chosen as shown. In order to develop a differential equation to relate $u(t)$ and $y(t)$, we need an intermediate variable. Let the output of the integrator on the right-hand side be $w(t)$. Then its input is $\dot{w}(t)$ and the input of the integrator on the left-hand side is $\ddot{w}(t)$. Using these variables, we can obtain from the summer on the left-hand side the following equation

$$\ddot{w}(t) = -2\dot{w}(t) - 3w(t) + 10u(t) \tag{3.88}$$

From the summer on the right-hand side we have

$$y(t) = \dot{w}(t) + 4w(t) \tag{3.89}$$

In order to develop an equation to relate $u(t)$ and $y(t)$, we must eliminate $w(t)$ from (3.88) and (3.89). Differentiations of (3.88) and (3.89) yield

Figure 3.26   *Continuous-time basic block diagram*

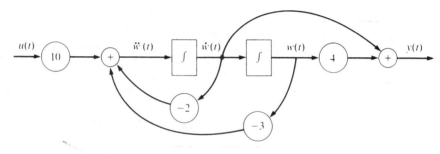

$$\ddot{w}(t) = -2\dot{w}(t) - 3w(t) + 10u(t)$$

$$w^{(3)}(t) + 2\ddot{w}(t) + 3\dot{w}(t) = 10\dot{u}(t)$$

$$y(t) = \dot{w}(t) + 4w(t)$$

$$\dot{y}(t) = \ddot{w}(t) + 4\dot{w}(t)$$

and

$$\ddot{y}(t) = w^{(3)}(t) + 4\ddot{w}(t)$$

Using these equations, it is straightforward to verify

$$\ddot{y}(t) + 2\dot{y}(t) + 3y(t) = 10\dot{u}(t) + 40u(t) \tag{3.90}$$

This second-order differential equation describes the basic block diagram in Figure 3.26. Compare this equation with (3.30).

Although every basic block diagram can be described by an LTIL differential equation, developing such an equation is not necessarily a simple task, as the preceding example showed. However, it is simple and straightforward to develop a state-variable equation, which will be introduced in Chapter 10, to describe any basic block diagram. From the state-variable equation, we can then develop a differential equation to describe the diagram.

EXERCISE 3.7.1

Find an LTIL differential equation to describe the block diagram in Figure 3.8(a) with unit-time delay elements replaced by integrators.

[**ANSWER:** $\ddot{y} + 2\dot{y} + 3y = 10u$]

The previous examples showed that LTIL differential equations do arise from LTI lumped systems. Now let us discuss the general form of LTIL differential equations. Consider

nth order LTIL differential equation

$$a_n y^{(n)}(t) + a_{n-1} y^{(n-1)}(t) + \cdots + a_1 y^{(1)}(t) + a_0 y(t)$$
$$= b_m u^{(m)}(t) + b_{m-1} u^{(m-1)}(t) + \cdots + b_1 u^{(1)}(t) + b_0 u(t) \tag{3.91}$$

where a_i and b_i are real constants, not necessarily all nonzero, $a_n \neq 0$,

$$y^{(i)}(t) := \frac{d^i}{dt^i} y(t) \quad \text{for } i = 1, 2, \ldots$$

and $u^{(i)}(t)$ are similarly defined. For $i = 1, 2$, we also write

$$\dot{y}(t) := y^{(1)}(t) := \frac{d}{dt}y(t)$$

$$\ddot{y}(t) := y^{(2)}(t) := \frac{d^2}{dt^2}y(t)$$

Equation (3.91) is called an nth order LTIL differential equation.

In the discrete-time case, we require $m \leq n$ in (3.35) for that equation to be causal. If $m > n$, $y[k]$ will depend on the future input $u[i]$ with $i > k$ and (3.35) is noncausal. Now consider this problem for (3.91): Assuming, for convenience, $n = 0$ and $m = 1$, then

$$y(t) = \dot{u}(t) = \frac{d}{dt}u(t)$$

This is a (pure) differentiator. If the differentiation is defined as

$$y(t) = \frac{d}{dt}u(t) = \lim_{\Delta \to 0} \frac{u(t + \Delta) - u(t)}{\Delta} \tag{3.92}$$

with $\Delta > 0$, then the output $y(t)$ depends on the future input $u(t + \Delta)$ and the differentiator is not causal. However, if we define the differentiation as

$$y(t) = \frac{d}{dt}u(t) := \lim_{\Delta \to 0} \frac{u(t) - u(t - \Delta)}{\Delta} \tag{3.93}$$

then $y(t)$ does not depend on the future input and the differentiator is causal. Therefore, whether or not a differentiator—or, more generally, Equation (3.91) with $m > n$—is noncausal is open to argument.

However, we still require $m \leq n$ in (3.91). This restriction is imposed for reasons other than causality. In practice, signals are often corrupted by high frequency noise. If $m > n$, the noise will be amplified, as shown in Figure 3.19, and the system may become useless. Therefore, we require $n \geq m$ in (3.91). In conclusion, in the discrete-time case, we require $n \geq m$ in (3.35) for causality; in the continuous-time case, we require $n \geq m$ in (3.91) for the noise problem.

3.8   SUMMARY

1. Impulse sequences can be easily generated. Impulse or δ-functions, however, cannot be generated in practice. Many functions can be used to define the impulse. Because the value of the impulse at $t = 0$ is not uniquely defined, the impulse is not a function in the usual sense (a function, by definition, must be uniquely defined at every t).

2. The impulse response is, by definition, the output of an initially relaxed

system excited by an impulse input. If the system is causal, then the impulse response is identically zero before the application of the impulse.

3. Convolutions are developed using only linearity (homogeneity and additivity) and time invariance. Therefore, every linear time-invariant system can be described by a convolution.

4. Graphical computation of convolutions involves four steps: flipping, shifting, multiplication, and summation or integration. Convolution with an impulse achieves a shifting. See Figure P3.23.

5. A discrete-time system is called an FIR system if its impulse response has a finite number of nonzero entries; it is called an IIR system if its impulse response has an infinite number of nonzero entries. The moving-average filter is an FIR system, and is widely used to compute the trend of sequences that change rapidly.

6. If a system is linear, time invariant, and lumped, then it can be described by a difference (differential) equation as well as a convolution. The difference equation is preferable to the convolution because it describes not only the zero-state response but also the zero-input response. Furthermore, it requires less computation. However, the difference equation cannot be used to describe distributed systems.

7. The difference equation is an algebraic equation and its solution can be obtained by direct substitution.

8. In the continuous-time case, the impulse response equals the differentiation of the unit step response. Although the unit step response can be easily measured, differentiation is rarely used in practice because of unavoidable high-frequency noise in measurement. Therefore, it is not simple to obtain impulse responses by measurement and differentiation.

PROBLEMS

3.1 Consider an LTI discrete-time system with impulse response sequence $h[k] = 1/4$, $k = 0, 1, 2, 3$. What is its zero-state output excited by the input $u[k]$, $k \geq 0$? Is the system FIR or IIR? Compute $y[k]$, for $k = 0, 1, \ldots, 20$, if $u[k] = 1$ for $k = 0, 1, 2$, and $u[k] = 0$ for $k > 2$.

3.2 Find the output $y[k]$ in Problem 3.1 using the graphical method discussed in Figure 3.4.

3.3 Consider an LTI discrete-time system with impulse response sequence $h[k] = e^{0.2k}$, $k = 0, 1, 2, \ldots$. What is the zero-state response $y[k]$ of the system excited by the input $u[k]$? Is the system an FIR or IIR system? Compute $y[k]$, $k = 0, 1, 2, 3$, if $u[k] = e^{-0.2k}$, $k = 0, 1, 2, 3, \ldots$.

3.4 Consider an LTI discrete-time system with the impulse-response sequence shown in Figure P3.4. What is the zero-state response $y[k]$ of the system? Is the system an FIR or an IIR system? Compute $y[k]$, $k = 0, 1, 2, \ldots$,

Figure P3.4

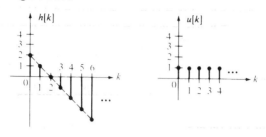

excited by the input shown, by direct substitution and by graphical method. Are the results the same?

3.5 Show that for an FIR system of length N, that is, $h[k] = 0$ for $k \geq N$, if $u[k] = 0$ for $k > k_0$, then $y[k] = 0$ for $k \geq k_0 + N$.

3.6 Consider an LTI discrete-time system with impulse response $h[0] = 0$ and $h[k] = 1$, for $k = 1, 2, \ldots$. Find a difference equation to describe the system.

3.7 Repeat Problem 3.6 if $h[k] = e^k$, $k = 0, 1, 2, \ldots$ and if $h[0] = 0$, $h[k] = e^{k-1}$, $k = 1, 2, \ldots$.

3.8 Consider the difference equation

$$y[k+1] + a_0 y[k] = b_1 u[k+1] + b_0 u[k]$$

If one multiplication requires 1 μs (microsecond = 10^{-6} second) and one addition or subtraction requires 0.4 μs, what conditions must be imposed on b_1 and on the sampling period in order to process the equation in real time without any time delay?

3.9 (a) Show

$$1 + a + \cdots + a^k = \frac{1 - a^{k+1}}{1 - a}$$

(b) Show, if $|a| < 1$, then

$$\sum_{k=0}^{\infty} a^k = 1 + a + \cdots + a^k + \cdots = \frac{1}{1 - a}$$

(c) What is the summation in (b) if $|a| > 1$? Is it true that the summation always equals infinity if $|a| = 1$? If not, give a counterexample.

3.10 Consider (3.36) or

$$y[k] + a_{n-1}y[k-1] + \cdots + a_1y[k-n+1] + a_0y[k-n]$$
$$= b_m u[k+m-n] + b_{m-1}u[k+m-n-1] + \cdots$$
$$+ b_1 u[k-n+1] + b_0 u[k-n]$$

Given $u[k]$, $k \geq 0$, in order to compute $y[k]$, $k \geq 0$, what are the required initial conditions? If $u[k] = 0$ for $k < 0$, what are the required initial conditions?

3.11 In Problem 3.10, set $m = n - 1$ and rename the coefficients to yield

$$y[k] + \alpha_1 y[k-1] + \alpha_2 y[k-2] + \cdots + \alpha_n y[k-n]$$
$$= \beta_1 u[k-1] + \beta_2 u[k-2] + \cdots + \beta_n u[k-n]$$

It can be reduced to, if some of the coefficients are zero,

$$y[k] + \alpha_1 y[k-1] + \cdots + \alpha_i y[k-i]$$
$$= \beta_1 u[k-1] + \beta_2 u[k-2] + \cdots + \beta_j u(k-j)$$

Show that this equation describes a causal system whether or not $i \geq j$ or $j \geq i$. This equation is widely used in the control literature.

3.12 Suppose a discrete-time system is described by

$$y[k+1] + 0.5y[k] = 2u[k+1] \qquad \text{for } k = -1, 0, 1, 2, \ldots$$

what is the impulse response sequence of the system? Recall that the impulse response is the output excited by zero initial conditions $y[-1] = 0$ and $u[k] = \delta[k]$.

3.13 Consider

$$y[k+1] + a_0 y[k] = b_1 u[k+1] + b_0 u[k]$$

What is its impulse response sequence $h[k]$? Show that $h[0] = 0$ if and only if $b_1 = 0$.

3.14 Which of the following difference equations are recursive equations? Which describe FIR systems?

(a) $2y[k+2] - y[k+1] - 4y[k] = u[k+1] - 3u[k]$

(b) $y[k] = u[k] - u[k-2] - u[k-3]$

3.15 Show that the nonrecursive equation

$$y[k] = 0.2(u[k] + u[k-1] + u[k-2] + u[k-3] + u[k-4])$$

is basically the same as the recursive equation

$$y[k] - y[k-1] = 0.2(u[k] - u[k-5])$$

Which equation is simpler from the computational point of view?

3.16 Suppose we borrow $10,000 to buy a house and are required to pay it off in thirty years, or 360 equal monthly payments. The unpaid debt will carry a 1% monthly interest rate. What is the monthly payment? What percentage of the first payment goes to pay off the debt?

※ 3.17 Find difference equations to describe the two basic block diagrams in Figure P3.17

※ 3.18 Show that the basic block diagram in Figure P3.18 can be described by

$$y[k] + a_0 y[k-1] = b_1 u[k] + b_0 u[k-1]$$

3.19 Compute the following

(a) $\int_0^5 \sin 2t \, \delta(t-3) dt$ $\overset{t-t_0}{=}$ $\sin 6$ $\notin$

$\delta(t)$ $\quad$ $t=3$ $\quad$ 3

(b) $\int_0^5 \sin 2t \, \delta(t-10) dt$ $= 0$

(c) $\int_0^\infty t^2 \delta(t-3) dt$ $= 3^2$

(d) $\int_0^3 t^3 \delta(t-3) dt$ $\qquad$ 3

3.20 Consider an LTI continuous-time system with impulse response $h_1(t) = \cos 2t$. What is its convolution description? Use correct limits of

Figure P3.17

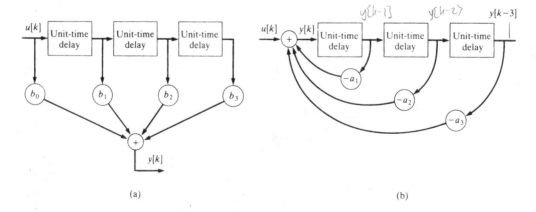

(a) $\qquad\qquad\qquad\qquad\qquad\qquad\qquad\qquad\qquad$ (b)

Figure P3.18

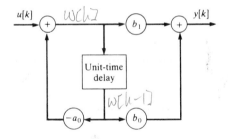

u[k] + $w[k]$ b_1 + y[k]

Unit-time delay

$w[k-1]$

$-a_0$ b_0

integration $(0, 0+, 0-, t, t+, \text{ or } t-)$ in the equation. It is assumed that the input contains no impulses.

3.21 Repeat Problem 3.20 for $h_2(t) = 2\delta(t) + \cos 2t$.

3.22 Compute the convolution integral, using the graphical method discussed in Figure 3.14, for $h_i(t)$ and $u_i(t)$, $i = 1, 2$, given in Figure P3.22. Can you induce that if $h(t) = 0$ for $t < 0$ and $t > P$, and $u(t) = 0$ for $t < 0$ and $t > Q$, then their convolution $y(t)$ has the property $y(t) = 0$ for $t < 0$ and $t > P + Q$? Compare this with Example 3.2.5 for the discrete-time case.

3.23 (a) Find the convolution of

$$y(t) = \int_{-\infty}^{\infty} f_1(t - \tau)f_2(\tau)d\tau$$

where f_i, $i = 1, 2$, are given in Figure P3.23. Note that $f_2(t) = \delta(t - 2)$ is an impulse. Verify that $y(t)$ is given as shown. We see that the convolution of a function and an impulse at t_0 simply shifts the function to t_0.

Figure P3.22

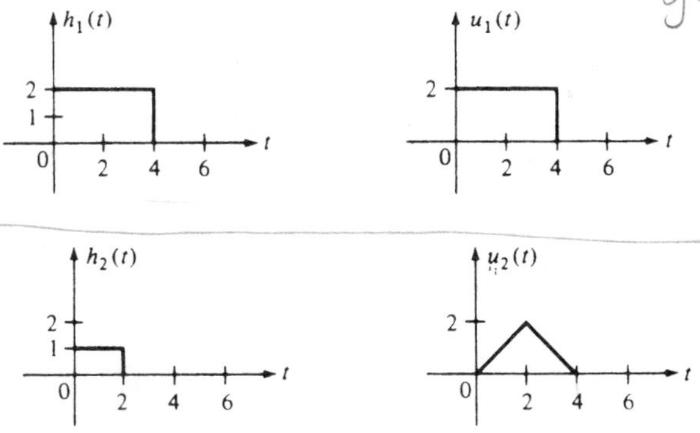

$h_1(t)$

2
1
0   2   4   6   → t

$u_1(t)$

2
0   2   4   6   → t

$h_2(t)$

2
1
0   2   4   6   → t

$u_2(t)$

2
0   2   4   6   → t

$y(t) = \begin{cases} 4t & 0 \le t \le 4 \\ 4(8-t) & 4 \le t \le 8 \\ 0 & t > 8 \end{cases}$

$y(t) = \begin{cases} \frac{t^2}{2} & 0 \le t < 2 \\ 4 - (t^2 - 6t + 10), & 2 < t \le 4 \\ 0.5(6 - t^2), & 4 < t \le 6 \\ 0 & t > 6 \end{cases}$

① $\int_0^t t\, dt$

② $\int_{t-2}^t t\, dt + \int_2^t (4-t)\, dt$ ③ $\int_{t-2}^4 (4-t)\, dt$

Figure P3.23

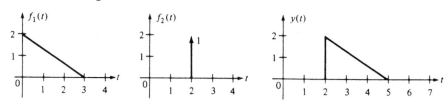

(b) Repeat (a) for $f_2(t) = \delta(t - 1) - 2\delta(t - 6)$.

(c) Repeat (a) for $f_2(t) = \delta(t - 1) + \delta(t - 3)$.

(d) Compute

$$\int_{-\infty}^{\infty} f_1(t) f_2(t)\, dt$$

for the preceding three cases. This is the sifting property in (3.50).

3.24 Find the convolutions of the two functions shown in Figure P3.24 with $P = 10$, $P = 2$, and $P = 1$.

3.25 Find a differential equation to describe a system with impulse response $h(t) = 2\sin 3t$, for $t \geq 0$.

3.26 Find a differential equation to describe a system with impulse response $\delta(t) + 2e^{-2t}$, for $t \geq 0$.

3.27 Find a differential equation to describe the network shown in Figure P3.27.

Figure P3.24

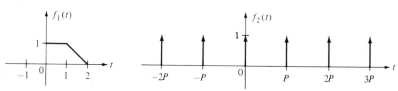

Figure P3.27

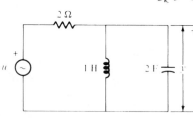

$$i_R = i_L + i_C$$

$$\frac{u(t) - y(t)}{R} = i_L + C\frac{dy}{dt}$$

$$\frac{u' - y'}{R} = \frac{di_L}{dt} + C\frac{d^2y}{dt^2}$$

$$\frac{u' - y'}{R} = \frac{y}{L} + Cy''$$

$$u' - y' = R(\dots)$$

$$u' = \frac{R}{L}y(t) + RCy''(t) + y'(t)$$

3.28 Find a differential equation to describe the network shown in Figure P3.28.

3.29 Consider the pendulum system shown in Figure P3.29, where the input is the horizontal force u and the output is the angular displacement θ. It is assumed that there is no friction in the hinge and no air resistance of the mass. Find a differential equation to describe the system. Is the equation a linear equation? If θ is small so that $\sin \theta$ and $\cos \theta$ can be approximated, respectively, by θ and 1, what is its differential equation description?

3.30 Figure P3.30 shows a model of an accelerometer. The model consists of a block with mass m connected to a box through two springs. The box is filled with liquid to create viscous friction with coefficient f. The box is rigidly attached to, for example, an airplane. Let u and z be the displacements of the box and block. Because the block is, roughly speaking, floating inside the box, we do not have $u = z$. Define $y = z - u$. It is the displacement of the block with respect to the box. Consider u to be the input and y to be the output. Find a differential equation to describe them.

3.31 Let $h[k]$ be the impulse response of an LTI discrete-time system (the out-

Figure P3.28

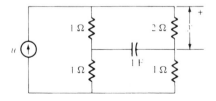

Figure P3.29

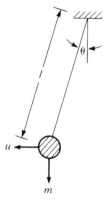

Figure P3.30

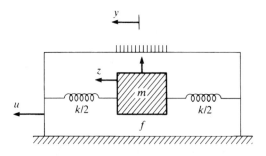

put excited by an impulse sequence) and $y_q[k]$ be the unit step response (the output excited by a unit step input sequence). Show

$$h[k] = y_q[k] - y_q[k - 1]$$

Compare this result with (3.65) for the continuous-time case.

3.32 Find differential equations to describe the basic block diagrams in Figure P3.17 with unit-time delay elements replaced by integrators.

3.33 Find a differential equation to describe the basic block diagram in Figure P3.18 with the unit-time delay element replaced by an integrator.

3.34 Find differential equations to describe the two continuous-time basic block diagrams shown in Figure P3.34. Are they the same?

3.35 Find difference equations to describe the two basic block diagrams shown in Figure P3.34 with the integrators replaced by unit-time delay elements.

3.36 Develop a differential equation to describe the basic block diagram shown in Figure P3.36.

$$x'''(t) = y(t)$$
$$w'''(t) = u(t)$$

Figure P3.34

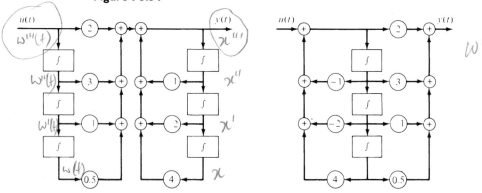

$$\left[0.5\,w(t) + w'(t) + 3w''(t) + 2w'''(t) \right] + 4x(t) + 2x'(t) + x''(t) = x'''(t)$$

Figure P3.36

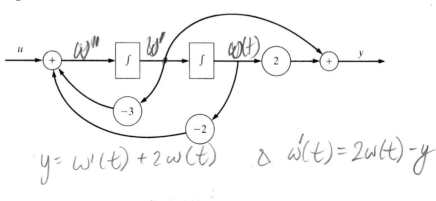

$$y = \omega'(t) + 2\omega(t) \qquad \& \quad \dot{\omega}'(t) = 2\omega(t) - y$$

$$\therefore \quad \underline{y''(t) + 3y'(t) + 2y(t) = u'(t) + 2u(t)}$$

$$\Rightarrow \quad 0.5\omega'(t) + \omega''(t) + 3u(t) + 2u'(t) + 4x'(t) + 2x''(t) +$$
$$+ y(t) = y'(t)$$

$$\Rightarrow \quad 0.5\omega''(t) + u(t) + 3u'(t) + 2u''(t) + 4x''(t) + 2y(t) +$$
$$+ y'(t) = y''(t)$$

$$\Rightarrow \quad 0.5\,u(t) + u'(t) + 3u''(t) + 2u'''(t) + 4y(t) + 2y'(t)$$
$$+ y''(t) = y'''(t)$$

$$\Rightarrow \quad 2u'''(t) + 3u''(t) + u'(t) + 0.5u'(t) = y'''(t) - y''(t)$$
$$- 2y'(t) - 4y(t).$$

USING **MATLAB**

MATLAB can be used to carry out the operations discussed in this chapter. In MATLAB, an array of data can be represented as a row vector. Consider the following two arrays

$$h = [2 \quad -1 \quad 3] \qquad u = [1 \quad 2 \quad -4 \quad 2 \quad 0]$$

Their discrete convolution can be obtained by the command

```
y = conv(h,u)
```

The result is

```
y = 2 3 -7 14 -14 6 0
```

Note that if the length of h is P and the length of u is Q, then the length of y is $P + Q - 1$. As discussed in Exercise 3.2.2, the coefficients of the product of two polynomials is the convolution of the coefficients of the two polynomials. Thus, the product of

$$h(x) = 2x^2 - x + 3 \qquad \text{and} \qquad u(x) = x^4 + 2x^3 - 4x^2 + 2x + 0$$

can be obtained by typing

```
conv ([2 -1 3],[1 2 -4 2 0])
```

The solution is $[2 \ 3 \ -7 \ 14 \ -14 \ 6 \ 0]$ or

$$h(x)u(x) = 2x^6 + 3x^5 - 7x^4 + 14x^3 - 14x^2 + 6x + 0$$

Deconvolution is the inverse operation of convolution and can be used to carry out polynomial division. For example, $u(x)$ divided by $h(x)$ can be carried out by typing

```
[q,r] = deconv (u,h)
```

which yields

```
q = 0.5 1.25 -2.125
r = 0 0 0 -3.875 6.375
```

where q is the quotient and r the remainder. Thus, we have

$$\frac{x^4 + 2x^3 - 4x^2 + 2x + 0}{2x^2 - x + 3} = 0.5x^2 + 1.25x - 2.125 + \frac{-3.875x + 6.375}{2x^2 - x + 3}$$

Difference equations can be solved by direct substitution and can be carried out using MATLAB. In MATLAB, DO loops are stated by `for . . . end`. Note that the index of variables in MATLAB cannot assume 0 or a negative number. Therefore, in order to solve the difference equation in Example 3.4.7 for $y[k]$, $k = 0, 1, \ldots, 9$, we restate the problem as follows: Find

$$y(k + 2) = 0.5y(k) - 1.5u(k + 1) + u(k)$$

for $k = 1, 2, \ldots, 10$ due to initial conditions $y(1) = 2$ and $y(2) = 1$ and input $u(1) = 0$, $u(2) = 0$, and $u(k + 2) = 1$ for $k \geq 1$. The following statements

```
y(1) = 2;y(2) = 1;u(1) = 0;u(2) = 0;
for k = 1:10;
u(k + 2) = 1
y(k + 2) = 0.5 * y(k) - 1.5 * u(k + 1) + u(k)
end
```

will eventually yield

```
u = 0 0 1 1 1 1 1 1 1 1 1 1
y = 2 1 1 -1 0 -1 -0.5 -1 -0.75 -1 -0.875 -1
```

Thus, difference equations can easily be solved using MATLAB.

The convolution integral in (3.56) must be discretized before we can compute it on a digital computer. Consider (3.56) with $t = kT$:

$$y(kT) = \int_{\tau=0}^{kT} h(kT - \tau)u(\tau)d\tau$$

which can be approximated by

$$y(kT) \approx T\sum_{i=0}^{k} h((k - i)T)u(iT)$$

where T is called the integration step size. This equation is a discrete convolution and can be computed using the command `conv` in MATLAB. Generally, the smaller the value of T, the more accurate the result. For example, to compute the convolution of $h(t) = e^{-0.1t}$ and $u(t) = \sin 2t$ from $t = 0$ to 5 with $T = 0.1$, we type

```
t = 0:0.1:50;
h = exp (-0.1 * t);u = sin (2 * t);
y = 0.1 * conv (h,u);
plot (y)
```

The result is as shown in Figure M3.1. Note that we use 501 points of h and u; thus, the output y has $501 + 501 - 1 = 1001$ points. The output y is actually the convolution of the first 501 points of h and u and the next 500 points of 0 for h and u. Therefore, only the first half of the plot yields the convolution of $h(t)$ and $u(t)$.

MATLAB PROBLEMS

Use MATLAB to compute the following problems.

M3.1 Compute the discrete convolution of $\{1, 2, 3, 2, 1\}$ and $\{-1, 2, -3 -2\}$. What is the length of the resulting sequence?

M3.2 Compute 10 points of the discrete convolution of $h[k] = 2 + 3k$ and $u[k] = 1$, for $k = 0, 1, 2, \ldots$. Use 6 points for $h[k]$ and $u[k]$. Then use 10 points for $h[k]$ and $u[k]$. Are the first 10 points of both computations the same? Which one is correct?

M3.3 Consider the three polynomials

$$a(s) = 2s^4 + 3s^2 - 9s + 2$$
$$b(s) = s^5 - 4s^4 + 2s^3 + s^2 - s - 2$$

and

$$c(s) = s^2 + 2s + 1$$

Compute
(a) $a(s)b(s)$
(b) $a(s)b(s)c(s)$

Figure M3.1

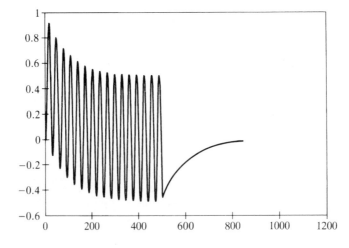

    (c) $b(s)/c(s)$
    (d) $a(s)/b(s)$

M3.4 Compute $y(k)$ from $k = 0$ to 10 of the difference equation (3.30) excited by a unit step input sequence and $y(-1) = 1$ and $y(-2) = 3$.

M3.5 Compute the convolution integral of the two functions in Figure 3.14 from $t = 0$ to 4. If the integration step size is chosen as $T = 1$, is the result close to the one in Figure 3.14(e)? Repeat the computation for $T = 0.5$ and $T = 0.1$. Remember to use enough points to obtain the correct solution.

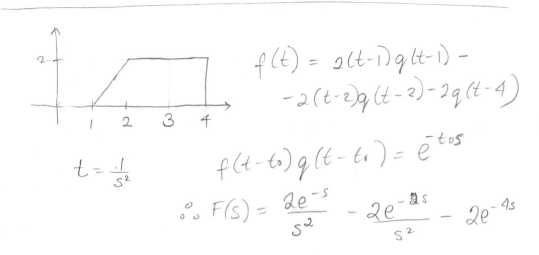

$$f(t) = 2(t-1)q(t-1) - 2(t-2)q(t-2) - 2q(t-4)$$

$$t = \frac{1}{s^2}$$

$$f(t-t_0)q(t-t_0) = e^{-t_0 s}$$

$$\therefore F(s) = \frac{2e^{-s}}{s^2} - \frac{2e^{-2s}}{s^2} - 2e^{-4s}$$

4

The Laplace Transform and Continuous-Time System Analysis

82

4.1 INTRODUCTION

Linear time-invariant lumped (LTIL) memoryless systems, continuous or discrete time, can be described by linear algebraic equations and can be studied using only algebraic manipulations: addition, subtraction, multiplication, and division. LTIL systems with memory are described by convolutions, differential or difference equations. These equations involve integrations and/or differentiations and cannot be studied *directly* using only algebraic manipulation. However, they can be so studied after some transformations. In this chapter, we will introduce such a transformation, called the Laplace transform, for continuous-time systems. The transformation for discrete-time systems will be studied in the next chapter.

Consider an LTIL system described by the nth-order differential equation

$$a_n y^{(n)}(t) + a_{n-1} y^{(n-1)}(t) + \cdots + a_1 y^{(1)}(t) + a_0 y(t)$$
$$= b_m u^{(m)}(t) + \cdots + b_1 u^{(1)}(t) + b_0 u(t)$$

One important problem in analysis is how to find the output $y(t)$ excited by some input $u(t)$ and initial conditions. This problem can be solved using the following methods.

Direct Method This method first finds the so-called complementary solution with unknown coefficients and sums it with the particular solution. Then, determine the unknown coefficients by matching the initial conditions. This method is often referred to as the classical method and can be found, for example, in Reference [20]. This method will not be discussed in this text.

State-variable Method This method transforms the nth-order differential equation into a set of n simultaneous first-order differential equations, called the state-variable equation. This equation can be studied more systematically than the *n*th-order differential equation and will be discussed in Chapter 10.

In addition to the preceding two methods, we can also solve the equation by using the Laplace transform method. This method consists of three steps:

1. Apply the Laplace transform to transform the differential equation into an *algebraic* equation. In the transformation, the initial conditions are automatically imbedded in the algebraic equation.
2. Carry out *algebraic* manipulations.
3. Compute the inverse Laplace transform to obtain the solution of the original differential equation.

Although the procedure appears to be long-winded, it is actually simpler to do than the direct and state-variable methods. Step 1 employs a transformation formula and involves a table lookup. Step 2 involves simple algebraic manipulations. Step 3 first carries out a partial fraction expansion and then uses a table. Thus, the method, although indirect, is simple and straightforward.[1] By using the Laplace transform, LTIL systems with memory can be studied using algebraic methods, just as LTIL memoryless systems are studied. The only difference is that the latter involves manipulation of real numbers while the former employs rational functions with real coefficients.

In addition to its simplicity, many design techniques in circuits, filters, and control systems have been developed in the Laplace transform domain. Therefore, the Laplace transform is a very important, if not *the* most important tool, in system analysis and design.

4.2 THE LAPLACE TRANSFORM

A function $f(t)$ that is defined for all t in $(-\infty, \infty)$ is called a positive-time function if it is zero for $t < 0$; a negative-time function if it is zero for $t > 0$. In the study of LTI systems, the initial time can be assumed, as discussed in Chapter 2,

---

[1]The Laplace transform transforms operations of calculus (differentiation and integration) into algebraic operations. This is similar to the logarithmic operation that transforms multiplications into additions. Using the formula

$$\log a \times b = \log a + \log b$$

the product of a and b can be obtained as

$$a \times b = \log^{-1}[\log a + \log b]$$

Thus, the multiplication can be replaced by one addition. However, we must use a table of logarithms three times to find $\log a$, $\log b$ and $\log^{-1}[\log a + \log b]$.

to be zero and the input is applied for $t \geq 0$. Therefore, the signals we will encounter in system analysis can be assumed to be positive-time signals.

The positive-time functions we will encounter in this chapter will be limited to step functions, ramp functions, sinusoidals (see Figure 1.6), exponential functions and their linear combinations. As discussed in Chapter 2, the characteristics of an LTI system can be obtained from the response of the system excited by *any* input signal. Therefore, we can use simple functions such as step or sinusoidal functions in system analysis. If we limit ourselves to the aforementioned signals, then their Laplace transforms will be rational functions and their manipulation will become very simple.

In addition to the aforementioned functions, we will also encounter the impulse discussed in Section 3.5.1. Because of possible inclusion of the impulse, care must be taken in defining the Laplace transform.

Definition 4.1

Consider a function $f(t)$ defined for all t. The Laplace transform pair is defined as

$$F(s) := \mathcal{L}[f(t)] := \int_{0-}^{\infty} f(t)e^{-st}dt \tag{4.1}$$

and

$$f(t) := \mathcal{L}^{-1}[F(s)] := \frac{1}{2\pi j}\int_{c-j\infty}^{c+j\infty} F(s)e^{st}ds \qquad \text{for } t \geq 0 \tag{4.2}$$

where s is a complex variable, called the Laplace-transform variable, $j := \sqrt{-1}$, and c is a real number to be discussed later. $F(s)$ is called the Laplace transform of $f(t)$, and $f(t)$ is the inverse Laplace transform of $F(s)$.

The Laplace transform is defined for the positive-time part of $f(t)$; the negative-time part is completely disregarded. However, if c in (4.2) is properly chosen, then (4.2) yields $f(t) = 0$ for $t < 0$. Therefore, we often assume $f(t) = 0$ for $t < 0$, and the Laplace transform is said to be defined only for positive-time functions. We first use a simple example to illustrate the use of (4.1) and discuss the problem that arises in its computation.

EXAMPLE **4.2.1**

Consider $f(t) = e^{2t}$. The function increases exponentially to infinity as $t \to \infty$. Its Laplace transform is

$$F(s) := \int_{0-}^{\infty} e^{2t}e^{-st}dt = \int_{0-}^{\infty} e^{-(s-2)t}dt = \frac{-1}{(s-2)}e^{-(s-2)t}\bigg|_{t=0-}^{\infty}$$

$$= \frac{-1}{s-2}\left[e^{-(s-2)t}\bigg|_{t=\infty} - e^{-(s-2)t}\bigg|_{t=0-} \right] \tag{4.3}$$

The value of $e^{-(s-2)t}$ is 1 at $t = 0-$ for all s. However the value of $e^{-(s-2)t}$ may be zero, infinity, or undefined at $t = \infty$. To see this, we express the complex variable s in rectangular coordinate as $s = \sigma + j\omega$, where $\sigma = Re\ s$ and $\omega = Im\ s$ are, respectively, the real part and the imaginary part of s. Then we have

$$e^{-(s-2)t} = e^{-(\sigma-2)t}e^{-j\omega t} = e^{-(\sigma-2)t}[\cos \omega t - j \sin \omega t]$$

If $\sigma = 2$, the function reduces to $\cos \omega t - j \sin \omega t$, whose real and imaginary parts may assume any value between 1 and -1 as $t \to \infty$. Thus, it is not defined as $t \to \infty$. If $\sigma < 2$, the exponential function $e^{-(\sigma-2)t}$ approaches infinity as $t \to \infty$. If $\sigma > 2$, the exponential function approaches 0 as $t \to \infty$. Thus, we conclude

$$e^{-(s-2)t} = \begin{cases} \pm\infty & \text{for } \sigma = Re\ s < 2 \\ \text{undefined} & \text{for } \sigma = Re\ s = 2 \\ 0 & \text{for } \sigma = Re\ s > 2 \end{cases}$$

for any $\omega = Im\ s$. Therefore, (4.3) is not defined for $\sigma = Re\ s \le 2$. However, if $\sigma = Re\ s > 2$, then (4.3) reduces to

$$F(s) = \mathcal{L}[e^{2t}] = \frac{1}{(s - 2)} \tag{4.4a}$$

This is the Laplace transform of e^{2t}.

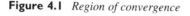

The Laplace transform in (4.4a) is, strictly speaking, defined only for $Re\ s > 2$. The region $Re\ s > 2$, shown in Figure 4.1, is called the *region of convergence*. The region of convergence is needed in computing the inverse Laplace transform of $1/(s - 2)$ by using (4.2). If c is chosen to be larger than 2, then (4.2) yields

$$\frac{1}{2\pi j}\int_{c-j\infty}^{c+j\infty} \frac{1}{s - 2}e^{st}ds = \begin{cases} e^{2t} & \text{for } t \ge 0 \\ 0 & \text{for } t < 0 \end{cases}$$

Figure 4.1 *Region of convergence*

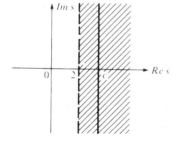

If c is chosen to be smaller than 2, then (4.2) will yield a negative-time function. In general, (4.2) may yield a positive-time, a negative-time, or a two-sided function, depending on how c is chosen. However, in application, we rarely use (4.2) to compute the inverse Laplace transform. We simply require the inverse Laplace transform to be a positive-time function and the inverse is to be obtained from a table; for example,

$$e^{2t} q(t) \longleftrightarrow \frac{1}{s-2} \qquad \text{or simply} \qquad e^{2t} \longleftrightarrow \frac{1}{s-2} \qquad \textbf{(4.4b)}$$

where $q(t)$ is the unit step function. Because we only study positive-time signals in this chapter, the step function is often dropped. Thus, the formula in (4.2) will not be used further. Furthermore, we will disregard the region of convergence when using (4.4), and consider (4.4) to be defined for all s except at $s = 2$.

In the preceding example, we have $e^{-(s-2)t} = 1$ at $t = 0$, $0-$, or $0+$; therefore, there seems to be no reason to use $0-$ as the lower limit of the integration in (4.1). This is indeed the case if $f(t)$ does not contain an impulse at $t = 0$. If $f(t)$ contains an impulse at $t = 0$, then the use of $0-$ is essential.

EXAMPLE 4.2.2

Let $f(t) = \delta(t)$ be the impulse defined in Figure 3.10. Its Laplace transform is, using the sifting property of the impulse,

$$F(s) = \mathcal{L}[\delta(t)] = \int_{0-}^{\infty} \delta(t) e^{-st} dt = e^{-st} \big|_{t=0} = 1$$

or

$$\delta(t) \longleftrightarrow 1$$

In this computation, if the lower limit of (4.1) is replaced by 0, then the result may be (as was discussed in Section 3.5.1) 0, 0.5, or 1, depending on which function in Figure 3.10 is used to define the impulse. Therefore, the use of $0-$ as the lower limit of (4.1) is essential.

This example shows one reason for using $0-$ in Definition 4.1. There is another important reason for using $0-$, and that will be discussed later.

Now we shall extend e^{2t} to e^{-at}, where a may be real or complex. This is important because almost all Laplace transforms studied in this text can be developed from the Laplace transform of e^{-at}.

EXAMPLE 4.2.3

Let $f(t) = e^{-at}$, where a is a real or complex constant. The Laplace transform of $f(t)$ is

$$F(s) = \int_{0-}^{\infty} e^{-at} e^{-st} dt = \int_{0-}^{\infty} e^{-(s+a)t} dt$$

$$= \frac{-1}{s+a} \left[e^{-(s+a)t} \bigg|_{t=\infty} - 1 \right]$$

If $e^{-(s+a)t} |_{t=\infty} = 0$, then $\quad = \dfrac{-1}{s+a} \Big[0 - 1 \Big] = -\dfrac{1}{s+a}(-1)$

$$F(s) = \mathcal{L}[e^{-at}] = \frac{1}{s+a} \tag{4.5a}$$

or

$$\boxed{\; e^{-at} \quad \text{for } t \ge 0 \longleftrightarrow \frac{1}{s+a} \;} \tag{4.5b}$$

This is a Laplace transform pair.

We now discuss some special cases of e^{-at}. If $a = 0$, then $f(t) = e^{-at} = 1$ for $t \ge 0$. It is the unit step function $q(t)$. Thus, we have

$$\boxed{\; q(t) \longleftrightarrow \frac{1}{s} \;}$$

If $a = j\omega$, a pure imaginary number, then $f(t) = e^{-j\omega t}$ is a complex exponential and its Laplace transform is

$$\boxed{\; e^{-j\omega t} q(t) \longleftrightarrow \frac{1}{s+j\omega} \quad \text{or simply} \quad e^{-j\omega t} \longleftrightarrow \frac{1}{s+j\omega} \;}$$

Note that signals in the real world are all real valued; however, it is convenient to introduce complex-valued functions in system and signal analysis, as you will see later.

The Laplace transforms of the functions studied in the preceding examples are all rational functions of s. This is not always the case, however. In fact, given a function, the following situations may occur:

1. Its Laplace transform does not exist. In other words, there is no region of convergence. The functions e^{t^2} and e^{e^t} are such examples. These two functions are mathematically contrived and do not arise in practice. It is

fair to say that all functions encountered in practice are Laplace transformable. Therefore, when using the Laplace transform, we usually pay no
attention to the existence problem.

2. Its Laplace transform exists but cannot be put into closed form. Most randomly generated time functions belong to this type.

3. Its Laplace transform exists and can be put into closed form but not as a
rational function of s. For example, the Laplace transforms of $1 / \sqrt{\pi t}$ and
$(\sinh at)/t$ are respectively $1 / \sqrt{s}$ and $0.5 \ln [(s + a)/(s - a)]$. They are irrational functions of s. See Reference [36].

4. Its Laplace transform exists and can be put into closed form as a rational
function of s.

Only functions that belong to the last class are studied in this chapter.
Although this is a very limited class of functions, they are very important in
engineering.

EXERCISE 4.2.1

Find the Laplace transforms of e^{-t} and e^{10t}.

[**ANSWERS:** $1/(s + 1)$, $1/(s - 10)$.]   ■

*4.2.1   Region of Convergence

We saw in (4.3) that the region of convergence is essential in computing
Laplace transforms. However, once Laplace transforms are computed, the
regions of convergence can be disregarded in their application, as will now be
explained.
 The region of convergence is required in computing the inverse Laplace
transform in (4.2). The constant c in (4.2) must be chosen so that the contour of
integration lies entirely inside the region of convergence. If the region of convergence is not specified, the inverse Laplace transform is not unique. In order
to see this nonuniqueness, we introduce the following two-sided or bilateral
Laplace transform:[2]

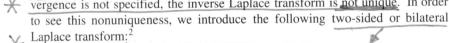

$$F(s) := \mathcal{L}_2[f(t)] := \int_{-\infty}^{\infty} f(t)e^{-st}dt \qquad (4.6)$$

The transform is defined for both the positive-time and negative-time parts of
$f(t)$. The subscript 2 stands for two-sided. If $f(t) = 0$ for $t < 0$, (4.6) reduces
to (4.1). In contrast to (4.6), (4.1) is called the one-sided, or unilateral Laplace

---

[2]The two-sided Laplace transform is rarely used in practice. Therefore, we discuss only the signifi-
cance of its region of convergence.

transform or, more often, simply the Laplace transform. The formula for the two-sided inverse Laplace transform is identical to (4.2). We show in the following that the two-sided Laplace transforms of two different functions may be identical. Indeed the two-sided Laplace transform of the positive-time function

$$f_1(t) = \begin{cases} e^{2t} & \text{for } t \geq 0 \\ 0 & \text{for } t < 0 \end{cases}$$

is

$$F_1(s) := \mathcal{L}_2[f_1(t)] = \frac{1}{s-2}$$

for *Re s* > 2, as computed in (4.4). The two-sided Laplace transform of the negative-time function

$$f_2(t) = \begin{cases} 0 & \text{for } t \geq 0 \\ -e^{2t} & \text{for } t < 0 \end{cases}$$

is

$$F_2(s) = \mathcal{L}_2[f_2(t)] = -\int_{-\infty}^{0} e^{2t} e^{-st} dt = \frac{-1}{2-s} e^{(2-s)t} \Big|_{-\infty}^{0}$$

$$= \frac{1}{s-2} \left[e^0 - e^{(2-s)t} \Big|_{t=-\infty} \right] = \frac{1}{s-2}$$

for *Re s* < 2, because $e^{(2-s)t} = 0$ as $t \to -\infty$ only if $(2 - Re\ s) > 0$ or *Re s* < 2. We see that the two-sided Laplace transforms of $f_1(t)$ and $f_2(t)$ are identical.

Although the two-sided Laplace transforms of $f_1(t)$ and $f_2(t)$ are identical, their regions of convergence are different, one in the region *Re s* > 2, the other in *Re s* < 2. Therefore, in the two-sided Laplace transforms, the specification of the region of convergence is essential. If the region of convergence is not specified, there is no one-to-one correspondence between a time function and its two-sided Laplace transform, and its inverse Laplace transform is not uniquely defined.

This is, however, not the case in the (one-sided) Laplace transform. In this case, all time functions are implicitly assumed to be positive time and all inverse Laplace transforms are required to be positive time. Under these assumptions, there is a one-to-one correspondence between the Laplace transform and its inverse. Therefore, in the (one-sided) Laplace transform, even if no region of convergence is specified, no ambiguity will arise. Thus, *once the*

Laplace transform of a function is computed, its region of convergence usually does not arise again in the application.⎤ For example, in using $\mathcal{L}[e^{2t}] = 1/(s-2)$, we pay no attention to its region of convergence, and consider $1/(s-2)$ as a rational function of s defined everywhere except at $s = 2$.

4.3   PROPERTIES OF THE LAPLACE TRANSFORM

7.5

In this section, we will discuss some properties of the Laplace transform. The application of these properties will also be discussed. The Laplace transform domain is often referred to as the *s*-domain.

● **Linearity** The Laplace transform is a linear operator; that is, if

$$F_1(s) = \mathcal{L}[f_1(t)] \quad \text{and} \quad F_2(s) = \mathcal{L}[f_2(t)]$$

then, for any real or complex constants α_1 and α_2,

$$\boxed{\mathcal{L}[\alpha_1 f_1(t) + \alpha_2 f_2(t)] = \alpha_1 F_1(s) + \alpha_2 F_2(s)}$$

The Laplace transform is defined through an integration. Because the integration is a linear operator, so is the Laplace transform.

EXAMPLE 4.3.1

We use this property to compute the Laplace transform of $\sin \omega_0 t$. Using Euler's identity

$$\cos \omega t = \frac{1}{2}\left(e^{j\omega t} + e^{-j\omega t}\right)$$

$$\boxed{\sin \omega_0 t = \frac{1}{2j}(e^{j\omega_0 t} - e^{-j\omega_0 t})}$$

we have

$$\mathcal{L}[\sin \omega_0 t] = \frac{1}{2j}(\mathcal{L}[e^{j\omega_0 t}] - \mathcal{L}[e^{-j\omega_0 t}])$$

$$= \frac{1}{2j}\left(\frac{1}{s - j\omega_0} - \frac{1}{s + j\omega_0}\right) = \frac{\omega_0}{s^2 + \omega_0^2}$$

EXERCISE 4.3.1

Verify

$$\boxed{\mathcal{L}[\sin \omega_0 t] = \frac{\omega_0}{s^2 + \omega_0^2}}$$

$$\mathcal{L}[\cos \omega_0 t] = \frac{s}{s^2 + \omega_0^2}$$

EXERCISE 4.3.2

Compute the Laplace transforms of the following positive-time functions.

(a) $e^{-2t} - 4e^{3t}$ $\mathcal{L}[1] = \dfrac{1}{s}$

(b) $2 - 2e^t + 0.5 \sin 4t$

[**ANSWERS:** (a) $-(3s + 11)/(s + 2)(s - 3)$; (b) $-2(s + 16)/s(s - 1)(s^2 + 16)$.] ∎

● **Shifting in the s-domain (Multiplication by e^{-at} in the time domain)** If $\mathcal{L}[f(t)] = F(s)$, then

   $$\boxed{\mathcal{L}[e^{-at}f(t)] = F(s + a)}$$ ✳

for any real or complex constant a.

By definition, we have

$$\mathcal{L}[e^{-at}f(t)] = \int_{0-}^{\infty} e^{-at}f(t)e^{-st}dt = \int_{0-}^{\infty} f(t)e^{-(s+a)t}dt = F(s + a)$$

This establishes the formula. Using the formula, we have

$$\boxed{\mathcal{L}[e^{-at}\sin \omega_0 t] = \frac{\omega_0}{(s + a)^2 + \omega_0^2}}$$

and

$$\boxed{\mathcal{L}[e^{at}\cos \omega_0 t] = \frac{(s - a)}{(s - a)^2 + \omega_0^2}}$$

EXERCISE 4.3.3

Find the Laplace transforms of the following functions

(a) $e^{-t}\sin 4t$

(b) $e^{3t}\sin (t - 30°)$

[**ANSWERS:** (a) $\dfrac{4}{(s + 1)^2 + 16}$; (b) $\dfrac{0.866}{(s - 3)^2 + 1} - \dfrac{0.5(s - 3)}{(s - 3)^2 + 1}$.] ∎

● **Differentiation in the s-domain (Multiplication by t in the time domain)** If $\mathcal{L}[f(t)] = F(s)$, then

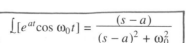 $$\boxed{\mathcal{L}[t f(t)] = -\frac{d}{ds}F(s)}$$ ✳   (4.7)

The differentiation of (4.1) with respect to s yields

$$\left[\frac{d}{ds}F(s) = \frac{d}{ds}\mathcal{L}[f(t)] = \frac{d}{ds}\int_{0-}^{\infty}f(t)e^{-st}dt = \int_{0-}^{\infty}f(t)[\frac{d}{ds}e^{-st}]dt = -\int_{0-}^{\infty}tf(t)e^{-st}dt\right.$$

which is, by definition, the Laplace transform of $-tf(t)$. This establishes the formula.

Using the formula and the fact $\mathcal{L}[q(t)] = 1/s$, where $q(t)$ is the unit step function, we can readily establish

$$\mathcal{L}[t] = \frac{-d}{ds}(\frac{1}{s}) = \frac{1}{s^2}$$

$$\mathcal{L}[t^2] = \frac{-d}{ds}(\frac{1}{s^2}) = \frac{2}{s^3}$$

and, in general,

$$\mathcal{L}[t^n] = \frac{n!}{s^{n+1}}$$

where $n! := 1 \cdot 2 \cdot 3 \ldots n$.

● **Differentiation in the time domain (Multiplication by s in the s-domain)** If $\mathcal{L}[f(t)] = F(s)$, then

$$\mathcal{L}[\frac{d}{dt}f(t)] =: \mathcal{L}[\dot{f}(t)] = s\mathcal{L}[f(t)] - f(0-) = sF(s) - f(0-) \qquad (4.8a)$$

$$\mathcal{L}[\frac{d^2}{dt^2}f(t)] =: \mathcal{L}[\ddot{f}(t)] = s^2F(s) - sf(0-) - \dot{f}(0-) \qquad (4.8b)$$

and, in general,

$$\mathcal{L}[\frac{d^n}{dt^n}f(t)] =: \mathcal{L}[f^{(n)}(t)]$$
$$= s^nF(s) - s^{n-1}f(0-) - s^{n-2}f^{(1)}(0-) - \cdots - f^{(n-1)}(0-) \qquad (4.8c)$$

where

$$f^{(k)}(0) := \frac{d^k}{dt^k}f(t)\bigg|_{t=0} \qquad \text{for } k = 0, 1, 2, \ldots,$$

and

$$\dot{f}(t) := f^{(1)}(t) \qquad \text{and} \qquad \ddot{f}(t) := f^{(2)}(t)$$

The first formula can be established using integration by parts. By definition, we have

$$\mathcal{L}[\frac{d}{dt}f(t)] = \int_{0-}^{\infty}[\frac{d}{dt}f(t)]e^{-st}dt = f(t)e^{-st}\Big|_{t=0-}^{\infty} - \int_{0-}^{\infty}f(t)\frac{d}{dt}[e^{-st}]dt$$

$$= 0 - f(0-) + s\int_{0-}^{\infty}f(t)e^{-st}dt = sF(s) - f(0-)$$

where we have assumed $f(t)e^{-st} = 0$ at $t = \infty$. This is true only if $Re\ s$ is larger than a certain value. For example, if $f(t) = e^{2t}$, then we need $Re\ s > 2$. In other words, (4.8a) holds only in the region of convergence of $f(t)$. However, in application, this restriction is completely disregarded. To show (4.8b), we apply (4.8a) to yield

$$\mathcal{L}[\frac{d}{dt}\dot{f}(t)] = s\mathcal{L}[\dot{f}(t)] - \dot{f}(0-) = s[sF(s) - f(0-)] - \dot{f}(0-)$$

$$= s^2F(s) - sf(0-) - \dot{f}(0-)$$

Proceeding forward, we can show (4.8c).

Here are two examples to illustrate the application of (4.8) and to discuss the reason for using $f(0-)$ rather than $f(0)$ or $f(0+)$.

EXAMPLE 4.3.2

In the preceding section, we showed $\mathcal{L}[\delta(t)] = 1$ and $\mathcal{L}[q(t)] = 1/s$, where $\delta(t)$ is the impulse and $q(t)$ is the unit step function. We also established in (3.62) that $\delta(t) = dq(t)/dt$. As was discussed in Figure 1.5(b) and Section 1.2, we have $q(0-) = 0$, $q(0+) = 1$. The value of $q(t)$ at $t = 0$, however, can be 0, 1, 0.5, or any other number. Now we use (4.8a) to compute

$$\mathcal{L}[\delta(t)] = \mathcal{L}[\frac{d}{dt}q(t)] = s\mathcal{L}[q(t)] - q(0-) = s \cdot \frac{1}{s} - 0 = 1$$

which is the same as the one computed from the definition. In this example, if we use $q(0)$ or $q(0+)$, then the result will be incorrect. This shows that the use of 0– in (4.8) is essential.

EXAMPLE 4.3.3

Consider

$$f(t) = \begin{cases} 2\cos\omega_0 t & \text{for } t \geq 0 \\ 0 & \text{for } t < 0 \end{cases}$$

Because of $f(0-) = 0$ and $f(0+) = 2$, the function $f(t)$ has a discontinuity at $t = 0$. Thus, we have, for $t \geq 0,$[3]

$$\dot{f}(t) = [f(0+) - f(0-)]\delta(t) - 2\omega_0 \sin \omega_0 t$$
$$= 2\delta(t) - 2\omega_0 \sin \omega_0 t$$

$\cos \phi = 1$

This can also be obtained as, using $q(t)$,

$$\frac{d}{dt}[f(t)\, q(t)] = \dot{f}(t)\, q(t) + f(t)\, \dot{q}(t) = -2\omega_0 \sin \omega_0 t\, q(t) + f(t)\delta(t)$$

$$= -2\omega_0 \sin \omega_0 t + f(0)\delta(t) = -2\omega_0 \sin \omega_0 t + 2\delta(t)$$

Now we use two methods to compute $\mathcal{L}[\dot{f}(t)]$. By definition, we have

$$\mathcal{L}[\dot{f}(t)] = \mathcal{L}[2\delta(t) - 2\omega_0 \sin \omega_0 t] = 2 - 2\omega_0 \cdot \frac{\omega_0}{s^2 + \omega_0^2} = \frac{2s^2}{s^2 + \omega_0^2}$$

Next, we use (4.8a) to compute

$$\mathcal{L}[\dot{f}(t)] = s\mathcal{L}[f(t)] - f(0-) = s \cdot \frac{2s}{s^2 + \omega_0^2} - 0 = \frac{2s^2}{s^2 + \omega_0^2}$$

which is the same as the one obtained from the definition.

For positive-time functions, we have $f(t) = 0$ for $t < 0$ which implies $f^{(i)}(0-) = 0$, for $i = 0, 1, 2, \ldots$. In this case, (4.8) reduces to

$$\mathcal{L}\left[\frac{d^n}{dt^n} f(t)\right] = s^n F(s) \tag{4.9}$$

Thus, differentiation in the time domain is equivalent to multiplication by s in the Laplace-transform domain. This relationship will be denoted by

$$\frac{d}{dt} \longleftrightarrow s$$

This is an important relationship.

For positive-time functions, we have $f^{(i)}(0-) = 0$ for $i = 1, 2, \ldots$. Then why do we include them in (4.8)? The reason is that the Laplace transform will be used to study LTIL differential equations with nonzero initial conditions.

---

[3]If $f(t) = 2 \cos \omega_0 t$ for *all* t, then $\dot{f}(t) = -2\omega_0 \sin \omega_0 t$ for all t. In this case, $\dot{f}(t)$ does not contain an impulse at $t = 0$.

Thus, we must use (4.8) rather than (4.9). This issue will become clear in Section 4.5.

Integration in the time domain (Division by s in the s-domain) If $L[f(t)] = F(s)$, then

$$L[\int_{0-}^{t} f(\tau)d\tau] = \frac{1}{s}F(s)$$

To establish this formula, we define

$$g(t) := \int_{0-}^{t} f(\tau)d\tau$$

and $G(s) := L[g(t)]$. Clearly, we have $g(0-) = 0$ and $f(t) = dg(t)/dt$. The Laplace transform of $f(t)$ is, using (4.8a),

$$F(s) = L[f(t)] = L[\frac{d}{dt}g(t)] = sG(s) - g(0-) = sG(s)$$

Thus, we have

$$G(s) = L[\int_{0-}^{t} f(\tau)d\tau] = \frac{1}{s}F(s)$$

This establishes the formula. This formula states that integration in the time domain is equivalent to multiplication by $1/s$ in the s-domain, denoted as

$$\int_{0-}^{t} \longleftrightarrow \frac{1}{s}$$

This is in contrast to $d/dt \longleftrightarrow s$ for differentiation.

EXAMPLE 4.3.4

We use the integration formula to reestablish (4.7). Because the Laplace transform of the unit step function $q(t)$ is $1/s$, we have

$$L[\int_{0-}^{t} q(\tau)d\tau] = L[t] = \frac{1}{s}L[q(t)] = \frac{1}{s} \cdot \frac{1}{s} = \frac{1}{s^2}$$

and

$$L[\int_{0-}^{t} \tau d\tau] = L[\frac{1}{2}t^2] = \frac{1}{s^3}$$

or

$$\mathcal{L}[t^2] = \frac{2}{s^3}$$

In general, we have

$$\mathcal{L}[t^n] = \frac{n!}{s^{n+1}}$$

EXERCISE 4.3.4

Find the Laplace transform of the function

$$f(t) = e^{2t} + 2e^{-2t} - t^2 \qquad \text{for } t \geq 0$$

[**ANSWER:** $(3s^4 - 2s^3 - 2s^2 + 8)/s^3(s^2 - 4)$.]

Convolution   The zero-state response of every linear time-invariant continuous-time system can, as was discussed in Section 3.5.2, be described by

$$y(t) = \int_0^t h(t - \tau)u(\tau)d\tau$$

where $h(t)$ is the impulse response of the system. If the system is causal, then $h(t) = 0$ for $t < 0$. Thus, $h(t)$ is a positive-time function. The output $y(t)$ excited by $u(t)$, for $t \geq 0$, can be considered as zero for $t < 0$ if the system is causal and initially relaxed. Thus, the three functions $h(t)$, $u(t)$, and $y(t)$ are all positive-time functions. We use $H(s)$, $U(s)$, and $Y(s)$ to denote their Laplace transforms. The substitution of the convolution into the definition of $Y(s)$ yields

$$Y(s) = \int_{0-}^{\infty} y(t)e^{-st}dt = \int_{0-}^{\infty} [\int_0^t h(t - \tau)u(\tau)d\tau]e^{-st}dt$$

Based on the discussion in Section 3.5.2 and the fact $h(t - \tau) = 0$, for $\tau > t$, we may change the limits of the integration inside the brackets from 0 to 0− and from t to ∞. We then change the order of integrations to yield

$$Y(s) = \int_{0-}^{\infty} [\int_{0-}^{\infty} h(t - \tau)e^{-s(t-\tau)}dt]u(\tau)e^{-s\tau}d\tau \qquad (4.10)$$

The integration inside the brackets becomes, after the substitution of $v := t - \tau$,

$$\int_{-\tau}^{\infty} h(v)e^{-sv}dv = \int_{0-}^{\infty} h(v)e^{-sv}dv = H(s)$$

where we have used again the fact that $h(v) = 0$ for $v < 0$. Thus, (4.10) becomes

$$Y(s) = \int_{0-}^{\infty} H(s)u(\tau)e^{-s\tau}d\tau = H(s)\int_{0-}^{\infty} u(\tau)e^{-s\tau}d\tau$$

or

$$\boxed{Y(s) = H(s)U(s)}$$ (4.11)

Thus, the Laplace transform transforms a convolution integral into an algebraic equation. The function $H(s)$, the Laplace transform of the impulse response, is called the *transfer function* of the system. Thus, if a system is initially relaxed, the Laplace transform of the output is simply equal to the product of the transfer function and the Laplace transform of the input. This is an algebraic relation and is very important in system analysis and design.

Table 4.1 lists some of the frequently used Laplace transform pairs. Table 4.2 lists properties of the Laplace transform. Some of the properties have not yet been discussed and will be discussed in later sections of this chapter.

4.4 INVERSE LAPLACE TRANSFORM—PARTIAL FRACTION EXPANSION

The computation of $f(t)$ from its Laplace transform $F(s)$ is called the inverse Laplace transform. The inverse Laplace transform of $F(s)$ can be computed from (4.2). Its computation, however, is complicated and is rarely used in engineering. In this section, we will discuss only the method of table lookup. The method is applicable only if $F(s)$ is a rational function of s.

Consider the rational function

$$F(s) = \frac{N(s)}{D(s)}$$

where $N(s)$ and $D(s)$ are two polynomials. We use "deg" to denote the degree of a polynomial. Depending on the relative degrees of $N(s)$ and $D(s)$, we have the following definitions:

Table 4.1 *Laplace Transform Pairs*

$f(t), t \geq 0$	$F(s)$
$\delta(t)$	1
1 or $q(t)$	$\dfrac{1}{s}$
t	$\dfrac{1}{s^2}$
t^n (n: positive integer)	$\dfrac{n!}{s^{n+1}}$
e^{-at} (a: real or complex)	$\dfrac{1}{s+a}$
te^{-at}	$\dfrac{1}{(s+a)^2}$
$t^n e^{-at}$	$\dfrac{n!}{(s+a)^{n+1}}$
$\sin \omega_0 t$	$\dfrac{\omega_0}{s^2 + \omega_0^2}$
$\cos \omega_0 t$	$\dfrac{s}{s^2 + \omega_0^2}$
$t \sin \omega_0 t$	$\dfrac{2\omega_0 s}{(s^2 + \omega_0^2)^2}$
$t \cos \omega_0 t$	$\dfrac{s^2 - \omega_0^2}{(s^2 + \omega_0^2)^2}$
$e^{-at} \sin \omega_0 t$	$\dfrac{\omega_0}{(s+a)^2 + \omega_0^2}$
$e^{-at} \cos \omega_0 t$	$\dfrac{s+a}{(s+a)^2 + \omega_0^2}$

$$F(s) \text{ improper} \longleftrightarrow \deg N(s) > \deg D(s) \longleftrightarrow F(\infty) = \pm\infty$$

$$F(s) \text{ proper} \longleftrightarrow \deg N(s) \leq \deg D(s) \longleftrightarrow |F(\infty)| = k < \infty$$

$$F(s) \text{ strictly proper} \longleftrightarrow \deg N(s) < \deg D(s) \longleftrightarrow F(\infty) = 0$$

$$F(s) \text{ biproper} \longleftrightarrow \deg N(s) = \deg D(s) \longleftrightarrow F(\infty) = k \neq 0 < \infty$$

For example, the rational functions

$$\frac{s^2 + 1}{s + 1} \qquad s^2 + 1 \qquad \text{and} \qquad \frac{s^{10}}{s^9 + s^8 + {}^7 - 10}$$

are improper. The rational functions

Table 4.2 *Properties of the Laplace transform*

Property	Time function	Laplace transform
Linearity	$\alpha_1 f_1(t) + \alpha_2 f_2(t)$	$\alpha_1 F_1(s) + \alpha_2 F_2(s)$
s-shifting	$e^{-at} f(t)$	$F(s + a)$
Time shifting	$f(t - T)q(t - T)$, for $T \geq 0$	$e^{-sT} F(s)$
s-differentiation	$tf(t)$	$\dfrac{-d}{ds} F(s)$
Time differentiation	$\dfrac{d}{dt} f(t)$	$sF(s) - f(0-)$
Time integration	$\displaystyle\int_{0-}^{t} f(\tau)d\tau$	$\dfrac{F(s)}{s}$
Time scaling	$f(at)$, for $a > 0$	$\dfrac{1}{a} F(\dfrac{s}{a})$
Convolution	$\displaystyle\int_{0}^{t} h(t - \tau)u(\tau)d\tau$	$H(s)U(s)$
Final value	$f(\infty) = \lim_{s \to 0} sF(s)$	(if $sF(s)$ has no open RHP poles)
Initial value	$f(0+) = \lim_{s \to \infty} sF(s)$	(if $F(s)$ is strictly proper)

$$2 \qquad \frac{s}{s + 3} \qquad \frac{1}{(s + 2)(s - 3)} \qquad \frac{s^3 - 1}{s^{10}}$$

are proper. The first two are also biproper, while the last two are strictly proper. Thus, proper rational functions include both strictly proper and biproper rational functions. If $F(s)$ is biproper, so is $F^{-1}(s) = D(s)/N(s)$.

The properness of a rational function $F(s)$ can also be determined from the value of $F(s)$ at $s = \infty$. It is clear that $F(s)$ is improper if and only if $F(\infty) = \pm\infty$; proper if and only if $F(\infty)$ is a finite nonzero or zero constant; biproper if and only if $F(\infty)$ is finite and nonzero; and strictly proper if and only if $F(\infty) = 0$. In Table 4.1, the first $F(s)$ is biproper; the rest are strictly proper.

EXERCISE 4.4.1

Classify the following rational functions.

$$s^2 + 1 \qquad 2 \qquad \frac{1}{s + 1} \qquad \frac{s^2 - 1}{s + 1} \qquad \frac{s - 1}{3s + 1}$$

How many of them are proper rational functions?

[**ANSWERS:** Improper, biproper, strictly proper, improper, biproper; 3.]

Let us consider the type of time functions whose Laplace transforms are improper rational functions. Since $\mathcal{L}[\delta(t)] = 1$, we have, using (4.9),

$$\mathcal{L}[\frac{d^n}{dt^n}\delta(t)] = s^n \mathcal{L}[\delta(t)] = s^n \qquad \text{for } n = 0, 1, 2, \ldots \qquad \textbf{(4.12)}$$

In other words, the Laplace transforms of the derivatives of an impulse are improper rational functions. The impulse is not an ordinary function, and the meaning of its derivatives is not obvious. Furthermore, it is not possible to generate an impulse, not to mention its derivatives, in practice. Therefore, we rarely encounter improper rational functions in engineering. For this reason, we'll study only proper rational functions in this section.

We introduce now the concepts of poles and zeros. Consider the proper rational function

$$F(s) = \frac{N(s)}{D(s)}$$

where $N(s)$ and $D(s)$ are polynomials with real coefficients. It is assumed that $N(s)$ and $D(s)$ have no common factors.[4] In this case, $N(s)$ and $D(s)$ are said to be coprime. Under this assumption, the roots of $D(s)$ are called the *poles* of $F(s)$, and the roots of $N(s)$ are called the *zeros* of $F(s)$. For example, the proper rational function

$$F(s) = \frac{(s-1)^2(s+3)}{(s^2-1)^2(s^2+2s+2)} = \frac{(s-1)^2(s+3)}{(s-1)^2(s+1)^2(s^2+2s+2)}$$

$$= \frac{s+3}{(s+1)^2(s+1+j1)(s+1-j1)}$$

has four poles, $-1, -1, -1-j1$, and $-1+j1$, and one zero, -3. Note that $s = 1$ is neither a pole nor a zero of $F(s)$ because the common factor $(s-1)^2$ should be canceled. A pole is called simple if it appears only once; it is called repeated if it appears twice or more. Thus, -1 is a repeated pole with multiplicity 2, and $-1 \pm j1$ are simple poles. Because $F(s)$ is assumed to have real coefficients, if a complex number $c - jd$ is a pole or zero, its complex conjugate $c + jd$ must also be a pole or zero.

Consider the proper rational function $F(s) = N(s)/D(s)$. If $N(s)$ and $D(s)$ have no common factors, the *degree* of $F(s)$ is defined as the degree of $D(s)$. For example, the rational function

---

[4]If two polynomials $D(s)$ and $N(s)$ can be written as $D(s) = \bar{D}(s)R(s)$ and $N(s) = \bar{N}(s)R(s)$, then the polynomial $R(s)$ is called a common factor of $D(s)$ and $N(s)$. For example, $D(s) = 2s + 6$ and $N(s) = 6s - 2$ have 2 or any real number as their common factor. Such a common factor, a polynomial of degree 0, is called a trivial common factor. We study here only nontrivial common factors, polynomials of degree 1 or higher.

$$F(s) = \frac{(s+1)}{s^2} = \frac{(s+1)R(s)}{s^2R(s)} =: \frac{N(s)}{D(s)}$$

has degree 2 for any polynomial $R(s)$. We cannot define the degree of $F(s)$ as the degree of $D(s)$ without requiring the coprimeness of $D(s)$ and $N(s)$. Otherwise, the degree of $D(s)$ and, consequently, the degree of $F(s)$ can be any number by introducing a common factor $R(s)$. Clearly, the number of the poles of $F(s)$ equals the degree of $F(s)$.

EXERCISE 4.4.2

What are the degrees of the following rational functions?

$$\frac{s^2 - 3s + 2}{(s-2)(s-1)} \qquad \frac{s^3 - 1}{s^3(s-1)} \qquad \frac{1}{s^3 + 6s^2 + s}$$

[**ANSWERS:** 0, 3, 3.]

With this preliminary, we are ready to discuss the computation of the inverse Laplace transform of a proper rational function $F(s)$. The basic idea is to expand $F(s)$ as a summation of terms whose inverse Laplace transforms are available in a table. We then use the table to find the inverse Laplace transform. To be more specific, we expand $F(s)$ as

$$F(s) = k_0 + F_1(s) + F_2(s) + \cdots + F_m(s) \qquad (4.13)$$

where k_0 is a constant and $F_i(s)$ is strictly proper and assumes the forms in Table 4.1. Because $F_i(\infty) = 0$ for all i, we have

$$k_0 = F(\infty)$$

Thus, the constant k_0 can be readily obtained. Note that k_0 is zero if $F(s)$ is strictly proper and nonzero if $F(s)$ is biproper. We then use Table 4.1 to find the inverse of $F_i(s)$ and the inverse Laplace transform of $F(s)$ is

$$f(t) = k_0\delta(t) + f_1(t) + f_2(t) + \cdots + f_m(t) \qquad \text{for } t \geq 0$$

This completes the computation of the inverse Laplace transform of $F(s)$. Thus, the major computation involves the expansion in (4.13). Two possibilities exist in the expansion:

1. We can exclusively use terms of the form $1/(s + a)$, $1/(s + a)^2$, ..., called linear factors, where a may be real or complex. In this case, our computation may involve complex numbers.

2. We can use linear factors and quadratic factors such as $b/[(s + c)^2 + b^2]$ and $(s + c)/[(s + c)^2 + b^2], \ldots$, where b and c are all real numbers. In this case, we can avoid complex numbers in our computation.

4.4.1 Expansion in Terms of Linear Factors

We use an example to illustrate the procedure. Consider the following rational function of degree 4.

$$F(s) = \frac{2s^4 + s^3 - 2s}{s^4 + 7s^3 + 18s^2 + 20s + 8}$$

It is a biproper rational function. The first step in partial fraction expansion is to factor the denominator of $F(s)$. This is certainly not a simple task if its degree is 3 or higher. Unfortunately, this factorization step cannot be avoided. If the denominator of $F(s)$ is not given in factored form or cannot be easily factored by hand, we may call on a digital computer (see the last section of this chapter). Therefore, we shall assume that every $F(s)$ is given in factored form, such as

$$F(s) = \frac{2s^4 + s^3 - 2s}{(s + 1)(s + 2)^3} \tag{4.14}$$

The function has a simple pole at -1 and a repeated pole with multiplicity 3 at -2. We can expand (4.14) as

$$F(s) = k_0 + \frac{k_1}{s + 1} + \frac{k_2}{s + 2} + \frac{k_3}{(s + 2)^2} + \frac{k_4}{(s + 2)^3} \tag{4.15}$$

We can also expand it as

$$F(s) = k_0 + \frac{k_1}{s + 1} + \frac{k_2 + k_3 s + k_4 s^2}{(s + 2)^3} \tag{4.16}$$

This form, however, is not as desirable as (4.15) because, unlike the inverse Laplace transform of every term of (4.15), the inverse Laplace transform of the last term of (4.16) is not available in Table 4.1. Therefore, we expand $F(s)$ in (4.15). Once all k_i in (4.15) are computed, then the inverse Laplace transform of $F(s)$ in (4.14) is, using Table 4.1,

$$f(t) = k_0 \delta(t) + k_1 e^{-t} + k_2 e^{-2t} + k_3 t e^{-2t} + \frac{k_4}{2} t^2 e^{-2t} \tag{4.17}$$

Now consider the computation of k_i in (4.15). Equation (4.15) is an identity; it holds for almost every s, including $s = \infty$. In order to illustrate the basic ideas, we will discuss a number of methods of computing k_i. Before proceeding, we set $s = \infty$. Then, (4.15) becomes

$$F(\infty) = k_0 + k_1 \cdot 0 + k_2 \cdot 0 + k_3 \cdot 0 + k_4 \cdot 0$$

or

$$k_0 = F(\infty) = \frac{2}{1} = 2 \qquad (4.18)$$

It is simply the ratio of the coefficients associated with s^4 in $F(s)$. Thus, k_0 can be easily computed.

Method I The multiplication of $(s + 1)(s + 2)^3$ to (4.15) yields

$$2s^4 + s^3 - 2s = 2(s + 1)(s + 2)^3 + k_1(s + 2)^3 + k_2(s + 1)(s + 2)^2 + k_3(s + 1)(s + 2) + k_4(s + 1)$$

which can be arranged as

$$2s^4 + s^3 - 2s - 2(s^4 + 7s^3 + 18s^2 + 20s + 8) = -13s^3 - 36s^2 - 42s - 16$$
$$= (k_1 + k_2)s^3 + (6k_1 + 5k_2 + k_3)s^2 + (12k_1 + 8k_2 + 3k_3$$
$$+ k_4)s + (8k_1 + 4k_2 + 2k_3 + k_4) \qquad (4.19)$$

Equating the coefficients of the like power of s yields

$$k_1 + k_2 = -13$$
$$6k_1 + 5k_2 + k_3 = -36$$
$$12k_1 + 8k_2 + 3k_3 + k_4 = -42$$
$$8k_1 + 4k_2 + 2k_3 + k_4 = -16$$

or, in matrix form,

$$\begin{bmatrix} 1 & 1 & 0 & 0 \\ 6 & 5 & 1 & 0 \\ 12 & 8 & 3 & 1 \\ 8 & 4 & 2 & 1 \end{bmatrix} \begin{bmatrix} k_1 \\ k_2 \\ k_3 \\ k_4 \end{bmatrix} = \begin{bmatrix} -13 \\ -36 \\ -42 \\ -16 \end{bmatrix} \qquad (4.20)$$

This equation can be solved using the methods discussed in Appendix A. If we use MATLAB, the following statements

```
a = [1 1 0 0; 6 5 1 0; 12 8 3 1; 8 4 2 1];
b = [-13; -36; -42; -16];
a \ b
```

yields $k_1 = 3$, $k_2 = -16$, $k_3 = 26$, and $k_4 = -28$. Once k_i are computed, the inverse Laplace transform of (4.14) is given as shown in (4.17).

Method 2 Equation (4.15) is an identity; it holds for every s except at the poles of $F(s)$. Other than $k_0 = 2$, there are four unknowns in (4.15). Now we shall choose four different values of s to yield four equations. These s values are to be chosen for the convenience of computation. If we choose $s = 0, 1, 2, -3$, then (4.15) becomes

$$s = 0 \qquad 0 = 2 + k_1 + \frac{k_2}{2} + \frac{k_3}{4} + \frac{k_4}{8}$$

$$s = 1 \qquad \frac{1}{2 \cdot 27} = 2 + \frac{k_1}{2} + \frac{k_2}{3} + \frac{k_3}{9} + \frac{k_4}{27}$$

$$s = 2 \qquad \frac{32 + 8 - 4}{3 \cdot 64} = 2 + \frac{k_1}{3} + \frac{k_2}{4} + \frac{k_3}{16} + \frac{k_4}{64}$$

$$s = -3 \qquad \frac{162 - 27 + 6}{(-2)(-1)^3} = 2 + \frac{k_1}{-2} + \frac{k_2}{-1} + \frac{k_3}{1} + \frac{k_4}{-1}$$

poles & zeros

Note that, other than -1 and -2, these s values can be chosen arbitrarily. These equations can be arranged in matrix form as

$$
\begin{bmatrix}
1 & 1/2 & 1/4 & 1/8 \\
1/2 & 1/3 & 1/9 & 1/27 \\
1/3 & 1/4 & 1/16 & 1/64 \\
-1/2 & -1 & 1 & -1
\end{bmatrix}
\begin{bmatrix}
k_1 \\ k_2 \\ k_3 \\ k_4
\end{bmatrix}
=
\begin{bmatrix}
-2 \\ -107/54 \\ -29/16 \\ 137/2
\end{bmatrix}
\qquad \textbf{(4.21)}
$$

If we use MATLAB, the following statements

```
a = [1  1/2  1/4  1/8; 1/2  1/3  1/9  1/27; 1/3  1/4
     1/16  1/64; -1/2  -1  1  -1];
b = [-2; -107/54; -29/16; 137/2];
a \ b
```

yields $k_1 = 3$, $k_2 = -16$, $k_3 = 26$, and $k_4 = -28$. The result is the same as the one obtained in Method 1. Note that we can also use the polynomial equation (4.19) to obtain four equations by choosing four different values of s. In using (4.19), we can choose any s, including the poles of $F(s)$. If we use the rational-function equation (4.15), then we cannot choose the poles of $F(s)$ as s.

Method 3 (Using formula) The multiplication of $(s + 1)$ and (4.15) yields

$$F(s)(s + 1) = k_0(s + 1) + k_1 + \frac{k_2(s + 1)}{s + 2} + \frac{k_3(s + 1)}{(s + 2)^2}$$

$$+ \frac{k_4(s + 1)}{(s + 2)^3} \qquad \textbf{(4.22)}$$

If we set $s = -1$, then every term, except k_1, on the right-hand-side of (4.22) is zero. Thus, we have

$$k_1 = F(s)(s + 1)\big|_{s=-1} = \frac{2s^4 + s^3 - 2s}{(s + 2)^3}\bigg|_{s=-1}$$

$$= \frac{2 \cdot 1 - 1 + 2}{1} = 3 \tag{4.23}$$

The multiplication of $(s + 2)^3$ and (4.15) yields

$$F(s)(s + 2)^3 = k_0(s + 2)^3 + \frac{k_1(s + 2)^3}{s + 1} + k_2(s + 2)^2$$

$$+ k_3(s + 2) + k_4 \tag{4.24}$$

which implies

$$k_4 = F(s)(s + 2)^3\big|_{s=-2} = \frac{2s^4 + s^3 - 2s}{s + 1}\bigg|_{s=-2}$$

$$= \frac{2 \cdot 16 - 8 + 4}{-2 + 1} = -28 \tag{4.25}$$

The differentiation of (4.24) yields *(cuz we don't have* $\frac{\dots}{(s+2)^2}$ *or* $\frac{\dots}{(s+2)}$ *in (4.14). ↓ actual function.)*

$$\frac{d}{ds}[F(s)(s + 2)^3] = 3k_0(s + 2)^2 + \frac{d}{ds}\left[\frac{k_1(s + 2)^3}{s + 1}\right]$$

$$+ 2k_2(s + 2) + k_3 \tag{4.26}$$

which implies

$$k_3 = \frac{d}{ds}[F(s)(s + 2)^3]\bigg|_{s=-2} = \frac{d}{ds}\left[\frac{2s^4 + s^3 - 2s}{s + 1}\right]\bigg|_{s=-2}$$

$$= \frac{(s + 1)(8s^3 + 3s^2 - 2) - (2s^4 + s^3 - 2s)}{(s + 1)^2}\bigg|_{s=-2}$$

$$= -(8 \cdot (-8) + 12 - 2) - (2 \cdot 16 - 8 + 4) = 26 \tag{4.27}$$

The differentiation of (4.26) yields

$$\frac{d^2}{ds^2}[F(s)(s + 2)^3] = 6k_0(s + 2) + \frac{d^2}{ds^2}\left[\frac{k_1(s + 2)^3}{s + 1}\right] + 2k_2$$

which implies

$$k_2 = \frac{1}{2} \frac{d^2}{ds^2} [F(s)(s+2)^3] \Big|_{s=-2}$$

which yields $k_2 = -16$ after lengthy computation. This is a different method of computing k_i.

From the preceding derivation, we can state the general formula for computing k_i. If

$$F(s) = k_0 + \sum_i \frac{k_i}{s - a_i} + \frac{c_1}{s - b} + \frac{c_2}{(s-b)^2} + \cdots + \frac{c_m}{(s-b)^m} \qquad (4.28)$$

that is, $F(s)$ has a number of simple poles at a_i and a repeated pole with multiplicity m at b, then we have

$$k_0 = F(\infty) \qquad (4.29a)$$
$$k_i = F(s)(s - a_i) \mid_{s=a_i} \qquad (4.29b)$$

and

$$c_i = \frac{1}{(m-i)!} \frac{d^{m-i}}{ds^{m-i}} [F(s)(s-b)^m] \Big|_{s=b} \qquad (4.29c)$$

for $i = m, m - 1, \ldots, 2, 1$. These formulas can be used to compute coefficients in partial fraction expansion. The formula in (4.29c) requires differentiation of rational functions and is complicated. A different set of formula can be developed from the polynomial equation in (4.19). It is obtained by differentiating the polynomial and requires less computation than the one in (4.29). However, the formulation is more complex. See Reference [6].

Method 4   This method first uses Method 3 to compute the coefficients associated with simple poles and the highest power of repeated poles and then uses Method 2 to compute the remaining coefficients. For example, consider (4.15). Using Method 3, we can easily compute $k_0 = 2$, $k_1 = 3$, $k_4 = -28$, and (4.15) becomes

$$\frac{2s^4 + s^3 - 2s}{(s+1)(s+2)^3} = 2 + \frac{3}{s+1} + \frac{k_2}{s+2} + \frac{k_3}{(s+2)^2} - \frac{28}{(s+2)^3}$$

There are still two unknowns to be solved. We choose arbitrarily $s = 0$ and $s = 1$ to yield

$$s = 0: \qquad 0 = 2 + 3 + \frac{k_2}{2} + \frac{k_3}{4} - \frac{28}{8}$$

and

$$s = 1: \qquad \frac{1}{2 \cdot 27} = 2 + \frac{3}{2} + \frac{k_2}{3} + \frac{k_3}{9} - \frac{28}{27}$$

These two equations can be simplified as

$$\frac{k_2}{2} + \frac{k_3}{4} = \frac{-3}{2}$$

$$\frac{k_2}{3} + \frac{k_3}{9} = \frac{-66}{27}$$

Its solution is $k_2 = -16$ and $k_3 = 26$.

In this section, we have discussed four methods of carrying out partial fraction expansions. In general, Method 4 requires the least computation and is preferred by this author.

EXERCISE 4.4.3

Find the inverse Laplace transforms of

(a) $F(s) = \dfrac{s - 1}{s(s + 1)}$

(b) $F(s) = \dfrac{s^3 + 1}{s(s + 1)(s + 2)}$

[**ANSWERS:** (a) $-1 + 2e^{-t}$, for $t \geq 0$; (b) $\delta(t) + 0.5 - 3.5e^{-2t}$, for $t \geq 0$.]   ■

EXERCISE 4.4.4

Find the inverse Laplace transforms of

(a) $\dfrac{s - 1}{s(s + 1)^2}$

(b) $\dfrac{s + 1}{s^3(s - 1)}$

[**ANSWERS:** (a) $-1 + e^{-t} + 2te^{-t}$, (b) $2e^t - 2 - 2t - \dfrac{1}{2}t^2$, for $t \geq 0$.]   ■

Complex Poles The preceding discussion is equally applicable whether a pole is real or complex. If it is complex, the computation will involve complex numbers. Furthermore, at the end of the partial fraction expansion, some additional manipulation will be required to transform complex-valued functions into real-valued functions. This is illustrated by an example.

Consider the transfer function

$$F(s) = \frac{s}{(s + 1)(s + 1 - j2)(s + 1 + j2)} \tag{4.30}$$

All poles are simple and distinct. We expand it as

$$F(s) = k_0 + \frac{k_1}{s + 1} + \frac{k_2}{s + 1 - j2} + \frac{k_3}{s + 1 + j2}$$

with, using (4.29a) and (4.29b),

$$k_0 = F(\infty) = 0$$

$$k_1 = F(s)(s + 1) \mid_{s=-1} = \left.\frac{s}{(s + 1 - j2)(s + 1 + j2)}\right|_{s=-1} = \frac{-1}{4}$$

$$k_2 = F(s)(s + 1 - j2) \mid_{s=-1+j2} = \frac{-1 + j2}{(j2)(j4)} = \frac{1 - j2}{8}$$

and

$$k_3 = F(s)(s + 1 + j2) \mid_{s=-1-j2} = \frac{-1 - j2}{(-j2)(-j4)} = \frac{1 + j2}{8}$$

In fact, once k_2 is computed, the computation of k_3 is unnecessary. It must equal the complex conjugate of k_2, that is, if $k_2 = a + jb$, then $k_3 = k_2^*$ $:= a - jb$. Using these k_i and Table 4.1, the inverse Laplace transform of (4.30) can be readily obtained as

$$f(t) = \frac{-1}{4}e^{-t} + \frac{1 - j2}{8}e^{-(1-j2)t} + \frac{1 + j2}{8}e^{-(1+j2)t} \tag{4.31}$$

for $t \geq 0$. It consists of three terms, two of which are complex-valued functions.

The coefficients of $F(s)$ in (4.30) are all real; therefore, the inverse Laplace transform of $F(s)$ must be a real-valued function. This implies that the two complex-valued functions in (4.31) can be combined into a real-valued function. We shall do so in a moment. We digress at this point to discuss the transformation of a complex number into polar form. The polar form of $x := a + jb$ is

$$x = re^{j\theta}$$

where $r = \sqrt{a^2 + b^2}$ and $\theta = \tan^{-1}(b/a)$. For example, if $x = -2 + j1$, then

$$x = \sqrt{4 + 1}\,e^{j\tan^{-1}[1/(-2)]} = \sqrt{5}\,e^{j\tan^{-1}(-0.5)}$$

Because $\tan(-26.5°) = -0.5$ and $\tan 153.5° = -0.5$, one may incorrectly write x as $\sqrt{5}\,e^{-j26.5°}$. The correct x, however, should be

$$x = \sqrt{5}\,e^{j153.5°}$$

as can be seen in Figure 4.2. In the complex plane, x is a vector with real part -2 and imaginary part 1, as shown. Thus, its phase is $153.5°$ rather than $-26.5°$. In computing the polar form of a complex number, it is advisable to draw a rough graph to ensure that you obtain the correct phase.

Using the preceding procedure we can now express k_2 and k_3 in (4.31) as

$$k_2 = \frac{\sqrt{5}}{8}e^{-j63.5°} = \frac{\sqrt{5}}{8}e^{-j1.1^{\text{radian}}} = k_3^*$$

The substitution of these into (4.31) yields

$$f(t) = -0.25e^{-t} + \frac{\sqrt{5}}{8}e^{-j1.1}e^{-(1-j2)t} + \frac{\sqrt{5}}{8}e^{j1.1}e^{-(1+j2)t} \qquad (4.32)$$

which can be written as

$$f(t) = -0.25e^{-t} + \frac{\sqrt{5}}{8}e^{-t}\left[e^{j(2t-1.1)} + e^{-j(2t-1.1)}\right]$$

Using

$$\cos\alpha = \frac{e^{j\alpha} + e^{-j\alpha}}{2}$$

we obtain

real - valued function →

$$f(t) = -0.25e^{-t} + \frac{\sqrt{5}}{4}e^{-t}\cos(2t - 1.1) \qquad (4.33)$$

$\Rightarrow \omega \wedge$ *for* $\cos(\omega t + \theta)$ *must be (both) in* RADIANS.

Figure 4.2 *Polar form*

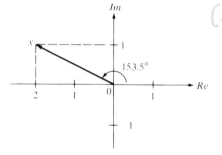

This is a real-valued function. Because the unit of ω in cos (ωt + θ) is radians per second, we must express the phase θ in radians rather than in degrees, as we did in (4.33). Otherwise ωt and θ have different units and cannot be directly added.

EXERCISE 4.4.5

Find the inverse Laplace transform of

$$F(s) = \frac{s - 1}{(s + 1)(s^2 + 2s + 5)} \qquad (4.34)$$

[**ANSWER:** $f(t) = -0.5e^{-t} + 0.5e^{-t}\cos 2t + 0.5e^{-t}\sin 2t$, for $t \geq 0$.] ▪

4.4.2 Expansion in Terms of Linear and Quadratic Factors— Real Computation

In the expansion in terms of linear factors, if $F(s)$ has complex poles, then the expansion will generally involve complex numbers. In this subsection, the complex poles will not be factored and will be retained as a quadratic term with real coefficients. By so doing, we can avoid using complex numbers in the expansion. For example, consider the strictly proper rational function

$$F(s) = \frac{s}{(s + 1)(s^2 + 2s + 5)} \qquad (4.35)$$

The quadratic factor has a pair of complex conjugate poles. Rather than factoring it as $(s + 1 + 2j)(s + 1 - 2j)$, we use the formula

$$s^2 + bs + c = s^2 + bs + (\frac{b}{2})^2 - (\frac{b}{2})^2 + c = (s + \frac{b}{2})^2 + (c - \frac{b^2}{4})$$

to express it as

$$s^2 + 2s + 5 = (s^2 + 2s + 1) - 1 + 5 = (s + 1)^2 + 2^2$$

Now we may expand (4.35) as

$$F(s) = \frac{k_1}{s + 1} + \frac{k'_2 s + k'_3}{(s + 1)^2 + 2^2} \qquad (4.36)$$

This expansion, however, is not as convenient as the following expansion

$$F(s) = \frac{k_1}{s + 1} + \frac{k_2(s + 1)}{(s + 1)^2 + 2^2} + \frac{k_3 \cdot 2}{(s + 1)^2 + 2^2} \qquad (4.37)$$

because (4.36) requires some additional manipulation before we can use Table 4.1. Every term in (4.37), on the other hand, has already assumed one of the forms in Table 4.1. Therefore the expansion in (4.37) is preferable. The k_1 can be computed as

$$k_1 = F(s)(s+1) \mid_{s=-1} = \frac{s}{s^2 + 2s + 5} \mid_{s=-1} = \frac{-1}{4}$$

To compute k_2 and k_3, we choose two real values of s other than -1 in (4.37) to yield two equations. If $s = 0$ and 1, then (4.37) becomes

$$0 = \frac{k_1}{1} + \frac{k_2}{5} + \frac{2k_3}{5}$$

and

$$\frac{1}{2 \cdot 8} = \frac{k_1}{2} + \frac{2k_2}{8} + \frac{2k_3}{8}$$

which can be simplified, after the substitution of $k_1 = -1/4$, as

$$k_2 + 2k_3 = \frac{5}{4}$$
$$k_2 + k_3 = -2k_1 + \frac{1}{4} = \frac{3}{4}$$

The solution is $k_3 = 1/2$ and $k_2 = 1/4$. Thus, the inverse Laplace transform of (4.35) is, from (4.37),

$$f(t) = \frac{-1}{4}e^{-t} + \frac{1}{4}e^{-t}\cos 2t + \frac{1}{2}e^{-t}\sin 2t \qquad \textbf{(4.38)}$$

for $t \geq 0$.

We now compare this result with the one in (4.33). In order to do so, we need the following trigonometric identity

$$\boxed{a \cos \theta + b \sin \theta = r(\frac{a}{r}\cos \theta + \frac{b}{r}\sin \theta)} \qquad \textbf{(4.39a)}$$

where $r = \sqrt{a^2 + b^2}$. Define $\phi = \tan^{-1}(b/a)$. Then we have $\cos \phi = a/r$ and $\sin \phi = b/r$. Thus, (4.39a) can be written as

$$a \cos \theta + b \sin \theta = r(\cos \theta \cos \phi + \sin \theta \sin \phi) = r \cos (\theta - \phi) \qquad \textbf{(4.39b)}$$

Using this formula, (4.38) can be written as

$$f(t) = -0.25e^{-t} + \frac{1}{4}e^{-t}(\cos 2t + 2 \sin 2t) \qquad \textbf{(4.40)}$$
$$= -0.25e^{-t} + \frac{1}{4}e^{-t}[\sqrt{5} \cos(2t - \phi)]$$

where $\phi = \tan^{-1}(2/1) = 63.5° = 1.1$ radians. Thus, (4.38) is identical to (4.33).

The procedure for the expansion in (4.37) is equally applicable to (4.36). The expansion in (4.36) is useful in the so-called parallel realization that will be discussed in Chapter 10. The discussion in this section can be extended to repeated quadratic terms. The basic idea is the same.

EXERCISE 4.4.6

Find the inverse Laplace transform of

$$F(s) = \frac{s - 1}{(s + 1)(s^2 + 2s + 5)}$$

which was studied in (4.34). Which method do you think is simpler?

[**ANSWER:** $f(t) = -0.5e^{-t} + 0.5e^{-t} \cos 2t + 0.5e^{-t} \sin 2t.$] ∎

4.5 LTIL DIFFERENTIAL EQUATIONS

3

In this section, we shall apply the Laplace transform to solve LTIL differential equations. Because every LTIL system can be described by such an equation, solving differential equations is an important part of system analysis.

Consider the mechanical system studied in Figure 3.20. The system is described by the following second-order LTIL differential equation, as derived in (3.73),

$$m\ddot{y}(t) + k_2\dot{y}(t) + k_1 y(t) = u(t)$$

where $\ddot{y}(t) := d^2y(t)/dt^2$ and $\dot{y}(t) := dy(t)/dt$. For easy discussion, we assume $m = 2$, $k_2 = 3$, and $k_1 = 1$. Then, the equation becomes

$$2\ddot{y}(t) + 3\dot{y}(t) + y(t) = u(t) \qquad \textbf{(4.41)}$$

In system analysis, we may ask: What is the response $y(t)$ excited by the unit step input $u(t) = q(t)$ and the initial conditions $y(0-) = -1$ and $\dot{y}(0-) = 1$. To solve this problem, we may apply the Laplace transform to (4.41) to yield, using (4.8),

$$2[s^2 Y(s) - sy(0-) - \dot{y}(0-)] + 3[sY(s) - y(0-)] + Y(s) = U(s) \quad \text{(4.42)}$$

This can be grouped as

$$(2s^2 + 3s + 1)Y(s) = 2sy(0-) + 2\dot{y}(0-) + 3y(0-) + U(s)$$

$$= (2s+3)y(0^-) + 2\dot{y}(0^-)$$

which implies

$$Y(s) = \frac{(2s+3)y(0-) + 2\dot{y}(0-)}{2s^2 + 3s + 1} + \frac{1}{2s^2 + 3s + 1}U(s) \quad \text{(4.43)}$$

$$\underbrace{\hspace{3cm}}_{\text{zero-input response}} \qquad \underbrace{\hspace{3cm}}_{\text{zero-state response}}$$

We see that the output consists of two parts: One part is excited by initial conditions and the other by the input. The former is called the zero-input response and the latter the zero-state response. This is a general property of every linear system, as was discussed in Chapter 2. If $y(0-) = -1$, $\dot{y}(0-) = 1$ and $U(s) = 1/s$, then (4.43) becomes

$$Y(s) = \frac{-2s - 3 + 2}{2s^2 + 3s + 1} + \frac{1}{(2s^2 + 3s + 1)s} = \frac{-2s^2 - s + 1}{(2s + 1)(s + 1)s}$$

This output is in the Laplace-transform domain. To find its time response, we must compute its inverse Laplace transform. We expand $Y(s)$ as

$$Y(s) = \frac{-2s^2 - s + 1}{(2s + 1)(s + 1)s} = k_0 + \frac{k_1}{2s + 1} + \frac{k_2}{s + 1} + \frac{k_3}{s}$$

with, using (4.29),

$$k_0 = F(\infty) = 0$$

$$k_1 = \left. \frac{-2s^2 - s + 1}{(s + 1)s} \right|_{s=-0.5} = -4$$

$$k_2 = \left. \frac{-2s^2 - s + 1}{(2s + 1)s} \right|_{s=-1} = 0$$

and

$$k_3 = \left. \frac{-2s^2 - s + 1}{(2s + 1)(s + 1)} \right|_{s=0} = 1$$

Thus, we have

$$Y(s) = \frac{-4}{2s+1} + \frac{0}{s+1} + \frac{1}{s} = \frac{-2}{s+0.5} + \frac{1}{s}$$

and the inverse Laplace transform of $Y(s)$ is

$$y(t) = -2e^{-0.5t} + 1 \qquad \qquad \textbf{(4.44)}$$

for $t \geq 0$. This is the time response of the system and is plotted in Figure 4.3.

From this example, we can see that the application of the Laplace transform to solve LTIL differential equations is simple and straightforward. First we apply the Laplace transform to change a differential equation into an algebraic equation. Next we substitute the input and initial conditions into the algebraic equation and compute the output. This output is in the Laplace-transform domain. Its inverse Laplace transform yields the time response.

EXERCISE 4.5.1

Find the responses of

$$\dot{y}(t) - 2y(t) = u(t) \qquad \text{and} \qquad \dot{y}(t) - 2y(t) = \dot{u}(t) \qquad \textbf{(4.45)}$$

excited by $y(0-) = 2$ and $u(t) = q(t)$. Also plot their responses.

[**ANSWERS:** $y(t) = 2.5e^{2t} - 0.5$ and $y(t) = 3e^{2t}$, for $t \geq 0$.]   ■

Figure 4.3   *The response of the differential equation in (4.41)*

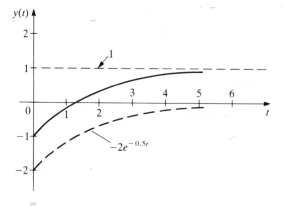

EXERCISE 4.5.2

Find the response of

$$\ddot{y}(t) + \dot{y}(t) - 2y(t) = \dot{u}(t) + 2u(t) \tag{4.46}$$

excited by $y(0-) = 1$, $\dot{y}(0-) = -1$ and $u(t) = 1$ for $t \geq 0$.

[**ANSWER:** $y(t) = -1 + \dfrac{2}{3}e^{-2t} + \dfrac{4}{3}e^{t}$, for $t \geq 0$. Note that $\dot{y}(t)$ is discontinuous at $t = 0$, thus $\dot{y}(0+) \neq \dot{y}(0-)$.] ∎

As was discussed in Section 4.2, one of the reasons for using 0– as the lower limit of (4.1) is to include the impulse $\delta(t)$ in the Laplace transform. Now we are ready to discuss one more reason for using 0–. If we use 0 as the lower limit in (4.1), then we must use the initial conditions $y(0)$ and $\dot{y}(0)$ in (4.8), rather than $y(0-)$ and $\dot{y}(0-)$. For the first equation in (4.45), we have $y(0-) = 2$ and $y(t) = 2.5e^{2t} - 0.5$, which implies $y(0+) = 2$. Thus, $y(t)$ is continuous at $t = 0$ and it is immaterial whether 0 or 0– is used. However, for the second equation in (4.45), we have $y(0-) = 2$ and $y(t) = 3e^{2t}$, which implies $y(0) = 3$. Thus, $y(t)$ is not continuous at $t = 0$. In this case, confusion may arise if we use 0 as the initial time. Thus, it is preferable to use 0– in (4.8). In conclusion, the Laplace transform is defined from 0– to ∞ in order to include wholly the impulse at $t = 0$ in the transform and to avoid possible confusion in initial conditions in solving LTIL differential equations.

4.6   ZERO-INPUT RESPONSE—CHARACTERISTIC POLYNOMIAL AND MODE

The response of LTIL systems, as was discussed in Chapter 2 and demonstrated in the preceding section, can be decomposed into the zero-input response and the zero-state response. In this section, we will discuss some general properties of the zero-input response.

Consider the system described by (4.41). The zero-input response of the system is, as derived in (4.43),

$$Y_{zi}(s) = \frac{(2s + 3)y(0-) + 2\dot{y}(0-)}{2s^2 + 3s + 1} \tag{4.47}$$

where the subscript zi denotes zero input. We call

$$D(s) := 2s^2 + 3s + 1 = 2(s + 0.5)(s + 1) \tag{4.48}$$

the denominator of Y_{zi}, the *characteristic polynomial* and its roots the *modes* of the system. Thus the modes of the system are -0.5 and -1. The zero-input response of the system excited by *any* initial conditions can always be expressed as

$$Y_{zi}(s) = \frac{2y(0-)s + [3y(0-) + 2\dot{y}(0-)]}{2(s + 0.5)(s + 1)} = \frac{k_1}{s + 0.5} + \frac{k_2}{s + 1} \qquad \text{(4.49a)}$$

where k_1 and k_2 depend on the initial conditions. Therefore, the zero-input response is, for $t \geq 0$,

$$y_{zi}(t) = k_1 e^{-0.5t} + k_2 e^{-t} \qquad \text{(4.49b)}$$

In other words, the zero-input response of (4.41) is always a linear combination of the two time functions $e^{-0.5t}$ and e^{-t}, which are the inverse Laplace transforms of $1/(s + 0.5)$ and $1/(s + 1)$. Thus, the form of the zero-input response excited by any initial conditions is completely determined by the modes of the system.

EXAMPLE 4.6.1

We show that for some initial conditions, k_1 or k_2 in (4.49) may become zero. If k_i is zero, we say that the corresponding mode is not excited. If $y(0-) = a$ and $\dot{y}(0-) = -a/2$, for any real a, then (4.49a) becomes

$$Y_{zi}(s) = \frac{2as + 3a - a}{(2s + 1)(s + 1)} = \frac{2a(s + 1)}{(2s + 1)(s + 1)} = \frac{2a}{2s + 1} = \frac{a}{s + 0.5}$$

which implies

$$y_{zi}(t) = ae^{-0.5t}$$

for $t \geq 0$. Therefore, if $\dot{y}(0-) = -y(0-)/2$, then the mode -1 or e^{-t} will not be excited. Similarly, it can be shown that if $\dot{y}(0-) = -y(0-)$, then the mode -0.5 or $e^{-0.5t}$ will not be excited. For other $y(0-)$ and $\dot{y}(0-)$, the two modes will always be excited.

EXERCISE 4.6.1

Find the modes and the general forms of the zero-input responses of the three LTIL differential equations in (4.45) and (4.46).

[ANSWERS: 2, $y_{zi}(t) = k_1 e^{2t}$; 2, $y_{zi}(t) = k_1 e^{2t}$; -2, 1, $y_{zi}(t) = k_1 e^{-2t} + k_2 e^{t}$ for $t \geq 0$.]

EXAMPLE **4.6.2**

Consider the automobile suspension system in Figure 3.21. As developed in Example 3.7.2, the system can be described by

$$\ddot{y}(t) + \frac{k_2}{m}\dot{y}(t) + \frac{k_1}{m}y(t) = \frac{1}{m}u(t)$$

Whenever the automobile hits a pothole, a vertical force $u(t)$ of a very short duration is applied to the system. Because the exact form of $u(t)$ is not known, it is simpler to assume that $u(t)$ excites nonzero initial conditions $y(0-)$ and $\dot{y}(0-)$ and then compute the zero-input response. Let $k_2/m = 2$, $k_1/m = 10$, $u = 0$, $y(0-) = 0.5$ and $\dot{y}(0-) = 0.4$. Then the application of the Laplace transform to the equation yields,

$$s^2 Y(s) - sy(0-) - \dot{y}(0-) + 2(sY(s) - y(0-)) + 10Y(s)$$
$$= (s^2 + 2s + 10)Y(s) - (s + 2)y(0-) - \dot{y}(0-) = 0 \quad \leftarrow y = 0$$

which implies

$$Y(s) = \frac{0.5(s + 2) + 0.4}{(s + 1)^2 + 9} = \frac{3k_1}{(s + 1)^2 + 9} + \frac{k_2(s + 1)}{(s + 1)^2 + 9}$$

with $k_1 = 0.3$ and $k_2 = 0.5$. Thus, we have

$$y(t) = 0.3e^{-t}\sin 3t + 0.5e^{-t}\cos 3t$$

This is plotted in Figure 4.4 with the solid line. We see that the response oscillates and takes about five seconds for the vertical displacement to return to zero. This happens when the shock absorber is worn out and the car is difficult to handle. If the shock absorber is replaced by a new one with $k_2/m = 7$, then the differential equation becomes, in the Laplace-transform domain,

$$s^2 Y(s) - sy(0-) - \dot{y}(0-) + 7(sY(s) - y(0-)) + 10Y(s)$$
$$= (s^2 + 7s + 10)Y(s) - (s + 7)y(0-) - \dot{y}(0-) = 0$$

which implies

$$Y(s) = \frac{0.5(s + 7) + 0.4}{(s + 2)(s + 5)} = \frac{k_1}{s + 2} + \frac{k_2}{s + 5}$$

with $k_1 = 2.9/3$ and $k_2 = -1.4/3$. Thus, we have

$$y(t) = \frac{2.9}{3}e^{-2t} - \frac{1.4}{3}e^{-5t}$$

Figure 4.4   *Responses of automobile suspension system*

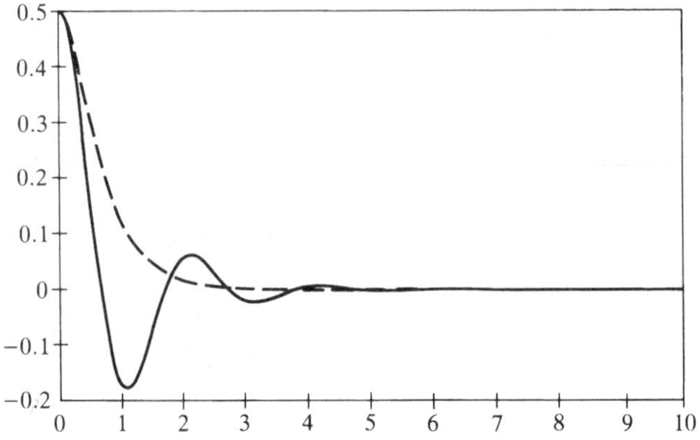

This is plotted in Figure 4.4 with the dotted line. We see that the response excited by the same initial conditions does not oscillate and dies out in about two seconds. Thus, the ride is smoother and the car is easier to handle. Therefore, if the shock absorbers of an automobile do not generate enough friction, they should be replaced.

For the general case, consider the *n*th-order LTIL differential equation

$$a_n y^{(n)}(t) + a_{n-1} y^{(n-1)}(t) + \cdots + a_1 y^{(1)}(t) + a_0 y(t)$$
$$= b_m u^{(m)}(t) + b_{m-1} u^{(m-1)}(t) + \cdots + b_1 u^{(1)}(t) + b_0 u(t) \qquad \textbf{(4.50)}$$

Define

$$D(p) := a_n p^n + a_{n-1} p^{n-1} + \cdots + a_1 p + a_0 \qquad \textbf{(4.51a)}$$

and

$$N(p) := b_m p^m + b_{m-1} p^{m-1} + \cdots + b_1 p + b_0 \qquad \textbf{(4.51b)}$$

where the variable p is the differentiator d/dt defined as

$$py(t) := \frac{d}{dt} y(t), \qquad p^2 y(t) := \frac{d^2}{dt^2} y(t), \qquad p^3 y(t) := \frac{d^3}{dt^3} y(t), \quad \textbf{(4.52)}$$

and so forth. Using this notation, (4.50) can be written as

$$D(p) y(t) = N(p) u(t) \qquad \textbf{(4.53)}$$

In the study of the zero-input response, we assume $u(t) \equiv 0$. In this case, (4.53) reduces to

$$D(p)y(t) = 0 \qquad (4.54)$$

This is called the homogeneous equation. Its response is excited entirely by initial conditions. The application of the Laplace transform to (4.54) yields, as in (4.47),

$$Y_{zi}(s) = \frac{I(s)}{D(s)}$$

where $D(s)$ is defined in (4.51a) with p replaced by s, and $I(s)$ is a polynomial of a degree smaller than that of $D(s)$. The coefficients of $I(s)$ are determined by initial conditions. We call $D(s)$ the characteristic polynomial of (4.53) because it governs the *free, unforced,* or *natural* response of (4.53). The roots of the polynomial $D(s)$ are called the *natural frequencies* or *modes*. We shall use the modes exclusively in this text for the following two reasons: First, we tend to associate frequency with oscillation, whereas the roots of $D(s)$ may not generate oscillation. Second, if $D(s) = s^2 + 2\zeta\omega_n s + \omega_n^2$, then ω_n is called the natural frequency in some control texts. Therefore, to avoid possible confusion, we shall call the roots of $D(s)$ the modes of the system. For example, if

$$D(s) = (s - 2)(s + 1)^3(s + 2 - j3)(s + 2 + j3)$$

then the modes are $2, -1, -1, -1, -2 + j3$, and $-2 - j3$. The root 2 and the complex roots $-2 \pm j3$ are simple modes and the root -1 is a repeated mode with multiplicity 3. Thus, for any initial conditions, $Y_{zi}(s)$ can be expanded as

$$Y_{zi}(s) = \frac{k_1}{s - 2} + \frac{k_2}{s + 2 - j3} + \frac{k_3}{s + 2 + j3} + \frac{c_1}{s + 1} + \frac{c_2}{(s + 1)^2} + \frac{c_3}{(s + 1)^3}$$

and its zero-input response is, using Table 4.1,

$$y_{zi}(t) = k_1 e^{2t} + k_2 e^{-(2-j3)t} + k_3 e^{-(2+j3)t} + c_1 e^{-t} + c_2 t e^{-t} + \frac{c_3}{2} t^2 e^{-t} \qquad (4.55a)$$

Thus, the general form of the zero-input response is determined by the modes of the system. In this expression, $k_2 e^{-(2-j3)t}$ and $k_3 e^{-(2+j3)t}$ are complex-valued functions. If we use the procedure in (4.32), then they can be combined into a real-valued function and (4.55a) becomes

$$y_{zi}(t) = k_1 e^{2t} + \bar{k}_2 e^{-2t} \sin(3t + \bar{k}_3) + c_1 e^{-t} + c_2 t e^{-t} + \frac{c_3}{2} t^2 e^{-t} \qquad (4.55b)$$

which can also be written as, using (4.39),

üi)

$$y_{zi}(t) = k_1 e^{2t} + \hat{k}_2 e^{-2t} \sin 3t + \hat{k}_3 e^{-2t} \cos 3t + c_1 e^{-t} \qquad \textbf{(4.55c)}$$

$$+ c_2 t e^{-t} + \frac{c_3}{2} t^2 e^{-t}$$

In (4.55b) and (4.55c), all coefficients are real. Any of these three expressions can be used to denote the zero-input response. We mention that, even though all zero-input responses can be expressed as in (4.55), the responses can be quite different for different initial conditions.

EXERCISE 4.6.2

Find the modes and the general form of the zero-input response of

$$D(p)y(t) = N(p)u(t)$$

where

$$D(p) = p^3 (p-2)^2 (p^2 + 4p + 8) \qquad \text{and} \qquad N(p) = 3p^2 - 10$$

[**ANSWERS:** $0, 0, 0, 2, 2, -2 + j2$, and $-2 - j2$; $y_{zi}(t) = k_1 + k_2 t + k_3 t^2 + k_4 e^{2t} + k_5 t e^{2t} + k_6 e^{-2t} \sin 2t + k_7 e^{-2t} \cos 2t$.] ■

4.7   ## ZERO-STATE RESPONSE—TRANSFER FUNCTION

14

Consider the differential equation (4.41) or

$$2\ddot{y}(t) + 3\dot{y}(t) + y(t) = u(t)$$

If all initial conditions are zero, the application of the Laplace transform yields

$$2s^2 Y(s) + 3sY(s) + Y(s) = U(s)$$

Thus, we have

$$Y(s) = \frac{1}{2s^2 + 3s + 1} U(s) =: H(s)U(s) \qquad \textbf{(4.56)}$$

where the rational function $H(s) = 1/(2s^2 + 3s + 1)$ is called the transfer function. It is the ratio of the Laplace transforms of the output and input with all initial conditions zero or

$$H(s) = \frac{Y(s)}{U(s)} \bigg|_{\text{initial conditions} = 0} \qquad \textbf{(4.57)}$$

This definition is consistent with the definition given in (4.11). Before proceeding, we mention that the transfer function can be obtained using the following methods:

1. The transfer function is, as developed in (4.11), the Laplace transform of the impulse response. Therefore, if the impulse response is available, its Laplace transform yields the transfer function.

2. It can be obtained from the zero-state response excited by *any* input, in particular, step or sinusoidal functions, by using (4.57).

3. It can be obtained from the differential equation description.

The first method is rarely used because of the difficulties in generating impulses and in measuring impulse responses. It is also a special case of the second method when the input is an impulse. We use examples to illustrate the second and third methods.

EXAMPLE 4.7.1

Consider an LTIL system. Its zero-state response excited by the input $u(t) = \sin 2t$, for $t \geq 0$, is measured as

$$y(t) = 2e^{-2t} + \sin 2t - 2 \cos 2t \qquad \text{for } t \geq 0$$

Now we use (4.57) to compute its transfer function. Clearly we have, using Table 4.1,

$$U(s) = \mathcal{L}[u(t)] = \frac{2}{s^2 + 4}$$

and

$$Y(s) = \mathcal{L}[y(t)] = \frac{2}{s + 2} + \frac{2}{s^2 + 4} - \frac{2s}{s^2 + 4} = \frac{2(s^2 + 4) + 2(s + 2) - 2s(s + 2)}{(s + 2)(s^2 + 4)}$$

$$= \frac{-2s + 12}{(s + 2)(s^2 + 4)}$$

Thus, the transfer function of the system is

$$H(s) = \frac{Y(s)}{U(s)} = \frac{\dfrac{-2s + 12}{(s + 2)(s^2 + 4)}}{\dfrac{2}{s^2 + 4}} = \frac{-s + 6}{s + 2} \qquad (4.58)$$

EXERCISE 4.7.1

The unit step response of a system is measured as

$$y(t) = 1 + e^{-t} - 2e^{-3t}$$

What is the transfer function of the system?

[**ANSWER:** $(5s + 3)/(s + 1)(s + 3)$.] ■

The previous example shows that the transfer function can be obtained by measurement. An important feature of the second method is that there is no need to know the internal structure of the system. We apply an input to the input terminal and measure the output at the output terminal. The transfer function can then be computed from the input and output. This is the case in theory, but it may not be so in practice. In reality, noise may enter into measurement and make the method difficult. If we know the structure of a system, we may use physical laws to develop a differential equation to describe the system. The transfer function can then be easily developed from the equation, as the following examples illustrate.

EXAMPLE 4.7.2

(**Mechanical system**) Consider the mechanical system in Figure 3.20. Using Newton's law, we developed in (3.73) the differential equation

$$m\ddot{y}(t) + k_2\dot{y}(t) + k_1 y(t) = u(t)$$

to describe the system. The application of the Laplace transform yields, assuming zero initial conditions,

$$ms^2 Y(s) + k_2 sY(s) + k_1 Y(s) = U(s)$$

or

$$(ms^2 + k_2 s + k_1)Y(s) = U(s)$$

Thus, the transfer function from u to y is

$$H(s) = \frac{Y(s)}{U(s)} = \frac{1}{ms^2 + k_2 s + k_1}$$

EXAMPLE 4.7.3

(**Electromechanical system**) Consider the field-controlled DC motor in Figure 3.22. It is described, as developed in (3.78), by

$$J\theta^{(3)}(t) + (f + \frac{JR_f}{L_f})\ddot{\theta}(t) + \frac{fR_f}{L_f}\dot{\theta}(t) = \frac{k_t}{L_f}u(t)$$

The application of the Laplace transform yields, assuming zero initial conditions,

$$Js^3\Theta(s) + (f + \frac{JR_f}{L_f})s^2\Theta(s) + \frac{fR_f}{L_f}s\Theta(s) = \frac{k_t}{L_f}U(s)$$

Thus, the transfer function from the input u to the output θ of the DC motor is

$$H(s) = \frac{\Theta(s)}{U(s)} = \frac{\dfrac{k_t}{L_f}}{Js^3 + (f + \dfrac{JR_f}{L_f})s^2 + \dfrac{fR_f}{L_f}s}$$

EXAMPLE 4.7.4

(**Electrical system**) Consider the network shown in Figure 3.13(b) and repeated in Figure 4.5(a). It is described, as developed in (3.74), by

$$2\dot{y}(t) + y(t) = u(t)$$

The application of the Laplace transform yields, assuming zero initial conditions,

$$(2s + 1)Y(s) = U(s)$$

Thus, the transfer function from u to y of the network is

$$H(s) = \frac{Y(s)}{U(s)} = \frac{1}{2s + 1} \tag{4.59}$$

We mention that for RLC networks, there is another method of computing transfer functions without computing differential equations. This method will be discussed in a later section.

Figure 4.5 *A network*

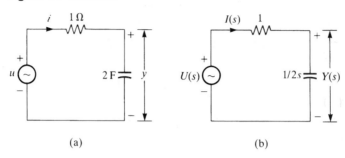

(a)   (b)

EXERCISE 4.7.2

(**Industrial process**) Find the transfer function from q_i to q_2 of the system studied in Figure 3.23.

[**ANSWER:** $H(s) = 1/(A_1 A_2 R_1 R_2 s^2 + (A_1 R_1 + A_2 R_2 + A_1 R_2)s + 1)$.] ■

As these examples illustrated, transfer functions can be readily obtained from differential equations. In general, if a system is described by

$$D(p)y(t) = N(p)u(t)$$

where $D(p)$ and $N(p)$ are defined as in (4.51), then the transfer function of the system is

$$H(s) = \frac{N(s)}{D(s)}$$

Note that once the transfer function of a system is obtained, the impulse response of the system can then be obtained by taking its inverse Laplace transform. For example, the impulse response of the transfer function in (4.58) or

$$H(s) = \frac{-s + 6}{s + 2} = \frac{-s - 2 + 8}{s + 2} = -1 + \frac{8}{s + 2}$$

is

$$h(t) = -\delta(t) + 8e^{-2t}$$

for $t \geq 0$. In conclusion, although the impulse response is defined as the output excited by an impulse, it is unnecessary to generate an impulse to find the impulse response. It can be obtained from the transfer function, which in turn can be obtained by analysis or by applying any input.

The transfer functions obtained in the preceding examples are all rational functions of s because the systems are all lumped. If an LTI system is distributed, the equation $Y(s) = H(s)U(s)$, as developed in (4.11), still holds. However, the transfer function $H(s)$ will not be a rational function of s.

4.7.1   Poles and Zeros of Proper Rational Transfer Functions

The zero-state response of a system is governed by its transfer function. Now we shall use it to compute the zero-state response. The transfer functions studied in this section are assumed to be proper rational functions. For this class of transfer functions, we may define poles and zeros. Consider the rational transfer function

$$H(s) = \frac{N(s)}{D(s)}$$

where $N(s)$ and $D(s)$ are polynomials with real coefficients. If $N(s)$ and $D(s)$ have no common factors, then as defined in Section 4.4, the roots of $D(s)$ and $N(s)$ are, respectively, the poles and zeros of the system described by $H(s)$. For example, the poles of the system described by (4.56), repeated in the following,

$$Y(s) = H(s)U(s) = \frac{1}{2s^2 + 3s + 1}U(s) = \frac{1}{2(s + 0.5)(s + 1)}U(s) \quad \textbf{(4.60)}$$

are −0.5 and −1. The system has no zero. Because of their importance, poles and zeros are defined once again using a different but equivalent definition.

Definition 4.2   A finite real or complex number λ is a pole of $H(s)$ if $|H(\lambda)| = \infty$, where $|\cdot|$ denotes the absolute value. It is a zero of $H(s)$ if $H(\lambda) = 0$.[5]

Consider the transfer function

$$H(s) = \frac{N(s)}{D(s)} = \frac{2s^3 + 6s^2 - 2s - 6}{(s - 1)(s + 2)(s + 1)^3}$$

To see whether or not $s = 1$ is a pole of $H(s)$, we compute $H(1)$:

$$H(1) = \frac{N(1)}{D(1)} = \frac{0}{0}$$

---

[5]We require λ to be finite in this definition. Otherwise, for the transfer function in (4.60), we have $H(s) \approx 1/2s^2$ for large s and $s = \infty$ is a repeated zero with multiplicity 2. In this text, we consider only finite poles and zeros unless stated otherwise.

It is not defined. However, using l'Hopital's rule,[6] we have

$$H(1) = \left.\frac{N(s)}{D(s)}\right|_{s=1} = \left.\frac{N'(s)}{D'(s)}\right|_{s=1} = \left.\frac{2(3s^2 + 6s - 1)}{5s^4 + 16s^3 + 12s^2 - 4s - 5}\right|_{s=1} = \frac{16}{24} \neq \infty$$

Thus, $s = 1$ is not a pole of $H(s)$. If we cancel the common factors of $H(s)$ to yield

$$H(s) = \frac{2(s + 3)(s - 1)(s + 1)}{(s - 1)(s - 2)(s + 1)^3} = \frac{2(s + 3)}{(s + 2)(s + 1)^2}$$

then we see immediately that $s = 1$ is not a pole. Clearly, $H(s)$ has one zero -3 and three poles $-2, -1$, and -1. The pole -2 is simple, and the pole -1 is repeated with multiplicity 2.

Every proper transfer function can be expressed as

$$H(s) = \frac{a(s - z_1)(s - z_2)\cdots}{(s - p_1)(s - p_2)(s - p_3)\cdots}$$

where z_i are zeros and p_i are poles. In addition to poles and zeros, we must specify a or the value of $H(s)$ at some finite s in order to determine $H(s)$ uniquely. This can be illustrated by an example.

EXAMPLE 4.7.5

A transfer function is known to have poles at $-2, 1, 1$, and $-1 \pm j3$ and zeros at 2 and -4. It is also known that $H(0) = -10$. What is the transfer function?

The transfer function is of the form

$$H(s) = \frac{a(s - 2)(s + 4)}{(s + 2)(s - 1)(s - 1)(s + 1 + j3)(s + 1 - j3)}$$

$$= \frac{a(s - 2)(s + 4)}{(s + 2)(s - 1)^2(s^2 + 2s + 10)}$$

Because $H(0) = -10$, we have

$$-10 = \frac{a(-8)}{2 \times 1 \times 10}$$

---

[6]In honor of Guillaume François Antoine l'Hopital (1661-1704). The rule was, in fact, a work of his teacher, Johann Bernoulli (1667-1748), a Swiss mathematician.

which implies

$$a = \frac{-10 \times 20}{-8} = 25$$

Thus, the transfer function is

$$H(s) = \frac{25(s + 2s - 8)}{(s + 2)(s - 1)^2(s^2 + 2s + 10)}$$

We now use the transfer function to compute the zero-state response. First, we compute the Laplace transform $U(s)$ of the input. Then, we multiply the transfer function $H(s)$ with $U(s)$ to yield the output $Y(s)$. The inverse Laplace transform of $Y(s)$ yields the zero-state response. This is illustrated by an example.

EXAMPLE 4.7.6

Find the unit step response of (4.60), that is, the zero-state response of (4.60) excited by $u(t) = 1$, for $t \geq 0$. The Laplace transform of $u(t)$ is $1/s$. Thus, we have

$$Y(s) = H(s)U(s) = \frac{1}{2(s + 0.5)(s + 1)} \cdot \frac{1}{s}$$

To find the time response, we carry out its partial fraction expansion as

$$Y(s) = \frac{0.5}{(s + 0.5)(s + 1)s} = \frac{k_1}{s + 0.5} + \frac{k_2}{s + 1} + \frac{k_3}{s}$$

with

$$k_1 = \left. \frac{0.5}{(s + 1)s} \right|_{s=-0.5} = \frac{0.5}{(0.5)(-0.5)} = -2$$

$$k_2 = \left. \frac{0.5}{(s + 0.5)s} \right|_{s=-1} = \frac{0.5}{(-0.5)(-1)} = 1$$

and

$$k_3 = \left. \frac{0.5}{(s + 0.5)(s + 1)} \right|_{s=0} = 1$$

Thus, the zero-state response is, by using Table 4.1,

$$y(t) = -2e^{-0.5t} + e^{-t} + 1 \qquad (4.61)$$

$$\underbrace{\qquad\qquad}_{\substack{\text{Due to poles} \\ \text{of } H(s)}} \quad \underbrace{\quad}_{\substack{\text{Due to poles} \\ \text{of } U(s)}}$$

for $t \geq 0$. The use of the Laplace transform to compute the zero-state response is indeed very simple.

────

This example reveals a very important property of the zero-state response. We see from (4.61) that the response consists of three terms. Two are the inverse Laplace transforms of $1/(s + 0.5)$ and $1/(s + 1)$, which are associated with the poles of the system. The remaining term is due to the step input. Therefore, the zero-state response of a system will consist of two parts. One part is due to the poles of the system, and the other is due to the applied input. In general, we have

> zero-state response = (terms due to poles of transfer function)
>   + (terms due to poles of input)

For example, consider

$$Y(s) = \frac{(s + 10)(s + 2)(s - 1)^2}{s^3(s - 2)(s + 3)(s + 2 - j2)(s + 2 + j2)} U(s)$$

It has simple poles at $2, -3$, and $-2 \pm j2$ and a repeated pole with multiplicity 3 at 0. It has simple zeros at -10 and -2 and a repeated zero at 1. Now, for any input $U(s)$, the zero-state response will be of the form

$$y(t) = k_1 + k_2 t + k_3 t^2 + k_4 e^{2t} + k_5 e^{-3t} + k_6 e^{-(2-j2)t} + k_7 e^{-(2+j2)t}$$
$$+ \text{(terms due to the poles of } U(s)) \qquad (4.62a)$$

or

$$y(t) = k_1 + k_2 t + k_3 t^2 + k_4 e^{2t} + k_5 e^{-3t} + \overline{k}_6 e^{-2t} \sin 2t + \overline{k}_7 e^{-2t} \cos 2t$$
$$+ \text{(terms due to the poles of } U(s)) \qquad (4.62b)$$

or

$$y(t) = k_1 + k_2 t + k_3 t^2 + k_4 e^{2t} + k_5 e^{-3t} + \hat{k}_6 e^{-2t} \sin (2t + \hat{k}_7)$$
$$+ \text{(terms due to the poles of } U(s)) \qquad (4.62c)$$

If $U(s)$ and $H(s)$ have no common poles—for example, if $U(s) = (s + 3)/(s + 1)(s + 2)$—then we have terms due to poles of $U(s) = k_8 e^{-t} + k_9 e^{-2t}$

If $U(s)$ and $H(s)$ have common poles—for example, if $U(s) = (s-3)/s(s-2)$—then we have

terms due to poles of $U(s) = k_8 t^3 + k_9 t e^{2t}$

Different inputs will generate different k_i. However, the form of (4.62) will always be the same. Note that if $U(s)$ contains a zero that is a pole of $H(s)$, then the coefficient associated with the pole will be zero and the pole is said to be not excited by the input. We mention that the zeros of $H(s)$ have no effect on the *form* of (4.62), but they do affect its coefficients and, consequently, its response, as the following example shows.

EXAMPLE 4.7.7

Consider

$$H_1(s) = \frac{2}{(s+1)(s+1+j1)(s+1-j1)}$$

$$H_2(s) = \frac{0.2(s+10)}{(s+1)(s+1+j1)(s+1-j1)}$$

$$H_3(s) = \frac{-0.2(s-10)}{(s+1)(s+1+j1)(s+1-j1)}$$

$$H_4(s) = \frac{10(s^2+0.1s+0.2)}{(s+1)(s+1+j1)(s+1-j1)}$$

$$H_5(s) = \frac{s}{(s+1)(s+1+j1)(s+1-j1)}$$

If we apply a unit step input or $U(s) = 1/s$, then their responses are all of the form

$$y(t) = k_1 e^{-t} + k_2 e^{-t} \sin(t+k_3) + k_4$$

but the set of k_i is different for each system. Their responses are plotted in Figure 4.6. The first four responses all approach 1 as $t \to \infty$; this is achieved by choosing $H_i(s)$ such that $H_i(0) = 1$ (see Example 4.7.8). The last response approaches zero because the pole of the input is not excited; it is canceled by the zero of $H_5(s)$.

In conclusion, the form of the zero-state response of a system is determined by the poles of the transfer function. The form of the zero-input response is determined by the modes of the characteristic polynomial.

Figure 4.6   *Unit step responses of systems*

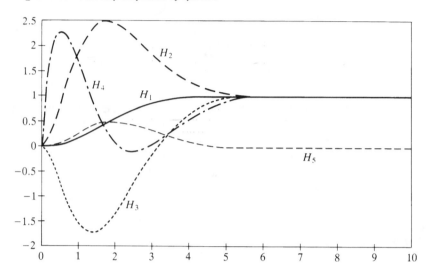

EXERCISE 4.7.3

Find the general form of the zero-state response of the system studied in Exercise 4.6.2.

[**ANSWER:** $k_1 + k_2 t + k_3 t^2 + k_4 e^{2t} + k_5 t e^{2t} + k_6 e^{-(2+j2)t} + k_7 e^{-(2-j2)t}$ + (terms due to the poles of $U(s)$).]   ∎

4.7.2   Time Responses of Modes and Poles

The response of any LTIL system can be decomposed as the zero-input response and zero-state response. The zero-input response is essentially dictated by the modes; the zero-state response is essentially dictated by the poles. Thus, poles and modes are important in system analysis. Here we will discuss their general time responses and their behavior as $t \to \infty$.

Poles can be real or complex, simple or repeated. They are often plotted on the complex plane or s-plane, as shown in Figure 4.7. The s-plane can be divided into three parts: the right half-plane (RHP), the left half-plane (LHP), and the pure imaginary or $j\omega$-axis. To avoid possible confusion of whether or not the RHP includes the $j\omega$-axis, we shall use the following convention: The *open* RHP is the RHP excluding the $j\omega$-axis, and the *closed* RHP is the RHP including the $j\omega$-axis. Because we will encounter a different complex plane in the next chapter, the open RHP of the Laplace complex plane is also called the open RH s-plane. If a pole lies inside the open LHP, then the pole has a negative real part; its imaginary part can be positive or negative. If a pole lies inside the closed LHP, then the pole has a negative or zero real part.

Figure 4.7 *Time responses of real and complex poles*

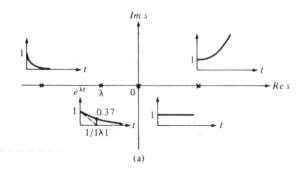

(a)

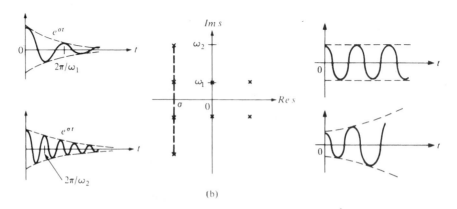

(b)

Poles and zeros are usually plotted on the *s*-plane by using small crosses and circles. Note that no zeros are plotted in Figure 4.7. For real poles, the time responses will increase or decrease exponentially as shown in Figure 4.7(a). The farther away a real pole is from the origin of the complex *s*-plane, the faster the rate of increase or decrease.

The time response of the complex pole $\sigma - j\omega$ is

$$e^{(\sigma - j\omega)t} = e^{\sigma t}e^{-j\omega t} = e^{\sigma t}(\cos \omega t - j \sin \omega t)$$

It is a complex-valued function of *t*. If a transfer function has exclusively real coefficients, then the complex conjugate of $\sigma - j\omega$ is also a pole and its time response is

$$e^{(\sigma + j\omega)t} = e^{\sigma t}(\cos \omega t + j \sin \omega t)$$

Furthermore, the coefficients associated with them must be complex conjugate. That is, if one coefficient is $k = re^{j\theta}$, the other is $k^* = re^{-j\theta}$. Therefore, we have

$$re^{j\theta}e^{(\sigma-j\omega)t} + re^{-j\theta}e^{(\sigma+j\omega)t} = re^{\sigma t}(e^{-j(\omega t-\theta)} + e^{j(\omega t-\theta)}) = 2re^{\sigma t}\cos(\omega t - \theta)$$

Figure 4.7(b) plots the response of two pairs of complex-conjugate poles with $2r = 1$ and $\theta = 0$. We see that the imaginary part of the poles governs the frequency of oscillation and the real part governs the rate of change of the envelope of the oscillation. Thus, from the locations of poles, the waveforms of their time responses can be readily sketched.

Now consider the time responses of poles as t approaches infinity. The time response of the real pole $1/(s + 0.01)$, which is in the open LHP, is

$$e^{-0.01t}$$

It clearly approaches zero as $t \to \infty$. The time response of the repeated pole $1/(s + 0.01)^2$ is, using Table 4.1,

$$te^{-0.01t}$$

It is the product of t, which goes to ∞, and $e^{-0.01t}$, which goes to 0 as $t \to \infty$. Therefore, it requires some computation to find its value as $t \to \infty$. We use l'Hopital's rule to compute

$$\lim_{t\to\infty} te^{-0.01t} = \lim_{t\to\infty} \frac{t}{e^{0.01t}} = \lim_{t\to\infty} \frac{1}{0.01e^{0.01t}} = 0$$

Thus, the time response $te^{-0.01t}$ approaches zero as $t \to \infty$. Similarly, we can show that

$$t^n e^{-0.01t} \to 0 \qquad \text{as } t \to \infty$$

for $n = 1, 2, 3, \ldots$. This is due to the fact that the exponential approaches zero at a rate much faster than the rate of t^n approaching infinity. Similarly, we have, for $\sigma < 0$,

$$t^n e^{\sigma t} \cos \omega t \to 0 \qquad \text{as } t \to \infty$$

for $n = 0, 1, 2, \ldots$. Thus we conclude that the time responses of open LHP poles, simple or repeated, all approach zero as $t \to \infty$. Using the same analysis, we can also conclude that the time responses of open RHP poles, simple or repeated, all approach infinity as $t \to \infty$.

The situation for pure imaginary poles is slightly more complex. If it is real and simple (that is, $1/s$), then its time response is a step function. Thus, it approaches a constant as $t \to \infty$. If it is real and repeated or $1/s^n$, $n = 2, 3, \ldots$, then its time response is t^{n-1}. Clearly, it approaches infinity as $t \to \infty$. The time response of simple and pure imaginary poles $1/(s^2 + \omega^2)$ is $\sin \omega t$, a sustained oscillation. If the pure imaginary poles are repeated, then its time response $t \sin \omega t$ approaches infinity oscillatorily. The preceding discussion is applicable to modes without any modification and is summarized in Table 4.3.

From this table, we make the following important conclusions:

1. The time response of a pole (mode), simple or repeated, approaches zero as $t \to \infty$ if, and only if, the pole (mode) lies inside the open LHP or has a negative real part.

2. The time response of a pole (mode) approaches a nonzero constant as $t \to \infty$ if, and only if, the pole (mole) is simple and located at the origin.

From these two statements, we can further conclude that the time response or the inverse Laplace transform of $F(s)$ approaches zero as $t \to \infty$ if and only if all poles of $F(s)$ lie inside the open LHP. The time response of $F(s)$ approaches a nonzero constant if and only if $F(s)$ has a simple pole at $s = 0$ and the remaining poles of $F(s)$ all lie inside the open LHP.

EXAMPLE 4.7.8

Consider the five systems in Example 4.7.7. The unit step responses of the systems are all of the form

$$Y_i(s) = H_i(s)U(s) = H_i(s)\frac{1}{s} = \frac{k_1}{s+1} + \frac{k_2}{s+1+j} + \frac{k_3}{s+1-j} + \frac{k_4}{s}$$

or, in the time domain,

$$y_i(t) = k_1 e^{-t} + k_2 e^{(-1-j)t} + k_3 e^{(-1+j)t} + k_4$$

for $t \geq 0$. Because all poles of $H_i(s)$ lie inside the open left half-plane, their responses approach zero as $t \to \infty$. Thus, we have

$$\lim_{t \to \infty} y_i(t) = k_4$$

Table 4.3 *Time responses of poles or modes as $t \to \infty$*

Poles or modes	Simple ($n = 1$)	Repeated ($n > 1$)
Open LHP	0	0
Open RHP	$\pm\infty$	$\pm\infty$
$j\omega$-axis ($1/s^n$)	A constant	∞
$j\omega$-axis ($1/(s^2 + a^2)^n$)	A sustained oscillation	$\pm\infty$

where k_4 can be computed using (4.29b) as

$$k_4 = Y_i(s)s\bigg|_{s=0} = H_i(s)\frac{1}{s}s\bigg|_{s=0} = H_i(s)\bigg|_{s=0} = H_i(0)$$

Clearly, we have

$$H_1(0) = \frac{2}{(0+1)(0+1+j)(0+1-j)} = \frac{2}{1+1} = 1$$

Similarly, we have $H_i(0) = 1$ for $i = 2, 3$, and 4 and $H_5(0) = 0$. Thus, the first four unit step responses approach 1 and the fifth approaches 0, as shown in Figure 4.6. Note that the pole of $U(s)$ is canceled by the zero of $H_5(s)$. Thus, the pole of $U(s)$ is not excited, and the unit step response of $H_5(s)$ approaches 0 as $t \to \infty$.

EXERCISE 4.7.4

Consider the following systems.

(a) $H_1(s) = \dfrac{s}{s^2 - 1}$

(b) $H_2(s) = \dfrac{1}{(s+1)(s+2)}$

(c) $H_3(s) = \dfrac{s}{(s+1)(s+2)}$

What values will their impulse responses and unit step responses approach?

[**ANSWERS:** (a) ∞, ∞; (b) $0, 0.5$; (c) $0, 0$.]

*4.8   LINEAR DIFFERENTIAL EQUATIONS WITH TIME-VARYING COEFFICIENTS

The Laplace transform is, as the preceding sections demonstrated, a very powerful tool in the study of linear time-invariant lumped systems. One may wonder, then, whether the transform can also be used to study other types of systems. This question will be answered here.

If a system is linear and time invariant, but it is not lumped, the equation $Y(s) = H(s)U(s)$ still holds. However, the transfer function $H(s)$ will be an irrational functions of s. Except for some very special cases, its manipulation is complicated. Therefore, the Laplace transform is not as widely used in LTI distributed systems as in LTI lumped systems.

If a system is linear and lumped, but it is not time invariant, then the Laplace transform cannot be used. For example, consider the first-order differential equation

$$\dot{y}(t) + a(t)y(t) = u(t)$$

If $a(t)$ is a constant, independent of time, then

$$\mathcal{L}[a(t)y(t)] = \mathcal{L}[ay(t)] = a\mathcal{L}[y(t)] = aY(s) \qquad \textbf{(4.63)}$$

and the differential equation can be transformed into

$$sY(s) - y(0-) + aY(s) = U(s)$$

or

$$(s + a)Y(s) = y(0-) + U(s)$$

This is an algebraic equation. This derivation hinges entirely on (4.63). If (4.63) does not hold, then the preceding derivation cannot be carried out.

Now consider the case where $a(t)$ is a time function. We assume $a(t) = e^{-t}$, for $t \geq 0$. If $y(t) = e^t$, for $t \geq 0$, then

$$\mathcal{L}[a(t)y(t)] = \mathcal{L}[1] = \frac{1}{s}$$

which is different from $\mathcal{L}[a(t)] = 1/(s + 1)$, $\mathcal{L}[y(t)] = 1/(s - 1)$, or their product. If $a(t) = y(t) = e^{-t}$, for $t \geq 0$, then

$$\mathcal{L}[a(t)y(t)] = \mathcal{L}[e^{-t}e^{-t}] = \frac{1}{s + 2}$$

which is again different from $\mathcal{L}[a(t)] = \mathcal{L}[y(t)] = 1/(s + 1)$ or their product. In other words, we do not have $\mathcal{L}[a(t)y(t)] = \mathcal{L}[a(t)]\mathcal{L}[y(t)]$; therefore, a linear time-varying differential equation cannot be transformed into an algebraic equation by applying the Laplace transform. For this reason, the Laplace transform is not used in the study of linear time-varying systems. The Laplace transform also is not used in the study of nonlinear systems.

We mention in passing that, if $A(s) = \mathcal{L}[a(t)]$, $Y(s) = \mathcal{L}[y(t)]$, and their regions of convergence include the $j\omega$-axis, then

$$\mathcal{L}[a(t)y(t)] \mid_{s=j\omega} = \frac{1}{2\pi}\int_{-\infty}^{\infty} A(j\alpha)Y(j\omega - j\alpha)d\alpha \qquad \textbf{(4.64)}$$

This is called the complex convolution, which will be discussed in Chapter 6.

*4.9   RLC NETWORK ANALYSIS

A network that consists of resistors, capacitors, inductors, current and/or voltage sources is called an RLC network. The network is an LTI system if all circuit elements are modeled as linear and time invariant. If the total number of inductors and capacitors is finite, the network is lumped. Clearly, all discussion in the preceding sections is directly applicable to such a network. For example, if we compute its differential equation description, then we can readily obtain its transfer function. In this section, we will discuss a different method of computing the transfer function, which does not involve computing the differential equation. It is carried out by applying the Laplace transform to each circuit element. This transforms the network into a "resistorlike" network. This new equivalent network can then be analyzed using algebraic methods.

The branch characteristic of an inductor with inductance L is, as discussed in (2.10),

$$v(t) = L\frac{di(t)}{dt} \quad \text{with } i(0-) = i_0 \tag{4.65}$$

with the polarities of current $i(t)$ and voltage $v(t)$ chosen as shown in Figure 4.8(a). Note that the initial inductor current $i(0-)$ must be specified; otherwise, the relationship between $v(t)$ and $i(t)$ is not unique. For example, the two different $i(t)$ in Figure 4.9 give the same $v(t)$. The application of the Laplace transform to (4.65) yields

$$V(s) = L[sI(s) - i(0-)] = LsI(s) - Li(0-) \tag{4.66}$$

This is an algebraic equation and can also be written as

$$I(s) = \frac{1}{Ls}V(s) + \frac{1}{s}i(0-) \tag{4.67}$$

Both (4.66) and (4.67) are illustrated in Figure 4.8. Note that, in the s-domain, $Li(0-)$ is an impulse with weight $Li(0-)$, whereas $i(0-)/s$ is a step function with

Figure 4.8   *An inductor*

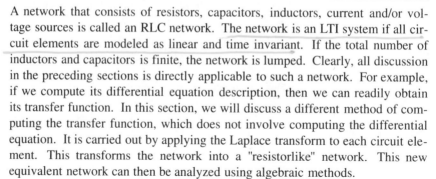

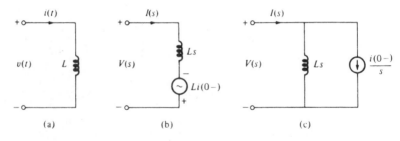

(a)   (b)   (c)

Figure 4.9 *Two different* i(t) *with the same* v(t)

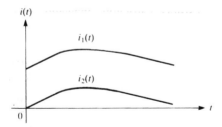

amplitude $i(0-)$. Therefore, the initial condition of the inductor will provide an impulse of voltage with the opposite polarity from $V(s)$ in Figure 4.8(b), but it will act as a constant current source in Figure 4.8(c).

The characteristic of a capacitor with capacitance C is, as discussed in (2.9),

$$v(t) = \frac{1}{C}\int_{0-}^{t} i(\tau)d\tau + v(0-)$$ **(4.68)**

with the polarities of $i(t)$ and $v(t)$ chosen as shown in Figure 4.10(a). The initial capacitor voltage $v(0-)$ must be specified. The application of the Laplace transform to (4.68) yields

$$V(s) = \frac{1}{Cs}I(s) + \frac{v(0-)}{s}$$ **(4.69)**

or

$$I(s) = CsV(s) - Cv(0-)$$ **(4.70)**

These relationships are shown in Figures 4.10(b) and (c). The remarks for Figures 4.8(b) and (c) are equally applicable to Figures 4.10(b) and (c). Either (b) or (c) can be used; they should be chosen to facilitate analysis.

Figure 4.10 *A capacitor*

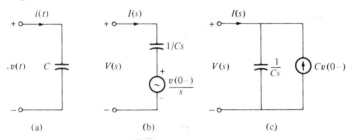

EXAMPLE 4.9.1

Consider the RLC network shown in Figure 4.11(a). Find the loop current excited by $i(0-) = 2$ amperes, $v(0-) = 3$ volts, and $u(t) = 1$ volt, for $t \geq 0$.

The network has one loop. If we use the equivalent s-domain circuit elements in Figures 4.8(c) and 4.10(c), then the resulting network will have three loops, and analysis will become very complicated. Therefore, we choose the elements in Figures 4.8(b) and 4.10(b), and the resulting network is shown in Figure 4.11(b). In this circuit, the total voltage source is $U(s) + 2 - 3/s$, and the total voltage drop is $2I(s) + sI(s) + (1/s)I(s)$. Therefore, we have

$$U(s) + 2 - \frac{3}{s} = (2 + s + \frac{1}{s})I(s)$$

which, after the substitution of $U(s) = 1/s$, can be simplified as

$$I(s) = \frac{2s - 2}{s^2 + 2s + 1} = \frac{2(s + 1) - 4}{(s + 1)^2} = \frac{-4}{(s + 1)^2} + \frac{2}{s + 1}$$

Thus, the loop current is

$$i(t) = -4te^{-t} + 2e^{-t} \qquad \text{for } t \geq 0$$

We see that the analysis is simple and straightforward.

EXERCISE 4.9.1

Consider the network shown in Figure 4.12. Find the voltage $v(t)$ due to $i(0-) = 3$ amperes, $v(0-) = 2$ volts, and $u(t) = 1$ ampere, for $t \geq 0$.

[**ANSWER:** $v(t) = -4te^{-t} + 2e^{-t}, t \geq 0$] ■

Figure 4.11 *Series network*

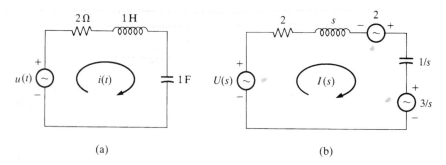

(a)   (b)

Figure 4.12 *Parallel network*

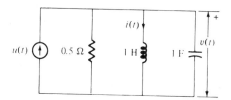

4.9.1   Transform Impedance

The advantage of using the Laplace transform in network analysis will become even more evident in the study of zero-state responses. If all initial voltage and current are zero, then we have

$$V(s) = LsI(s) \quad \text{(inductor)} \qquad = j\omega L * I \quad \textbf{(4.71a)}$$

$$V(s) = \frac{1}{Cs}I(s) \quad \text{(capacitor)} \qquad = \frac{1}{j\omega L} * I \quad \textbf{(4.71b)}$$

and

$$V(s) = RI(s) \quad \text{(resistor)} \qquad = R * I \quad \textbf{(4.71c)}$$

These can be written as $V(s) = Z(s)I(s)$, and $Z(s)$ is called the (transform) impedance. Thus, the impedances of the resistor, inductor, and capacitor are, respectively, R, Ls, and $1/Cs$. The equation $V(s) = Z(s)I(s)$ is an algebraic equation and is a generalization of Ohm's law $v = Ri$ in the Laplace-transform domain. If we consider $I(s)$ as an input and $V(s)$ as an output, then the impedance is the transfer function discussed in $(4.57)^7$. The concept of impedance is basic in network analysis. We emphasize once again that it describes only the zero-state response. In other words, when we use impedances, all initial conditions must be assumed to be zero.

The manipulation involving impedances is purely algebraic, identical to the manipulation of resistances. For example, the resistance of the series connection of two resistances R_1 and R_2 is $R_1 + R_2$; the resistance of the parallel connection of R_1 and R_2 is $R_1R_2/(R_1 + R_2)$. Similarly, the impedance of the series connection of two impedances $Z_1(s)$ and $Z_2(s)$ is $Z_1(s) + Z_2(s)$; the impedance of the parallel connection of $Z_1(s)$ and $Z_2(s)$ is

$$\frac{Z_1(s)Z_2(s)}{Z_1(s) + Z_2(s)}$$

---

^{7}If the input is a current and the output is a voltage, the transfer function is also called an impedance. It is a generalization of the concept of resistance. If the input is a voltage and the output is a current, then the transfer function is also called an admittance. It is a generalization of the concept of conductance.

as shown in Figure 4.13. The only difference is that now we deal with rational functions rather than real numbers, as in the resistive case.

EXAMPLE 4.9.2

The transfer function from u to y of the network shown in Figure 4.5(a) was computed in (4.59) from its differential equation description. Now we shall compute it using the concept of impedance. The equivalent resistorlike network of Figure 4.5(a) is shown in Figure 4.5(b). The impedance of the series connection of 1 and $1/2s$ is $1 + 1/2s$. Thus, the loop current $I(s)$ shown in Figure 4.5(b) equals

$$I(s) = \frac{U(s)}{1 + \dfrac{1}{2s}} = \frac{2s}{2s + 1} U(s)$$

and the output voltage equals

$$Y(s) = \frac{1}{2s} I(s) = \frac{1}{2s} \left(\frac{2s}{2s + 1} U(s) \right) = \frac{1}{2s + 1} U(s)$$

Therefore, the transfer function from u to y is

$$H(s) = \frac{Y(s)}{U(s)} = \frac{1}{2s + 1}$$

This is the same as (4.59).

Figure 4.13 *Series and parallel connection of two impedances*

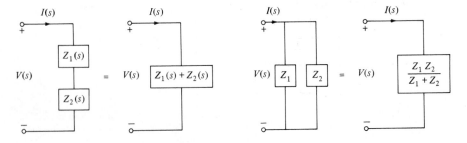

EXAMPLE **4.9.3**

Find the current $i(t)$ of the network shown in Figure 4.14(a). Its equivalent network in the s-domain is shown in Figure 4.14(b). The impedance of the series connection of the inductor and 2 Ω resistor is $3s + 2$. The impedance of the parallel connection of $1/2s$ and $(3s + 2)$ is

$$\frac{\frac{1}{2s}(3s + 2)}{\frac{1}{2s} + 3s + 2} = \frac{3s + 2}{1 + 6s^2 + 4s}$$

as shown in Figure 4.14(c). Thus, the current $I(s)$ is given by

$$I(s) = \frac{U(s)}{1 + \frac{3s + 2}{6s^2 + 4s + 1}} = \frac{6s^2 + 4s + 1}{6s^2 + 7s + 3}U(s)$$

If $u(t)$ is given, we can compute $U(s)$ and then $I(s)$. The inverse Laplace transform of $I(s)$ yields the current $i(t)$.

EXERCISE 4.9.2

Suppose the capacitor in Figure 4.15 is initially relaxed. Find the current $i(t)$ and voltage $y(t)$ excited by a unit step function.

[**ANSWERS:** $i(t) = \frac{1}{3} + \frac{1}{6}e^{-3t/4}$, $y(t) = \frac{1}{3} - \frac{1}{3}e^{-3t/4}$, $t \geq 0$.] ∎

Figure 4.14 *Network*

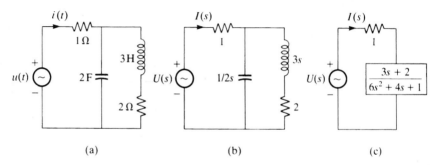

(a)　　　　　　(b)　　　　　　(c)

Figure 4.15 *Network*

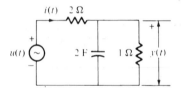

In the study of linear algebraic equations in Appendix A, we use (A.72) to illustrate that most discussions for equations with real coefficients are applicable to equations whose coefficients are rational functions. Now we shall show that (A.72) arises from an RLC network.

EXAMPLE 4.9.4

Consider the network shown in Figure 4.16(a). Its equivalent network is shown in Figure 4.16(b). Let the two loop currents be chosen as shown. Then, from the left-hand-side loop, we have

$$\text{Loop } \textcircled{1}: \quad U(s) = (2 + \frac{1}{0.5s} + 2s)I_1(s) - 2sI_2(s)$$

From the right-hand-side loop, we have

$$\text{Loop } \textcircled{2}: \quad -2sI_1(s) + (2s + \frac{1}{2s} + 1)I_2(s) = 0$$

Figure 4.16 *Network with two loops*

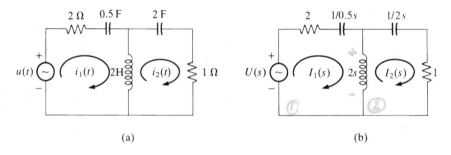

(a)   (b)

They can be written as

$$\frac{2s + 2 + 2s^2}{s}I_1(s) - 2sI_2(s) = U(s)$$

$$-2sI_1(s) + \frac{4s^2 + 2s + 1}{2s}I_2(s) = 0$$

This is the set of equations in (A.72).

4.10 TRANSFER-FUNCTION REPRESENTATION— COMPLETE CHARACTERIZATION

Every LTIL system can be described by a convolution, a differential equation, or a transfer function. Among these three descriptions, the transfer function is the most convenient to employ in analysis and design. For example, if we want to compute the response of a system excited by input $u(t)$, then the response is simply the product of its transfer function and the Laplace transform of $u(t)$ or

$$Y(s) = H(s)U(s) \qquad (4.72)$$

The inverse Laplace transform of $Y(s)$ yields the response of the system. Therefore, in analysis and design, we often use the transfer function to represent a system, as shown in Figure 4.17(a). This is called the transfer-function, or block, representation. In Figure 4.17(a), the input and output are expressed in the time domain, but the system is represented by its transfer function, which is in the Laplace-transform domain. Therefore, strictly speaking, the representation is incorrect. If the input and output are in the time domain, the block should be represented by the impulse response, as shown in Figure 4.17(b). If the block is represented by a transfer function, then the input and output should also be in the Laplace-transform domain, as shown in Figure 4.17(c). However, for convenience, we often plot it as shown in Figure 4.17(a) by mixing the time and Laplace-transform domains.

Figure 4.17 *Transfer-function representation*

(a)   (b)   (c)

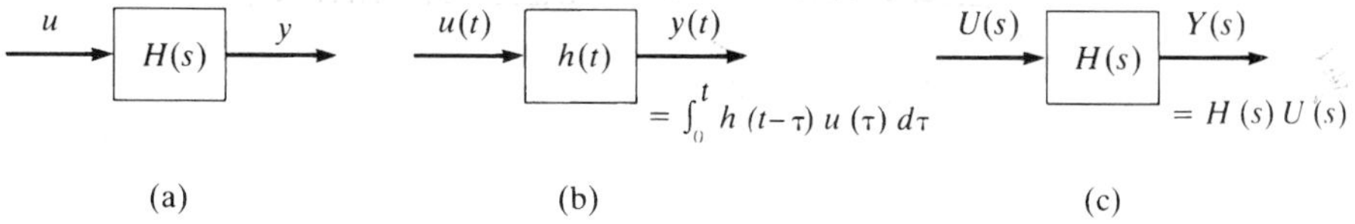

The advantage of using transfer functions can further be seen from the connection of two systems shown in Figure 4.18. The connections in Figures 4.18(a), (b), (c), and (d) are called, respectively, the tandem or cascade, parallel, negative feedback, and positive feedback connections. Let the two systems be described by

$$Y_1(s) = H_1(s)U_1(s) \qquad Y_2(s) = H_2(s)U_2(s) \tag{4.73}$$

and let the input and output of the connections be denoted by $u(t)$ and $y(t)$. In the tandem connection, we have $u(t) = u_1(t)$, $y(t) = y_2(t)$, and $y_1(t) = u_2(t)$. Using these and (4.73), we have

$$\begin{aligned} Y(s) &= Y_2(s) = H_2(s)U_2(s) = H_2(s)Y_1(s) \\ &= H_2(s)H_1(s)U_1(s) = H_1(s)H_2(s)U(s) \end{aligned} \tag{4.74}$$

Thus, the transfer function of the tandem connection is $H_1(s)H_2(s)$, the product of the two individual transfer functions. In the parallel connection, we have $u(t) = u_1(t) = u_2(t)$ and $y(t) = y_1(t) + y_2(t)$ which imply

$$\begin{aligned} Y(s) &= Y_1(s) + Y_2(s) = H_1(s)U(s) + H_2(s)U(s) \\ &= (H_1(s) + H_2(s))U(s) \end{aligned} \tag{4.75}$$

Thus, the transfer function of the parallel connection is $H_1(s) + H_2(s)$, the sum of the two individual transfer functions. In the negative feedback connection in Figure 4.18(c), we have $y(t) = y_1(t) = u_2(t)$ and $u_1(t) = u(t) - y_2(t)$. The substitution of (4.73) into these equations yields

Figure 4.18 *Connections of two systems*

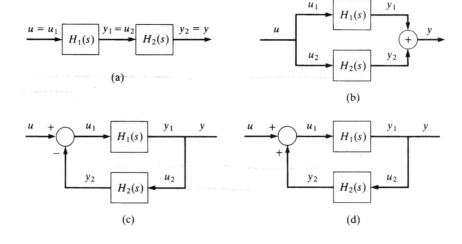

(a)

(b)

(c)

(d)

$$U_1(s) = U(s) - H_2(s)U_2(s) = U(s) - H_2(s)Y_1(s) = U(s) - H_2(s)H_1(s)U_1(s)$$

which implies

$$(1 + H_2(s)H_1(s))U_1(s) = U(s) \quad \text{or} \quad U_1(s) = \frac{1}{1 + H_1(s)H_2(s)}U(s)$$

Thus, we have

$$Y(s) = Y_1(s) = H_1(s)U_1(s) = \frac{H_1(s)}{1 + H_1(s)H_2(s)}U(s) \qquad \text{(4.76)}$$

which means that the transfer function of the negative feedback connection is $H_1(s)/(1 + H_1(s)H_2(s))$.

EXERCISE 4.10.1

Show that the transfer function from u to y of the positive feedback system in Figure 4.18(d) is

$$H(s) = \frac{Y(s)}{U(s)} = \frac{H_1(s)}{1 - H_1(s)H_2(s)} \qquad \text{(4.77)}$$

∎

The preceding derivation involves only *algebraic* manipulation; thus, the transfer functions of the connections can be easily obtained. If we use convolutions or differential equations, then the derivation involves integrations and differentiations and is very complicated. For this reason, the use of transfer functions is preferred to the use of convolutions and differential equations in analysis and design.

A question may be raised at this point. Whenever we use transfer functions, all initial conditions must be assumed to be zero and, therefore, zero-input responses or responses excited by initial conditions are completely disregarded. Can we do this? Consider a system described by the differential equation

$$D(p)y(t) = N(p)u(t)$$

where $D(p)$ and $N(p)$ are defined as in (4.51). If $D(p) = (p + 1)(p + 2)(p - 2)$, then, as was discussed in Section 4.6, the modes of the system are -1, -2, and 2 and the zero-input response is of the form

$$y_{zi}(t) = k_1 e^{-t} + k_2 e^{-2t} + k_3 e^{2t} \qquad \text{(4.78)}$$

for some constant k_i. The zero-state response of the system is governed by its transfer function

$$H(s) = \frac{N(s)}{D(s)}$$

and the poles of the systems are defined as the roots of $D(s)$ after canceling all common factors between $N(s)$ and $D(s)$. For example, if $N(s) = s^2 - 4$ and $D(s) = (s + 1)(s + 2)(s - 2)$, then

$$H(s) = \frac{s^2 - 4}{(s + 1)(s + 2)(s - 2)} = \frac{(s + 2)(s - 2)}{(s + 1)(s + 2)(s - 2)} = \frac{1}{(s + 1)}$$

and the system has only one pole at -1. Note that even though -2 and 2 are modes, they are not poles. In this case, the zero-state response excited by any input $u(t)$ is generally of the form

$$y_{zs}(t) = \hat{k}_1 e^{-t} + \text{(terms due the poles of } U(s)) \tag{4.79}$$

We see that, from the zero-state response in (4.79), there is no way to detect the existence of the modes -2 and 2. If initial conditions of the system are different from zero, then the zero-input response, because of e^{2t}, will approach infinity as $t \to \infty$ and the system will burn out or disintegrate. This phenomenon, however, cannot be detected from the transfer function. In this case, the system is said to be not completely characterized by its transfer function, and care must be exercised in using transfer functions to study systems.

If $N(s)$ and $D(s)$ have no common factors, then every root of $D(s)$ is a pole of $H(s) = N(s)/D(s)$. For example, if $D(s) = (s + 1)(s + 2)(s - 2)$ and $N(s) = s(s - 1)$, then the poles of the system are $-1, -2$, and 2 and the zero-state response of the system excited by the input $u(t)$ is of the form

$$y_{zs}(t) = \bar{k}_1 e^{-t} + \bar{k}_2 e^{-2t} + \bar{k}_3 e^{2t} + \text{(terms due the poles of } U(s)) \tag{4.80}$$

We see that the zero-input response in (4.78) is essentially included in the zero-state response and no essential information is lost in using the transfer function. In this case, the system is said to be completely characterized by its transfer function.

In conclusion, if $N(s)$ and $D(s)$ are coprime (have no common factors), then

$$\{\text{the set of the poles}\} = \{\text{the set of the modes}\} \tag{4.81}$$

and the system is *completely characterized* by its transfer function. In this case, there is no loss of essential information when using transfer functions to study systems. If $N(s)$ and $D(s)$ have common factors, say, $R(s)$, then the roots of $R(s)$ are modes but not poles of the system and

$$\{\text{the set of the poles}\} \subset \{\text{the set of the modes}\} \tag{4.82}$$

In this case, the roots of $R(s)$ are called the missing poles of the transfer function and the system is not completely characterized by its transfer function. If a system is not completely characterized by its transfer function, care must be exercised in using the transfer function to study the system.

EXAMPLE 4.10.1

Consider the network shown in Figure 3.24 and repeated in Figure 4.19. The input u and the output y are described by, as derived in (3.86),

$$\dot{y}(t) - 0.75y(t) = \dot{u}(t) - 0.75u(t)$$

or

$$(p - 0.75)y(t) = (p - 0.75)u(t) \tag{4.83}$$

The mode of the system is the root of $(s - 0.75)$, or 0.75. Therefore, its zero-input response is

$$y_{zi}(t) = k_1 e^{0.75t}$$

where k_1 depends on the initial voltage of the capacitor in Figure 4.19. We see that if the initial voltage is different from zero, then the response will approach infinity as $t \to \infty$.

Figure 4.19 *A network*

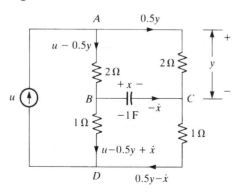

The transfer function of the system is

$$H(s) = \frac{s - 0.75}{s - 0.75} = 1$$

Because of the common factor, the transfer function reduces to 1. Thus, the system has no pole and the zero-state response is $y(t) = u(t)$, for all t. This system is not completely characterized by its transfer function because the mode 0.75 does not appear as a pole of the system. In other words, the transfer function has missing pole 0.75.

If we use the transfer function to study the system, we would conclude that the system is satisfactory. But in fact, the system is unacceptable because if the voltage of the capacitor becomes nonzero due to whatever reason, the response will grow without bound and the system will either saturate or burn out. Thus, the system is useless in practice.

The existence of the missing pole in Figure 4.19 can be easily explained from the structure of the network. Because of the symmetry of the four resistors, if the initial voltage of the capacitor is zero, its voltage will remain zero no matter what input current is applied. Therefore, the removal of the capacitor will not affect the zero-state response of the system. Thus, the system has a superfluous component as far as the input and output are concerned. This type of system should not be built in practice.

In RLC networks, capacitors and inductors are called energy storage elements because they can store energy in their electric and magnetic fileds. Roughly speaking, if the degree of the transfer function of an RLC network is less than the total number of capacitors and inductors, then the network is not completely characterized by its transfer function. This is the case for the network in Figure 4.19.

EXAMPLE 4.10.2

Consider the network shown in Figure 4.20(a). It has an LC loop connected in series with the current source $u(t)$. Because the current $i(t)$ shown in the figure always equals $u(t)$, as far as the output voltage $y(t)$ is concerned, the network can be reduced to the one in Figure 4.20(b). Thus, the transfer functions of both networks in Figure 4.20(a) and (b) equal

$$H_1(s) = \frac{Y_1(s)}{U_1(s)} = \frac{1 \times \dfrac{1}{2s}}{1 + \dfrac{1}{2s}} = \frac{1}{2s + 1} \tag{4.84}$$

Figure 4.20 *Networks*

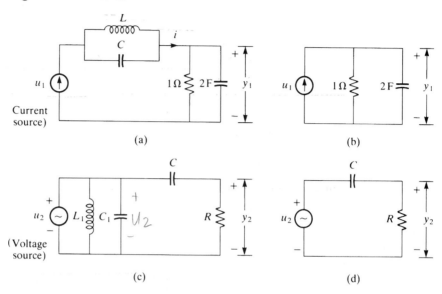

(a)   (b)

(c)   (d)

This transfer function has degree one. The network in Figure 4.20(a) has three energy storage elements; therefore, it is not completely characterized by its transfer function. The network in Figure 4.20(b) has one energy storage element, which is equal in number to the degree of its transfer function. Therefore, the network is completely characterized by its transfer function.

EXERCISE 4.10.2

Consider the networks shown in Figures 4.20(c) and (d). The one in Figure 4.20(c) has a capacitor and an inductor connected in parallel with the voltage source $u(t)$. Show that the transfer functions of both networks equal

$$H_2(s) = \frac{Y_2(s)}{U_2(s)} = \frac{R}{R + \dfrac{1}{Cs}} = \frac{RCs}{RCs + 1} \tag{4.85}$$

Are they completely characterized by their transfer functions?

[**ANSWERS:** The network in Figure 4.20(c) is not completely characterized by (4.85); the network in Figure 4.20(d) is completely characterized by (4.85).] ∎

As can be seen from the networks in Figures 4.19 and 4.20, if a system is not completely characterized by its transfer function, then the system contains unnecessary components. Certainly, it is not desirable to design such systems. Therefore, most physical systems are designed to be completely characterized by their transfer functions. If a system is completely characterized by its transfer function, we may disregard its zero-input response and use its transfer function in analysis and design.

4.10.1   Loading Problem

The advantage of using transfer functions is most evident in studying composite systems, or connections of two or more systems, as shown in Figure 4.18. One problem, however, may arise in connecting two systems. This can be illustrated by an example.

EXAMPLE 4.10.3

The network in Figure 4.21(a) is the tandem connection of the two networks in Figures 4.20(b) and (d). We compute the transfer functions from u to y_1 and from u to y_2 shown. The impedance of the parallel connection of $1/(2s + 1)$ and $R + 1/Cs$ is

$$Z(s) = \frac{\dfrac{1}{2s + 1} \times (R + \dfrac{1}{Cs})}{\dfrac{1}{2s + 1} + R + \dfrac{1}{Cs}} = \frac{RCs + 1}{2RCs^2 + (RC + C + 2)s + 1}$$

Thus, the voltage y_1 equals

$$Y_1(s) = \frac{RCs + 1}{2RCs^2 + (RC + C + 2)s + 1} U(s)$$

Figure 4.21   *Connection of two networks*

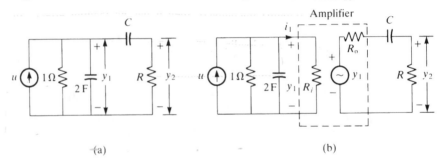

(a)   (b)

and the voltage y_2 equals   *voltage divider*

$$Y_2(s) = \frac{R}{R + \dfrac{1}{Cs}} Y_1(s) = \frac{RCs}{2RCs^2 + (RC + C + 2)s + 1} U(s)$$

Thus, the transfer functions from u to y_1 and from u to y_2 are, respectively,

$$\overline{H}_1(s) = \frac{Y_1(s)}{U(s)} = \frac{RCs + 1}{2RCs^2 + (RC + C + 2)s + 1} \tag{4.86}$$

and

$$H(s) = \frac{Y_2(s)}{U(s)} = \frac{RCs}{2RCs^2 + (RC + C + 2)s + 1} \tag{4.87}$$

The transfer functions of the two networks in Figures 4.20(b) and (d) were computed in (4.84) and (4.85). Their product is

$$H_1(s)H_2(s) = \frac{1}{2s + 1} \cdot \frac{RCs}{RCs + 1} = \frac{RCs}{2RCs^2 + (RC + 2)s + 1}$$

which is different from $H(s)$ in (4.87). Thus, the transfer function of the tandem connection of two systems is not necessarily equal to the product of the two individual transfer functions, as developed in (4.74).

In developing the transfer functions for various connections in Figure 4.18, it is implicitly assumed that the transfer function of each system does not change after connection. This may not be the case in practice. For example, the transfer function from u_1 to y_1 of the network in Figure 4.20(b) is $1/(2s + 1)$. The transfer function, however, becomes the one in (4.86), which is different from the original transfer function and depends on what is connected to it. Different connections yield different transfer functions. This motivates the following definition: If the transfer function of a system changes after it is connected to another system, the connection is said to have a *loading effect*. If the transfer function remains unchanged, then it has no loading effect. In using transfer functions to study various connections of systems, we must make sure that connections have no loading effects. Otherwise, the transfer functions developed in Equations (4.74) through (4.77) cannot be used.

In RLC networks, the loading problem can often be eliminated by using isolating amplifiers. Consider the connection shown in Figure 4.21(b), where we have inserted an amplifier with gain 1 between the two networks. Ideally, the input resistance R_i of the amplifier should be infinity, and the output resistance R_o should be zero. If $R_i = \infty$, the current i_1 shown is zero, and the transfer function from u to y_1 is $1/(2s + 1)$, no matter what is connected to the output of the amplifier. Similarly, if $R_o = 0$, the transfer function of the second network is $RCs/(RCs + 1)$, no matter what is connected to the input of the amplifier. Thus, the amplifier is called an *isolating amplifier* or *buffer* and the transfer function of the tandem connection in Figure 4.21(b) is

$$\frac{1}{2s + 1} \cdot 1 \cdot \frac{RCs}{RCs + 1} = H_1(s)H_2(s)$$

In general, if no isolating amplifiers are used, RLC networks cannot be easily separated into blocks, each represented by a transfer function. The basic block diagrams in Figures 3.25 and 3.26 are built from operational amplifiers, which have properties close to isolating amplifiers and, therefore, have no loading problem. The loading problem of electromechanical systems usually can be taken care of by properly selecting inputs and outputs. For a detailed discussion, see Reference [8].

4.11   THE FINAL-VALUE AND INITIAL-VALUE THEOREMS

Some of the properties of the Laplace transform have not yet been discussed. We shall do so in the remainder of this chapter.

Consider a positive-time function $f(t)$. The value of $f(t)$ at $t = \infty$ or, more precisely, the value of $f(t)$ as t approaches infinity is called the final value of $f(t)$. To find the final value of $f(t)$, we may simply set $t = \infty$ and compute its value. For example, if $f(t) = 1 + e^{-t}$, then $f(\infty) = 1$ because e^{-t} approaches 0 as $t \to \infty$. On the other hand, if $f(t) = 1 + e^t$, then $f(\infty) = \infty$. If $f(t) = 1 + \sin t$, then $f(t)$ oscillates between 0 and 2 as $t \to \infty$. Therefore, $f(\infty)$ is not defined.

Consider a Laplace transform $F(s)$. In some applications, we may be asked to find the final value of $f(t)$. If we first compute the inverse Laplace transform of $F(s)$, the final value of $f(t)$ can then be obtained by setting $t = \infty$. Alternatively, we may use the following theorem without computing the inverse Laplace transform.

The final-value theorem   Let $F(s)$ be the Laplace transform of $f(t)$ and let $F(s)$ be a proper rational function. Then $f(t)$ approaches a zero or nonzero constant as $t \to \infty$ if and only if all poles of $sF(s)$ have negative real parts. Under this condition, we have

$$\lim_{t \to \infty} f(t) = \lim_{s \to 0} sF(s) \tag{4.88}$$

Consider first the condition of the theorem. Using the linearity property of the Laplace transform and Table 4.3, we can readily conclude that the inverse Laplace transform of $F(s)$ approaches zero if and only if all poles of $F(s)$ lie inside the open left half s-plane or, equivalently, have negative real parts. The inverse of $F(s)$ approaches a nonzero constant if and only if $F(s)$ has a simple pole at $s = 0$, and the rest of the poles of $F(s)$ have negative real parts. These two statements are combined as stated in the theorem. If the condition of the theorem is not met, (4.88) does not hold.

EXAMPLE 4.11.1

Consider $f(t) = e^t$. Clearly, we have $f(\infty) = \infty$. Its Laplace transform is $1/(s - 1)$ and

$$\lim_{s \to 0} sF(s) = \lim_{s \to 0} \frac{s}{s - 1} = 0$$

which is not equal to $f(\infty)$. The formula in (4.88) does not hold because $sF(s) = s/(s - 1)$ has one pole with a positive real part. Consider $f(t) = \sin 2t$. Its value is not defined as $t \to \infty$. However, we have

$$\lim_{s \to 0} sF(s) = \lim_{s \to 0} s \cdot \frac{2}{s^2 + 4} = 0$$

The formula in (4.88) again does not hold because $sF(s)$ has poles $\pm j2$ that have zero real part.

We now prove the final-value theorem. From (4.8a), we have

$$sF(s) - f(0-) = \mathcal{L}[\dot{f}(t)] = \int_{0-}^{\infty} \frac{df(t)}{dt} e^{-st} dt$$

If $s \to 0$, the equation becomes

$$\lim_{s \to 0} sF(s) - f(0-) = \lim_{s \to 0} \int_{0-}^{\infty} \frac{df(t)}{dt} e^{-st} dt$$

$$= \int_{0-}^{\infty} \frac{df(t)}{dt} dt = \lim_{t \to \infty} f(t) - f(0-)$$

which reduces to (4.88) after removing $f(0-)$ from both sides. Note that if all poles of $sF(s)$ have negative real parts, then $f(t)$ decreases exponentially and its integration from 0– to ∞ converges. Thus the last equality in the above equation holds. Note also that the final-value theorem can also be established using partial fraction expansion, as in Example 4.7.8.

Recall that the poles of $sF(s)$ are defined as the roots of the denominator of $sF(s)$ after canceling out all common factors. Therefore, in order to apply the final-value theorem, we must check first whether or not there are common factors between the numerator and denominator of $sF(s)$. If yes, we cancel out the common factors and then check whether or not all roots of the remaining denominator have negative real parts. If $F(s)$ is given in factored form, these steps can be carried out easily.

EXERCISE 4.11.1

Find $f(\infty)$ if $F(s)$ is given by

(a) $\dfrac{1}{s-1}$

(b) $\dfrac{s-2}{s(s+1)}$

(c) $\dfrac{s+5}{s(s+2-j1)(s+2+j1)}$

(d) $\dfrac{1}{(s+j)(s-j)}$

(e) $\dfrac{s+10}{(s+2)(s+3)(s+4)}$

(f) $\dfrac{s-1}{s(s^2-1)}$

[**ANSWERS:** (a) ∞; (b) -2; (c) 1; (d) not defined; (e) 0; (f) 1. Note that there is one common factor between the numerator and denominator in (f).] ∎

If $F(s)$ is not in factored form, the employment of the final-value theorem is not necessarily simple. For example, consider

$$F_1(s) = \frac{N_1(s)}{D_1(s)} = \frac{1}{s^3 + 2s^2 + 9s + 68}$$

and

$$F_2(s) = \frac{N_2(s)}{D_2(s)} = \frac{s^3 + s^2 - 2}{s^4 + 2s^3 + 2s^2 - 2s - 3}$$

The poles of $sF_1(s)$ are clearly the roots of $D_1(s)$. However, the poles of $sF_2(s)$ are not necessarily the roots of $D_2(s)$ because of the possible existence of common factors between $D_2(s)$ and $N_2(s)$. Therefore, before applying the final-value theorem, we must check whether or not two polynomials have common factors. Once this step is completed, we then check whether all roots of the remaining polynomial have negative real parts. If the degree of a polynomial is three or higher, as in $D_i(s)$ for $i = 1$ and 2, the computation of the roots is not a simple task by hand. (Fortunately, a method called the Routh test can be used to check this property without solving for the roots; it will be discussed in Chapter 8.) Thus, the employment of the final-value theorem is not simple if $F(s)$ is not given in factored form.

We now introduce a dual property of the final-value theorem. Let $F(s) = N(s)/D(s)$ be the Laplace transform of a positive-time function $f(t)$, where $N(s)$ and $D(s)$ are polynomials. Recall that $F(s)$ is strictly proper if deg $D(s) > $ deg $N(s)$.

The initial-value theorem Let $F(s)$ be the Laplace transform of $f(t)$ and let $F(s)$ be a rational function. If $F(s)$ is strictly proper, then

$$f(0+) = \lim_{s \to \infty} sF(s)$$

An example can illustrate the necessity of the condition. Consider

$$\mathcal{L}[f(t)] := \mathcal{L}[\delta(t) - e^{-2t}] = 1 - \frac{1}{s+2} = \frac{s+1}{s+2} =: F(s)$$

It is not strictly proper. Because $\delta(t) = 0$ for $t \neq 0$, we have $f(0+) = -1$; whereas we have $\lim_{s \to \infty} sF(s) = \infty$. Thus, the theorem does not hold if $F(s)$ is not strictly proper. One may suggest to modify the theorem as

$$f(0) = \lim_{s \to \infty} sF(s) \tag{4.89}$$

This is not acceptable because $\delta(t)$ is not defined at $t = 0$. Furthermore, we may encounter the case $f(0-) \neq f(0+)$ and ambiguity would arise in (4.89).

We now establish the theorem. Equation (4.8a) implies

$$sF(s) = f(0-) + \int_{0-}^{0+} e^0 \frac{df(t)}{dt}\, dt + \int_{0+}^{\infty} e^{-st}\dot{f}(t)dt$$

$$= f(0-) + [f(0+) - f(0-)] + \int_{0+}^{\infty} e^{-st}\dot{f}(t)dt$$

$$= f(0+) + \int_{0+}^{\infty} e^{-st}\dot{f}(t)dt$$

The exponential e^{-st} tends to zero as $s \to \infty$ for every t in the range $(0+, \infty)$ (this is not so if 0 is included). Hence, we have

$$\lim_{s \to \infty} sF(s) = \lim_{s \to \infty} [f(0+) + \int_{0+}^{\infty} e^{-st}\dot{f}(t)dt] = f(0+)$$

This establishes the theorem.

EXERCISE 4.11.2

Find $f(0+)$ if $F(s) = \mathcal{L}[f(t)] = (-s + 3)/(s - 1)$. Does the initial-value theorem hold?

[**ANSWERS:** $f(0+) = 2$. No.]   ∎

EXERCISE 4.11.3

Find $f(0+)$ if

$$F(s) = \frac{1}{s^3 + 2s + 1} \qquad \text{and} \qquad F(s) = \frac{2s + 1}{s^2 - 1}$$

[**ANSWERS:** 0, 2.]   ∎

The application of the initial-value theorem is fairly simple. It depends only on the relative degrees of $N(s)$ and $D(s)$. From their degrees, we can readily establish the following:

1. If $\deg D(s) - \deg N(s) \le 0$, the initial-value theorem is not applicable.
2. If $\deg D(s) - \deg N(s) \ge 2$, $f(0+) = 0$
3. If $\deg D(s) - \deg N(s) = 1$, $f(0+)$ is a nonzero constant and equals the ratio of the leading coefficients of $D(s)$ and $N(s)$.

4.12   FURTHER PROPERTIES OF THE LAPLACE TRANSFORM

This section will discuss some more properties of the Laplace transform. The Laplace transforms encountered in the preceding sections are all rational functions of s. Because this will no longer be the case in this section, it is appropriate to separate this discussion from the preceding sections.

Time delay   Let $f(t)$ be a positive-time function and let T be a positive constant. Then $f(t - T)$ is a shift to the right (delay) of $f(t)$ by T seconds as shown in Figure 4.22(b). Because $f(t) = 0$ for $t < 0$, the value of $f(t - T)$ between 0 and T is zero and $f(t - T)$ is still a positive-time function. If T is negative, then $f(t - T)$ is a shift to the left (advance) of $f(t)$ by T seconds, as shown in Figure 4.22(c). Note that $f(t - T)$, for $T < 0$, is not a positive-time function.

Consider a positive-time function $f(t)$ and a positive T. In order to stress the fact that $f(t - T) = 0$ for $t < T$, it is convenient to use the unit step function $q(t - T)$ defined in (1.10). Then, the delay of $f(t)$ by T seconds can be written simply as $f(t - T)q(t - T)$. For example, if

Figure 4.22 *Positive-time functions*

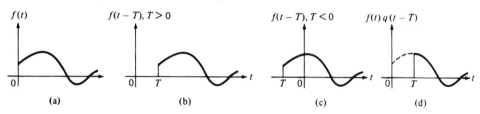

$$f(t) = \begin{cases} e^{-2t} & \text{for } t \geq 0 \\ 0 & \text{for } t < 0 \end{cases}$$

then the delay of $f(t)$ by T seconds can be simply written as $e^{-2(t-T)}q(t-T)$. If we do not use $q(t)$, then it must be written as

$$f(t - T) = \begin{cases} e^{-2(t-T)} & \text{for } t \geq T \\ 0 & \text{for } t < T \end{cases}$$

This is inconvenient. Therefore, we shall use $q(t - T)$ in this section.

EXERCISE 4.12.1

Let $f(t)$ be a positive-time function. Plot $f(t-T)q(t)$, $f(t-T)q(t-T)$ and $f(t)q(t-T)$. If $f(t)$ is not a positive-time function, are the plots of $f(t-T)q(t)$ and $f(t-T)q(t-T)$ the same?

[**ANSWERS:** The first two plots are as shown in Figure 4.22(b); the third is as shown in Figure 4.22(d). They are different.] ∎

Let $F(s)$ be the Laplace transform of a positive-time function $f(t)$. Then, we have

$$\mathcal{L}[f(t-T)q(t-T)] = e^{-sT}F(s) \qquad \text{for } T \geq 0 \tag{4.90}$$

To show this, we write

$$\mathcal{L}[f(t-T)q(t-T)] = \int_{0-}^{\infty} f(t-T)q(t-T)e^{-st}dt = \int_{T-}^{\infty} f(t-T)e^{-s(t-T)}e^{-sT}dt$$

$$= e^{-sT}\int_{0-}^{\infty} f(v)e^{-sv}dv = e^{-sT}F(s)$$

where we have changed the variable $v := t - T$. This establishes (4.90). Because of the factor e^{-sT}, the Laplace transform of $f(t - T)$ is *not* a rational function of s. Clearly, if $f(t) = \delta(t)$, then $F(s) = 1$, and

$$\mathcal{L}[\delta(t - t_0)] = e^{-st_0} \tag{4.91}$$

This can also be easily obtained from (4.1) by using the sifting property of the impulse.

Now we shall apply (4.90) to compute the Laplace transform of the pulse defined by

$$p_T(t) = \begin{cases} 1 & \text{for } 0 \le t < T \\ 0 & \text{elsewhere} \end{cases} \tag{4.92}$$

which is shown in Figure 4.23. It is clear that

$$p_T(t) = q(t) - q(t - T)$$

Thus, we have, using (4.90),

$$\begin{aligned} P_T(s) &= \mathcal{L}[p_T(t)] = \mathcal{L}[q(t)] - \mathcal{L}[q(t - T)] \\ &= \frac{1}{s} - e^{-sT}\frac{1}{s} = \frac{1}{s}(1 - e^{-sT}) \end{aligned} \tag{4.93}$$

EXAMPLE 4.12.1

Compute the response of the system with transfer function $1/(s + 2)$ excited by the pulse input shown in Figure 4.23. The output clearly equals

$$Y(s) = H(s)U(s) = \frac{1}{s + 2} \cdot \frac{1 - e^{-sT}}{s} \tag{4.94}$$

Figure 4.23   *A pulse*

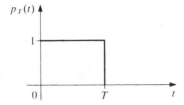

We compute

$$\frac{1}{s(s+2)} = \frac{0.5}{s} - \frac{0.5}{s+2}$$

Thus, (4.94) becomes

$$Y(s) = \frac{0.5}{s} - \frac{0.5}{s} \cdot e^{-sT} - \frac{0.5}{s+2} + \frac{0.5}{s+2} \cdot e^{-sT} \qquad \textbf{(4.95)}$$

Its inverse Laplace transform is

$$y(t) = 0.5q(t) - 0.5q(t-T) - 0.5e^{-2t}q(t) + 0.5e^{-2(t-T)}q(t-T)$$

EXAMPLE 4.12.2

Consider the positive feedback system shown in Figure 4.24(a). The impulse response of the unit-time delay element is $\delta(t-1)$; thus, its transfer function is e^{-s}, as shown in Figure 4.24(b). The transfer function of the tandem connection of a and e^{-s} is ae^{-s}. Thus, the transfer function from u to y is, using (4.77),

$$H(s) = \frac{ae^{-s}}{1 - ae^{-s}}$$

This is an irrational function of s, and the feedback system is a distributed system.

EXERCISE 4.12.2

Verify that the transfer function of the negative-feedback system in Figure 4.24(c) is

Figure 4.24 *Feedback systems*

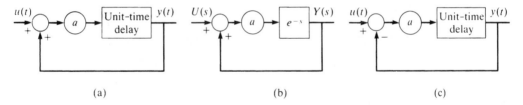

(a)   (b)   (c)

$$H(s) = \frac{ae^{-s}}{1 + ae^{-s}}$$

Is it a lumped system?

[**ANSWER:** No.]   ■

In general, irrational transfer functions are difficult to manipulate. The only exception is those involving e^{-st_0}, as Example 4.12.1 and the following example show.

EXAMPLE **4.12.3**

The unit step response of the system in Figure 4.24(a) with transfer function $H(s) = ae^{-s}/(1 - ae^{-s})$ is

$$Y(s) = H(s)\frac{1}{s} = \frac{1}{s}\frac{ae^{-s}}{1 - ae^{-s}}$$

Using

$$\frac{1}{1-x} = 1 + x + x^2 + x^3 + \cdots$$

we expand $Y(s)$ as

$$Y(s) = \frac{1}{s}ae^{-s}[1 + ae^{-s} + a^2e^{-2s} + a^3e^{-3s} + \cdots]$$

$$= \frac{1}{s}[ae^{-s} + a^2e^{-2s} + a^3e^{-3s} + \cdots]$$

The inverse Laplace transform of e^{-Ts}/s is $q(t - T)$. Thus, we have

$$y(t) = aq(t - 1) + a^2q(t - 2) + a^3q(t - 3) + \cdots$$

which is plotted in Figures 4.25(a) and (b) for $a = 1$ and $a = 0.5$. Note that these responses can also be obtained directly using the argument in Example 3.2.2.

Although the final- and initial-value theorems are stated for rational functions, they are also applicable here. The initial value of $y(t)$ is

$$y(0+) = \lim_{s \to \infty} sY(s) = \lim_{s \to \infty} s \cdot \frac{ae^{-s}}{s(1 - ae^{-s})} = 0$$

Figure 4.25 *Unit step responses*

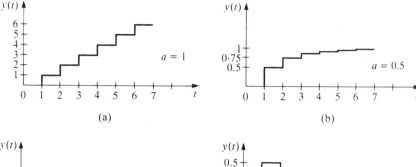

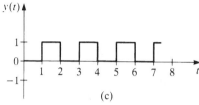
(a)

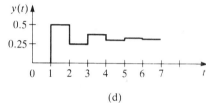
(b)

(c)

(d)

where we have used $e^{-s} = 0$, as $s \to \infty$. The final value of $y(t)$ for $a = 1$ does not approach a constant and the final-value theorem cannot be used. The final value of $y(t)$ for $a = 0.5$ approaches a constant and equals

$$y(\infty) = \lim_{s \to 0} sY(s) = \lim_{s \to 0} s \cdot \frac{ae^{-s}}{s(1 - ae^{-s})} = \frac{a}{1 - a} = \frac{0.5}{1 - 0.5} = 1$$

as shown in Figure 4.25(b).

EXERCISE 4.12.3

Verify that the unit step responses of the negative feedback system shown in Figure 4.24(c) are as shown in Figure 4.25(c) and (d) for $a = 1$ and $a = 0.5$. Use the final-value theorem to verify, for $a = 0.5$, $y(\infty) = 0.5/(1 + 0.5) = 0.33$. ∎

Periodic functions A positive-time function $f(t)$ is said to be periodic with period P if

$$f(t + P) = f(t) \qquad \text{for all } t \geq 0$$

as shown in Figure 4.26. Let $f_P(t)$ be the first period of $f(t)$, that is

$$f_P(t) = \begin{cases} f(t) & \text{for } 0 \leq t < P \\ 0 & \text{for elsewhere} \end{cases}$$

Figure 4.26 *A periodic function*

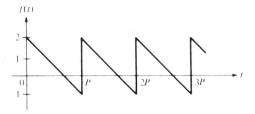

and let

$$F_P(s) = \mathcal{L}[f_P(t)]$$

If the first period of $f(t)$ is $f_P(t)$, the second period is $f_P(t - P)$, the third period is $f_P(t - 2P)$, and so forth. Thus, $f(t)$ can be expressed as

$$f(t) = f_P(t) + f_P(t - P) + f_P(t - 2P) + \cdots \qquad \text{(4.96)}$$

The application of the Laplace transform to (4.96) yields, using (4.90),

$$\begin{aligned} F(s) &= F_P(s) + e^{-sP}F_P(s) + e^{-2sP}F_P(s) + \cdots \\ &= (1 + e^{-sP} + e^{-2sP} + \cdots)F_P(s) \end{aligned} \qquad \text{(4.97)}$$

Using the formula

$$1 + r + r^2 + \cdots + r^n + \cdots = \frac{1}{1 - r}$$

for $|r| < 1$, we can write (4.97) as

$$F(s) = F_P(s)\frac{1}{1 - e^{-sP}} \qquad \text{(4.98)}$$

This is the Laplace transform of the periodic function $f(t)$. Strictly speaking, the closed form is valid only if $|e^{-sP}| < 1$. In its application, this restriction is often disregarded.

Time scaling   On computer simulation, it is sometimes desirable to increase or decrease the speed of simulation. For example, it may take hours for a space shuttle to circle the earth once; however, a complete circular orbit can be generated in a matter of minutes on a computer screen. This is achieved by a change of time scale. For example, the $f(t)$ in Figure 4.27(a) can be stretched or compressed, as shown in Figures 4.27(b) and (c) by choosing a different a in $f(at)$. If $a < 1$, the function is stretched in time; if $a > 1$, the function is

Figure 4.27 *Change of time scale*

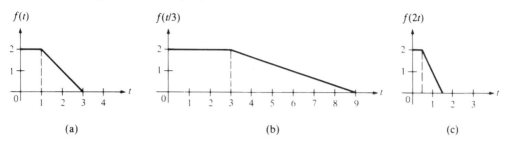

compressed in time. A negative a is not permitted because it will change $f(at)$ into a negative-time function.

Let $F(s)$ be the Laplace transform of $f(t)$. Then we have

$$\mathcal{L}[f(at)] = \frac{1}{a} F(\frac{s}{a}) \quad \text{for } a > 0 \tag{4.99}$$

This formula can be established as follows. The Laplace transform of $f(at)$ is, using (4.1),

$$F_a(s) := \mathcal{L}[f(at)] = \int_{0-}^{\infty} f(at)e^{-st}dt$$

which becomes, after defining $\tau = at$,

$$F_a(s) = \int_{0-}^{\infty} f(\tau)e^{s\tau/a}d(\tau/a) = \frac{1}{a}\int_{0-}^{\infty} f(\tau)e^{-s\tau/a}d\tau = \frac{1}{a}F(\frac{s}{a})$$

This establishes (4.99).

4.13   SUMMARY

1. The Laplace transform of a time function is defined as an integration from 0− to infinity. The reasons for using 0− rather than 0 are (a) to include the impulse at $t = 0$ wholly in the transform and (b) to avoid possible confusion regarding initial conditions. If a time function does not contain an impulse at $t = 0$ and is continuous at $t = 0$, then there is no difference between using 0−, 0, or 0+ in defining the Laplace transform.

2. The Laplace transform transforms a convolution integral into an algebraic equation. The Laplace transform of the impulse response is called the transfer function. If a system is linear and time invariant but is distributed, then the transfer function is an irrational function of s. In this case, generally, not much advantage is gained in using the Laplace transform. How-

ever, if a system is linear, time invariant, and lumped, the transfer function is a rational function of s. Mathematically, manipulation of rational functions is the same as manipulation of real numbers. Thus, for this class of systems, the use of transfer functions in analysis and design becomes fairly simple.

3. The Laplace transform is not used if a system is time varying or nonlinear.

4. The region of convergence in the Laplace transform can often be disregarded in the study of causal systems and positive-time signals.

5. The response of a system with transfer function $H(s)$ excited by an input can be obtained by partial fraction expansion and a table lookup. This procedure will yield a closed-form solution and reveal general properties of the system. However, the procedure is not a good method for computing the response on a digital computer. A better method is to transform $H(s)$ into a state-variable equation and then to carry out integration directly. This will be discussed in Chapter 10.

6. If a system is described by the differential equation

$$D(p)y(t) = N(p)u(t)$$

where $D(p)$ and $N(p)$ are defined as in (4.51), then the zero-input response of the system is dictated by the modes or the roots of the characteristic polynomial $D(s)$ and the zero-state response is dictated by the poles of the transfer function $H(s) = N(s)/D(s)$. The set of modes is not necessarily equal to the set of poles due to the possible existence of common factors between $N(s)$ and $D(s)$. To find the poles of a transfer function, we must cancel the common factors between the denominator and numerator. Only then will the roots of the remaining denominator be the poles of the transfer function.

7. The time response of a pole (mode) approaches 0 as $t \to \infty$ if, and only if, the pole (mode) lies inside the open left half-plane. It approaches infinity if the pole (mode) lies inside the open right half-plane. If the pole (mode) is on the imaginary axis, its response approaches infinity only if the pole (mode) is repeated.

8. The transfer-function representation of systems is convenient for analysis and design. When using transfer functions, we completely disregard zero-input responses, or responses excited by nonzero initial conditions. If a system is completely characterized by its transfer function, then the zero-input response is essentially included in the zero-state response and no essential information is lost in using the transfer function.

9. When using the final-value and initial-value theorems of the Laplace transform, we must check the applicability conditions. Otherwise, the theorems may yield erroneous results.

PROBLEMS

4.1 Find the Laplace transform of e^{-2t}. What is its region of convergence? Does it include the imaginary axis?

4.2 Find the two-sided Laplace transforms of

$$f_1(t) = \begin{cases} e^{-3t} & \text{for } t \geq 0 \\ 0 & \text{for } t < 0 \end{cases}$$

and

$$f_2(t) = \begin{cases} 0 & \text{for } t \geq 0 \\ -e^{-3t} & \text{for } t < 0 \end{cases}$$

Do they have the same transform? What are their regions of convergence?

4.3 Find the Laplace transform of $e^{-0.2t} \sin(2t + 45°)$ for $t \geq 0$.

4.4 Find the Laplace transforms of the following.

(a) $e^{-2t} - 3e^{-3t}$

(b) $t + t^2 e^{-t} + \sin t$

(c) $1 + t \sin 3t$

4.5 (a) Compute the Laplace transform of $\sin at \sin bt$.

(b) Compute the Laplace transform of $e(t) = (1 + 0.5\sin t) \sin 1000t$. A rough sketch of $e(t)$ is shown in Figure P4.5. What are the values of a, b, c, and d on the plot? (The signal is called an amplitude-modulated signal.)

4.6 Find the inverse Laplace transforms of the following.

Figure P4.5

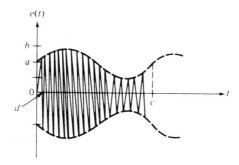

(a) $\dfrac{s}{(s^2 - 1)}$

(b) $\dfrac{s^3 + 1}{s(s - 1)(s + 1)}$

(c) $\dfrac{s^2 + 4s - 5}{s(s^2 - 1)}$

4.7 Using only real computation, find the inverse Laplace transforms of the following.

(a) $\dfrac{1}{s(s^2 + 2s + 2)}$

(b) $\dfrac{s^3 + 1}{s(s^2 + 2s + 2)}$

4.8 Express the complex numbers $1 + j2$, $-1 + j2$, $-1 - j2$, and $1 - j2$ in polar forms.

4.9 Repeat Problem 4.7 using only linear factors. Combine the terms with complex numbers into real terms. Are these results the same as those found for Problem 4.7?

4.10 Find the inverse Laplace transforms of the following.

(a) $\dfrac{s^2 + 1}{(s + 2)^3}$

(b) $\dfrac{s + 1}{(s - 1)^3(s + 2)}$

4.11 Show that if $f(t)$ is a real-valued function, its Laplace transform $F(s)$ is a function of s with real coefficients. [Hint: Show $F^*(s) = F(s^*)$ and then consider real s].

4.12 The vertical movement of the landing of an aircraft can be modeled as shown in Figure 3.21 and described by

$$m\ddot{y}(t) + k_2\dot{y}(t) + k_1 y(t) = u(t)$$

We may assume that $u(t)$ becomes a unit step function with magnitude mg when the aircraft touches down on the runway. If $k_2/m = 2$ and $k_1/m = 10$, plot $y(t)$ as a function of t.

4.13 Repeat Problem 4.12 for $k_2/m = 7$ and $k_1/m = 10$.

4.14 Consider the differential equation

$$\ddot{y}(t) + 4y(t) = u(t)$$

Compute $y(t)$ excited by $y(0-) = 2$, $\dot{y}(0-) = -1$, and $u(t) = 1$ for $t \geq 0$. What are its zero-state response and zero-input response? Plot $y(t)$ as a function of t.

4.15 Compute $y(t)$ in Problem 4.14 excited by $y(0-) = 0$, $\dot{y}(0-) = 0$, and $u(t) = \delta(t)$. Is $y(t)$ the impulse response of the system?

4.16 Find the solution of the differential equation

$$\ddot{y}(t) - 3\dot{y}(t) + 2y(t) = \dot{u}(t) + u(t)$$

excited by $\dot{y}(0-) = 1$, $y(0-) = 0$, and $u(t) = 1$ for $t \geq 0$. Is $y(t)$ continuous at $t = 0$? Is $\dot{y}(t)$ continuous at $t = 0$?

4.17 In Problem 4.16, find $y(0-)$ and $\dot{y}(0-)$ so that the output $y(t)$ excited by $u(t) = e^{-2t}$ is of the form e^{-2t}.

4.18 Consider the system described by the differential equation in Problem 4.16.

(a) What are the modes of the system? What is the general form of its zero-input response?

(b) What is its transfer function? What are its poles and zeros? What is the general form of its zero-state response?

(c) Is the set of the modes of the system equal to the set of the poles?

4.19 Answer the questions in Problem 4.18 for the system described by

$$\ddot{y}(t) + 3\dot{y}(t) + 2y(t) = \dot{u}(t) + u(t)$$

4.20 Consider a system with the same set of poles and modes as $\{0, 0, 1, 1, 1\}$. What are the general forms of the zero-input response and the zero-state response?

4.21 Consider an LTIL system. The zero-state response of the system excited by $u(t) = \sin 2t$ is measured as

$$y(t) = 1 + 3 \sin 2t - e^{-t}$$

(a) What is its transfer function? What is its impulse response?

(b) What is the unit step response of the system?

4.22 Consider the heating system shown in Figure P4.22. Let $\theta(t)$ be the temperature and $q(t)$ be the heat pumping into the chamber. Because of imperfect insulation, we assume that the temperature can be described by the first-order differential equation

$$\dot{\theta}(t) + 0.1\theta(t) = q(t)$$

(a) What is the transfer function from q to θ of the system?

(b) If no heat is applied and if the temperature is 80 degrees, how long will it take for the temperature to decrease to 70 degrees?

Figure P4.22

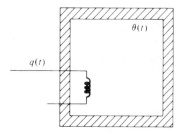

4.23 Consider the network shown in Figure P4.23. It is called a ladder net-
      work.

   (a) Find the transfer function from u to y of the network.

   (b) Let the output y be 1 volt. Compute the voltage of each node from the
       output to the input. What is the transfer function of the network?
       (This technique can be used to compute the transfer function of ladder
       networks.)

4.24 Consider the network shown in Figure P4.24(a) with $v(0-) = 0$.

   (a) Find $v(t)$ after the switch is connected to A at $t = 0$.

Figure P4.23

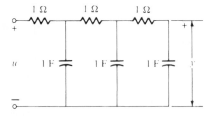

Figure P4.24

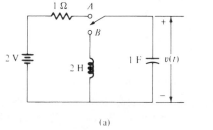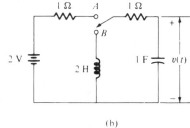

(a)   (b)

(b) At $t = 20$ seconds, the switch is moved to B. Find $v(t)$ for $t \geq 20$. Will $v(t)$ approach zero as $t \to \infty$?

4.25 Answer the questions in Problem 4.24 for the network shown in Figure P4.24(b). Will $v(t)$ approach zero as $t \to \infty$? Give a reason why this answer is different from the one given in Problem 4.24.

4.26 Consider the network shown in Figure P4.26. Find the voltage $v(t)$ excited by the voltage source and the initial conditions $i(0) = -1$ A (ampere) and $v(0) = -2$ V.

4.27 Find the impedances of the networks in Figure P4.27.

4.28 Verify that $y(t) = 2e^{0.5t^2}$ is the solution of the time-varying differential equation

$$\dot{y}(t) - ty(t) = 0$$

excited by the initial condition $y(0) = 2$. Can you solve the equation by using the Laplace transform?

4.29 Consider a system described by $D(p)y(t) = N(p)u(t)$ with

$$D(p) = p^3(p + 1)^2(p + 3) \qquad N(p) = (p - 1)(p + 2)$$

Figure P4.26

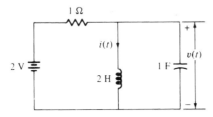

Figure P4.27

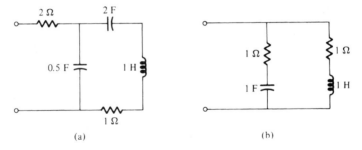

(a)   (b)

What are the modes, zeros, and poles of the system? What are the general forms of its zero-input and zero-state responses? Is the system completely characterized by its transfer function?

4.30 Repeat Problem 4.29 for

$$D(p) = p^3(p + 1)^2(p + 3) \qquad N(p) = (p + 1)(p + 3)$$

4.31 Let $H_1 = 1/s(s + 1)$ and $H_2 = (s - 1)/(s + 3)$. Compute the transfer functions of their tandem, parallel, positive-feedback, and negative-feedback connections, as shown in Figure 4.18.

4.32 Find the transfer functions of the networks shown in Figure P4.32. Are the networks completely characterized by their transfer functions?

4.33 Consider the basic block diagram studied in Problem 3.36. What is its transfer function? Is the block diagram completely characterized by its transfer function?

4.34 Consider the two hydraulic tanks shown in Figure P4.34. Use the procedure discussed in Example 3.7.5 to develop a block diagram for the system. What is its overall transfer function from q_i to q_2? Is it the product of two individual transfer functions? Is there a loading problem in the system? Can the overall transfer function of the system in Figure 3.23 be written as the product of the two individual transfer functions? Give your reasons.

4.35 Let the Laplace transforms of $f(t)$ be

(a) $\dfrac{s^3 - 10}{s(s^2 + 2s + 2)}$

(b) $\dfrac{s^2 - 10}{s^2(s^2 + 2s + 2)}$

(c) $\dfrac{s - 10}{s(s^2 - 2s + 2)}$

Do $f(t)$ approach finite constants as $t \to \infty$? What are $f(\infty)$?

4.36 Find the initial values of $f(t)$ in Problem 4.35.

Figure P4.32

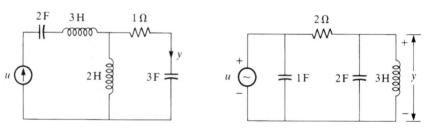

Figure P4.34

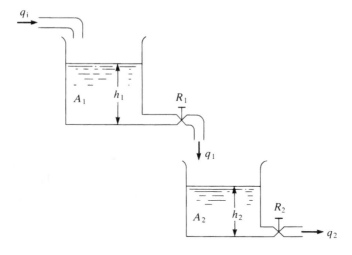

4.37 Find the Laplace transforms of the functions shown in Figure P4.37. Are they rational functions of s?

4.38 Find the Laplace transforms of the periodic functions shown in Figure P4.38.

Figure P4.37

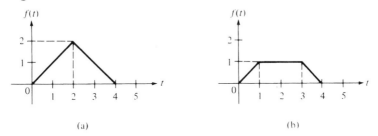

(a)   (b)

Figure P4.38

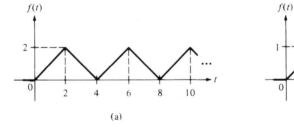

(a)   (b)

Figure P4.40

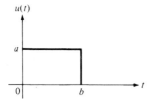

4.39 Find the Laplace transform of $f(t) = |\sin t|$ for $t \geq 0$.

4.40 Find the zero-state response of a system with transfer function $1/s\,(s + 2)$ excited by the application of the pulse input shown in Figure P4.40. If $a = 1$, what is b in order for the response to reach 5 as $t \to \infty$? If $b = 1$, what is a in order for the response to reach 5 as $t \to \infty$?

USING **MATLAB**

MATLAB can be used to carry out the operations discussed in this chapter. Consider a transfer function such as

$$F(s) = \frac{N(s)}{D(s)} = \frac{2s^2 - s + 5}{s^3 - 7s - 6} \tag{M4.1}$$

Its denominator and numerator are expressed as row vectors as

```
num = [2 -1 5];den = [1 0 -7 -6];
```

To find the roots of `den`, the command

```
roots(den)
```

yields

```
ans = 3.0000
     -2.0000
     -1.0000
```

The denominator has three real roots at 3, −2, and −1. If we want to express a transfer function as

$$H(s) = k \frac{(s - z(1))(s - z(2)) \cdots}{(s - p(1))(s - p(2))(s - p(3)) \cdots}$$

called the zero-pole-gain form, the command

```
[z,p,k] = tf2zp(num,den)
```

where `tf2zp` stands for transfer function to zero pole, yields

```
z = 0.2500 + 1.5612i
    0.2500 - 1.5612i
p = 3.0000
   -2.0000
   -1.0000
k = 2
```

Thus, we have

$$F(s) = \frac{2s^2 - s + 5}{s^3 - 7s - 6} = 2 \frac{(s - 0.25 - 1.5612i)(s - 0.25 + 1.5612i)}{(s - 3)(s + 2)(s + 1)}$$

The partial fraction expansion can be carried out in MATLAB by calling residue. For example, if we type

```
[r,p,k]=residue(num,den)
```

then the output is

```
r = 1.0000
    3.0000
   -2.0000
p = 3.0000
   -2.0000
   -1.0000
k = [ ]
```

This means that $H(s)$ can be expanded as

$$H(s) = \frac{2s^2 - s + 5}{s^3 - 7s - 6} = \frac{1}{s - 3} + \frac{3}{s + 2} + \frac{-2}{s + 1} + k$$

where k is a polynomial and is zero in this example. We mention that residue can be used only if $H(s)$ has simple poles. To find the unit step response of $H(s)$ from $t = 0$ to 10 seconds and with the output computed or printed every 0.1 second, we type

```
t = 0:0.1:10;
y =  step(num,den,t);
plot(t,y)
```

then the response will appear on the screen. If we type

```
step(num,den)
```

then the response will appear on the monitor. In this case, the program will select automatically the final time and the time interval of printout.

The inverse Laplace transform of $H(s)$ is, by definition, the impulse response of $H(s)$. Thus, command impulse can be used to compute the inverse Laplace transform. For example, the commands

```
t = 0:0.2:10;
h = impulse(num,den,t)
```

compute the inverse Laplace transform of $H(s)$ from $t = 0$ to 10 seconds with the inverse computed every 0.2 second. If command impulse is not available, as in the *Student Edition of MATLAB*, we may use command step to compute the step response of $sH(s)$. For example, the inverse Laplace transform of $H(s)$ in (M4.1) can also be obtained by typing

```
step([2 -1 5 0],den)
```

The step responses in Figure 4.6 are obtained by the following commands

```
t = 0:0.1:10;
n1 = 2;d = [1 3 4 2];
n2 = [0.2 2];n3 = [-0.2 2];
n4 = [10 1 2];n5 = [1 0];
y1 = step(n1,d,t);
y2 = step(n2,d,t);
y3 = step(n3,d,t);
y4 = step(n4,d,t);
y5 = step(n5,d,t);
plot(t,y1,t,y2,t,y3,t,y4,t,y5)
```

Now consider some numerical problems. The roots of a polynomial is very sensitive to parameter variations. For example, the commands

```
roots([1 5 11 13 8 2]) and
roots([1 5 11 13 8 2.001])
```

yield, respectively, the roots $(-1, -1, -1, -1 \pm i)$ and the roots $(-1.0005 \pm i, -1.0997, -0.9497 \pm 0.0866i)$. The two polynomials differ in only one coefficient and the difference is smaller than 0.1%, but their five roots are quite different. This high sensitivity occurs not only in polynomials with repeated roots but also in polynomials with distinct roots. A well-known example is the polynomial with $1, 2, \ldots, 19$, and 20 as roots, or

$$p = (s - 1)(s - 2) \cdots (s - 19)(s - 20)$$

The following commands

```
k = 0:20;r(k) = k;
p = poly (r); (Compute the coefficients of the polynomial p.)
roots(p) (Compute the 20 roots of p.)
```

yield $(1, 2, 3, 4, 5, 6, 6.9999, 8.0003, 8.9983, 10.0059, \ldots, 18.0079, 18.9983, 20.0002)$. We see that roots of polynomials are indeed very sensitive to parameter variations. Therefore, in numerical computation, we should avoid, if possible, computing roots of polynomials. If there is no way to avoid it, we must use the most reliable method. Presently, the most reliable method of computing the roots of a polynomial is to change the polynomial into a matrix and then use LINPACK or EISPACK to compute the eigenvalues of the matrix. MATLAB uses this procedure to compute roots of polynomials.

Numerically, using partial fraction expansion to compute inverse Laplace transforms and system responses is not desirable, because it requires computing roots of polynomials. A better method is to transform a transfer function into a state-space equation (using command `tf2ss`), to discretize the continuous-time state equation into a discrete-time equation (using command `c2d`), and then to compute the response (see Chapter 10). Command `step` in MATLAB takes this procedure, and generally yields more accurate result than the one obtained by using partial fraction expansion.

MATLAB PROBLEMS

Use MATLAB to compute the following.

M4.1 Find the roots of

$$D(s) = 2s^5 + 3s^4 - 3s^3 + 4s^2 + 5s + 1$$

Express $D(s)$ in factor form.

M4.2 Express

$$F(s) = \frac{N(s)}{D(s)} \tag{M4.2}$$

with $N(s) = s^3 + 2s + 1$ and $D(s)$ in Problem M4.1 in the zero-pole-gain form.

M4.3 Use command `residue` to carry out the partial fraction expansion of

$$F(s) = \frac{s^2 + 2s + 3}{s(s+1)(s+2)} \tag{M4.3}$$

M4.4 Repeat Problem M4.3 for

$$F(s) = \frac{s^2 + 2s + 3}{s(s+1)(s+2)^2} \tag{M4.4}$$

Do you encounter any problem?

M4.5 Repeat Problem M4.3 for $F(s)$ in Problem M4.2.

M4.6 (a) Use command `impulse` to find the inverse Laplace transforms of $F(s)$ in (M4.2), (M4.3), and (M4.4).

(b) Use command `step` to repeat (a).

5

The z-Transform and Discrete-Time System Analysis

5.1   INTRODUCTION

This chapter will discuss the counterpart of the Laplace transform, called the z-transform, for discrete-time systems. Every linear time-invariant discrete-time system with memory can be described by a discrete convolution. If the system is lumped as well, then it can also be described by an LTIL difference equation, such as

$$2y\,[k+2] + 3y\,[k+1] + y\,[k] = u\,[k+2] + u\,[k+1] - u\,[k] \qquad \textbf{(5.1)}$$

for $k = -2, -1, 0, 1, 2, \ldots$, or

$$2y\,[k] + 3y\,[k-1] + y\,[k-2] = u\,[k] + u\,[k-1] - u\,[k-2] \qquad \textbf{(5.2)}$$

for $k = 0, 1, 2, \ldots$. The response of either equation is excited by an input $u\,[k]$ and some initial conditions. By convention, we assume that the input is applied at $k = 0$ and $u\,[k] = 0$ for $k < 0$ and the initial conditions are $y\,[-1]$ and $y\,[-2]$. Because difference equations are basically algebraic equations, their solutions can be obtained by direct substitution. For example, if we write (5.1) as

$$y\,[k+2] = \frac{1}{2}(-3y\,[k+1] - y\,[k] + u\,[k+2] + u\,[k+1] - u\,[k])$$

then the response $y\,[k]$ due to any $u\,[k]$ for $k \geq 0$ and $y\,[-1]$ and $y\,[-2]$ can be computed recursively by substituting $k = -2, -1, 0, 1, \ldots$. The solution, however, is not in closed form and is difficult to develop from the solution general properties of the system. This difficulty will not arise if we use the z-transform to study the system. Furthermore, a number of design techniques have been developed in the z-transform domain, making the z-transform an important tool in the study of discrete-time systems.

Unlike differential equations, which have only one form, difference equations may assume the form in (5.1) or (5.2). For easy reference, we call (5.1) the *advanced form* and (5.2) the *delayed form*. The advanced form is widely used in control and filter design; the delayed form is widely used in estimation, identification, and digital signal processing. Although both forms are basically equivalent, they differ somewhat in their use of the z-transform. Both forms will be discussed here.

5.2   THE Z-TRANSFORM

A discrete-time signal $f[k] := f(kT)$, where k is an integer ranging from $(-\infty, \infty)$, is called a positive-time sequence if $f[k] = 0$ for $k < 0$; it is called a negative-time sequence if $f[k] = 0$ for $k > 0$. In the study of LTI discrete-time systems, the initial time instant can be assumed, as was discussed in Chapter 2, to be 0 and the input sequence is applied for $k \geq 0$. Therefore, signals encountered in system analysis are mostly positive-time sequences.

Definition 5.1   Consider a discrete-time sequence $f[k] := f(kT)$ defined for all integer k and $T > 0$. The z-transform is defined as

$$F(z) := \mathscr{Z}[f[k]] := \sum_{k=0}^{\infty} f[k]z^{-k} \tag{5.3}$$

and the inverse z-transform is

$$f[k] := \mathscr{Z}^{-1}[F(z)] := \frac{1}{2\pi j} \oint F(z)z^{k-1}dz \quad \text{for } k \geq 0 \tag{5.4}$$

where z is a complex variable, called the z-transform variable, and the small circle in (5.4) is a contour to be discussed later. $F(z)$ is called the z-transform of $f[k]$, and $f[k]$ is called the inverse z-transform of $F(z)$.

The z-transform is defined for the positive-time part of $f[k]$; the negative-time part is completely disregarded. However, if the contour of integration in (5.4) is properly chosen, then (5.4) yields $f[k] = 0$, for $k = -1, -2, -3, \ldots$. Therefore, we often assume $f[k] = 0$ for $k < 0$, and the z-transform is said to be defined only for positive-time sequences.

The z-transform of $f[k]$ is defined as an infinite power series of z^{-1}. In this series, z^{-i} can be interpreted as indicating the ith sampling instant. In other words, z^0 indicates the initial time instant $k = 0$; z^{-1} indicates the time instant $k = 1$, and, in general, z^{-i} indicates the time instant $k = i$. In this sense, the z-transform merely expresses the sequence as a function of z^{-1} with z^{-i} indicating its time instant. In this interpretation, there is not much difference between a sequence and its z-transform. For convenience, we call z^{-1} the unit-time delay element.

The z-transforms we will encounter in this chapter, however, can also be expressed in closed form as rational functions of z. They will be developed using the formula

$$1 + r + r^2 + r^3 + \cdots = \sum_{k=0}^{\infty} r^k = \frac{1}{1-r}, \quad |r| < 1 \quad (5.5)$$

where r is a real or complex constant with absolute value smaller than 1, denoted as $|r| < 1$ (see Problem 3.9). If $|r| > 1$, the power series in (5.5) diverges to ∞. If $r = 1$, the sum is ∞. If $r = -1$, the sum could be 1 or -1 and is therefore undefined. Thus, the condition $|r| < 1$ is essential in (5.5). This is the equation we will use to develop closed-form z-transforms.

EXAMPLE 5.2.1

Consider the sequence $f[k] = b^k$ for all integer k, where b is a real or complex number. Its z-transform is

$$F(z) = \sum_{k=0}^{\infty} f[k]z^{-k} = \sum_{k=0}^{\infty} b^k z^{-k} = \sum_{k=0}^{\infty} (bz^{-1})^k$$

If $|bz^{-1}|$ is smaller than 1, then the infinite power series converges and, using (5.5), can be expressed as

$$F(z) = \mathcal{Z}[b^k] = \frac{1}{1 - bz^{-1}} = \frac{z}{z - b} \quad (5.6)$$

This is the z-transform of $f[k] = b^k$. Strictly speaking, (5.6) holds only if $|bz^{-1}| < 1$ or $|b| < |z|$. The region $|b| < |z|$, as shown in Figure 5.1, is called the *region of convergence*. In application, we often disregard the region of convergence, and consider $z/(z - b)$ to be defined everywhere except at $z = b$. This will be explained in the next section.

The z-transform of b^k is $1/(1 - bz^{-1})$ or $z/(z - b)$. Unlike the Laplace transform, where we use exclusively positive powers of s, both positive and negative powers of z are used in the z-transform. To differentiate these two forms, we use the notations

$$F(z) = \frac{z}{z - b} \quad \text{and} \quad F^-(z^{-1}) = \frac{1}{1 - bz^{-1}}$$

Figure 5.1 *Region of convergence*

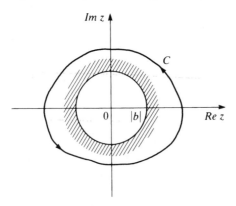

Note that, if $F(z)$ is defined as shown, then we have $F(z^{-1}) = z^{-1}/(z^{-1} - b)$, which is different from $F^-(z^{-1})$. Therefore, we must use different notations to denote the two different forms of the same function. We call $F(z)$ the positive-power form and $F^-(z^{-1})$ the negative-power form. Both forms will be used in this chapter. Before proceeding, consider some special cases of (5.6).

1. **Unit step sequence.** The unit step sequence $q[k]$ is defined as

$$q[k] = \begin{cases} 1 & \text{for } k = 0, 1, 2, \ldots \\ 0 & \text{for } k < 0 \end{cases}$$

It is the sequence in (5.6) with $b = 1$. Thus, its z-transform is

$$Q(z) := \mathcal{Z}[q[k]] = \sum_{k=0}^{\infty} z^{-k} = \frac{1}{1 - z^{-1}} = \frac{z}{z - 1} \qquad (5.7)$$

2. **Exponential sequence.** If $b = e^{aT}$, where a is a real or complex constant, then $f[k] = e^{akT}$. It increases exponentially if a is real and positive; it decreases exponentially if a is real and negative. If a is complex, then $f[k]$ is a complex-valued sequence. The z-transform of $f[k] = e^{akT}$ is

$$F(z) = \mathcal{Z}[e^{akT}] = \sum_{k=0}^{\infty} e^{akT} z^{-k} = \frac{1}{1 - e^{aT} z^{-1}} = \frac{z}{z - e^{aT}}$$

3. **Kronecker delta sequence.** Consider the impulse sequence or Kronecker delta sequence

$$\delta[k] = \begin{cases} 1 & \text{for } k = 0 \\ 0 & \text{for } k \neq 0 \end{cases}$$

Its z-transform is[1]

$$\Delta(z) := \mathcal{Z}[\delta[k]] = \sum_{k=0}^{\infty} \delta[k]z^{-k} = 1 \qquad \text{❋} \qquad (5.8a)$$

More generally, if the impulse sequence is shifted to $k = i$, where i is a positive integer, as

$$\delta[k - i] = \begin{cases} 1 & \text{for } k = i \\ 0 & \text{for } k \neq i \end{cases}$$

then we have

$$\mathcal{Z}[\delta[k - i]] = \sum_{i=0}^{\infty} \delta[k - i]z^{-k}$$

which, because $\delta[k - i]$ is zero except when $k = i$, reduces to

$$\mathcal{Z}[\delta[k - i]] = z^{-i} \qquad \text{❋} \qquad (5.8b)$$

Note that if i is a negative integer, then $\delta[k - i]$ is a negative-time sequence and its z-transform is zero. However, (5.8b) will be used for $i \geq 0$ and $i < 0$, such as $\mathcal{Z} \delta[k - 1]] = z^{-1}$, $\mathcal{Z} \delta[k + 1]] = z$, and $\mathcal{Z}[\delta[k + 5]] = z^5$. Strictly speaking $\mathcal{Z}[\delta[k - i]]$ for $i < 0$ should be defined using the two-sided z-transform, that will be discussed in the next section. Here we may simply interpret z as the unit-time advance operator (shifting to the left) and z^{-1} as the unit-time delay operator (shifting to the right). For example, $\delta[k - 3]$ is the shifting of $\delta[k]$ to the right 3 sampling periods and its z-transform is z^{-3}; $\delta[k + 3]$ is the shifting of $\delta[k]$ to the left 3 sampling periods and its z-transform is z^3.

EXERCISE 5.2.1

Find the z-transforms of the following.

(a) $f[k] = 2^k$ $\qquad \frac{z}{z-2}$

(b) $f[k] = (-2)^k$ $\qquad \frac{z}{z+2}$

(c) $f[k] = 1$, for $k = 3, 4, 5, \ldots$, and $f[k] = 0$, for $k < 3$

$$q[k-3] \quad \overbrace{\frac{z^{-3}}{1-z^{-1}}} = \frac{1}{z^2(z-1)}$$

---

[1] If we define $0^0 := 1$, then we have $\delta[k] = b^k$, with $b = 0$, $k = 0, 1, 2, \ldots$. Thus, the z-transform of the sequence is, from (5.6), $z/(z - b)|_{b=0} = z/z = 1$.

(d) $f[k] = \delta[k-2] - 2\delta[k-5]$

(e) $f[k] = -3\delta[k+4] + 1\delta[k+2] - 2\delta[k] + 1.5\delta[k-1]$

[**ANSWERS:** (a) $z/(z-2) = 1/(1-2z^{-1})$, (b) $z/(z+2) = 1/(1+2z^{-1})$, (c) $z^{-3}/(1-z^{-1}) = 1/z^2(z-1)$, (d) $z^{-2} - 2z^{-5} = (z^3 - 2)/z^5$, (e) $-3z^4 + z^2 - 2 + 1.5z^{-1}$.] ∎

The z-transforms of the sequences in the preceding example and exercise are all rational functions of z. This is, however, not always the case. In fact, given a positive-time sequence, the following situations may occur.

1. Its z-transform does not exist. In other words, there is no region of convergence. The sequences e^{k^2} and e^{e^k} for $k = 0, 1, 2, \ldots$, are such examples. These sequences are mathematically contrived and do not arise in practice. It is fair to say that all sequences encountered in practice are z-transformable. Therefore, when using the z-transform, we usually pay no attention to the existence problem.

2. Its z-transform exists but cannot be put into closed form. Most randomly generated sequences of infinite length belong to this type.

3. Its z-transform exists and can be put into closed form but not as a rational function of z. For example, the z-transforms of

$$f_1[k] = \begin{cases} 0 & \text{for } k = 0 \\ 1/k & \text{for } k = 1, 2, 3, \ldots \end{cases}$$

and

$$f_2[k] = \begin{cases} 0 & \text{for } k = 0, 2, 4, \ldots \\ 1/k! & \text{for } k = 1, 3, 5, \ldots \end{cases}$$

are

$$F_1(z) = z^{-1} + \frac{1}{2}z^{-2} + \frac{1}{3}z^{-3} + \cdots = -\ln(1 - z^{-1})$$

and

$$F_2(z) = z^{-1} + \frac{1}{3!}z^{-3} + \frac{1}{5!}z^{-5} + \cdots = \sinh z^{-1}$$

They are not rational functions of z.

4. Its z-transform exists and can be put into closed form as a rational function of z.

In this chapter, we study only sequences that belong to the last class. Although this is a very limited class of sequences, it is the most important in engineering.

*5.2.1   Region of convergence

As discussed in (5.6), the power series $\sum_{k=0}^{\infty} b^k z^{-k}$ converges to $F(z) = z/(z - b)$ only if $|z| > |b|$. The region $|z| > |b|$ is called the region of convergence. This region of convergence is needed if the inverse z-transform of $F(z)$ is to be computed from (5.4). The contour of integration in (5.4) must be chosen to lie entirely inside the region of convergence, as shown in Figure 5.1. If the region of convergence is not specified, the inverse z-transform of $F(z)$ is not uniquely defined. In order to see this nonuniqueness, we introduce the two-sided or bilateral z-transform

$$F(z) = \mathcal{Z}_2[f[k]] := \sum_{k=-\infty}^{\infty} f[k]z^{-k} \tag{5.9}$$

This transform is defined for both the positive-time and negative-time parts of $f[k]$. The subscript 2 stands for two-sided. If $f[k] = 0$ for $k < 0$, then (5.9) reduces to (5.3). In contrast to (5.9), (5.3) is called the one-sided or unilateral z-transform or, more often, simply the z-transform. The formula for the inverse two-sided z-transform is the same as (5.4).

Now consider the two different sequences shown in Figure 5.2. The sequence $f_1[k]$ in Figure 5.2(a) is a positive-time sequence and its two-sided z-transform reduces to the (one-sided) z-transform as

$$F_1(z) = \mathcal{Z}_2[f_1[k]] = \mathcal{Z}[f_1[k]] = \frac{z}{z - 1}$$

with the region of convergence $|z| > 1$. The two-sided z-transform of $f_2[k]$ in Figure 5.2(b) is

Figure 5.2 *Two different sequences with the same two-sided z-transform*

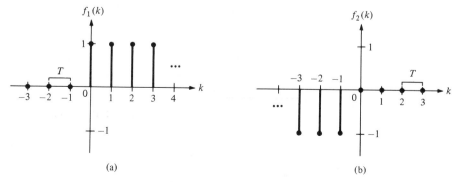

(a)

(b)

$$F_2(z) = \sum_{k=-\infty}^{\infty} f_2[k]z^{-k} = \sum_{k=-\infty}^{-1}(-1)z^{-k} = -\sum_{k=1}^{\infty} z^k$$

$$= -z\sum_{k=0}^{\infty} z^k = -z\left(\frac{1}{1-z}\right) = \frac{z}{z-1}$$

with the region of convergence $|z| < 1$. In other words, we have

$$\frac{z}{z-1} = \begin{cases} 1 + z^{-1} + z^{-2} + z^{-3} + \cdots & \text{(positive time)} \\ -z - z^2 - z^3 - z^4 - \cdots & \text{(negative time)} \end{cases} \tag{5.10}$$

We see that both $f_1[k]$ and $f_2[k]$ have the identical two-sided z-transform; however, they have different regions of convergence. Thus, in the two-sided z-transform, the region of convergence must be specified; otherwise, there is no one-to-one correspondence between a time sequence and its two-sided z-transform.

This is, however, not the case in the (one-sided) z-transform. In this case, all sequences are assumed to be positive time. Under this assumption, there is a unique relationship between every time sequence and its z-transform, and the inverse z-transform can be obtained from a table lookup. Therefore, in the (one-sided) z-transform, even if no region of convergence is specified, no ambiguity will arise. Thus, in employing the z-transform, we usually pay no attention to the region of convergence. This situation is identical to the continuous-time case.

5.2.2   From the Laplace transform to the z-transform

The Laplace transform is defined for (positive-time) continuous-time signals whereas the z-transform is defined for (positive-time) discrete-time sequences. If we apply the Laplace transform directly to a sequence, then the result is zero, that is, $\mathcal{L}[f(kT)] = 0$. Thus, the sequence must be modified before we can establish the relationship between the two transforms.

Consider the impulse $\delta(t)$ defined in Figure 3.10. It is zero everywhere except at $t = 0$. If we shift it to $t = kT$, then it becomes $\delta(t - kT)$. Now consider the function

$$\delta(t) + \delta(t - T) + \delta(t - 2T) + \cdots$$

It is a continuous-time function defined for all t, but it assumes nonzero values only at discrete-time instants $t = 0, T, 2T, \ldots$. Using this idea, we can express the discrete-time sequence $f(kT)$, $k = 0, 1, 2, \ldots$, as

$$f_s(t) := \sum_{k=0}^{\infty} f(kT)\delta(t - kT) \tag{5.11}$$

It is a continuous-time signal because it is defined at every $t \geq 0$. The continuous-time signal $f_s(t)$ is zero everywhere except at $t = kT$. At time instant $t = kT$, $f_s(t)$ has an impulse with weight $f(kT)$. Therefore, it is reasonable to consider $f_s(t)$ as a continuous-time representation of the discrete-time sequence $f(kT)$.

The Laplace transform of $f_s(t)$ is, using (4.91),

$$F_s(s) := \mathcal{L}[f_s(t)] = \sum_{k=0}^{\infty} f(kT)\mathcal{L}[\delta(t - kT)] = \sum_{k=0}^{\infty} f(kT)e^{-kTs} \qquad (5.12)$$

If we define $z = e^{Ts}$, then (5.12) becomes

$$F_s(s)\Big|_{z=e^{Ts}} = \sum_{k=0}^{\infty} f(kT)z^{-k}$$

Its right-hand side is, by definition, the z-transform of $f(kT)$. Thus, the z-transform of $f(kT)$ is the Laplace transform of $f_s(t)$ with the substitution $z = e^{Ts}$ or

$$\boxed{\mathcal{Z}[f(kT)] = \mathcal{L}[f_s(t)]\Big|_{z=e^{Ts}}}$$

This establishes the relationship between the Laplace transform and the z-transform.

We discuss now the implication of

$$z = e^{Ts} \quad \text{or} \quad s = \frac{1}{T}\ln z \qquad (5.13)$$

where ln stands for the natural logarithm. If $s = 0$, then $z = e^0 = 1$; that is, the origin of the s-plane is mapped into the point $z = 1$ in the z-plane. If $s = j\pi/2T$, then

$$z = e^{j\pi/2} = \cos\frac{\pi}{2} + j\sin\frac{\pi}{2} = j1$$

That is, the point A on the s-plane in Figure 5.3(a) is mapped into the point A on the z-plane in Figure 5.3(b). The point B or $s = j\pi/T$ is mapped into $z = e^{j\pi} = -1$. Similarly, the imaginary axis of the s-plane from $\omega = 0$ to $\omega = 2\pi/T$ is mapped into the unit circle on the z-plane, and the strip with slanted lines on the s-plane is mapped inside the unit circle, as shown in Figure 5.3. The mapping, however, is not one to one. The imaginary axis from $\omega = 2\pi/T$ to $\omega = 4\pi/T$ will again be mapped into the unit circle on the z-plane. In general, the imaginary axis from $\omega = 2\pi k/T$ to $2\pi(k + 1)/T$, for any integer k, will all be mapped into the unit circle. In fact, the entire left half–s-plane will be mapped

Figure 5.3 *Mapping between the s-plane and the z-plane*

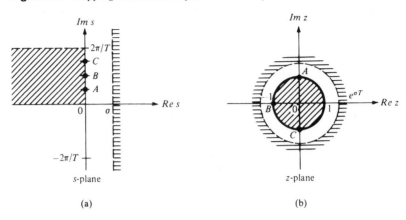

inside the unit circle on the z-plane. This is an important property and will be useful to subsequent discussion. Note that $z = 0$ on the z-plane corresponds to the negative infinity ($s = -\infty$) on the s-plane.

The region of convergence in the z-transform can also be obtained from that of the Laplace transform using the mapping in (5.13). The region of the convergence of a positive-time function is, as discussed in (4.3) and (4.4), of the form $Re\ s > \sigma$ as shown in Figure 5.3(a). If $s = \sigma + j\omega$, then

$$z = e^{Ts} = e^{(\sigma + j\omega)T} = e^{\sigma T}e^{j\omega T}$$

which implies

$$|z| = |e^{\sigma T}||e^{j\omega T}| = |e^{\sigma T}|$$

for all ω. It is a circle with radius $e^{\sigma T}$ in the z-plane, as shown in Figure 5.3(b). Thus, the vertical line passing through $Re\ s = \sigma$ in the s-plane is mapped into the circle with radius $e^{\sigma T}$ in the z-plane. Now, if $Re\ s > \sigma$, then $|z| > |e^{\sigma T}|$. Thus, the region of convergence $Re\ s > \sigma$ in the Laplace transform is mapped into the region outside the circle with radius $b := e^{\sigma T}$. This is the region of convergence for the positive-time sequence discussed in (5.6). Thus, the regions of convergence of the Laplace and z-transforms are related by (5.13).

5.3   **PROPERTIES OF THE Z-TRANSFORM**

In this section, we will discuss some properties of the z-transform and the application of these properties.

Linearity The z-transform is a linear operator. Let $f_i[k] := f_i(kT)$, $i = 1, 2$, be any two positive-time sequences with the same sampling period T. If

$$F_i(z) = \mathscr{Z}[f_i[k]] \quad \text{for } i = 1,2$$

then, for any real or complex constants α_1 and α_2,

$$\mathscr{Z}[\alpha_1 f_1[k] + \alpha_2 f_2[k]] = \alpha_1 \mathscr{Z}[f_1[k]] + \alpha_2 \mathscr{Z}[f_2[k]]$$
$$= \alpha_1 F_1(z) + \alpha_2 F_2(z)$$

The z-transform is defined through a summation. Because the summation is a linear operator, so is the z-transform.

EXAMPLE 5.3.1

We use this property to compute the z-transform of $\sin k\omega_0 T$. We use Euler's identity to write

$$\sin k\omega_0 T = \frac{e^{jk\omega_0 T} - e^{-jk\omega_0 T}}{2j}$$

The z-transform of $e^{\pm jk\omega_0 T}$ is, using (5.6),

$$\mathscr{Z}[e^{\pm jk\omega_0 T}] = \mathscr{Z}[(e^{\pm j\omega_0 T})^k] = \frac{z}{z - e^{\pm j\omega_0 T}} = \frac{1}{1 - e^{\pm j\omega_0 T} z^{-1}}$$

Therefore, we have

$$\mathscr{Z}[\sin k\omega_0 T] = \frac{1}{2j} \left[\frac{z}{z - e^{j\omega_0 T}} - \frac{z}{z - e^{-j\omega_0 T}} \right] = \frac{(e^{j\omega_0 T} - e^{-j\omega_0 T})z}{2j(z^2 - (e^{j\omega_0 T} + e^{-j\omega_0 T})z + 1)}$$

$$= \frac{(\sin \omega_0 T)z}{z^2 - 2(\cos \omega_0 T)z + 1} = \frac{(\sin \omega_0 T)z^{-1}}{1 - 2(\cos \omega_0 T)z^{-1} + z^{-2}}$$

EXERCISE 5.3.1

Show

$$\mathscr{Z}[\cos k\omega_0 T] = \frac{z(z - \cos \omega_0 T)}{z^2 - 2(\cos \omega_0 T)z + 1} = \frac{1 - (\cos \omega_0 T)z^{-1}}{1 - 2(\cos \omega_0 T)z^{-1} + z^{-2}}$$

EXERCISE 5.3.2

Find the z-transforms of the following.

(a) $f_1[k] = \begin{cases} 2^k & k = 0, 2, 4, \ldots \\ 0 & k = 1, 3, 5, \ldots \end{cases}$

(b) $f_2[k] = \begin{cases} 0 & k = 0, 2, 4, \ldots \\ 2^k & k = 1, 3, 5, \ldots \end{cases}$

Note that they can be written as $(2^k \pm (-2)^k)/2$.

[**ANSWERS:** (a) $z^2/(z^2 - 4) = 1/(1 - 4z^{-2})$, (b) $2z/(z^2 - 4) = 2z^{-1}/(1 - 4z^{-2})$.] ∎

● **Multiplication by k in the time domain**   Let $F(z)$ be the z-transform of $f[k]$. Then we have

$$\boxed{\mathcal{Z}[kf[k]] = -z\frac{dF(z)}{dz}} \qquad (5.14)$$

This can be established as follows. The differentiation of

$$F(z) = \sum_{k=0}^{\infty} f[k]z^{-k}$$

with respect to z yields

$$\frac{dF(z)}{dz} = \sum_{k=0}^{\infty} - kf[k]z^{-k-1}$$

which becomes, after the multiplication of $-z$ on both sides,

$$-z\frac{dF(z)}{dz} = \sum_{k=0}^{\infty} kf[k]z^{-k}$$

Its right-hand side is, by definition, the z-transform of $kf[k]$. This establishes (5.14).

EXAMPLE 5.3.2

Find the z-transform of kb^k. The z-transform of b^k is $z/(z - b)$. Thus, we have

$$\mathcal{Z}[kb^k] = -z\frac{d}{dz}\left(\frac{z}{z - b}\right) = -z \cdot \frac{(z - b) - z}{(z - b)^2} = \frac{bz}{(z - b)^2} = \frac{bz^{-1}}{(1 - bz^{-1})^2}$$

EXERCISE 5.3.3

Verify

$$\mathscr{Z}[k^2 b^k] = \frac{b(z+b)z}{(z-b)^3} = \frac{b(1+bz^{-1})z^{-1}}{(1-bz^{-1})^3}$$

∎

• **Multiplication by a^k in the time domain** Let $F(z)$ be the z-transform of $f[k]$. Then we have

$$\mathscr{Z}[a^k f[k]] = F(\frac{z}{a})$$

This follows directly from

$$\mathscr{Z}[a^k f[k]] = \sum_{k=0}^{\infty} f[k]a^k z^{-k} = \sum_{k=0}^{\infty} f[k](\frac{z}{a})^{-k} = F(\frac{z}{a})$$

EXERCISE 5.3.4

Show

$$\mathscr{Z}[(\cos k\omega_o T)f[k]] = \frac{1}{2}(F(e^{-j\omega_o T}z) + F(e^{j\omega_o T}z))$$

and

$$\mathscr{Z}[(\sin k\omega_o T)f[k]] = \frac{1}{2j}(F(e^{-j\omega_o T}z) - F(e^{j\omega_o T}z))$$

∎

Table 5.1 lists some of the often used z-transform pairs. Both the positive-and negative-power forms are listed.

• **Time delay (right shift)** Consider a sequence $f[k] := f(kT)$, as shown in Figure 5.4(a). It is not necessarily a positive-time sequence. We plot in Figure 5.4(b) $f[k-1]$, which is the shifting of $f[k]$ one sampling period to the right or the delay of $f[k]$ one sampling period. The z-transform of $f[k]$ is

$$F(z) = \sum_{k=0}^{\infty} f[k]z^{-k} = f[0] + f[1]z^{-1} + f[2]z^{-2} + \cdots \qquad (5.15)$$

$$\sin(A+B) = \sin A \cos B + \cos A \sin B$$
$$\sin(B-A) = \sin B \cos A - \cos B \sin A$$

Table 5.1 *z-Transform pairs*

$f[k]$	$F(z)$	$F^-(z^{-1})$
$\delta[k]$	1	1
$\delta[k-n]$	z^{-n}	z^{-n}
$q[k]$	$\dfrac{z}{z-1}$	$\dfrac{1}{1-z^{-1}}$
e^{-akT}	$\dfrac{z}{z-e^{-aT}}$	$\dfrac{1}{1-e^{-aT}z^{-1}}$
b^k	$\dfrac{z}{z-b}$	$\dfrac{1}{1-bz^{-1}}$
kb^k	$\dfrac{bz}{(z-b)^2}$	$\dfrac{bz^{-1}}{(1-bz^{-1})^2}$
$k^2 b^k$	$\dfrac{b(z+b)z}{(z-b)^3}$	$\dfrac{b(1+bz^{-1})z^{-1}}{(1-bz^{-1})^3}$
$\sin k\omega_0 T$	$\dfrac{(\sin \omega_0 T)z}{z^2 - 2(\cos \omega_0 T)z + 1}$	$\dfrac{(\sin \omega_0 T)z^{-1}}{1 - 2(\cos \omega_0 T)z^{-1} + z^{-2}}$
$\cos k\omega_0 T$	$\dfrac{(z - \cos \omega_0 T)z}{z^2 - 2(\cos \omega_0 T)z + 1}$	$\dfrac{1 - (\cos \omega_0 T)z^{-1}}{1 - 2(\cos \omega_0 T)z^{-1} + z^{-2}}$
$b^k \sin k\omega_0 T$	$\dfrac{b(\sin \omega_0 T)z}{z^2 - 2b(\cos \omega_0 T)z + b^2}$	$\dfrac{b(\sin \omega_0 T)z^{-1}}{1 - 2b(\cos \omega_0 T)z^{-1} + b^2 z^{-2}}$
$b^k \cos k\omega_0 T$	$\dfrac{(z - b\cos \omega_0 T)z}{z^2 - 2b(\cos \omega_0 T)z + b^2}$	$\dfrac{1 - b(\cos \omega_0 T)z^{-1}}{1 - 2b(\cos \omega_0 T)z^{-1} + b^2 z^{-2}}$

see pg. 243 for proof

$$b^{-k} q[-k-1] = \frac{bz}{1-bz} = \frac{-z}{z-\frac{1}{b}} \quad \text{if } |z| \leq \frac{1}{|b|}$$

It is defined for the positive-time part, and $f[-1]$, $f[-2]$, ... do not appear in $F(z)$. The z-transform of $f[k-1]$ is

$$\mathscr{Z}[f[k-1]] = \sum_{k=0}^{\infty} f[k-1]z^{-k} = f[-1] + f[0]z^{-1} + f[1]z^{-2} + \cdots$$

which, after substituting (5.15), becomes

$$\mathscr{Z}[f[k-1]] = f[-1] + z^{-1}F(z) \tag{5.16a}$$

This can also be obtained as follows: Delaying $f[k]$ one sampling period is equivalent to multiplying $F(z)$ by z^{-1}. It also shifts $f[-1]$ to $k = 0$, as shown in

Figure 5.4 *Shifting of time sequences*

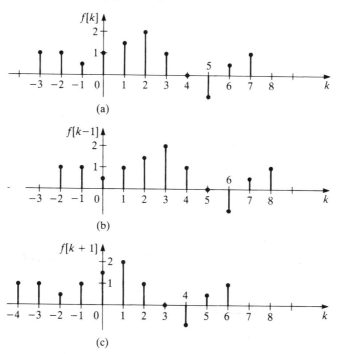

(a)

(b)

(c)

Figure 5.4(b). Thus, the z-transform of $f[k-1]$ must include $f[-1]$, as in (5.16a). Using the same argument, we have

$$\mathscr{Z}[f[k-2]] = f[-2] + f[-1]z^{-1} + z^{-2}F(z) \qquad (5.16b)$$

and, in general,

$$\mathscr{Z}[f[k-i]] = z^{-i}F(z) + \sum_{n=1}^{i} f[-n]z^{-i+n} \qquad (5.16c)$$

for any positive integer i.

- **Time advance (left shift)** Consider the sequence $f[k]$ shown in Figure 5.4(a). Figure 5.4(c) shows $f[k+1]$, which is the shifting of $f[k]$ to the left one sampling period or the advance of $f[k]$ one sampling period. Its z-transform is

$$\mathscr{Z}[f[k+1]] = \sum_{k=0}^{\infty} f[k+1]z^{-k} = f[1] + f[2]z^{-1} + f[3]z^{-2} + \cdots \qquad (5.17)$$

We see that $f[0]$ does not appear in the transform because it has been shifted to $k = -1$, as shown in Figure 5.4(c), and the z-transform is defined only for $k \geq 0$. Substituting (5.15) into (5.17) yields

$$\mathscr{Z}[f[k+1]] = z(F(z) - f[0]) \tag{5.18a}$$

This can also be obtained as follows: Advancing $f[k]$ one sampling period is equivalent to multiplying $F(z)$ by z. It also shifts $f[0]$ to $k = -1$, as shown in Figure 5.4(c). Thus, the z-transform of $f[k+1]$ should not contain $f[0]$, as in (5.18a). Using the same argument, we have

$$\mathscr{Z}[f[k+2]] = z^2(F(z) - f[0] - f[1]z^{-1}) \tag{5.18b}$$

and, in general,

$$\boxed{\mathscr{Z}[f[k+i]] = z^i\left(F(z) - \sum_{n=0}^{i-1} f[n]z^{-n}\right)} \tag{5.18c}$$

for any positive integer i.

- **Discrete convolution**   The zero-state response of every linear time-invariant discrete-time system can be described, as was discussed in Section 3.2.1, by a discrete convolution as

$$y[k] = \sum_{i=0}^{k} h[k-i]u[i] = \sum_{i=0}^{k} h[i]u[k-i] \tag{5.19}$$

where $h[k]$ is the impulse response of the system and is the output excited by the impulse sequence $\delta[k]$. If the system is causal, then $h[k] = 0$ for $k < 0$. Thus, $h[k]$ is a positive-time sequence. The output $y[k]$ excited by the input $u[k]$, for $k \geq 0$, is zero for $k < 0$ if the system is causal and initially relaxed. Thus, $u[k]$, $h[k]$, and $y[k]$ are all positive-time sequences. We use $U(z)$, $H(z)$, and $Y(z)$ to denote their z-transforms. The z-transform will transform the convolution summation into an algebraic equation. The substitution of (5.19) into the z-transform of $y[k]$ yields

$$Y(z) = \sum_{k=0}^{\infty} y[k]z^{-k} = \sum_{k=0}^{\infty}\left[\sum_{i=0}^{k} h[k-i]u[i]\right]z^{-k}$$

Because $h[k-i] = 0$ for $k < i$, the upper limit k in the inner summation can be replaced by ∞. The equation then becomes, after changing the order of summations,

$$Y(z) = \sum_{i=0}^{\infty}\left[\sum_{k=0}^{\infty} h[k-i]z^{-(k-i)}\right]u[i]z^{-i} \tag{5.20}$$

The summation inside the brackets becomes, after substituting $l := k - i$ and using $h[l] = 0$ for $l < 0$,

$$\sum_{k=0}^{\infty} h[k-i]z^{-(k-i)} = \sum_{l=-i}^{\infty} h[l]z^{-l} = \sum_{l=0}^{\infty} h[l]z^{-l} = H(z)$$

Thus, (5.20) becomes

$$Y(z) = \sum_{i=0}^{\infty} H(z)u(i)z^{-i} = H(z)\sum_{i=0}^{\infty} u(i)z^{-i}$$

or

$$Y(z) = H(z)U(z) \tag{5.21}$$

We see that the z-transform transforms the discrete convolution in (5.19) into an algebraic equation. The function $H(z)$ is called the discrete-time transfer function of the system. It is the z-transform of the impulse sequence. Thus, if a system is initially relaxed, the z-transform of the output simply equals the product of the transfer function and the z-transform of the input. Equation (5.21) is similar to (4.11) and is of fundamental importance in system analysis and design.

To conclude this section, we list some properties of the z-transform in Table 5.2. The last two properties will be developed in a later section.

Table 5.2 *Properties of the z-transform*

Property	Time sequence	z-transform
Linearity	$\alpha_1 f_1[k] + \alpha_2 f_2[k]$	$\alpha_1 F_1(z) + \alpha_2 F_2(z)$
Multiplication by k	$kf[k]$	$-z\dfrac{dF(z)}{dz}$
Multiplication by a^k	$a^k f[k]$	$F(\dfrac{z}{a})$
Time delay $(i > 0)$	$f[k-i]$	$z^{-i}F(z) + \displaystyle\sum_{n=1}^{i} f[-n]z^{-i+n}$
Time advance $(i > 0)$	$f[k+i]$	$z^{i}[F(z) - \displaystyle\sum_{n=0}^{i-1} f[n]z^{-n}]$
Convolution	$\displaystyle\sum_{i=0}^{k} h[k-i]u[i]$	$H(z)U(z)$
Final value	$\lim_{k\to\infty} f[k] = \lim_{z\to 1} (z-1)F(z)$	(If $(z-1)F(z)$ has no pole on or outside the unit circle)
Initial value	$f[0] = \lim_{z\to\infty} F(z)$	(If $F(z)$ is proper)

✓ don't do

5.4   INVERSE z-TRANSFORM—DIRECT DIVISION

Go to
5.4.1.

The process of computing $f[k]$ from its z-transform $F(z)$ is called the inverse z-transform and is denoted as

$$f[k] = \mathcal{Z}^{-1}[F(z)]$$

It can be carried out using direct division, using partial fraction expansion, or using a computer to compute the inversion formula in (5.4). The first two methods are applicable only if $F(z)$ is a rational function of z; these methods will be discussed in this chapter. The last method will be discussed in Chapter 7.

Consider the rational function

$$F(z) = \frac{-2z^4 + 5z^3 + z^2 - 6z + 3}{z^2 - z - 2} = -2z^2 + 3z + \frac{3}{z^2 - z - 2} \quad (5.22)$$

It is an improper rational function. We use direct division to compute

$$
\require{enclose}
\begin{array}{r}
-2z^2 + 3z \qquad\qquad\quad +3z^{-2} + 3z^{-3} + 9z^{-4} \\
z^2 - z - 2\,\enclose{longdiv}{-2z^4 + 5z^3 + \; z^2 - 6z + 3} \\
\underline{-2z^4 + 2z^3 + 4z^2} \\
3z^3 - 3z^2 - 6z + 3 \\
\underline{3z^3 - 3z^2 - 6z} \\
0 \quad + 0 + 3
\end{array}
$$

$$
\begin{array}{r}
3 - 3z^{-1} - 6z^{-2} \\
\underline{3z^{-1} + 6z^{-2}} \\
3z^{-1} - 3z^{-2} - 6z^{-3} \\
\underline{9z^{-2} + 6z^{-3}} \\
9z^{-2} - 9z^{-3} - 18z^{-4} \\
15z^{-3} + 18z^{-4}
\end{array}
$$

Then $F(z)$ can be expressed as

$$F(z) = -2z^2 + 3z + 0 + 0 \cdot z^{-1} + 3z^{-2} + 3z^{-3} + 9z^{-4} + \cdots$$

If we use (5.8), then the inverse z-transform of $F(z)$ is

$$
\begin{aligned}
f[k] = &-2\delta[k+2] + 3\delta[k+1] + 0 \cdot \delta[k] + 0 \cdot \delta[k-1] \\
&+ 3\delta[k-2] + 3\delta[k-3] + 9\delta[k-4] + \cdots
\end{aligned}
$$

or

$$\{f[-2], f[-1], f[0], f[1], f[2], f[3], f[4], \ldots\} = \{-2, 3, 0, 0, 3, 3, 9, \ldots\}$$

This is not a positive-time sequence because nonzero entries appear at $k = -2$ and $k = -1$. Strictly speaking, $F(z)$ is not the z-transform of $f[k]$ because the z-transform is defined for $f[k]$ with $k \geq 0$. However, we shall accept $f[k]$ to be the inverse z-transform of $F(z)$.

If $F(z)$ is proper, then we can use direct division to express it exclusively as a negative power series of z and the inverse z-transform of $F(z)$ is a positive-time sequence. For example, consider the proper rational function $3/(z^2 - z - 2)$. It is the proper part of the one in (5.22) and the preceding direct division applies. Thus, we have

$$\frac{3}{z^2 - z - 2} = 0 + 0 \cdot z^{-1} + 3z^{-2} + 3z^{-3} + 9z^{-4} + \dots$$

and its inverse z-transform is $\{f[0], f[1], f[2], f[3], \dots\} = \{0, 0, 3, 3, \dots\}$ or

$$f[k] = 3\delta[k - 2] + 3\delta[k - 3] + 9\delta[k - 4] + \dots$$

We mention that '$3/(z^2 - z - 2)$ can also be expressed, using direct division, exclusively as a positive power of z as

$$
-2 - z + z^2 \overline{\smash{\big)}\
\begin{array}{l}
-\dfrac{3}{2} + \dfrac{3}{4}z - \dfrac{9}{8}z^2 \\[2mm]
3 \\[1mm]
3 + \dfrac{3}{2}z - \dfrac{3}{2}z^2 \\[1mm]
\hline
-\dfrac{3}{2}z + \dfrac{3}{2}z^2 \\[1mm]
-\dfrac{3}{2}z - \dfrac{3}{4}z^2 + \dfrac{3}{4}z^3 \\[1mm]
\hline
\dfrac{9}{4}z^2 - \dfrac{3}{4}z^3 \\[1mm]
\dfrac{9}{4}z^2 + \dfrac{9}{8}z^3 - \dfrac{9}{8}z^4 \\[1mm]
\hline
\end{array}
}
$$

Thus, we have

$$\frac{3}{z^2 - z - 2} = -\frac{3}{2} + \frac{3}{4}z - \frac{9}{8}z^2 + \dots$$

This is a positive power series of z and denotes a negative-time sequence. Since we are studying only positive-time sequences, this sequence is not the one we are looking for.

In conclusion, every proper rational function has a positive-time sequence as its inverse z-transform. The sequence can be obtained by direct division. In the division, the polynomials must be arranged in the descending powers of z; otherwise, a negative-time sequence will be obtained.

If a rational function $F(z)$ is improper, then it is not possible to express it *exclusively* as a negative power series of z and $F(z)$ cannot be the z-transform of a positive-time sequence. Thus, an improper $F(z)$ is also called a noncausal rational function. We will study mostly proper rational functions here. The direct division method can also be applied to the negative power form of rational functions. We omit its discussion.

EXERCISE 5.4.1

Find the inverse z-transform of

$$F(z) = \frac{z^2 - 1}{z^3 - 1}$$

[**ANSWER:** 0, 1, 0, −1, 1, 0, −1, 1, 0, ..., or $\delta[k-1] - \delta[k-3] + \delta[k-4] + \ldots$.] ∎

Pole-zero excess and time delay   Consider a rational function

$$F(z) = \frac{N(z)}{D(z)} = \frac{b_m z^m + b_{m-1} z^{m-1} + \cdots + b_1 z + b_0}{a_n z^n + a_{n-1} z^{n-1} + \cdots + a_1 z + a_0} \tag{5.23}$$

where $a_n \neq 0$, $b_m \neq 0$, and the remaining coefficients may or may not be zero. We assume $N(z)$ and $D(z)$ are coprime (have no common factors). Then, as in the analog case, every root of $N(z)$ is defined as a zero and every root of $D(z)$ is defined as a pole of $F(z)$. Clearly, $F(z)$ has n poles and m zeros. Define

$$r := \text{number of poles} - \text{number of zeros} = n - m$$

It is called the *pole-zero excess* of $F(z)$. For example, the rational function

$$F(z) = \frac{3}{z^2 - z - 2} = \frac{3}{(z+1)(z-2)} \tag{5.24a}$$

has poles −1 and 2 and no zero and has a pole-zero excess of 2. Clearly, $F(z)$ is improper if $r < 0$, biproper if $r = 0$, strictly proper if $r > 0$, and proper if $r \geq 0$. If $F(z)$ is proper and if we use direct division to express it as

$$F(z) = \frac{b_m}{a_n} z^{m-n} + \cdots = \frac{b_m}{a_n} z^{-r} + \cdots$$

then its inverse z-transform is

$$f[0] = 0, \quad f[1] = 0, \quad \cdots, \quad f[r-1] = 0, \quad f[r] = \frac{b_m}{a_n} \neq 0$$

In other words, if $F(z)$ has pole-zero excess r, then its time sequence will start to appear at $k = r$. Thus, there is a delay of r sampling period. This simple property is worth remembering.

In the discrete-time case, rational functions in both positive-power and negative-power forms are widely used. For example, the positive-power rational function in (5.24a) can be expressed in negative-power form as

$$F(z) = \frac{3z^{-2}}{1 - z^{-1} - 2z^{-2}} =: F^-(z^{-1}) = \frac{3z^{-2}}{(1 + z^{-1})(1 - 2z^{-1})} \qquad \textbf{(5.24b)}$$

$$= \frac{3(z^{-1} - 0)(z^{-1} - 0)}{-2(z^{-1} + 1)(z^{-1} - \frac{1}{2})} =: \frac{N^-(z^{-1})}{D^-(z^{-1})}$$

As $F(z)$, we may define the roots of $N^-(z^{-1})$ as zeros and the roots of $D^-(z^{-1})$ as poles. Then, $F^-(z^{-1})$ has two zeros at 0 and two poles at -1 and $1/2$. They are different from those of $F(z)$. Thus, in the discrete-time case, it is important to specify $F(z)$ or $F^-(z^{-1})$ in discussing poles and zeros. This will be discussed further in Section 5.7.1.

The negative-power rational function $F^-(z^{-1})$ has two poles and two zeros; thus, its pole-zero excess is 0, which is different from the one of $F(z)$. Thus, the pole-zero excess of $F^-(z^{-1})$ cannot be used to specify the time delay of the inverse z-transform of $F^-(z^{-1})$.

Now we develop an alternative form of $F^-(z^{-1})$. Multiplying z^{-n} to the numerator and denominator of $F(z)$ in (5.23) yields

$$F^-(z^{-1}) = F(z) = \frac{z^{m-n}(b_m + b_{m-1}z^{-1} + \cdots + b_0 z^{-m})}{a_n + a_{n-1}z^{-1} + \cdots + a_1 z^{-n+1} + a_0 z^{-n}}$$

which can be written as

$$F^-(z^{-1}) = \frac{z^{-r}(\beta_0 + \beta_1 z^{-1} + \cdots + \beta_q z^{-q})}{\alpha_0 + \alpha_1 z^{-1} + \cdots + \alpha_p z^{-p}} \qquad \textbf{(5.25)}$$

with $r := n - m$, $\beta_0 = b_m \neq 0$, $\alpha_0 = a_n \neq 0$, $p > 0$, and $q > 0$. Other than a_n and b_m, some or all of b_i and a_j can be zero, therefore p in (5.25) can be larger than, equal to, or smaller than q. Independent of p and q, the inverse z-transform of $F^-(z^{-1})$ is a positive-time sequence if and only if $r \geq 0$ and a nonzero entry starts to appear at $k = r$. Thus (5.25) provides some explicit information and is widely used to represent negative-power transfer functions.

5.4.1   Partial-Fraction Expansion and Table Lookup

This section discusses a different method of computing inverse z-transforms. This method consists of two steps: partial fraction expansion and a table lookup. The expansion depends heavily on the table that is used, We will use Table 5.1, which consists of positive-power and negative-power z-transform pairs. Let us begin with an example that illustrates the procedure, which closely follows the one presented in Section 4.4. We first use the negative-power form and then the positive-power form.

EXAMPLE 5.4.1

$$F(z) = \frac{3}{z^2 - z - 2}$$

Find the inverse z-transform of the rational function in (5.24).

Method 1 Using negative-power z-transform pairs, we express $F^-(z^{-1})$ as a sum of the terms in Table 5.1 as

$$F^-(z^{-1}) = \frac{3z^{-2}}{(1 + z^{-1})(1 - 2z^{-1})} \tag{5.26}$$

$$= k_0 + k_1 \frac{1}{1 - (-1)z^{-1}} + k_2 \frac{1}{1 - 2z^{-1}}$$

If $z^{-1} = \infty$, then we have

$$k_0 = F^-(z^{-1})\Big|_{z^{-1} = \infty} = \frac{3}{-2} = -1.5$$

Multiplying $(1 + z^{-1})$ and (5.26) yields

$$F^-(z^{-1})(1 + z^{-1}) = k_0(1 + z^{-1}) + k_1 + k_2 \frac{1 + z^{-1}}{1 - 2z^{-1}}$$

which implies

$$k_1 = F^-(z^{-1})(1 + z^{-1})\Big|_{z^{-1} = -1} = \frac{3(-1)^2}{1 + 2} = \frac{3}{3} = 1$$

Similarly, we have

$$k_2 = F^-(z^{-1})(1 - 2z^{-1})\Big|_{z^{-1} = 0.5} = \frac{3(0.5)^2}{1 + 0.5} = \frac{0.75}{1.5} = 0.5$$

Thus, $F^-(z^{-1})$ can be expanded as

$$F^-(z^{-1}) = -1.5 + \frac{1}{1 - (-1)z^{-1}} + 0.5 \frac{1}{1 - 2z^{-1}}$$

and its inverse z-transform is, using Table 5.1,

$$f[k] = -1.5\delta[k] + (-1)^k + 0.5 \cdot 2^k$$

for $k = 0, 1, 2, \ldots$. For example, we have

$$
\begin{aligned}
f[0] &= -1.5 + (-1)^0 + 0.5 \cdot 2^0 = -1.5 + 1 + 0.5 = 0 \\
f[1] &= 0 + (-1)^1 + 0.5 \cdot 2^1 = -1 + 0.5 \cdot 2 = 0 \\
f[2] &= 0 + (-1)^2 + 0.5 \cdot 2^2 = 1 + 0.5 \cdot 4 = 3
\end{aligned}
$$

and so forth.

Method 2 Using positive-power z-transform pairs, we express $F(z)$ as

$$F(z) = \frac{3}{(z+1)(z-2)} = k_0 + k_1 \frac{z}{z+1} + k_2 \frac{z}{z-2} \qquad (5.27)$$

If $z = 0$, then (5.27) becomes

$$-\frac{3}{-2} = k_0 - k_1 \cdot 0 + k_2 \cdot 0$$

Thus, $k_0 = -1.5$. Multiplying $(z+1)/z$ and (5.27) yields

$$F(z)\frac{z+1}{z} = k_0 \frac{z+1}{z} + k_1 + k_2 \frac{z+1}{z-2}$$

which implies

$$k_1 = \left. F(z)\frac{z+1}{z} \right|_{z=-1} = \left. \frac{3}{(z-2)z} \right|_{z=-1} = \frac{3}{(-1-2)(-1)} = 1$$

Similarly, we have

$$k_2 = \left. F(z)\frac{z-2}{z} \right|_{z=2} = \left. \frac{3}{(z+1)z} \right|_{z=2} = \frac{3}{(2+1)(2)} = 0.5$$

Thus, $F(z)$ can be expanded as

$$F(z) = \frac{3}{z^2 - z - 2} = 1.5 + \frac{z}{z-(-1)} + 0.5 \cdot \frac{z}{z-2}$$

and its inverse z-transform is

$$f[k] = -1.5\delta[k] + (-1)^k + (0.5)2^k$$

for $k = 0, 1, 2, \ldots$.

$\cancel{\cancel{\times}}$ **Method 3** Expanding $F(z)/z$, instead of $F(z)$, as a sum of terms $1/(z - b)$, such as

$$\frac{F(z)}{z} = \sum \frac{k_i}{z - b_i} \tag{5.28}$$

allows the procedure in the Laplace transform part to be applied without any modification. Once k_i are computed, we then multiply (5.28) by z to yield

$$F(z) = \sum k_i \frac{z}{z - b_i}$$

and the inverse z-transform of $F(z)$ can then be obtained using Table 5.1. For this example, we expand

$$\frac{F(z)}{z} = \frac{3}{z(z + 1)(z - 2)} = \frac{k_1}{z} + \frac{k_2}{z + 1} + \frac{k_3}{z - 2}$$

with

$$k_1 = \left. \frac{3}{(z + 1)(z - 2)} \right|_{z=0} = -1.5$$

$$k_2 = \left. \frac{3}{z(z - 2)} \right|_{z=-1} = \frac{3}{(-1)(-3)} = 1$$

$$k_3 = \left. \frac{3}{z(z + 1)} \right|_{z=2} = \frac{3}{2 \cdot 3} = 0.5$$

We then multiply z to the equation to yield

$$F(z) = k_1 + \frac{k_2 z}{z + 1} + \frac{k_3 z}{z - 2} = -1.5 + \frac{z}{z + 1} + 0.5 \frac{z}{z - 2}$$

Thus, the inverse z-transform of $F(z)$ is

$$f[k] = -1.5\delta[k] + (-1)^k + 0.5 \cdot 2^k$$

for $k = 0, 1, 2, \ldots$.

All three methods yield the same result, and any of them can be used to compute inverse z-transforms.

EXERCISE 5.4.2

Use the preceding three methods to compute the inverse z-transform of

$$F(z) = \frac{6z}{z^2 - 9}$$

[**ANSWER:** $f[k] = 3^k - (-3)^k$, for $k = 0, 1, 2, \ldots$] ∎

We give one more example that employs the various ideas discussed in Section 4.4. We discuss only the positive-power form.

EXAMPLE 5.4.2   *COMPLEX POLES*

Consider the rational function

$$F(z) = \frac{z - 1}{(z + 2)(z^2 + 2z + 4)}$$

Its denominator has the linear factor $(z + 2)$ and the quadratic factor $(z^2 + 2z + 4)$. Two methods will be used to compute its inverse z-transform. In the first method, the quadratic factor will be further factored as linear factors with complex poles. Thus, the computation will involve complex numbers. In the second method, the quadratic factor will be retained and complex numbers will be avoided in the computation.

Method I: Using linear factors   We factor the quadratic factor as

$$z^2 + 2z + 4 = (z + 1)^2 + 3 = (z + 1 - j\sqrt{3})(z + 1 + j\sqrt{3}) = (z - re^{j\theta})(z - re^{-j\theta})$$

where $r = \sqrt{1 + 3} = 2$ and $\theta = \tan^{-1}\sqrt{3}/(-1) = 120^o = 2.09$ radians. Note that θ is not equal to -60°, as can be readily verified by drawing a graph such as the one in Figure 4.2. In the multiplication and division of complex numbers, it is easier to use the polar form; in addition and subtraction, it is simpler to use the rectangular form. In the following, both forms will be used. We expand

$$\frac{F(z)}{z} = \frac{z - 1}{z(z + 2)(z - 2e^{j2.09})(z - 2e^{-j2.09})}$$

$$= \frac{k_1}{z} + \frac{k_2}{z + 2} + \frac{k_3}{z - 2e^{j2.09}} + \frac{k_4}{z - 2e^{-j2.09}} \qquad \textbf{(5.29)}$$

where

$$k_1 = \left.\frac{z-1}{(z+2)(z+2z+4)}\right|_{z=0} = \frac{-1}{8}$$

$$k_2 = \left.\frac{z-1}{z(z^2+2z+4)}\right|_{z=-2} = \frac{-3}{(-2)\cdot4} = \frac{3}{8}$$

$$k_3 = \left.\frac{z-1}{z(z+2)(z+1+j\sqrt{3})}\right|_{z=-1+j\sqrt{3}} = \frac{-2+j\sqrt{3}}{(-1+j\sqrt{3})(1+j\sqrt{3})(j\,2\sqrt{3})}$$

$$= \frac{-2+j\sqrt{3}}{(-1-3)(j\,2\sqrt{3})} = \frac{-\sqrt{3}-2j}{8\sqrt{3}} = 0.19e^{-j2.29}$$

and

$$k_4 = k_3^* = 0.19e^{j2.29}$$

We then write (5.29) as

$$F(z) = k_1 + k_2\frac{z}{z+2} + k_3\frac{z}{z-2e^{j2.09}} + k_3^*\frac{z}{z-2e^{-j2.09}}$$

Therefore, its inverse is, for $k = 0, 1, 2, \ldots$,

$$f[k] = k_1\delta[k] + k_2(-2)^k + k_3(2e^{j2.09})^k + k_3^*(2e^{-j2.09})^k$$

$$= \frac{-1}{8}\delta[k] + \frac{3}{8}(-2)^k + (0.19)\cdot2^ke^{j(2.09k-2.29)} + (0.19)2^ke^{-j(2.09k-2.29)}$$

$$= \frac{-1}{8}\delta[k] + \frac{3}{8}(-2)^k + (0.19)\cdot2^k[e^{j(2.09k-2.29)} + e^{-j(2.09k-2.29)}]$$

$$= \frac{-1}{8}\delta[k] + \frac{3}{8}(-2)^k + (0.38)\cdot2^k\cos(2.09k-2.29) \tag{5.30}$$

This completes the computation. Because of the existence of complex poles, this method involves complex numbers. If we do not factor the quadratic factor, then complex numbers can be avoided in the computation. This is discussed in the next method.

Method 2: Using linear and quadratic factors   The inverse z-transform can be obtained from Table 5.1; therefore, we shall first express the quadratic factor in one of the forms given in Table 5.1. We write

$$z^2 + 2z + 4 = z^2 - 2b(\cos\omega_0 T)z + b^2$$

Therefore, we have $b = 2$ and $\cos\omega_0 T = -1/2$, which implies $\omega_0 T = 2.09$ radians and $\sin\omega_0 T = \sqrt{3}/2$. Now we expand

$$F(z) = \frac{z-1}{(z+2)(z^2+2z+4)}$$

$$F(z) = \bar{k}_0 + \bar{k}_1\frac{z}{z+2} + \bar{k}_2\frac{b(\sin \omega_0 T)z}{z^2+2z+4} + \bar{k}_3\frac{(z - b\cos \omega_0 T)z}{z^2+2z+4}$$

$$= \bar{k}_0 + \bar{k}_1\frac{z}{z+2} + \bar{k}_2\frac{\sqrt{3}\,z}{z^2+2z+4} + \bar{k}_3\frac{(z+1)z}{z^2+2z+4} \tag{5.31}$$

Every term assumes one of the forms in Table 5.1. If $z = 0$, then

$$\bar{k}_0 = \frac{-1}{2\cdot 4} = \frac{-1}{8}$$

Multiplying $(z+2)/z$ and then setting $z = -2$ yield

$$\bar{k}_1 = F(z)\frac{z+2}{z}\bigg|_{z=-2} = \frac{z-1}{z(z^2+2z+4)}\bigg|_{z=-2} = \frac{-2-1}{(-2)(4-4+4)} = \frac{3}{8}$$

If $z = -1$, then (5.31) becomes

$$\frac{-2}{(-1+2)(1-2+4)} = \bar{k}_0 + \bar{k}_1\frac{-1}{-1+2} + \bar{k}_2\frac{\sqrt{3}(-1)}{1-2+4} + \bar{k}_3\cdot 0$$

which implies

$$\bar{k}_2 = -\sqrt{3}\left[\frac{-2}{3} + \frac{1}{2}\right] = \frac{\sqrt{3}}{6} = \frac{1}{2\sqrt{3}}$$

Similarly, if we set $z = 1$, then $\bar{k}_3$ can be solved as

$$\bar{k}_3 = \frac{-1}{4}$$

Therefore, we have

$$F(z) = \frac{-1}{8} + \frac{3}{8}\frac{z}{z+2} + \frac{1}{2\sqrt{3}}\frac{b(\sin \omega_0 T)z}{z^2+2z+4} - \frac{1}{4}\frac{(z - b\cos \omega_0 T)z}{z^2+2z+4}$$

and, using Table 5.1,

$$f[k] = \frac{-1}{8}\delta[k] + \frac{3}{8}(-2)^k + \frac{1}{2\sqrt{3}}2^k \sin 2.09k - \frac{1}{4}2^k \cos 2.09k \tag{5.32}$$

for $k = 0, 1, 2, \ldots$. This completes the computation. No complex numbers arise in this computation.

EXERCISE 5.4.3

Verify that (5.30) equals (5.32). ∎

EXERCISE 5.4.4

Find the inverse z-transform of

$$F(z) = \frac{z + 2}{(z - 1)(z^2 + 2z + 4)}$$

[**ANSWER:** $-0.5\delta[k] + \dfrac{3}{7} - \dfrac{5\sqrt{3}}{42} 2^k \sin 2.09k + \dfrac{1}{14} 2^k \cos 2.09k$,

for $k = 0, 1, 2, \ldots$.] ∎

*5.4.2   Rational functions with repeated poles at z = 0

We now discuss a special case of inverse z-transforms which has no analog counterpart. Consider a proper rational function $F(z)$ with a repeated pole at $z = 0$, such as

$$F(z) = \frac{1}{z^n} \cdot F_1(z) \tag{5.33}$$

Although the procedure discussed in the preceding section could be used to compute its inverse z-transform, there is a simpler method. We first disregard the poles at $z = 0$ and compute the inverse z-transform of $F_1(z)$. Although $F(z)$ is proper, $F_1(z)$ may not be proper. Therefore, the inverse z-transform of $F_1(z)$ may not be a positive-time sequence. Suppose we have

$$\mathcal{Z}^{-1}[F_1(z)] = f_1[k] \quad \text{for } k = -n, -n + 1, \ldots, 0, 1, 2, 3, \ldots$$

Because z^{-1} is a unit-time delay element, z^{-n} will introduce a delay of n sampling periods. Thus, we have

$$\mathcal{Z}^{-1}[F(z)] = f_1[k - n] \tag{5.34}$$

for $k \geq 0$. This can be illustrated by an example.

EXAMPLE 5.4.3

Consider

$$F(z) = \frac{2z^2 + z + 3}{z^3(z - 0.5)}$$

It has a pole at $z = 0.5$ and a repeated pole at $z = 0$ with multiplicity 3. Define

$$F_1(z) := \frac{2z^2 + z + 3}{z - 0.5}$$

We expand it as

$$F_1(z) = \frac{2z^2 + z + 3}{z - 0.5} = k_{-1}z + k_0 + k_1\frac{z}{z - 0.5}$$

$$= \frac{k_{-1}z^2 - (0.5k_{-1} + k_0 + k_1)z - 0.5k_0}{z - 0.5}$$

Equating the coefficients of like powers of z yields

$$k_{-1} = 2 \qquad -0.5k_0 = 3 \qquad \text{and} \qquad 0.5k_{-1} + k_0 + k_1 = -1$$

Thus, we have $k_{-1} = 2$, $k_0 = -6$, $k_1 = 4$, and

$$F_1(z) = 2z - 6 + 4\frac{z}{z - 0.5}$$

The inverse z-transform of $2z - 6$ is, as discussed in (5.8),

$$\mathscr{Z}^{-1}[2z - 6] = 2\delta[k + 1] - 6\delta[k] \tag{5.35}$$

for all integers k or at least for $k \geq -1$. The inverse z-transform of $z/(z - 0.5)$ is

$$\mathscr{Z}^{-1}[\frac{z}{z - 0.5}] = \begin{cases} (0.5)^k & \text{for } k = 0, 1, 2, \ldots \\ 0 & \text{for } k < 0 \end{cases}$$

$$= 0.5^k q[k] \qquad \text{for all integers } k$$

where $q[k]$ is the unit step sequence. The first form cannot be combined with (5.35) because of the difference in the ranges of k. Thus, we must use the second form and the inverse z-transform of $F_1(z)$ is

$$f_1[k] := \mathscr{Z}^{-1}[F_1(z)] = 2\delta[k + 1] - 6\delta[k] + 4(0.5)^k q[k]$$

for all integers k or at least for $k \geq -1$. Note that $f_1[k]$ is not a positive-time sequence. Because the inverse z-transform of $F(z)$ is simply the delay of three sampling periods of $f_1[k]$, we replace k by $k - 3$ in $f_1[k]$ to yield

$$f[k] = \mathscr{Z}^{-1}[\frac{2z^2 + z + 3}{z^3(z - 0.5)}]$$

$$= f_1[k - 3] = 2\delta[k - 2] - 6\delta[k - 3] + 4(0.5)^{k-3} q[k - 3]$$

for all integers k or, simply, for $k \geq 0$. Note that this is a positive-time sequence. Because $F(z)$ has a pole-zero excess of 2, we have $f[0] = f[1] = 0$.

EXAMPLE 5.4.4

To find the inverse z-transform of $1/(z - 2)$, we write

$$F(z) = \frac{1}{z - 2} = \frac{1}{z} \cdot \frac{z}{z - 2} = \frac{1}{z} F_1(z)$$

The inverse z-transform of $F_1(z)$ is

$$f_1[k] = 2^k q[k] \qquad \text{for all } k$$

Thus, the inverse z-transform of $1/(z - 2)$ is

$$f[k] = 2^{k-1} q[k-1], \qquad \text{for all } k \qquad \text{(5.36a)}$$

Delay due to $\frac{1}{z}$

or $f[0] = 0$, $f[1] = 1$, $f[2] = 2$, $f[3] = 2^2 = 4$, ... This can also be written as

$$f[k] = \begin{cases} 0 & \text{for } k = 0 \\ 2^{k-1} & \text{for } k \geq 1 \end{cases} \qquad \text{(5.36b)}$$

An alternative way of computing the inverse z-transform of $1/(z - 2)$ is to express it as

$$\frac{1}{z - 2} = k_0 + k_1 \frac{z}{z - 2}$$

with $k_0 = -0.5$ (obtained by setting $z = 0$) and $k_1 = -k_0 = 0.5$ (obtained by setting $z = \infty$). Thus, the inverse z-transform of $1/(z - 2)$ is, using Table 5.1,

$$f[k] = -0.5\delta[k] + 0.5 \cdot 2^k \qquad \text{(5.36c)}$$

for $k = 0, 1, 2, \dots$.

EXERCISE 5.4.5

Verify that the three expressions in (5.36) all denote the same sequence.

EXERCISE 5.4.6

Find the inverse z-transforms of the following

(a) $\dfrac{2}{z + 0.1}$

(b) $\dfrac{1}{z^2(z - 0.5)}$

(c) $\dfrac{z^3 - 1}{z^5(z - 1)}$

[**ANSWERS:** (a) $2(-0.1)^{k-1} q[k - 1]$; (b) $(0.5)^{k-3} q[k - 3]$; (c) $\delta[k - 3] + \delta[k - 4]$
$+ \delta[k - 5] = q[k - 3] - q[k - 6]$.] $\blacksquare$

The inverse z-transform or the time sequence of every pole other than $z = 0$
has infinitely many nonzero entries. Poles at the origin or $z = 0$ introduce either
delay or impulse sequences that are identically zeros for almost all k. Thus,
poles at the origin are called *trivial poles*. Trivial poles in digital systems have
no counterpart in analog systems.

Poles at $z = 0$ introduce delay; zeros at $z = 0$ introduce advance. For example, if

$$f[k] = \mathscr{Z}^{-1}\left[\frac{z^n N(z)}{D(z)}\right] \qquad f_1[k] = \mathscr{Z}^{-1}\left[\frac{N(z)}{D(z)}\right]$$

then $f[k] = f_1[k + n]$. However, partial fraction expansion of $z^n N(z)/D(z)$ is
no more complex than that of $N(z)/D(z)$; therefore, there is no reason to compute $f_1[k]$ first, as in (5.33).

5.5   LTIL DIFFERENCE EQUATIONS

In this section, we shall apply the z-transform to solve linear difference equations with constant real coefficients, or LTIL difference equations. Because
every LTIL discrete-time system can be described by such an equation, solving
difference equations is an important part of system analysis. As was discussed
earlier, every difference equation can be solved by direct substitution without
resorting to any method. Such a solution, however, will not be in closed form,
and it would be difficult to abstract from that solution the general property of
the system. The use of the z-transform, however, will provide such information.
Therefore, the z-transform is an important tool in discrete-time system analysis.

Consider a system described by the second-order delayed-form difference
equation

$$2y[k] + 3y[k - 1] + y[k - 2] = u[k] + u[k - 1] - u[k - 2] \qquad \textbf{(5.37a)}$$

or its equivalent advanced-form

$$2y[k+2] + 3y[k+1] + y[k] = u[k+2] + u[k+1] - u[k] \qquad \textbf{(5.37b)}$$

The output of the system is excited by initial conditions $y[-1]$ and $y[-2]$ and an input $u[k]$ applied at $k = 0$. The procedure of using the z-transform to compute the response of (5.37) is identical to the analog case. It consists of the following three steps: (1) Apply the z-transform to (5.37) to yield an algebraic equation. (2) Substitute the initial conditions and the z-transform of the input into the algebraic equation and compute, using algebraic manipulation, the output $Y(z)$. This output is in the z-transform domain. (3) Compute the inverse z-transform of $Y(z)$ to yield the time response. Although this procedure is simple and straightforward, some differences will arise when using the two forms in (5.37). This is discussed in the following examples.

EXAMPLE 5.5.1

(**Delayed form**) Find the response of (5.37a) excited by $y[-1] = 2$, $y[-2] = -1$, and a unit step sequence $u[k] = 1$ for $k \geq 0$ and $u[k] = 0$ for $k < 0$. Applying (5.16) to (5.37a) and using $u[-1] = u[-2] = 0$ yield

$$2y[k] + 3y[k-1] + y[k-2] = u[k] + u[k-1] - u[k-2]$$

$$2Y(z) + 3(y[-1] + z^{-1}Y(z)) + (y[-2] + y[-1]z^{-1} + z^{-2}Y(z))$$
$$= U(z) + z^{-1}U(z) - z^{-2}U(z)$$

This can be grouped as

$$(2 + 3z^{-1} + z^{-2})Y(z) = -3y[-1] - y[-2] - y[-1]z^{-1} + (1 + z^{-1} - z^{-2})U(z)$$

which implies

$$Y(z) = \frac{-3y[-1] - y[-2] - y[-1]z^{-1}}{2 + 3z^{-1} + z^{-2}} + \frac{1 + z^{-1} - z^{-2}}{2 + 3z^{-1} + z^{-2}}U(z) \qquad \textbf{(5.38)}$$

$$\underbrace{\qquad\qquad\qquad\qquad}_{\text{zero-input response}} \qquad \underbrace{\qquad\qquad\qquad}_{\text{zero-state response}}$$

We see that the output consists of two parts: One part is excited by the initial conditions, and the other is excited by the input. The former is the zero-input response, and the latter is the zero-state response. This is a general property of every linear system, as was discussed in Chapter 2. If $y[-2] = -1$, $y[-1] = 2$, and $U(z) = z/(z-1) = 1/(1 - z^{-1})$, then (5.38) becomes

$$Y(z) = \frac{-5 - 2z^{-1}}{2 + 3z^{-1} + z^{-2}} + \frac{1 + z^{-1} - z^{-2}}{2 + 3z^{-1} + z^{-2}} \cdot \frac{1}{1 - z^{-1}}$$

which can be simplified as

$$Y(z) = \frac{-4 + 4z^{-1} + z^{-2}}{2(1 + 0.5z^{-1})(1 + z^{-1})(1 - z^{-1})} \tag{5.39}$$

To find the time response of $Y(z)$, we expand it as

$$Y(z) = k_0 + \frac{k_1}{1 + 0.5z^{-1}} + \frac{k_2}{1 + z^{-1}} + \frac{k_3}{1 - z^{-1}}$$

with

$$k_0 = Y(z)\Big|_{z^{-1} = \infty} = 0$$

$$k_1 = Y(z)(1 + 0.5z^{-1})\Big|_{z^{-1} = -1/0.5} = \frac{-4 + 4z^{-1} + z^{-2}}{2(1 + z^{-1})(1 - z^{-1})}\Big|_{z^{-1} = -2} = \frac{-4 - 8 + 4}{2(-1) \cdot 3} = \frac{4}{3}$$

$$k_2 = Y(z)(1 + z^{-1})\Big|_{z^{-1} = -1} = \frac{-4 + 4z^{-1} + z^{-2}}{2(1 + 0.5z^{-1})(1 - z^{-1})}\Big|_{z^{-1} = -1} = \frac{-4 - 4 + 1}{2(0.5) \cdot 2} = \frac{-7}{2}$$

and

$$k_3 = Y(z)(1 - z^{-1})\Big|_{z^{-1} = 1} = \frac{-4 + 4z^{-1} + z^{-2}}{2(1 + 0.5z^{-1})(1 + z^{-1})}\Big|_{z^{-1} = 1} = \frac{-4 + 4 + 1}{2(1.5) \cdot 2} = \frac{1}{6}$$

Thus, the inverse z-transform of $Y(z)$ is, using Table 5.1,

$$y[k] = k_1(-0.5)^k + k_2(-1)^k + k_3(1)^k = \frac{4}{3}(-0.5)^k - \frac{7}{2}(-1)^k + \frac{1}{6} \tag{5.40}$$

for $k \geq 0$. This is the output of the system excited by a unit step sequence and initial condition $y[-1] = 2$ and $y[-2] = -1$.

EXAMPLE 5.5.2

(**Advanced form**) We can recompute the response of the system in Example 5.5.1 using the advanced-form difference equation. The application of the z-transform to (5.37b) yields, using (5.18),

$$2z^2(Y(z) - y[0] - y[1]z^{-1}) + 3z(Y(z) - y[0]) + Y(z) \tag{5.41}$$
$$= z^2(U(z) - u[0] - u[1]z^{-1}) + z(U(z) - u[0]) - U(z)$$

This can be arranged as

$$Y(z) = \frac{2z^2 y[0] + 2zy[1] + 3zy[0] - z^2 u[0] - zu[1] - zu[0]}{2z^2 + 3z + 1} \quad \textbf{(5.42)}$$

$$+ \frac{z^2 + z - 1}{2z^2 + 3z + 1} U(z)$$

The first term after the equal sign contains $y[0]$, $y[1]$, $u[0]$, and $u[1]$, and it is not clear whether the term represents the zero-input response. In order to compute (5.42), we must first compute $y[0]$ and $y[1]$. From (5.37b) and using $u[k] = 0$ for $k < 0$, we have, by setting $k = -2$ and $k = -1$,

$k = -2:$ $2y[0] - u[0] = -3y[-1] - y[-2]$ **(5.43a)**

and

$k = -1:$ $2y[1] + 3y[0] - u[1] - u[0] = -y[-1]$ **(5.43b)**

Substituting these into (5.42) yields

$$Y(z) = \underbrace{\frac{z^2(-3y[-1] - y[-2]) + z(-y[-1])}{2z^2 + 3z + 1}}_{\text{zero-input response}} + \underbrace{\frac{z^2 + z - 1}{2z^2 + 3z + 1} U(z)}_{\text{zero-state response}} \quad \textbf{(5.44)}$$

It turns out that the first term after the equal sign in (5.42) depends only on the initial conditions and represents the zero-input response. Equation (5.44) is identical to (5.38) except that it is in positive-power form and it can likewise be decomposed as the zero-state response and the zero-input response. If $y[-1] = 2$, $y[-2] = -1$, and $U(z) = z/(z - 1)$, then (5.44) becomes

$$Y(z) = \frac{-5z^2 - 2z}{2z^2 + 3z + 1} + \frac{z(z^2 + z - 1)}{(2z^2 + 3z + 1)(z - 1)} = \frac{z(-4z^2 + 4z + 1)}{2(z + 0.5)(z + 1)(z - 1)}$$

We expand $Y(z)/z$ as

$$\frac{Y(z)}{z} = \frac{-4z^2 + 4z + 1}{2(z + 0.5)(z + 1)(z - 1)} = \frac{k_1}{z + 0.5} + \frac{k_2}{z + 1} + \frac{k_3}{z - 1} \quad \textbf{(5.45)}$$

with

$$k_1 = \left. \frac{-4z^2 + 4z + 1}{2(z + 1)(z - 1)} \right|_{z=-0.5} = \frac{-1 - 2 + 1}{2(0.5)(-1.5)} = \frac{-2}{-1.5} = \frac{4}{3}$$

$$k_2 = \left. \frac{-4z^2 + 4z + 1}{2(z + 0.5)(z - 1)} \right|_{z=-1} = \frac{-4 - 4 + 1}{2(-0.5)(-2)} = \frac{-7}{2}$$

and

$$k_3 = \left.\frac{-4z^2 + 4z + 1}{2(z + 0.5)(z + 1)}\right|_{z=1} = \frac{-4 + 4 + 1}{2(1.5)(2)} = \frac{1}{6}$$

Substituting these into (5.45) and then multiplying the whole equation by z yield

$$Y(z) = \frac{4}{3}\frac{z}{z + 0.5} - \frac{7}{2}\frac{z}{z + 1} + \frac{1}{6}\frac{z}{z - 1}$$

Thus, the inverse z-transform of $Y(z)$ is

$$y[k] = \frac{4}{3}(-0.5)^k - \frac{7}{2}(-1)^k + \frac{1}{6}(1)^k$$

for $k \geq 0$. This is the same as the one obtained in Example 5.5.1 using the delayed-form difference equation.

The procedure for solving LTIL difference equations using the z-transform can be summarized as follows: First, we apply the z-transform to change a difference equation into an algebraic equation. Next, we substitute the initial conditions and the input into the algebraic equation and compute the output using algebraic manipulation. This output is in the z-transform domain. Its inverse z-transform yields the time response.

This procedure can be applied to both delayed-form and advanced-form difference equations. However, for the advanced form, we must use the original difference equation to compute $y[0]$ and $y[1]$, as in (5.43). No such computation is needed for the delayed form. Therefore, it is simpler to use delayed-form difference equations to compute responses.

EXERCISE 5.5.1

Find the response of

$$y[k] + y[k - 1] - 2y[k - 2] = u[k - 1] + 2u[k - 2]$$

excited by $y[-1] = -0.5$, $y[-2] = 0.25$, and $u[k] = 1$, for $k = 0, 1, 2, \ldots$.

[**ANSWER:** $y[k] = k + (-2)^k$, $k = 0, 1, 2, \ldots$.]

EXERCISE 5.5.2

Find the time responses of

$$y[k+1] - 2y[k] = u[k] \quad \text{and} \quad y[k+1] - 2y[k] = u[k+1]$$

excited by $y[-1] = 1$ and $u[k] = 1$, for $k = 0, 1, 2, \ldots$.

[**ANSWERS:** $y[k] = 3 \cdot 2^k - 1$ and $y[k] = 4 \cdot 2^k - 1$, for $k = 0, 1, 2, \ldots$.]   ■

5.6   ZERO-INPUT RESPONSE— CHARACTERISTIC POLYNOMIAL

The response of LTIL discrete-time systems can be decomposed, as was discussed in Chapter 2 and demonstrated in the preceding section, as the zero-input response and the zero-state response. In this section, we will discuss some general properties of the zero-input response.

Consider the system described by the difference equation (5.37). If the input $u[k]$ is identically zero, then the response excited by initial conditions $y[-1]$ and $y[-2]$ can be expressed, as computed in (5.38) and (5.44), as

$$
Y_{zi}(z) = \frac{-3y[-1] - y[-2] - y[-1]z^{-1}}{2 + 3z^{-1} + z^{-2}}
$$
$$
= \frac{(-3y[-1] - y[-2])z^2 - y[-1]z}{2z^2 + 3z + 1} \tag{5.46}
$$

where the subscript zi denotes zero input. We call

$$D(z) := 2z^2 + 3z + 1 = 2(z + 0.5)(z + 1)$$

the denominator of $Y_{zi}(z)$, the *characteristic polynomial*, and we call the roots of $D(z)$ the *modes* of the system. Thus, the modes of this system are -0.5 and -1. The zero-input response of (5.37) excited by *any* initial condition can always be expressed as

$$
Y_{zi}(z) = \frac{(-3y[-1] - y[-2])z^2 - y[-1]z}{2(z + 0.5)(z + 1)} = \frac{k_1 z}{z + 0.5} + \frac{k_2 z}{z + 1} \tag{5.47}
$$

where k_1 and k_2 depend on the initial conditions. Therefore, the zero-input response is of the form

$$y_{zi}[k] = k_1(-0.5)^k + k_2(-1)^k$$

for $k = 0, 1, 2, \ldots$. In other words, the zero-input response of (5.37) is always a linear combination of the two time sequences $(-0.5)^k$ and $(-1)^k$ that are the inverse z-transform of $z/(z + 0.5)$ and $z/(z + 1)$. Thus, the form of the zero-input response excited by any initial condition is completely determined by the modes of the system.

Similarly, we may also define

$$D^-(z^{-1}) := 2 + 3z^{-1} + z^{-2} = (2 + z^{-1})(1 + z^{-1})$$

as the characteristic polynomial and its roots -2 and -1 as the modes of the system. The modes defined here will be the inverses of the modes defined for $D(z)$. Unless stated otherwise, characteristic polynomials and modes will be those defined for $D(z)$.

EXAMPLE 5.6.1

We show that, for some initial conditions, k_1 or k_2 in (5.47) may be zero. If k_i is zero, the corresponding mode is not excited. If $y[-1] = a$ and $y[-2] = -2a$ for any real a, then (5.47) becomes

$$Y_{zi}(z) = \frac{-az^2 - az}{2(z + 0.5)(z + 1)} = \frac{-az(z + 1)}{2(z + 0.5)(z + 1)} = \frac{-az}{2(z + 0.5)}$$

which implies

$$y_{zi}[k] = -0.5a\,(-0.5)^k \quad = -\tfrac{1}{2}\,a\,(-0.5)^k$$

for $k \geq 0$. Thus, for the set of initial conditions, the mode -1 or $(-1)^k$ is not excited. Similarly, it can be shown that if $y[-1] = -y[-2]$, then the mode -0.5 or $(-0.5)^k$ will not be excited. For other $y[-1]$ and $y[-2]$, the two modes will always be excited.

To extend the preceding discussion to the general case, consider the nth-order delayed-form LTIL difference equation

$$a_n y[k] + a_{n-1} y[k-1] + \cdots + a_1 y[k-n+1] + a_0 y[k-n]$$
$$= b_m u[k+m-n] + b_{m-1} u[k+m-n-1] \tag{5.48a}$$
$$+ \cdots + b_1 u[k-n+1] + b_0 u[k-n]$$

or its advanced form

$$a_n y[k+n] + a_{n-1} y[k+n-1] + \cdots + a_1 y[k+1] + a_0 y[k]$$
$$= b_m u[k+m] + b_{m-1} u[k+m-1] + \cdots + b_1 u[k+1] + b_0 u[k] \tag{5.48b}$$

Define

$$D^-(p^{-1}) := a_n + a_{n-1}p^{-1} + \cdots + a_1 p^{-n+1} + a_0 p^{-n} \tag{5.49a}$$

$$N^-(p^{-1}) := b_m p^{m-n} + b_{m-1}p^{m-n-1} + \cdots + b_1 p^{1-n} + b_0 p^{-n} \tag{5.49b}$$

$$D(p) := a_n p^n + a_{n-1}p^{n-1} + \cdots + a_1 p + a_0 \tag{5.50a}$$

and

$$N(p) := b_m p^m + a_{m-1}p^{m-1} + \cdots + b_1 p + b_0 \tag{5.50b}$$

where the variable p^{-1} is the unit-time delay operator defined as

$$p^{-1}y[k] = y[k-1], \qquad p^{-2}y[k] = y[k-2], \tag{5.51a}$$
$$p^{-3}y[k] = y[k-3], \ldots$$

and the variable p is the unit-time advance operator defined as

$$py[k] = y[k+1], \qquad p^2 y[k] = y[k+2], \tag{5.51b}$$
$$p^3 y[k] = y[k+3], \ldots$$

Using these notations, (5.48) can be written as

$$D^-(p^{-1})y[k] = N^-(p^{-1})u[k] \tag{5.52a}$$

and

$$D(p)y[k] = N(p)u[k] \tag{5.52b}$$

In the study of the zero-input response, we assume $u[k] \equiv 0$. In this case, (5.52) reduces to

$$D^-(p^{-1})y[k] = 0 \quad \text{and} \quad D(p)y[k] = 0 \tag{5.53}$$

 These are called the homogeneous equations. Their solutions are excited exclusively by initial conditions. The application of the z-transform to the first equation in (5.53) yields

$$a_n Y(z) + a_{n-1}(z^{-1}Y(z) + y[-1]) + a_{n-2}(z^{-2}Y(z) + y[-1]z^{-1} + y[-2]) + \cdots$$
$$+ a_1(z^{-n+1}Y(z) + y[-1]z^{-n+2} + \cdots + y[-n+1]) + a_0(z^{-n}Y(z)$$
$$+ y[-1]z^{-n+1} + \cdots y[-n+1]z^{-1} + y[-n]) = 0$$

which can be written as

$$Y(z) =$$

$$\frac{(a_{n-1} + a_{n-2}z^{-1} + \cdots + a_0 z^{-n+1})y[-1] + \cdots + (a_1 + a_0 z^{-1})y[-n+1] + a_0 y[-n]}{a_n + a_{n-1}z^{-1} + \cdots + a_1 z^{-n+1} + a_0 z^{-n}}$$

$$=: \frac{I^{-}(z^{-1})}{D^{-}(z^{-1})} \tag{5.54a}$$

where $I^{-}(z^{-1})$ is a polynomial of z^{-1} depending on initial conditions and $D^{-}(z^{-1})$ is the polynomial defined in (5.49a) with p^{-1} replaced by z^{-1}. If we apply the z-transform to the second equation in (5.53), then the output will depend on $y[0]$, $y[1]$, $\ldots$, which can be expressed in terms of $y[-1]$, $y[-2]$, $\ldots$, by using $D(p)y[k] = 0$. Thus, eventually we have

$$Y(z) = \frac{I(z)}{D(z)} \tag{5.54b}$$

where $I(z)$ is a polynomial of z depending on initial conditions and $D(z)$ is the polynomial defined in (5.49b) with p replaced by z. We call $D(z)$ the characteristic polynomial of (5.52b) because it governs the *free, unforced,* or *natural* response of (5.52). The roots of $D(z)$ are called the *natural frequencies* or *modes*. We use the modes exclusively in this text. For example, if

$$D(z) = (z - 2)(z + 1)^3(z + 2 - j3)(z + 2 + j3)$$

then its modes are 2, -1, -1, -1, $-2 + j3$, and $-2 - j3$. The root 2 and the complex roots $-2 \pm j3$ are simple modes, and the root -1 is a repeated mode with multiplicity 3. We express the complex modes in polar form as $-2 \pm j3 = 3.6e^{\pm j123.7°} = 3.6e^{\pm j2.16\text{rad}}$. Then, for any initial conditions, the zero-input response is always of the form

$$y_{zi}[k] = k_1 2^k + k_2(3.6e^{j2.16})^k + k_3(3.6e^{-j2.16})^k \tag{5.55a}$$
$$+ c_1(-1)^k + c_2 k(-1)^k + c_3 k^2(-1)^k$$

for some constants k_i and c_i. If all coefficients of $D(z)$ are real, then $k_2 = k_3^*$ and the second and third terms after the equal sign can be combined into real terms as

$$y_{zi}[k] = k_1 2^k + \overline{k}_2(3.6)^k \sin(2.16k + \overline{k}_3) + c_1(-1)^k \tag{5.55b}$$
$$+ c_2 k(-1)^k + c_3 k^2(-1)^k$$

or

$$y_{zi}[k] = k_1 2^k + \hat{k}_2(3.6)^k \sin 2.16k + \hat{k}_3(3.6)^k \cos 2.16k \tag{5.55c}$$
$$+ c_1(-1)^k + c_2 k(-1)^k + c_3 k^2(-1)^k$$

with all real coefficients. This is similar to (4.55) in the analog case. Note that even though all zero-input responses can be expressed as in (5.55), their time responses can be quite different for different initial conditions.

There is one situation that arises in the discrete-time case does not arise in the continuous-time case. The mode at $s = 0$ with multiplicity n in the continuous-time case will generate the time response t^{n-1}. The mode at $z = 0$ with multiplicity n in the discrete-time case will introduce a delay, as was discussed in (5.33), or generate $\delta[k - i]$, $i = 0, 1, \ldots, n$, which are identically zero for $k > n$. Therefore, the mode at $z = 0$ can be considered as a *trivial* mode.[2] As $D(z)$, we may call $D^-(z^{-1})$ the characterestic polynomial in negative power form. Its roots may be called the modes. These modes, however, are the reciprocal of those defined for $D(z)$. To avoid confusion, we use $D(z)$ and its roots exclusively unless stated otherwise.

EXERCISE 5.6.1

Find the modes and the form of the zero-input response of

$$D(p)y[k] = N(p)u[k]$$

where

$$D(p) = (p - 2)^2(p^2 + 4p + 8) \quad \text{and} \quad N(p) = 3p^2 - 10$$

[**ANSWERS:** 2, 2, $-2 + j2$, and $-2 - j2$; and $y_{zi}[k] = k_1(2)^k + k_2 k(2)^k + k_3(-2 + j2)^k + k_4(-2 - j2)^k = k_1(2)^k + k^2 k(2)^k + k_3(-2.83)^k \sin(2.36 + k_4).$] ∎

5.7   ZERO-STATE RESPONSE—TRANSFER FUNCTION

Consider the system described by (5.37) or

$$2y[k] + 3y[k - 1] + y[k - 2] = u[k] + u[k - 1] - u[k - 2]$$

and its equivalent form

$$2y[k + 2] + 3y[k + 1] + y[k] = u[k + 2] + u[k + 1] - u[k]$$

If all initial conditions are zero, the application of the z-transform yields, as was derived in (5.38) and (5.44),

---

[2] The mode at $z = 0$ in the discrete-time case corresponds to the mode at $s = -\infty$ in the continuous-time case. This infinite mode generates an impulse or its derivatives at $t = 0$. Therefore, the mode $s = -\infty$ can be considered as a trivial mode in the continuous-time case.

$$Y(z) = \frac{1 + z^{-1} - z^{-2}}{2 + 3z^{-1} + z^{-2}} U(z) = \frac{z^2 + z - 1}{2z^2 + 3z + 1} U(z)$$

We call

$$H^-(z^{-1}) := \frac{1 + z^{-1} + z^{-2}}{2 + 3z^{-1} + z^{-2}} \qquad \text{(5.56a)}$$

the negative-power-form transfer function of the system and

$$H(z) = \frac{z^2 + z - 1}{2z^2 + 3z + 1} \qquad \text{(5.56b)}$$

the positive-power-form transfer function or, simply, the transfer function of the system. To differentiate them from the analog case, they are also called the *discrete-time, discrete,* or *digital* transfer functions. They are the ratio of the z-transforms of the output and input when all initial conditions are zero, or

$$H^-(z^{-1}) = H(z) = \left.\frac{Y(z)}{U(z)}\right|_{\text{Initial conditions}=0} \qquad \text{(5.57)}$$

This definition is consistent with the definition given in (5.21).

The transfer function of a system can be obtained from

1. its impulse response or the response excited by the Kronecker delta sequence $\delta[k]$;
2. its zero-state response excited by any input, in particular, step or sinusoidal sequence;
3. its difference equation description.

The first method follows simply from the development of (5.21). It is also a special case of the second method because, if the input is $\delta[k]$, then $U(z) = 1$ and $Y(z) = H(z)$. An example illustrates the second method.

EXAMPLE 5.7.1

Consider an LTIL system. Its zero-state response excited by the input $u[k] = \sin 2k$, for $k = 0, 1, 2, \ldots$ is measured as

$$y[k] = 2e^{-2k} + \sin 2k - 2 \cos 2k \qquad \text{for } k \geq 0$$

In order to find the transfer function of the system, we compute, using Table 5.1,

$$U(z) = \mathcal{Z}[\sin 2k] = \frac{\sin 2}{z^2 - 2(\cos 2)z + 1}$$

and

$$Y(z) = \mathscr{Z}[y[k]] = 2 \cdot \frac{z}{z - e^{-2}} + \frac{\sin 2}{z^2 - 2(\cos 2)z + 1} - \frac{2(z - \cos 2)z}{z^2 - 2(\cos 2)z + 1}$$

Thus, the transfer function of the system is, using (5.57) and after some length manipulation,

$$H(z) = \frac{Y(z)}{U(z)} = \frac{(-2 \cos 2 + \sin 2 + 2e^{-2})z + [2 - (\sin 2 + 2 \cos 2)e^{-2}]}{\sin 2(z - e^{-2})}$$

The substitution of the values $e^{-2} = 0.135$, $\sin 2 = \sin 115° = \sin 65° = 0.9$, and $\cos 2 = \cos 115° = -\cos 65° = -0.42$ into the equation yields

$$H(z) = \frac{2.233z + 2.213}{z - 0.135}$$

This is the transfer function of the system. The inverse z-transform of $H(z)$ yields the impulse response of the system.

How do we obtain transfer functions from difference equations? Consider a system described by

$$D^-(p^{-1})y[k] = N^-(p^{-1})u[k] \tag{5.58a}$$

where $D^-(p^{-1})$ and $N^-(p^{-1})$ are defined as in (5.49). Recall that $p^{-i}y[k] = y[k-i]$. If all initial conditions are zero or, equivalently, $y[k] = 0$ for $k < 0$, then (5.16) reduces to

$$\mathscr{Z}[p^{-i}y[k]] = \mathscr{Z}[y[k-i]] = z^{-i}Y(z)$$

Thus, if all initial conditions are zero, the application of the z-transform to (5.58a) yields

$$D^-(z^{-1})Y(z) = N^-(z^{-1})U(z)$$

and the transfer function of the system is

$$H^-(z^{-1}) = \frac{N^-(z^{-1})}{D^-(z^{-1})} \tag{5.58b}$$

EXAMPLE 5.7.2

Consider the basic block diagram shown in Figure 3.7. Its difference equation description is, as derived in (3.25),

$$y[k] - y[k-1] + 2y[k-2] - 3y[k-3] = u[k]$$

If all initial conditions are zero, the application of the z-transform yields

$$Y(z) - z^{-1}Y(z) + 2z^{-2}Y(z) - 3z^{-3}Y(z) = U(z)$$

or

$$(1 - z^{-1} + 2z^{-2} - 3z^{-3})Y(z) = U(z)$$

Thus, the transfer function of the diagram from $u[k]$ to $y[k]$ is

$$H^-(z^{-1}) := \frac{Y(z)}{U(z)} = \frac{1}{1 - z^{-1} + 2z^{-2} - 3z^{-3}} = \frac{z^3}{z^3 - z^2 + 2z - 3}$$

Now consider a system described by the advanced-form difference equation

$$D(p)y[k] = N(p)u[k] \tag{5.59a}$$

where $D(p)$ and $N(p)$ are defined as in (5.50). Even though all initial conditions are zero, we do not have

$$\mathscr{Z}[p^i y[k]] = \mathscr{Z}[y[k+i]] = z^i Y(z)$$

What we have is

$$\mathscr{Z}[p^i y[k]] = \mathscr{Z}[y[k+i]] = z^i[Y(z) - \sum_{n=0}^{i-1} y[n]z^{-n}]$$

However, if all initial conditions are zero, by applying the z-transform to (5.59a) and after some manipulation, we will eventually obtain

$$D(z)Y(z) = N(z)U(z)$$

Thus, the transfer function of the system is

$$H(z) = \frac{Y(z)}{U(z)} = \frac{N(z)}{D(z)} \tag{5.59b}$$

In conclusion, transfer functions can be easily obtained from delayed-form or advanced-form difference equations.

EXAMPLE 5.7.3

Consider the basic block diagram shown in Figure 3.7. Its difference equation description is, as derived in (3.26),

$$y[k + 3] - y[k + 2] + 2y[k + 1] - 3y[k] = u[k + 3]$$

Applying the z-transform and carrying out some manipulation yield

$$z^3 Y(z) - z^2 Y(z) + 2zY(z) - 3Y(z) = z^3 U(z)$$

or

$$(z^3 - z^2 + 2z - 3)Y(z) = z^3 U(z)$$

Thus, the transfer function of the block diagram is

$$H(z) = \frac{Y(z)}{U(z)} = \frac{z^3}{z^3 - z^2 + 2z - 3}$$

EXERCISE 5.7.1

Find the transfer functions of the systems described by the following

   (a) $3y[k] - 4y[k - 1] + 2y[k - 2] = u[k] + 3u[k - 2]$

   (b) $y[k] - 4y[k - 2] = u[k - 1] - 3u[k - 2]$

[**ANSWERS:** (a) $(1 + 3z^{-2})/(3 - 4z^{-1} + 2z^{-2})$, (b) $(z^{-1} - 3z^{-2})/(1 - 4z^{-2})$.]   ■

EXERCISE 5.7.2

Show that if all initial conditions are zero, the z-transform of

$$y[k + 2] - 4y[k] = u[k + 1] - 3u[k]$$

is

$$z^2 Y(z) - 4Y(z) = zU(z) - 3U(z)$$

What is its transfer function?

[**ANSWER:** $(z-3)/(z^2-4)$]   ■

The transfer functions obtained in the preceding examples and exercises are all rational functions of z. This is so because the systems are lumped. If a system is linear and time-invariant but is distributed, the equation $Y(z) = H(z)U(z)$ still holds. However, the transfer function $H(z)$ is an irrational function of z, such as $\ln(1 - z^{-1})$ or $\sin z$. In the remainder of this section, we study only rational transfer functions.

5.7.1   Poles and zeros of proper rational transfer functions

The zero-state response of a system is governed by its transfer function. We now discuss the general form of zero-state responses. But, first, the concepts of poles and zeros should be introduced. Consider the discrete-time rational transfer function

$$H(z) = \frac{N(z)}{D(z)}$$

where $N(z)$ and $D(z)$ are polynomials with real coefficients, and $\deg N(z) \le \deg D(z)$. If $N(z)$ and $D(z)$ have no common factors, then the roots of $D(z)$ and $N(z)$ are, respectively, the poles and zeros of $H(z)$. For example, the system with transfer function

$$H(z) = \frac{3z + 12}{2z^3 - 3.5z - 1.5} = \frac{3(z + 4)}{2(z + 0.5)(z + 1)(z - 1.5)} \tag{5.60a}$$

has three poles at -0.5, -1, and 1.5 and one zero at -4. As in the continuous-time case, the pole and zero can be defined as follows: A finite real or complex number λ is a pole of $H(z)$ if $|H(\lambda)| = \infty$. It is a zero if $|H(\lambda)| = 0$. Poles and zeros can also be defined for $H^-(z^{-1})$. For example, if

$$H^-(z^{-1}) = H(z) = \frac{3z + 12}{2z^3 - 3.5z - 1.5} = \frac{3z^{-2} + 12z^{-3}}{2 - 3.5z^{-2} - 1.5z^{-3}}$$

$$= \frac{3z^{-2}(1 + 4z^{-1})}{2(1 + 0.5z^{-1})(1 + z^{-1})(1 - 1.5z^{-1})} \tag{5.60b}$$

$$= \frac{12z^{-2}(z^{-1} + 0.25)}{-1.5(z^{-1} + 2)(z^{-1} + 1)(z^{-1} - 0.6667)}$$

then $H^-(z^{-1})$ has three poles at -2, -1, and $1/1.5 = 0.6667$ and three zeros at 0, 0, and -0.25. These are different from the poles and zeros of $H(z)$.

In order to relate the poles and zeros of $H(z)$ with those of $H^-(z^{-1})$, we modify the definitions of poles and zeros as follows: A finite or infinite number λ is a pole (zero) of $H(z)$ if $|H(\lambda)| = \infty$ ($H(\lambda) = 0$). If we use this definition, then $H(z)$ in (5.60a) has a repeated zero with multiplicity 2 at $z = \infty$ because $H(z)$ can be approximated as $H(z) = 3/2z^2$ for $|z|$ large and $H(\infty) = 0$. If we consider both finite and infinite poles and zeros, then

$$\text{number of poles of } H(z) = \text{number of zeros of } H(z)$$
$$= \text{number of poles of } H^-(z^{-1}) = \text{number of zeros of } H^-(z^{-1})$$

and

$$\boxed{\text{pole (zero) of } H(z) = \frac{1}{\text{pole (zero) of } H^-(z^{-1})}}$$

Infinite poles and zeros are of no use other than establishing the preceding two properties. Thus, we study only finite poles and finite zeros.

EXERCISE 5.7.3

Find finite and infinite poles and zeros of

$$\frac{z}{(z - 0.1)(z + 2)} = \frac{z^{-1}}{(1 - 0.1z^{-1})(1 + 2z^{-1})}$$

Are their numbers all the same?

[**ANSWERS:** Poles of $H(z)$: 0.1, -2; zeros of $H(z)$: 0, ∞; poles of $H^-(z^{-1})$: 10, -0.5; zeros of $H^-(z^{-1})$: 0, ∞; yes.]   ∎

Now we will study the general property of the zero-state response. The output of a system with transfer function $H(z)$ excited by an input $u[k]$ can be computed as follows: First we compute the z-transform of $u[k]$. We then multiply $H(z)$ and $U(z)$ to yield $Y(z)$. The inverse z-transform of $Y(z)$ yields the zero-state response. This can be illustrated by an example.

EXAMPLE 5.7.4

Find the unit step response of the system described by (5.56) (that is, the response excited by the unit step sequence $u[k] = 1$, for $k = 0, 1, 2, \dots$). The transfer function of the system is $H(z) = (z^2 + z - 1)/(2z^2 + 3z + 1)$. The z-transform of $u[k]$ is $z/(z - 1)$. Thus, we have

$$Y(z) = H(z)U(z) = \frac{z^2 + z - 1}{2(z + 0.5)(z + 1)} \cdot \frac{z}{z - 1}$$

To find its time response, we compute

$$\frac{Y(z)}{z} = \frac{z^2 + z - 1}{2(z + 0.5)(z + 1)(z - 1)} = \frac{k_1}{z + 0.5} + \frac{k_2}{z + 1} + \frac{k_3}{z - 1}$$

with

$$k_1 = \frac{z^2 + z - 1}{2(z + 1)(z - 1)}\bigg|_{z=-0.5} = \frac{0.25 - 0.5 - 1}{2(0.5)(-1.5)} = \frac{-1.25}{-1.5} = \frac{5}{6}$$

$$k_2 = \frac{z^2 + z - 1}{2(z + 0.5)(z - 1)}\bigg|_{z=-1} = \frac{1 - 1 - 1}{2(-0.5)(-2)} = \frac{-1}{2}$$

and

$$k_3 = \frac{z^2 + z - 1}{2(z + 0.5)(z + 1)}\bigg|_{z=1} = \frac{1 + 1 - 1}{2(1.5)(2)} = \frac{1}{6}$$

Thus, we have

$$Y(z) = \frac{k_1 z}{z + 0.5} + \frac{k_2 z}{z + 1} + \frac{k_3 z}{z - 1}$$

and

$$y[k] = \underbrace{\frac{5}{6}(-0.5)^k - \frac{1}{2}(-1)^k}_{\substack{\text{Due to poles} \\ \text{of } H(z)}} + \underbrace{\frac{1}{6}1^k}_{\substack{\text{Due to pole} \\ \text{of } U(z)}} \tag{5.61}$$

for $k \geq 0$. This is the zero-state response of the system.

This example reveals a very important property of the zero-state response. We see from (5.61) that the response consists of three terms. Two are the inverse z-transform of $z/(z + 0.5)$ and $z/(z + 1)$, which are due to the poles of the system. The remaining term is due to the step input. In other words, the zero-state response consists of two parts. One part is due to the poles of the system, and the other is due to the applied input. This is, in fact, a general property of the zero-state response. For example, consider

$$H(z) = \frac{(z + 10)(z + 2)(z - 1)^2}{(z - 0.1)(z - 2)^2(z + 2 - j2)(z + 2 + j2)}$$

It has simple poles at 0.1 and $-2 \pm j2 = 2.83e^{\pm j135°} = 2.83e^{\pm j2.36\text{rad}}$ and a repeated pole with multiplicity 2 at 2. The zero-state response of $H(z)$ excited by any input $U(z)$ is of the form

$$y[k] = k_1(0.1)^k + k_2(2.83e^{j2.36})^k + k_3(2.83e^{-j2.36})^k + c_1 2^k + c_2 k(2)^k$$
$$+ \text{(terms due to the poles of } U(z)) \tag{5.62a}$$

or

$$y[k] = k_1(0.1)^k + \bar{k}_2(2.83)^k \cos(2.36k + \bar{k}_3) + c_1 2^k + c_2 k(2)^k$$
$$+ \text{(terms due to the poles of } U(z)) \tag{5.62b}$$

If $U(z)$ and $H(z)$ have no poles in common, such as $U(z) = z/(z - 0.5)(z + 0.5)$, then

$$\text{terms due to the poles of } U(z) = k_4(0.5)^k + k_5(-0.5)^k \tag{5.63a}$$

If $U(z)$ and $H(z)$ have poles in common, such as $U(z) = z/(z - 0.1)(z - 2)$, then

$$\text{terms due to the poles of } U(z) = k_4 k(0.1)^k + c_3 k^2(2)^k \tag{5.63b}$$

The forms in (5.62) and (5.63) are determined by the poles of $H(z)$ and $U(z)$ and are independent of the zeros of $H(z)$ and $U(z)$. The zeros, however, do affect k_i and, consequently, the response of the system. This is illustrated in the next example.

EXAMPLE 5.7.5

Consider the system in Example 5.7.4. If $U(z) = (z + 1)/(z - 1)$, then the zero-state response is

$$Y(z) = \frac{z^2 + z - 1}{2(z + 0.5)(z + 1)} \cdot \frac{z + 1}{z - 1} = \frac{z^2 + z - 1}{2(z + 0.5)(z - 1)}$$

To find its time response, we expand

$$\frac{Y(z)}{z} = \frac{z^2 + z - 1}{2z(z + 0.5)(z - 1)} = \frac{1}{z} - \frac{5}{6(z + 0.5)} - \frac{1}{3(z - 1)}$$

Thus, we have

$$Y(z) = 1 - \frac{5z}{6(z + 0.5)} - \frac{z}{3(z - 1)}$$

which implies, for $k \geq 0$,

$$y[k] = \delta[k] - \frac{5}{6}(-0.5)^k - \frac{1}{3}(1)^k$$

We see that the pole -1 of $H(z)$ is canceled by a zero of $U(z)$ and is not excited. Similarly, we can show that the input $U(z) = (z + 0.5)/(z - 2)$ will not excite the pole -0.5 and the input $U(z) = (z + 0.5)(z + 1)/z(z + 3)$ will not excite either pole.

EXAMPLE 5.7.6

Consider the following systems

$$H_1(z) = \frac{0.551}{(z + 0.9)(z - 0.8 + j0.5)(z - 0.8 - j0.5)}$$
$$= \frac{0.551}{z^3 - 0.7z^2 - 0.55z + 0.801} \tag{5.64a}$$

$$H_2(z) = \frac{z - 1}{z^3 - 0.7z^2 - 0.55z + 0.801} \tag{5.64b}$$

$$H_3(z) = \frac{5.51(z - 0.9)}{z^3 - 0.7z^2 - 0.55z + 0.801} \tag{5.64c}$$

$$H_4(z) = \frac{-5.51(z - 1.1)}{z^3 - 0.7z^2 - 0.55z + 0.801} \tag{5.64d}$$

and

$$H_5(z) = \frac{5.51(z^2 - 1.9z + 1)}{z^3 - 0.7z^2 - 0.55z + 0.801} \tag{5.64e}$$

We express the complex poles as $0.8 \pm j0.5 = 0.94e^{\pm j0.56\text{rad}}$. The unit step responses of the systems (outputs excited by $U(z) = z/(z - 1)$) are all of the form

$$y_i[k] = k_1(-0.9)^k + k_2(0.94)^k \sin(0.56k + k_3) + k_4(1)^k \tag{5.65}$$

for $k \geq 0$. These responses are plotted in Figure 5.5. The second response approaches zero as k approaches infinity because of the pole-zero cancellation of $(z - 1)$ and the pole of the input is not excited. All other responses approach

Figure 5.5   *Unit step responses of various systems with the same set of poles*

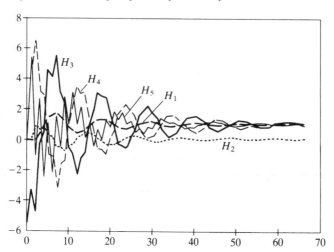

1 as k approaches infinity; this is achieved by choosing $H_i(z)$ so that $H_i(1) = 1$ (see Example 5.7.7). Although all responses can be expressed as in (5.65), they are quite different. Thus, zeros of transfer functions also affect responses of systems.

To conclude this section, we mention that the pole at $z = 0$, like the mode at $z = 0$, may introduce only a delay or an impulse sequence. Thus, the pole at $z = 0$ can be considered to be a trivial pole.

EXERCISE 5.7.4

What are the general forms of the zero-state responses of the systems with the following transfer functions?

$$H_1(z) = \frac{1}{(z + 0.4)^2(z - 2)} \quad \text{and} \quad H_2(z) = \frac{(z + 2 + j2)(z + 2 - j2)}{(z + 0.4)^2(z - 2)}$$

[**ANSWERS:** Both are $k_1(-0.4)^k + k_2k(-0.4)^k + k_3 2^k +$ (terms due to the poles of $U(z)$).]   ∎

5.7.2   Time responses of modes and poles

The response of LTIL difference equations can be decomposed as the zero-input response and the zero-state response. The form of the zero-input response is determined by the modes, and the form of the zero-state response is determined by the poles. This section discusses time responses of poles and modes and their behavior as $k \rightarrow \infty$.

Poles can be real or complex, simple or repeated. They are often plotted on the complex plane or z-plane, as shown in Figure 5.6. Unlike the s-plane, which is divided into the right and left half-planes and the imaginary axis, we shall divide the z-plane into three parts: the unit circle, the region outside the unit circle, and the region inside the unit circle. This division is expected in view of the mapping discussed in Figure 5.3. If a pole r, real or complex, is on the unit circle, then its magnitude equals 1 or $|r| = 1$. If the pole is inside the unit circle, then $|r| < 1$; if it is outside, then $|r| > 1$. Poles and zeros are usually plotted on the z-plane using small crosses and small circles.

The time sequence of a simple pole r is r^k. If the pole r is repeated with multiplicity 2, then its time response is kr^k. If it is repeated with multiplicity m, then its time response is of the form $k^{m-1}r^k$. It is clear that if the pole, simple or repeated, is outside the unit circle, or $|r| > 1$, then the response approaches infinity as $k \rightarrow \infty$. On the other hand, if the pole, simple or repeated, is inside the unit circle, or $|r| < 1$, then the response approaches 0 as $k \rightarrow \infty$. This is obvious for a simple r. If r is repeated with multiplicity 2, then the response is kr^k. As $k \rightarrow \infty$, it is the product of an infinity and a zero; therefore, some computation is needed to find its value. We use l'Hopital's rule to compute

$$\lim_{k \to \infty} kr^k = \lim_{k \to \infty} \frac{k}{r^{-k}} = \lim_{k \to \infty} \frac{-1}{r^{-k}\ln r} = 0$$

where we have used $d(r^{-k})/dk = r^{-k}(-\ln r)$ and, for $|r| < 1$, $r^{-k} \rightarrow \infty$ as $k \rightarrow \infty$. Thus, the time sequence kr^k approaches zero as $k \rightarrow \infty$. Similarly, we can show, for $|r| < 1$,

Figure 5.6 *Time response of poles and modes*

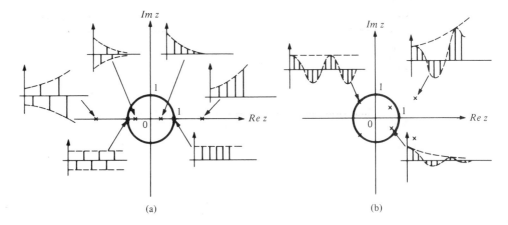

(a)   (b)

$$k^n r^k \to 0 \qquad as \ k \to \infty$$

for $n = 1, 2, 3, \ldots$. In conclusion, the time responses of poles, simple or repeated, that lie inside the unit circle approach zero as $k \to \infty$. Some of the responses are plotted in Figure 5.6(a).

If a pole is on the unit circle, or $|r| = 1$, the situations for simple and repeated poles are different. The time response of a simple pole on the unit circle is either a constant or a sinusoidal sequence; it will approach neither zero nor infinity. If the pole is repeated, then its response will approach infinity either monotonically or oscillatorily. This situation is similar to poles on the imaginary axis of the s-plane in the continuous-time case.

In the continuous-time case, if complex conjugate poles are expressed in the rectangular form as $\sigma \pm j\omega$, then the imaginary part ω determines the frequency of oscillation and the real part σ governs the envelope of the oscillation. In the discrete-time case, instead of using the rectangular form, we shall express complex conjugate poles in polar form as $re^{\pm j\theta}$. Then, from Table 5.1, we can see that the response will be of the form

$$r^k \cos k\theta \qquad or \qquad r^k \sin k\theta$$

or their linear combinations. Thus, the phase θ determines the frequency of oscillation and the amplitude r governs the envelope of oscillation. Because $|\theta| = |\omega_0 T| \le \pi$, the highest frequency of oscillation is limited to the range $(-\pi/T, \pi/T]$, where T is the sampling period. This is consistent with the discussion in Section 1.6. We plot in Figure 5.6(b) the responses of three pairs of complex-conjugate poles.

The preceding discussion is applicable to modes without any modification and is summarized in Table 5.3. From this table, we have the following important conclusions:

1. The time response of a pole (mode), simple or repeated, approaches zero as $k \to \infty$ if, and only if, the pole (mode) lies inside the unit circle or its magnitude is smaller than 1.

2. The time response of a pole (mode) approaches a nonzero constant as $k \to \infty$ if, and only if, the pole (mode) is simple and located at $z = 1$.

Table 5.3 *Time responses of poles and modes as $k \to \infty$*

Poles or modes	Simple ($n = 1$)	Repeated ($n = 2, 3, \ldots$)
Inside the unit circle	0	0
Outside the unit circle	$\pm\infty$	$\pm\infty$
$1/(z - 1)^n$	A constant	∞
$1/[(z - e^{j\theta})(z - e^{-j\theta})]^n$	A sustained oscillation	$\pm\infty$

From these two statements, we can further conclude that the time response or the inverse z transform of $F(z)$ approaches 0 as $k \to \infty$ if and only if all poles of $F(z)$ lie inside the unit circle. The time response of $F(z)$ approaches a nonzero constant if and only if $F(z)$ has a simple pole at $z = 1$ and the remaining poles of $F(z)$ all lie inside the unit circle.

EXAMPLE 5.7.7

Consider a system described by

$$y[k + 3] - 1.6y[k + 2] - 0.76y[k + 1] - 0.08y[k] = u[k + 2] - 4u[k]$$

Its characteristic polynomial is

$$D(z) = z^3 - 1.6z^2 - 0.76z - 0.08 = (z - 2)(z + 0.2)^2$$

Thus, the system has modes 2, -0.2, and -0.2 and its zero-input response has the form

$$y_{zi}[k] = k_1(2)^k + k_2(-0.2)^k + k_3 k(-0.2)^k$$

Because the mode 2 is outside the unit circle, its response 2^k increases with k without bound. Thus, if $k_1 \neq 0$, then the zero-input response of the system approaches infinity as $k \to \infty$.

The transfer function of the system is

$$H(z) = \frac{z^2 - 4}{z^3 - 1.6z^2 - 0.76z - 0.08} = \frac{(z + 2)(z - 2)}{(z - 2)(z + 0.2)^2} = \frac{z + 2}{(z + 0.2)^2}$$

Thus, the system has a repeated pole at -0.2 with multiplicity 2. The unit step response of the system is

$$Y(z) = H(z)U(z) = \frac{z + 2}{(z + 0.2)^2} \cdot \frac{z}{z - 1}$$

$Y(z)$ has two poles at -0.2, which lie inside the unit circle, and a simple pole at 1. Thus, $y[k]$ approaches a nonzero constant as $k \to \infty$. This constant can be computed as follows: We expand $Y(z)$ as

$$Y(z) = k_1 \frac{z}{z + 0.2} + k_2 \frac{z}{(z + 0.2)^2} + k_3 \frac{z}{z - 1}$$

The responses of the first two terms approach zero as $k \to \infty$. Thus, we have

$$y[k] \to k_3(1)^k = k_3 \qquad \text{as } k \to \infty$$

The constant k_3 equals

$$k_3 = Y(z)\frac{z-1}{z}\Big|_{z=1} = H(z)U(z)\frac{z-1}{z}\Big|_{z=1} = H[1]$$

$$= \frac{z+2}{(z+0.2)^2}\Big|_{z=1} = \frac{3}{(1.2)^2} = 2.08$$

Thus, the unit step response of the system approaches 2.08 as $k \to \infty$. This is essentially the final-value theorem that will be discussed in Section 5.9.

EXERCISE 5.7.5

What are the unit step response and the zero-input response excited by any initial conditions, as $k \to \infty$, of the system described by

$$(p-0.5)(p+0.8)(p+0.2+j0.5)(p+0.2-j0.5)y[k] = (p-1)(p+2)u[k]$$

[**ANSWERS:** $0, 0.$]   ■

To conclude this section, note that, although the z-transform is a very powerful tool in the study of linear time-invariant lumped discrete-time systems, it is rarely used in the study of linear time-varying systems. The reason is the same as in the continuous-time case (see Section 4.8). If a system is linear time invariant but is distributed, the equation $Y(z) = H(z)U(z)$ still holds. However, $H(z)$ is not a rational function of z and its manipulation becomes complicated. Therefore, again, the z-transform is rarely used in the study of LTI distributed systems.

5.8   TRANSFER-FUNCTION REPRESENTATION— COMPLETE CHARACTERIZATION

Every LTIL system can be described by a convolution, a difference equation, or a transfer function. Of these three descriptions, the transfer function is the one most often used in analysis and design. For example, if we want to compute the response of a system excited by an input $u[k]$, then the output is simply the product of the transfer function and the z-transform of $u[k]$ or

$$Y(z) = H(z)U(z)$$

The inverse z-transform of $Y(z)$ yields the response of the system. Thus, we often use the transfer function to represent a system, as shown in Figure 5.7(a). In Figure 5.7(a), the input and output are expressed in the time domain whereas

Figure 5.7 *Transfer-function representation*

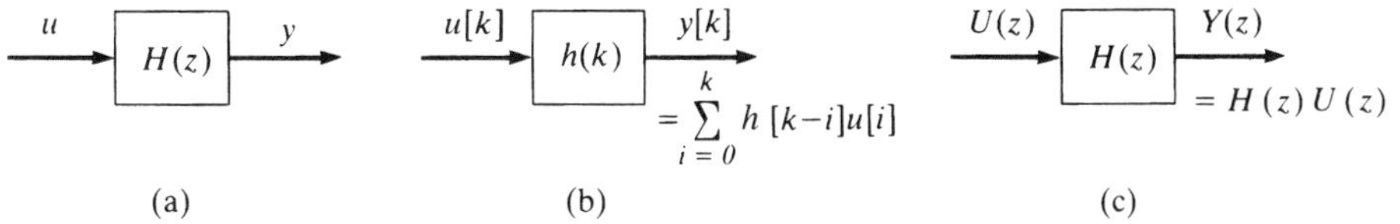

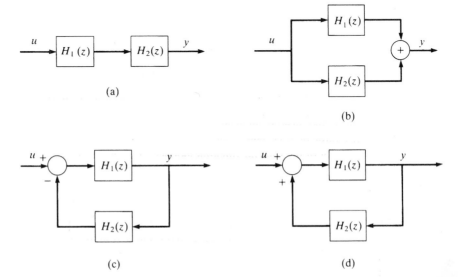

$$H(z) \qquad\qquad \begin{array}{c} u[k] \\ \longrightarrow \end{array} h(k) \begin{array}{c} y[k] \\ \longrightarrow \\ = \displaystyle\sum_{i=0}^{k} h\,[k-i]u[i] \end{array} \qquad\qquad U(z) \quad H(z) \quad \begin{array}{c} Y(z) \\ = H(z)\,U(z) \end{array}$$

(a)   (b)   (c)

the system is represented by its transfer function, which is in the z-transform domain. Therefore, strictly speaking, the representation is incorrect. The correct representations should be as shown in Figure 5.7(b) or (c). The former is in the time domain, and the latter is in the transform domain. However, for convenience, we often plot the transfer function as shown in Figure 5.7(a) by mixing the time and transform domains.

The advantage of using transfer functions in the discrete-time case is identical to the advantage of using them in continuous-time case (see Section 4.10). For example, the transfer function of the tandem connection of two systems with transfer functions $H_1(z)$ and $H_2(z)$, as shown in Figure 5.8(a), is $H_2(z)H_1(z)$. The transfer function of the parallel connection in Figure 5.8(b) is $H_1(z) + H_2(z)$. The transfer function of the negative feedback system shown in Figure 5.8(c) is

$$H(z) = \frac{H_1(z)}{1 + H_1(z)H_2(z)} \qquad\qquad (5.66)$$

Figure 5.8 *Connection of two systems*

(a)

(b)

(c)

(d)

Figure 5.9   *Feedback system and its transfer function*

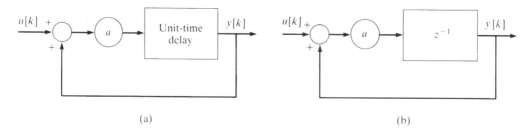

(a)   (b)

and the one of the positive feedback system shown in Figure 5.8(d) is

$$H(z) = \frac{H_1(z)}{1 - H_1(z)H_2(z)} \tag{5.67}$$

Their derivations are identical to the ones in the analog case and will not be repeated here. In the analog case, the connection of two systems may create the loading problem discussed in Section 4.10.1. In the digital case, we have a similar but less serious problem. If discrete-time systems are built with digital circuits, then the number of terminals connected to a point may have to be limited; otherwise, the loading problem will occur. If systems are implemented on a digital computer, the problem will not arise.

EXAMPLE 5.8.1

Consider the system shown in Figure 5.9(a). Its impulse response, as was discussed in Example 3.2.2, is

$$h[k] = a\delta[k-1] + a^2\delta[k-2] + \cdots = \sum_{i=1}^{\infty} a^i \delta[k-i] \tag{5.68}$$

The transfer function of the unit-time delay element is z^{-1}, as shown in Figure 5.9(b). The transfer function of the tandem connection of a and z^{-1} is az^{-1}. The transfer function of the positive feedback system is, using (5.67),

$$H(z) = \frac{az^{-1}}{1 - az^{-1}} = \frac{a}{z - a} \tag{5.69}$$

This representation is considerably simpler than the impulse response in (5.68). We mention that the discrete-time system in Figure 5.9(a) is a lumped system; its analog counterpart in Figure 4.24 is a distributed system, however.

Although transfer functions are convenient to employ in analysis and design, they describe only zero-state responses. In other words, whenever we use transfer functions, initial conditions are implicitly assumed to be zero or, equivalently, zero-input responses are completely disregarded. Can we do this? The answer is the same as in the analog case. Consider a discrete-time system described by the difference equation

$$D(p)y[k] = N(p)u[k]$$

where $D(p)$ and $N(p)$ are defined as in (5.50). If $D(p) = (p + 0.5)$ $(p - 1)(p - 2)$, then the modes of the system are -0.5, 1, and 2, and the zero-input response is of the form

$$y_{zi}[k] = k_1(-0.5)^k + k_2(1)^k + k_3(2)^k \qquad (5.70)$$

for some constant k_i. The zero-state response of the system is governed by its transfer function

$$H(z) = \frac{N(z)}{D(z)}$$

and the poles of the system are defined as the roots of $D(z)$ after canceling all common factors between $N(z)$ and $D(z)$. For example, if $D(z) = (z + 0.5)$ $(z - 1)(z - 2)$ and $N(z) = z^2 - 3z + 2$, then

$$H(z) = \frac{z^2 - 3z + 2}{(z + 0.5)(z - 1)(z - 2)} = \frac{(z - 1)(z - 2)}{(z + 0.5)(z - 1)(z - 2)} = \frac{1}{z + 0.5}$$

and the system has only one pole at -0.5. Note that, even though 1 and 2 are modes, they are not poles. In this case, the zero-state response excited by any input is of the form

$$y_{zs}[k] = \bar{k}_1(-0.5)^k + \text{(terms due to the poles of } U(z))$$

We see that the zero-state response does not contain 1^k and 2^k, which generally arise in the zero-input response. Thus, if $N(z)$ and $D(z)$ have common factors, then every pole is a mode but not every mode is a pole, which is denoted as

$$\{\text{the set of poles}\} \subset \{\text{the set of modes}\}$$

In this case, the system is said to have missing poles and the system is not completely characterized by its transfer function. For example, the system in Example 5.7.7 is not completely characterized by its transfer function because the system has three modes at 2, -0.2, and -0.2 but only two poles at -0.2 and

−0.2. The unit step response of the system approaches a constant and the system appears to be satisfactory. However, if the initial conditions of the system are different from zero, then the response approaches infinity and the system will burn out. Thus, if a system is not completely characterized by its transfer function, care must be exercised in using the transfer function to study the system.

On the other hand, if $D(z)$ and $N(z)$ are coprime (have no common factors), then every pole is a mode and every mode is a pole, which is denoted as

$$\{\text{the set of poles}\} = \{\text{the set of modes}\}$$

In this case, the zero-input response is essentially contained in the zero-state response. For example, if $D(z) = (z + 0.5)(z - 1)(z - 2)$ and $N(z) = z(z + 3)$, then the zero-state response of the system excited by any $u[k]$ is

$$y_{zs}[k] = \bar{k}_1(-0.5)^k + \bar{k}_2(1)^k + \bar{k}_3(2)^k \tag{5.71}$$
$$+ \text{ (terms due to the poles of } U(z))$$

We see that the zero-input response in (5.70) is essentially contained in (5.71). In this case, the system is said to be completely characterized by its transfer function and there is no loss of essential information in using the transfer function to study the system. This is identical to the analog case.

5.9   THE FINAL-VALUE AND INITIAL-VALUE THEOREMS

Some of the properties of the z-transform have not yet been discussed. We shall do so in this section.

Let $f[k]$, $k = 0, 1, 2, \ldots$, be a positive-time sequence. We call $f[0]$ the initial value and $f[k]$ as $k \to \infty$ the final value of the sequence. For example, the initial value and final value of $f[k] = 1 + 2^k$ are 2 and ∞. The initial value and final value of $f[k] = 1 + (0.2)^k$ are 2 and 1. The final value of $f[k] = 1 + \sin 2k$ is not defined because it oscillates between 0 and 2 and does not approach a constant. Let $F(z)$ be the z-transform of $f[k]$. It is assumed that $F(z)$ is a proper rational function of z. In this section, we shall express the final and initial values of $f[k]$ in terms of $F(z)$.

The Final-value theorem   Let $F(z)$ be the z-transform of a positive-time sequence $f[k]$, $k = 0, 1, 2, \ldots$, and let $F(z)$ be a proper rational function. If every pole of $(z - 1)F(z)$ has a magnitude smaller than 1 or, equivalently, lies inside the unit circle of the z-plane, then $f[k]$ approaches a constant (zero or nonzero), and

$$\lim_{k \to \infty} f[k] = \lim_{z \to 1} (z - 1)F(z) \tag{5.72}$$

We will discuss the condition of the theorem first. Using the linearity property of the z-transform and Table 5.3, we can readily conclude that the inverse

z-transform of $F(z)$ approaches 0 as $k \to \infty$ if and only if every pole of $F(z)$ lies inside the unit circle. If every pole of $F(z)$ lies inside the unit circle, so does every pole of $(z-1)F(z)$. The inverse of $F(z)$ approaches a nonzero constant if and only if $F(z)$ has a simple pole at $z = 1$ and the rest of the poles of $F(z)$ lie inside the unit circle. In this case, every pole of $(z-1)F(z)$ lies inside the unit circle because the simple pole at $z = 1$ is cancelled by $(z-1)$. Thus, under the condition of the theorem, $f[k]$ approaches a constant as $k \to \infty$. If the condition is not met, $f[k]$ does not approach a constant and the formula in (5.72) does not hold.

EXAMPLE 5.9.1

Consider $f[k] = 2^k$. Its final value is ∞. Its z-transform is $z/(z-2)$ and

$$\lim_{z \to 1} (z-1)F(z) = \lim_{z \to 1} \frac{(z-1)z}{z-2} = 0$$

which is not equal to $f[\infty]$. For this example, (5.72) does not hold because $(z-1)F(z)$ has a pole outside the unit circle.

EXAMPLE 5.9.2

Consider $f[k] = \sin k$. Its value as $k \to \infty$ is not defined. However, we have

$$\lim_{z \to 1} (z-1)F(z) = \lim_{z \to 1} (z-1)\frac{(\sin 1)z}{z^2 - 2(\cos 1)z + 1} = 0$$

For this example, (5.72) does not hold because $(z-1)F(z)$ has a pair of complex poles on the unit circle.

To derive the final-value theorem, let $f[k]$ be a positive-time sequence. Then we have

$$\mathscr{Z}[f[k]] = F(z) = \lim_{N \to \infty} \sum_{k=0}^{N} f[k]z^{-k} \tag{5.73}$$

and, using (5.18a),

$$\mathscr{Z}[f[k+1]] = z[F(z) - f[0]] = \lim_{N \to \infty} \sum_{k=0}^{N} f[k+1]z^{-k} \tag{5.74}$$

Subtracting (5.73) from (5.74) yields

$$(z - 1)F(z) - zf[0] = \lim_{N \to \infty} \sum_{k=0}^{N} [f[k+1]z^{-k} - f[k]z^{-k}] \qquad \textbf{(5.75)}$$

As $z \to 1$, the right-hand side of (5.75) reduces to $(f[N+1] - f[0])$. Thus, (5.75) implies

$$\lim_{z \to 1} (z - 1)F(z) - f[0] = \lim_{N \to \infty} f[N+1] - f[0]$$

This establishes the final-value theorem. Note that the final-value theorem can also be established using partial fraction expansion, as in Example 5.7.7.

EXERCISE 5.9.1

Find $f[\infty]$ if $F(z)$ is given by

(a) $\dfrac{3z}{(z-1)(z+1)}$

(b) $\dfrac{z+20}{(z+0.9)^{10}}$

(c) $\dfrac{z+1}{3(z^2-1)(z+0.9)}$

(d) $\dfrac{(2z+1)(z-10)}{z(z+2)}$

[**ANSWERS:** (a) Not defined; (b) 0; (c) 1/5.7, and note that there is a common factor; (d) $\pm\infty$.] ∎

Let us compare the conditions for the continuous-time final-value theorem and the discrete-time final-value theorem. The condition for the continuous-time case is that every pole of $sF(s)$ has a negative real part or, equivalently, lies inside the *open* left half-s-plane (left-half plant excluding the $j\omega$-axis). This region is transformed by (5.13) into the region inside the unit circle of the z-plane. The point $s = 0$ is transformed by (5.13) into $z = 1$. Thus, the discrete-time final-value theorem can be obtained from the continuous-time one.

Now we will turn to the initial-value theorem.

The Initial-value theorem Let $F(z)$ be the z-transform of a positive-time sequence $f[k]$, $k = 0, 1, 2, \ldots$, and let $F(z)$ be a proper rational function. Then we have

$$\boxed{f[0] = \lim_{z \to \infty} F(z)} \qquad \textbf{(5.76)}$$

This theorem follows from

$$F(z) = f[0] + f[1]z^{-1} + f[2]z^{-2} + \cdots$$

which becomes (5.76) by setting $z = \infty$.

EXERCISE 5.9.2

Find the initial values of $F(z)$ in Exercise 5.9.1.

[**ANSWERS:** (a) 0, (b) 0, (c) 0, (d) 2.] ■

5.10 SUMMARY

1. The z-transform of a sequence is defined as a power series of z^{-1}, with z^{-i} indicating its time instant. Such a power series is not necessarily expressible in closed form. It can be expressed as a rational function of z if the time sequence is a step, ramp, sinusoidal, or exponential sequence or a linear combination of those.

2. The two-sided z-transform and its inverse are not uniquely related unless the region of convergence is specified. However, the region of convergence can be completely disregarded if we study only causal systems and positive-time sequences.

3. The inverse z-transform can be obtained by direct division or by partial fraction expansion. In partial fraction expansion, we may use z-transform pairs in negative-power or positive-power form. We may also expand $F(z)/z$ and then multiply the resulting expansion by z. In this case, the procedure in the analog case can be used without any modification.

4. The z-transform will transform a discrete convolution into an algebraic equation. The z-transform of the impulse response, which can be easily generated in the discrete-time case, is called the (discrete-time, discrete, or digital) transfer function. If a system is linear and time invariant but is distributed, then the transfer function is an irrational function of z. In this case, not much advantage will be gained in using the z-transform. However, if a system is linear, time invariant, and lumped, then its transfer function is a rational function of z. Mathematically, the manipulation of rational functions is the same as the manipulation of real numbers. Thus, the study of this class of systems becomes fairly simple.

5. The z-transform is not used in the study of nonlinear or time-varying systems.

6. In the discrete-time case, the transfer function is expressed sometimes as $H(z)$ and sometimes as $H^-(z^{-1})$. It is important to specify the form; otherwise, confusion may arise in discussion. For example, the poles and

zeros of $H(z)$ are different from those of $H^-(z^{-1})$. In the continuous-time case, we use $H(s)$ exclusively.

7. In the discrete-time case, an LTIL system is causal if, and only if, its transfer function is a proper rational function of z. In the continuous-time case, transfer functions are required to be proper to avoid amplification of high-frequency noise.

8. The time response of a pole (mode) approaches zero as $k \to \infty$ if, and only if, the pole (mode) lies inside the unit circle in the z-plane. It approaches infinity if the pole (mode) lies outside the unit circle. If the pole (mode) is on the unit circle, its response approaches infinity only if the pole (mode) is repeated.

9. If a system is described by the difference equation

$$D(p)y[k] = N(p)u[k]$$

where $D(p)$ and $N(p)$ are defined as in (5.50), then the form of the zero-input response of the system is determined by the modes, or the roots of the characteristic polynomial $D(z)$. The zero-state response is determined by the poles of the transfer function $H(z) = N(z)/D(z)$. The set of modes is not necessarily equal to the set of poles due to the possible existence of common factors between $N(z)$ and $D(z)$. To find the poles of a transfer function, we must cancel all common factors between the denominator and numerator. Only then will the roots of the remaining denominator be the poles of the transfer function.

10. Consider a discrete-time system described by the difference equation

$$D(p)y[k] = N(p)u[k]$$

If $D(p)$ and $N(p)$ are coprime (have no common factors), then the system is completely characterized by its transfer function $H(z) = N(z)/D(z)$. In this case, there is no loss of essential information in using the transfer function to study the system.

11. When using the final-value theorem of the z-transform, we must check the applicability condition. Otherwise, the theorem may yield erroneous results.

12. Although advanced-form and delayed-form difference equations are equivalent forms, the delayed-form difference equations are more convenient for computing solutions because initial conditions appear directly in z-transformed equations. Positive-power and negative-power transfer functions are both used in practice. However, it seems simpler to use the former when discussing general properties of transfer functions. For example, the pole-zero excess of $H(z)$ is related to time delay but the pole-zero excess of $H^-(z^{-1})$ is not.

PROBLEMS

5.1 Find the z-transforms of the following sequences.

   (a) $3\delta[k - 2] + 2^k$

   (b) $\sin(4k + 1.1)$

   (c) $k^2(-2)^k$

   (d) $e^{0.1k}$

   (e) $e^{k-2}q[k - 2]$

   (f) $e^{0.1k}q[k - 2]$

5.2 Find a negative time sequence that has the same two-sided z-transform as the z-transform of 2^{k-1}, $k = 0, 1, 2, \ldots$.

5.3 (a) Find the z-transforms of the following sequences.

$$f[k] = k + \sin 2k$$
$$f[k] = k(-1)^k + k \cdot 1^k$$
$$f[k] = \begin{cases} 0 & \text{for } k = 1, 3, 5, \ldots \text{(odd)} \\ 0.5k & \text{for } k = 0, 2, 4, \ldots \text{(even)} \end{cases}$$

   (b) Verify

$$\mathscr{Z}[k^2 b^k] = \frac{b(z + b)z}{(z - b)^3}$$

5.4 Find the z-transforms of $q[k + 2]$, $q[k + 1]$, $q[k]$, $q[k - 1]$, and $q[k - 2]$, where $q[k]$ is the unit step sequence. Compute them from (5.3) and from (5.16) and (5.18).

5.5 Is the mapping in (5.13) a one-to-one mapping in the sense that for every z_0 there is only one s_0 in the s-plane that is mapped into z_0? If not, find three points in the s-plane that will be mapped into $z = 1 + j1$. In the continuous-time case, if a pole is real, then its time response will not oscillate. In the discrete-time case, if a pole is real, is it true that its time response will not oscillate?

5.6 Use the direct division method to find the inverse z-transforms of the following

   (a) $\dfrac{z^2}{z^2 + 3z - 4}$

   (b) $\dfrac{z + 2}{(z - 0.9)(z + 1)^3}$

   (c) $\dfrac{1}{(z - 0.8)(z^2 + 2z + 3)}$

5.7 What is the inverse z-transform of $z^2/(z + 1)$? Is it a positive-time sequence?

5.8 Use the partial fraction expansion method to find the inverse z-transforms of the functions in Problem 5.6. Use both positive-power and negative-power z-transform pairs.

(a)

(e)

5.9 Find the inverse z-transforms of the following.

(a) $F(z) = \dfrac{z - 1}{z^5(z + 1)(z - 2)}$

(b) $F(z) = \dfrac{z^3 + 2z + 1}{z^4(z + 2)}$

5.10 Find the solutions of these difference equations.

(a) $y[k] + 2y[k - 1] = u[k - 1] - u[k - 2]$ excited by
$y[-1] = -1$, $y[-2] = -1$, and $u[k] = e^{-k}$, for $k = 0, 1, 2, \ldots$.

(b) $y[k] + 0.8y[k - 1] + 0.16y[k - 2] = 2u[k - 2]$ excited by
$y[-1] = y[-2] = 0$, and $u[k] = 1$, for $k = 0, 1, 2, \ldots$.

(c) $y[k] + \sqrt{2}\,y[k - 1] + y[k - 2] = u[k - 2]$ excited by
$y[-1] = 2\sqrt{2}$, $y[-2] = 2$, and $u[k] = 0$, for $k = 0, 1, 2, \ldots$.

5.11 Find the solutions of these difference equations. They are advanced form of the delayed-form equations in Problem 5.10. Are the results the same? Which form is simpler to use?

(a) $y[k + 2] + 2y[k + 1] = u[k + 1] - u[k]$ excited by
$y[-1] = -1$, $y[-2] = -1$, and $u[k] = e^{-k}$, for $k = 0, 1, 2, \ldots$.

(b) $y[k + 2] + 0.8y[k + 1] + 0.16y[k] = 2u[k]$ excited by
$y[-1] = y[-2] = 0$ and $u[k] = 1$, for $k = 0, 1, 2, \ldots$.

(c) $y[k + 2] + \sqrt{2}\,y[k + 1] + y[k] = u[k]$ excited by
$y[-1] = 2\sqrt{2}$, $y[-2] = 2$, and $u[k] = 0$, for $k = 0, 1, 2, \ldots$.

5.12 Find the modes of the three difference equations in Problems 5.10 and 5.11.

5.13 Find the transfer functions of the three equations in Problems 5.10 and 5.11. Also, find their poles and zeros.

5.14 Is it possible to find a set of initial conditions and an input so that the output of the difference equation in Problem 5.10(b) is $y[k] = 1$, for $k = 0, 1, 2, \ldots$?

5.15 Consider the difference equation in Problem 5.10(b). Find inputs that will excite only one pole. Find an input that will not excite any pole.

5.16 (a) Find the poles and zeros of the following transfer functions

$$\dfrac{2(z - 1)}{z(z + 2)(z - 2e^{j30°})(z - 2e^{-j30°})}$$

$$\dfrac{z^2}{(z + 3)(z - 0.5)}$$

$$\dfrac{(2z - 1)}{(z + 2)(z - 0.5)}$$

(b) Express them as $H^-(z^{-1})$, and then find their poles and zeros. Establish their relationship by including poles and zeros at infinity.

5.17 What are the general forms of the zero-state responses of the systems with transfer functions in Problem 5.16?

5.18 Find advanced- and delayed-form difference equations to describe the systems with transfer functions in Problem 5.16.

5.19 (a) Consider an FIR filter of length N. What is the form of its transfer function? Where are its poles? Does it have any nontrivial poles?

(b) Consider a filter with the impulse response $h[k] = b^k$, for $k = 0, 1, \ldots, 99$, and $h[k] = 0$, for $k > 99$. It is an FIR filter of length 100. Show that its transfer function can be written as

$$H(z) = 1 + bz^{-1} + \cdots + b^{99}z^{-99} = \frac{1 - b^{100}z^{-100}}{1 - bz^{-1}}$$

Write both a nonrecursive and a recursive difference equation to describe the filter. Which equation requires less computation?

5.20 The rational function

$$H(z) = \frac{b_m z^m + b_{m-1} z^{m-1} + \cdots + b_0}{a_n z^n + a_{n-1} z^{n-1} + \cdots + a_0}$$

is proper if $n \geq m$ and $a_n \neq 0$. If $p \geq 0$ and $q \geq 0$, what are the conditions for

$$H^-(z^{-1}) = \frac{\beta_0 + \beta_1 z^{-1} + \cdots + \beta_q z^{-q}}{\alpha_0 + \alpha_1 z^{-1} + \cdots + \alpha_p z^{-p}}$$

to be proper or causal? Do you need $p \geq q$ or $p < q$?

5.21 Show that the time response of $1/(z - 0.01)^3$ will approach zero as $k \to \infty$.

5.22 (a) Find the frequency of oscillation of the inverse z-transform of $1/(z^2 - 4z + 9)$. Does it depend on the sampling period T?

(b) In the continuous-time case, the larger the imaginary part of a complex pole, the larger the frequency of oscillation. In the discrete-time case, is it true that the larger the phase of a complex pole, the larger the frequency of oscillation? Is there a limit on its frequency of oscillation?

5.23 Which of the following systems are completely characterized by their transfer functions?

(a) $y[k + 2] - y[k + 1] - 2y[k] = u[k + 1] - u[k]$

(b) $y[k] - y[k - 1] - 2y[k - 2] = u[k - 1] + u[k - 2]$

(c) $(p + 1)(p^2 + p + 4)y[k] = (p^2 - p + 4)u[k]$

5.24 Find the final value and initial value of the positive-time sequences with these z-transforms

(a) $\dfrac{z^3 - 1}{(z^2 - 1)(z + 0.5)}$

(b) $\dfrac{z^3 + 1}{(z^2 - 1)(z + 0.5)}$

(c) $\dfrac{z - 1}{(z^2 + 0.2z + 0.4)(z^2 - 0.2z + 0.4)}$

5.25 Let $F(z)$ be the z-transform of the positive-time sequence $f[k]$. It is assumed that

$$S := \sum_{k=0}^{\infty} f[k]$$

is finite. Show $S = F(1)$.

5.26 (a) Let $F(z)$ be the z-transform of the positive-time sequence $f[k]$. Define

$$g[k] := \sum_{i=0}^{k} f[i]$$

Is $g[k]$ a positive-time sequence?

(b) Show

$$g[k] := \sum_{i=0}^{k} q[k - i]f[i]$$

where $q[k]$ is the unit step sequence.

(c) Show

$$G(z) = \frac{z}{z - 1}F(z)$$

USING **MATLAB**

A discrete transfer function may be expressed in positive-power form $F(z)$ or negative-power form $F^-(z^{-1})$. If it is expressed in positive-power form, then many operations discussed in the previous chapters can be directly applied. For example, to express the improper transfer function in (5.22) as a sum of a polynomial and a strictly proper transfer function, the following commands

```
n = [-2 5 1 -6 3];d = [1 -1 -2];
[q,r] = deconv(n,d)
```

yield

```
q = -2 3 0
r = 0 0 0 0 3
```

This is what we have in (5.22). Note that deconv is the deconvolution discussed at the end of Chapter 3. To find the partial fraction expansion of the rational function in (5.29), we first use zp2tf to transform the zero-pole-gain form of (5.29) into the transfer function form and then call residue. The following commands

```
z = [1];p = [0 -2 2*exp(2.09*j) 2*exp(-2.09*j)];k = 1;
[n,d] = zp2tf(z,p,k);
[r,p,k] = residue(n,d)
```

yield

```
r = -0.1236 - 0.1437i = 0.1895e -2.2811i
    -0.1236 + 0.1437i
     0.3722
    -0.128
p = -0.9924 + 1.7364i
    -0.9924 - 1.7364i
    -2.0000
     0
k = 0
```

The values of r computed in Example 5.4.2 are $0.19e^{\pm 2.29j}$, $3/8 = 0.375$, and $-1/8 = -0.125$. These are very close to those computed using MATLAB. Note that both i and j are defined as $\sqrt{-1}$ in MATLAB.

The discrete counterparts of the commands step and impulse in the analog case are dstep and dimpulse, which are not available in the *Student Edition of MATLAB*. However, we may use command filter in the student edition to compute the impulse and step responses of discrete-time systems. Consider the difference equation

$$y[k] + a(2)y[k-1] + \cdots + a(p+1)y(k-p)$$
$$= b(1)u[k] + b(2)u[k-1] + \cdots + b(q+1)u[k-q]$$

or, in the z-transform domain and with zero initial conditions,

$$Y(z) = \frac{b(1) + b(2)z^{-1} + \cdots + b(q+1)z^{-q}}{1 + a(1)z^{-1} + \cdots + a(p+1)z^{-p}} U(z)$$

If we define

```
n = [b(1) b(2) . . . b(q + 1)]; d = [1 a(1) . . .
     a(p + 1)];
```

then command `filter(n,d,u)` will generate the zero-state response excited by `u`. Unlike the analog case, where `n` and `d` are the coefficients, in the descending power of *s*, of the numerator and denominator, the `n` and `d` in `filter` are the coefficients, in the ascending power of z^{-1}. Thus, we must use the negative-power form of transfer functions when using `filter`.

The inverse z-transform of a rational function is the same as the impulse response of the rational function, which is the output exited by the impulse sequence [1 0 0 0...]. To find the inverse z-transform of the rational function in Exercise 5.4.1, we write the rational function as

$$F(z) = \frac{z^2 - 1}{z^3 - 1} = \frac{z^{-1} - z^{-3}}{1 - z^{-3}} \tag{M5.1}$$

Then the following commands

```
n = [0 1 0 -1];d = [1 0 0 -1];
u = [1 zeros(1,8)]; (zeros(1,8) generates [0 0 0 0 0 0
0 0].)
y = filter(n,d,u)
```

generate

```
y = 0 1 0 -1 1 0 -1 1 0
```

which is the same as the answer in Exercise 5.4.1. Note that, if $F(z)$ in (M5.1) is written as

$$F(z) = \frac{z^2 - 1}{z^3 - 1} = \frac{0 \cdot z^3 + z^2 + 0 \cdot z - 1}{z^3 - 1} \tag{M5.2}$$

so that its numerator and denominator have the same highest power, then `n` and `d` of (M5.2) are `n = [0 1 0 -1]` and `d = [1 0 0 -1]`, which are the same as those for the negative-power rational function. In other words, if we require n and d to be of the same length for positive-power transfer functions, then they can be used in `filter(n,d,u)`.

To find the unit step responses of the five systems in (5.64) from $k = 0$ to 69, we type

```
d = [1 -0.7 -0.55 0.801];n1 = [0 0 0 0.551];
n2 = [0 0 1 -1];n3 = [0 0 5.51 -5.51*0.9];
n4 = [0 0 -5.51 5.51*1.1];n5 = [0 5.51 -5.51*1.9 5.51];
u = [ones(1,70)];
y1 = filter(n1,d,u);
y2 = filter(n2,d,u);
y3 = filter(n3,d,u);
y4 = filter(n4,d,u);
y5 = filter(n5,d,u);
k = 0:69;
plot(k,y1,k,y2,k,y3,k,y4,k,y5)
```

The result is plotted in Figure 5.5.

MATLAB PROBLEMS

Use MATLAB to compute the following.

M5.1 Express the improper rational function $(3z^4 + - 2z + 1)/(2z + 1)$ as a sum of a polynomial and a strictly proper rational function.

M5.2 Use command `filter` to find the impulse response and the step response of $F(z)$ in Example 5.4.3. Can you use command `residue` for this problem?

CHAPTER

Continuous-Time Signal Analysis

67

6.1 INTRODUCTION

10

(8)

In the preceding two chapters, we studied system analysis. In this and the following chapters, we shall study signal analysis. The most important part of signal analysis is finding the frequency content or frequency range of signals. This is fundamental in communication. For example, as Figure 6.1 shows, signals sent out by AM radio stations are limited to the frequency band from 540 KHz to 1600 KHz. FM radio stations, TV broadcasting, and microwave communication all take different frequency bands, as the figure shows. The frequency content of human voices, on the other hand, is generally limited to from 200 Hz to 4 KHz. How is human voice transmitted in the radio frequency band? To answer this, we must discuss the exact meaning of frequency content, which is the main topic of this chapter.

Signal analysis has many applications in practice. From the EKG waveform irregularities shown in Figure 1.1, we can detect a heart's defects. Each submarine generates its own distinct acoustic wave because of the rotation of the propellers, engine vibration, or vibration of the hull. This knowledge can be used in underwater detection. Africanized or "killer" bees and domestic bees are almost identical in physical size and appearance, and, until recently, the only way to tell them apart was by tedious examination under a microscope. Since it has been found that they beat their wings at different frequencies and, consequently, generate different signals, these signals are being used to identify the killer bees and control their spread. Signal analysis also can detect subtle noises in the depth of nuclear reactors; it has successfully pinpointed loose parts in reactor cooling systems, including nuts, bolts, and even tools left behind after maintenance. Another important application of signal analysis is the elimina-

Figure 6.1 *Various frequency bands*

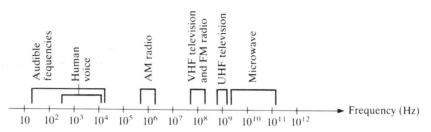

tion of some unwanted noise such as the roar of engines, whine of power transformers and industrial fans, and throb of heavy machinery. This type of noise is periodic and can be broken into basic components as shown in Figure 6.2(a). For example, a microphone in an airline cabin can pick up the whine of engines. A computer can analyze the noise and break down its basic components and then generate, through a loudspeaker, a mirror image of the noise, called antinoise, as shown in Figure 6.2(b). This will cancel the noise without affecting human conversation. Antinoise can be pumped into air ducts, so heavy insulation does not have to be installed. However, this antinoise technique cannot be used if the noise is random or has no repetitive pattern.

A signal is called a *deterministic* signal if it can be described without any uncertainty. This type of signal can be reproduced exactly and repeatedly. Signals studied in the preceding chapters are all of this type. A signal is called a *random* signal if it cannot be described with certainty before it actually occurs. For example, the outcome of rolling a die is a random signal, because the die may show any number between 1 and 6. The signals in Figures 1.1 and 6.2(a) are also random signals. These types of signals generally cannot be reproduced

Figure 6.2 *Decomposition and elimination of repetitive noise*

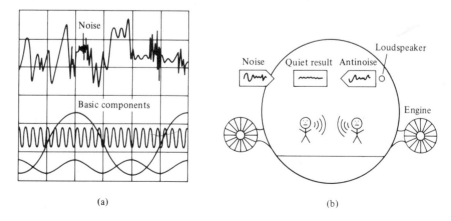

(a)   (b)

exactly and repeatedly. The study of random signals requires a knowledge of probability theory and stochastic processes, which are outside the scope of this text. Therefore, we study here only deterministic signals.

This chapter studies only continuous-time signals. Discrete-time signals will be studied in the next chapter. Signals in system analysis are all positive-time signals. Signals in this chapter, however, will be extended to negative time and be defined for all t in $(-\infty, \infty)$. This extension is needed, so that the frequency contents for periodic functions (including constants) can be easily interpreted. Signals in the real world are all real valued. The signals in this chapter, however, may assume complex numbers. This extension is needed so that signals can be represented in terms of complex exponentials (see Section 1.2, in particular, Figure 1.8). Before proceeding, let us discuss some classifications of signals.

6.1.1   Periodic and Aperiodic Signals

A time function, real or complex valued, is said to be periodic with period P if

$$f(t) = f(t + P) \tag{6.1}$$

for all t in $(-\infty, \infty)$. If (6.1) holds, then

$P = period$

$$f(t) = f(t + P) = f(t + 2P) = \cdots = f(t + nP)$$

for all t and for every positive integer n. Therefore, if $f(t)$ is periodic with period P, it is also periodic with period $2P$, $3P$, ... The smallest such P is called the *fundamental* period. The fundamental frequency is then defined as $2\pi/P$ in rad/s or $1/P$ in Hz. If a function is not periodic, it is called a *nonperiodic* or an *aperiodic* function.

$T = \dfrac{1}{f}$

As was discussed in Chapter 1, $\sin(\omega_0 t + \theta)$ is periodic with fundamental period $P = 2\pi/\omega_0$ and frequency ω_0. The signals sent out by radars are periodic sequences of pulses, as shown in Figure 6.3(a). The sawtooth waveform shown in Figure 6.3(b), which is used in television sets as a time base, is also periodic. The signals in Figure 1.1 and the one in the upper part of Figure 6.2(a) are aperiodic. So are the exponentially increasing or decreasing functions illus-

Figure 6.3 *Periodic signals*

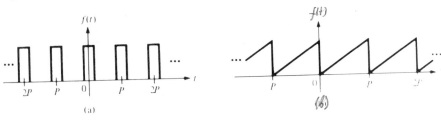

(a)   (b)

trated in Figure 1.7. The function $f(t) = 1$ for all t in $(-\infty, \infty)$ is periodic because (6.1) holds for every P. However, its fundamental frequency cannot be defined from $1/P$. A constant function is called a DC signal and is defined to have frequency 0. This can be justified by considering $f(t) = 1$ as the limit of $\cos(2\pi t/P)$ as $P \to \infty$. Thus, the frequency of $f(t) = 1$ is 0.

Consider two periodic functions $f_i(t)$ with period P_i, $i = 1,2$. The sum of $f_1(t) + f_2(t)$ may or may not be periodic depending on the relationship between P_1 and P_2. Recall that, if $f_i(t)$ is periodic with period P_i, then it is also periodic with period $n_i P_i$ for any positive integer n_i. Now, if we can find two integers n_1 and n_2 such that

re assg. 1

$$n_1 P_1 = n_2 P_2 \quad \text{or} \quad \frac{P_1}{P_2} = \frac{n_2}{n_1}$$

then $f_1(t)$ and $f_2(t)$ have a *common period*, and their sum will be periodic with period $n_1 P_1 = n_2 P_2$. If n_1 and n_2 are coprime (they have no integer other than 1 as a common factor), then $n_1 P_1 = n_2 P_2$ is the fundamental period. Thus, the condition for $f_1(t) + f_2(t)$ to be periodic is that the ratio of P_1/P_2 is a rational number or a ratio of two integers. Otherwise, $f_1(t)$ and $f_2(t)$ have no common period and their sum is not periodic.

EXAMPLE 6.1.1

The function $\sin 2t$ is periodic with period $P_1 = 2\pi/2 = \pi$. The function $\sin 2\pi t$ is a periodic function with period $P_2 = 2\pi/2\pi = 1$. Because

$$\frac{P_1}{P_2} = \frac{\pi}{1} = \pi$$

is not a rational number, $\sin 2t$ and $\sin 2\pi t$ do not have a common period. Therefore, $\sin 2t + \sin 2\pi t$ is not periodic.

The function $\sin 21t$ is periodic with period $P_3 = 2\pi/21$. Because

$$\frac{P_1}{P_3} = \frac{\pi}{2\pi/21} = \frac{21}{2}$$

is a rational number, the function $\sin 2t + \sin 21t$ is periodic with fundamental period $2P_1 = 21P_3 = 2\pi$ and its fundamental frequency is $2\pi/2\pi = 1$.

The fundamental frequency of sums of two or more sinusoidal functions can be computed without computing first their common periods. Consider two real numbers a and b. If a can be expressed as $a = nb$, where n is an integer, then b is called a *divisor* of a. In other words, b is a divisor of a if a is an integer multiple of b. For example, 0.01, 0.1, 0.2, 0.3, and 0.6 are divisors of 0.6, but 0.4, 0.5, and 6 are not. The greatest common divisor (GCD)[1] of two or more numbers is defined as the largest number that is a divisor of all numbers. For example, the GCD of 0.6 and 1.2 is 0.6. The GCD of 0.6, 1.2, and 2.1 is 0.3.

Consider $f(t) = \sin \omega_1 t + \sin \omega_2 t$. If ω_1 and ω_2 do not have a common divisor, then $f(t)$ is not periodic. If they have a common divisor, then $f(t)$ is periodic with their GCD as its fundamental frequency. For example, 2 and 2π have no common divisor, so $\sin 2t + \sin 2\pi t$ is not periodic. 2 and 21 have 1 as their GCD, so $\sin 2t + \sin 21t$ is periodic with fundamental frequency 1.

EXAMPLE 6.1.2

Consider

$$f_1(t) = \pi \sin 0.6t - 3 \cos 1.5t + 11 \sin 3t$$

The three frequencies 0.6, 1.5, and 3 have 0.3 as their GCD. Therefore, $f_1(t)$ is periodic with fundamental frequency 0.3 and fundamental period $2\pi/0.3$. Consider

$$f_2(t) = \sin 1.2\pi t - 2 \cos 3\pi t$$

The frequencies 1.2π and 3π have 0.6π as their GCD. Thus, $f_2(t)$ is periodic with fundamental frequency 0.6π and fundamental period $2\pi/0.6\pi = 1/0.3$.

EXERCISE 6.1.1

Which of the following are periodic functions? Find their fundamental frequencies and periods.

(a) $\sin 3t + \sin \pi t$

(b) $\pi \sin 20t - 3.1 \cos 21t$

(c) $\sin 0.8t + \pi \cos 2t - 21 \sin 1.2t$

(d) $1 - \cos \pi t + 2 \sin \pi t + \cos 3\pi t$

[**ANSWERS:** (a) No; (b) yes, 1, 2π; (c) yes, 0.4, 5π; (d) yes, π, 2.] ∎

---

[1]Conventionally, the greatest common divisor (GCD) is defined only for integers. We have extended it to real numbers. However, we require the quotient b/a to be an integer.

6.1.2   Energy and Power Signals

Signals in electrical systems are usually voltages or currents. If a voltage signal $v(t)$ is applied to a 1 Ω resistor, the current passing through the resistor is $v(t)/1$ and the power dissipated is $v(t)\,i(t) = v^2(t)$. Thus, the energy provided by the signal $v(t)$ over the time interval $[t_0, t_1]$ is

$$\int_{t_0}^{t_1} v^2(t)\,dt$$

Similarly, if a current signal $i(t)$ passes through a 1 Ω resistor, the voltage across the resistor is $i(t) \cdot 1$ and the power dissipated is $v(t)\,i(t) = i^2(t)$. Thus, the energy provided by the signal $i(t)$ over $[t_0, t_1]$ is

$$\int_{t_0}^{t_1} i^2(t)\,dt$$

Because of these facts, the energy of a real- or complex-valued signal $f(t)$, be it electrical, mechanical, or any type of signal, is customarily defined as

$$\int_{t_0}^{t_1} |f(t)|^2\,dt = \int_{t_0}^{t} f(t)\,f^*(t)\,dt$$

where the asterisk denotes the complex conjugate and $|\cdot|$ denotes the absolute value. If $f(t)$ is real, we have $f(t)\,f^*(t) = f^2(t)$.

A signal $f(t)$ is called an *energy signal* if its total energy in $(-\infty, \infty)$ is finite, that is,

also see pg. 348

$$E_\infty := \int_{-\infty}^{\infty} |f(t)|^2\,dt < \infty$$

The signal $e^{-2|t|}$, for all t, shown in Figure 6.4(a) is an energy signal, because

$$E_\infty = \int_{-\infty}^{\infty} (e^{-2|t|})^2\,dt = \int_{-\infty}^{0} e^{4t}\,dt + \int_{0}^{\infty} e^{-4t}\,dt = 2\int_{0}^{\infty} e^{-4t}\,dt$$

$$= \left.\frac{-2}{4} e^{-4t}\right|_{t=0}^{\infty} = \frac{-2}{4}[0-1] = \frac{1}{2}$$

A function $f(t)$ is said to be of finite duration if it is nonzero only over a finite time interval, that is, there exist t_1 and t_2 such that $f(t) = 0$, for $t < t_1$ and $t > t_2$. A function is bounded if there exists a finite M such that $|f(t)| \le M$, for all t. A bounded signal of finite duration is an energy signal because

Figure 6.4 *Signals*

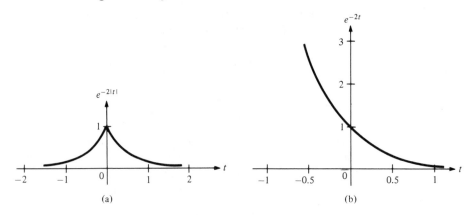

(a)   (b)

$$E_\infty = \int_{-\infty}^{\infty} |f(t)|^2 dt \leq M^2 \int_{t_1}^{t_2} dt = M^2(t_2 - t_1) < \infty$$

Most signals encountered in practice do not go to infinity (they are bounded) and must start somewhere and end somewhere (they are of finite duration); therefore, they are energy signals.

The average power of a signal over $[t_1, t_2]$ is defined as

$$\boxed{P_{av} := \frac{1}{t_2 - t_1} \int_{t_1}^{t_2} |f(t)|^2 dt}$$

A signal is called a *power signal* if the average power defined by

$$\boxed{P_\infty := \lim_{T \to \infty} \frac{1}{2T} \int_{-T}^{T} |f(t)|^2 dt} \quad = \lim_{T \to \infty} \frac{1}{2T} E_T \quad \textbf{(6.2)}$$

is *nonzero* and finite. An energy signal has a finite total energy; thus, we have $P_\infty = 0$. Therefore, no energy signal is a power signal. The total energy of a power signal must be infinity, otherwise P_∞ would be zero. Therefore, no power signal can be an energy signal. In conclusion, a signal can be a power signal or an energy signal, but it cannot be both. However, energy and power signals are not the only types of signals. Some signals are neither energy signals nor power signals.

EXAMPLE 6.1.3

The signal e^{-2t} in Figure 6.4(b) is not an energy signal because

$$E_T := \int_{-T}^{T} (e^{-2t})^2 dt = \int_{-T}^{T} e^{-4t} dt = \frac{-1}{4}[e^{-4T} - e^{4T}]$$

which approaches infinity as $T \to \infty$. Its average power is

$$P_\infty = \lim_{T \to \infty} \frac{1}{2T} E_T = \lim_{T \to \infty} \frac{e^{4T} - e^{-4T}}{8T} = \lim_{T \to \infty} \frac{e^{4T}}{8T} = \lim_{T \to \infty} \frac{4e^{4T}}{8} = \infty$$

where we have used l'Hopital's rule. Thus, e^{-2t} is neither an energy signal nor a power signal.

For periodic signals, the computation of the average power can be simplified. Clearly, we have

also see pg. 320 321

$$P_{av} := P_\infty = \lim_{T \to \infty} \frac{1}{2T} \int_{-T}^{T} |f(t)|^2 dt$$

$$= \frac{1}{P} \int_{-P/2}^{P/2} |f(t)|^2 dt = \frac{1}{P} \int_{0}^{P} |f(t)|^2 dt \qquad (6.3)$$

for a periodic signal $f(t)$ with period P. If $f(t)$ is bounded, P_∞ in (6.3) is finite. So, *every bounded and periodic signal is a power signal.* However, the converse is not true; that is, a power signal is not necessarily a bounded and periodic signal. For example, $\sin 2t + \sin 2\pi t$ is, as was discussed earlier, not periodic, but it can be shown to be a power signal. (See also Problem 6.3.)

EXAMPLE 6.1.4

Consider $f(t) = f(t + 1)$, for all t, with $f(t) = t^{-2/3}$ in [0, 1]. This is a periodic function with period 1. Its average power is

$$\int_{0}^{1} |f(t)|^2 dt = \int_{0}^{1} t^{-4/3} dt = -3t^{-1/3} \Big|_{t=0}^{t=1} = -3(1 - \infty) = \infty$$

It is not finite. Thus, the periodic signal is not a power signal. Note that the signal is not bounded.

EXAMPLE 6.1.5

Consider the signal $f(t) = A \sin(\omega_0 t + \theta)$. It is periodic with period $P = 2\pi/\omega_0$. Its average power is

$$P_\infty = \frac{1}{P}\int_{-P/2}^{P/2} A^2 \sin^2(\omega_0 t + \theta)dt = \frac{\omega_0}{2\pi}\int_{-\pi/\omega_0}^{\pi/\omega_0} \frac{A^2}{2}[1 - \cos 2(\omega_0 t + \theta)]dt$$

$$= \frac{\omega_0 A^2}{4\pi}[\frac{\pi}{\omega_0} + \frac{\pi}{\omega_0}] = \frac{A^2}{2}$$

where we have used

$$\int_{-\pi/\omega_0}^{\pi/\omega_0} \cos 2(\omega_0 t + \theta)dt = 0$$

 that is, the integration of a sinusoidal function over a complete period or its multiples is zero. Thus, the average power of a sinusoid with amplitude A equals $A^2/2$. It equals the power supplied by a DC voltage with amplitude $A/\sqrt{2}$ over a 1Ω resistor. Thus, a sinusoid with amplitude A is said to have an effective amplitude of $A/\sqrt{2}$.

rms

6.1.3   **Orthogonality of Complex Exponentials**

Consider the complex exponential

$$e^{j\omega t} = \cos \omega t + j \sin \omega t \qquad (6.4)$$

It is a complex-valued function with real part $\cos \omega t$ and imaginary part $\sin \omega t$. The amplitude of $e^{j\omega t}$ is 1 for all ω and for all t, as shown in Figure 1.8(b). As t increases, $e^{j\omega t}$ rotates along the unit circle counterclockwise for $\omega > 0$ and clockwise for $\omega < 0$. The complex exponential $e^{j\omega t}$ is periodic with fundamental frequency ω and period $2\pi/\omega$.
    Consider the set of complex exponentials

$$\phi_m(t) := e^{jm\omega_0 t} \qquad (6.5)$$

where $m = 0, \pm 1, \pm 2, \ldots$. For each integer m, the function $e^{jm\omega_0 t}$ is a periodic function with fundamental frequency $m\omega_0$ and fundamental period $P_m := 2\pi/m\omega_0$. If we define $P := 2\pi/\omega_0$, then $P = mP_m$ and any time interval of length P contains m complete cycles of $e^{jm\omega_0 t}$. Now we show

$$\int_{t_0}^{t_0 + P} e^{jm\omega_0 t}dt = = \int_{t_0}^{t_0 + P} (\cos m\omega_0 t + j \sin m\omega_0 t)\, dt = \begin{cases} P & \text{if } m = 0 \\ 0 & \text{if } m \neq 0 \end{cases} \qquad (6.6)$$

for any t_0. Indeed, if $m = 0$, then $e^{jm\omega_0 t} = 1$ for all t; thus, the integration equals P. If $m \neq 0$, then net areas of sinusoidals over complete periods are zero, as shown in Figure 6.5, and the integration is 0. This establishes (6.6).

Figure 6.5 *Net area of a sinusoid over complete periods*

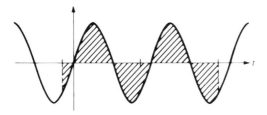

There is an important property of the set of periodic functions $e^{jm\omega_0 t}$ for $m = 0, \pm 1, \pm 2, \ldots$. Let the asterisk denote the complex conjugate. For example, we have

$$\phi_m^*(t) = (e^{jm\omega_0 t})^* = (\cos m\omega_0 t + j \sin m\omega_0 t)^* = \cos m\omega_0 t - j \sin m\omega_0 t = e^{-jm\omega_0 t}$$

Then, (6.6) implies

Orthogonality Property ✳

$$\int_{<P>} \phi_m(t)\phi_n^*(t)dt := \int_{t_0}^{t_0 + P} e^{j(m-n)\omega_0 t}dt = \begin{cases} P & \text{if } m = n \\ 0 & \text{if } m \neq n \end{cases} \tag{6.7}$$

for any t_0. The notation $<P>$ is used to denote integration over one period P. This is called the *orthogonality* property of the set $\phi_m(t)$.

Consider the linear combination of $\phi_m(t)$

$$f(t) = \sum_{m=-\infty}^{\infty} c_m e^{jm\omega_0 t}$$

where c_m are real or complex constants. It is clear that ω_0 is the GCD of all $m\omega_0$. Thus, for every set of c_m, $f(t)$ is periodic with fundamental period ω_0. Conversely, every periodic function with fundamental period ω_0 can be represented by such a linear combination for some c_m, as will be demonstrated in the next section.

6.2 FOURIER SERIES OF PERIODIC FUNCTIONS

Consider a periodic function $f(t)$ with fundamental period P and fundamental frequency

$$\boxed{\omega_0 = \frac{2\pi}{P}} = 2\pi f \Rightarrow f = \frac{1}{T}$$

We show that $f(t)$ can be expressed as

$$f(t) = \sum_{m=-\infty}^{\infty} c_m e^{jm\omega_0 t} \qquad (6.8)$$

with

c_m = discrete / line frequency spectrum

$$c_m := \frac{1}{P} \int_{t_0}^{t_0+P} f(t) e^{-jm\omega_0 t} dt =: \frac{1}{P} \int_{<P>} f(t) e^{-jm\omega_0 t} dt \qquad (6.9)$$

for $m = 0, \pm 1, \pm 2, \ldots$ and for any t_0. This is called the *complex exponential Fourier series*[2]. Note that both positive and negative frequencies appear in (6.8). If negative frequencies are not included in (6.8), then such an expression is not possible. To establish (6.9), we multiply $e^{-jn\omega_0 t}$ and (6.8) and then integrate the product over $[t_0, t_0 + P]$ to yield

$$\int_{t_0}^{t_0+P} f(t) e^{-jn\omega_0 t} dt = \sum_{m=-\infty}^{\infty} \int_{t_0}^{t_0+P} c_m e^{jm\omega_0 t} e^{-jn\omega_0 t} dt = \sum_{m=-\infty}^{\infty} c_m \int_{t_0}^{t_0+P} e^{j(m-n)\omega_0 t} dt$$

Because of the orthogonality property in (6.7), every integral on the right-hand side equals zero if $m \neq n$ and equals P if $m = n$. Thus, the infinity summation reduces to Pc_n, and the equation becomes

$$\int_{t_0}^{t_0+P} f(t) e^{-jn\omega_0 t} dt = Pc_n$$

This becomes (6.9) after renaming the index n as m. In using (6.9), t_0 is often chosen as 0 or $-P/2$ to yield

$$c_m = \frac{1}{P} \int_0^P f(t) e^{-jm\omega_0 t} dt = \frac{1}{P} \int_{-P/2}^{P/2} f(t) e^{-jm\omega_0 t} dt \qquad (6.10)$$

Here are some examples.

EXAMPLE 6.2.1

Consider the function

$$f(t) = 1 - 3\cos 0.6\pi t + 2\sin 1.2\pi t + \cos 2.1\pi t \qquad \text{for all } t \qquad (6.11)$$

---

[2]In honor of French mathematician Jean Baptiste Joseph Fourier (1768–1830).

The GCD of $\{0.6\pi, 1.2\pi, 2.1\pi\}$ is 0.3π. Thus, $f(t)$ is periodic with fundamental frequency $\omega_0 := 0.3\pi$ and fundamental period $P := 2\pi/0.3\pi = 20/3$. For this function, it is unnecessary to use (6.9) or (6.10) to compute the coefficients of the complex exponential Fourier series. We use the following formulas

$$\boxed{\sin \omega t = \frac{e^{j\omega t} - e^{-j\omega t}}{2j} \qquad \cos \omega t = \frac{e^{j\omega t} + e^{-j\omega t}}{2}}$$

to express (6.11) as

$$p(t) = 1 - \frac{3}{2}\left(e^{j0.6\pi t} + e^{-j0.6\pi t}\right) + \frac{2}{2j}\left(e^{j1.2\pi t} - e^{-j1.2\pi t}\right) + \frac{1}{2}\left(e^{2.1\pi t} + e^{-2.1\pi t}\right)$$

$$
\begin{aligned}
f(t) &= \frac{1}{2}e^{-j2.1\pi t} - \frac{2}{2j}e^{-j1.2\pi t} - \frac{3}{2}e^{-j0.6\pi t} \\
&\quad + 1 - \frac{3}{2}e^{j0.6\pi t} + \frac{2}{2j}e^{j1.2\pi t} + \frac{1}{2}e^{j2.1\pi t} \\
&= 0.5e^{-j7\omega_0 t} + je^{-j4\omega_0 t} - 1.5e^{-j2\omega_0 t} \\
&\quad + 1 - 1.5e^{j2\omega_0 t} - je^{j4\omega_0 t} + 0.5e^{j7\omega_0 t}
\end{aligned}
\tag{6.12}
$$

This is the complex exponential Fourier series of (6.11). The coefficients are

$$
\begin{aligned}
c_{-7} &= 0.5 \qquad c_{-4} = j \qquad c_{-2} = -1.5 \\
c_0 &= 1 \qquad c_2 = -1.5 \qquad c_4 = -j \qquad c_7 = 0.5
\end{aligned}
\tag{6.13}
$$

and the rest of c_m are zero. Even though $f(t)$ is a real-valued function, c_m can be real or complex. Note that the subscript m denotes frequency $m\omega_0$.

EXAMPLE 6.2.2

Consider the periodic function $f(t)$ shown in Figure 6.6(a). It is a train of rectangular pulses with amplitude 1, duration $2a$ and period P. Its fundamental frequency is $\omega_0 = 2\pi/P$. To express the function in the Fourier series, we compute

$$c_0 = \frac{1}{P}\int_{-P/2}^{P/2} f(t)dt := \frac{1}{P}\int_{-a}^{a} 1 dt = \frac{2a}{P} \tag{6.14a}$$

and

$$
\begin{aligned}
c_m &= \frac{1}{P}\int_{-P/2}^{P/2} f(t)e^{-jm\omega_0 t}dt = \frac{1}{P}\int_{-a}^{a} e^{-jm\omega_0 t}dt \\
&= \frac{1}{-jm\omega_0 P}[e^{-jm\omega_0 a} - e^{jm\omega_0 a}] = \frac{2\sin m\omega_0 a}{m\omega_0 P}
\end{aligned}
\tag{6.14b}
$$

Figure 6.6 *(a) A train of rectangular pulses and (b) its Fourier coefficients with* P = 4a

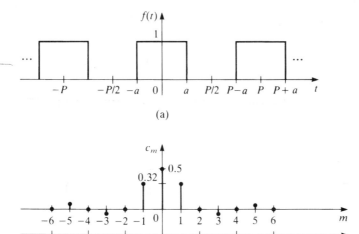

(a)

(b)

Thus, the complex exponential Fourier series of $f(t)$ is

$$f(t) = \sum_{m=-\infty}^{\infty} c_m e^{jm\omega_0 t}$$

with c_m computed in (6.14).

If $P = 4$ and $a = 1$, then (6.14) becomes

$$\omega_0 = \frac{2\pi}{P} = \frac{\pi}{2} \qquad c_0 = \frac{2a}{P} = 0.5$$

and

$$c_m = \frac{2 \sin m\omega_0 a}{m\omega_0 P} = \frac{2 \sin(m\pi/2)}{2\pi m} = \frac{\sin(m\pi/2)}{\pi m}$$

$\rightarrow$ for complex Exponential fourier series

or $c_1 = c_{-1} = 1/\pi$, $c_2 = c_{-2} = 0$, $c_3 = c_{-3} = -1/3\pi$, $c_4 = c_{-4} = 0$, $c_5 = c_{-5} = 1/5\pi, \ldots$. Hence, we have

$$f(t) = 0.5 + \frac{1}{\pi}(e^{j\omega_0 t} + e^{-j\omega_0 t}) - \frac{1}{3\pi}(e^{j3\omega_0 t} + e^{-j3\omega_0 t}) + \cdots \qquad \textbf{(6.15)}$$

The coefficients c_m happen to be all real. They are plotted in Figure 6.6(b) as a function of m and $\omega = m\omega_0$.

EXERCISE 6.2.1

Consider

$$f(t) = \sin 2\pi t - 2\cos 2.1\pi t$$

Is it periodic? What is its fundamental frequency? Express it in the complex exponential Fourier series.

[**ANSWERS:** Periodic, $\omega_0 = 0.1\pi$, $f(t) = -e^{-j21\omega_0 t} + 0.5je^{-j20\omega_0 t} - 0.5je^{j20\omega_0 t} - e^{j21\omega_0 t}$.] ∎

Not every periodic function can be expressed in the complex exponential Fourier series. Sufficient conditions for $f(t)$ to have the Fourier series are

1. $f(t)$ is absolutely integrable over one period, that is,

$$\int_0^P |f(t)|\, dt \le M < \infty$$

and

2. $f(t)$ has a finite number of discontinuities and a finite number of maxima and minima in one period.

These are called the *Dirichlet* conditions.[3] For example, consider the two periodic functions with fundamental period $P = 1$ shown in Figure 6.7. The first function is $1/(1 - t)$, whose integration over $[0, 1]$ is infinity. Thus, the function is not absolutely integrable over one period and its Fourier series does not exist. The second function is $\sin 1/t$; its rate of oscillation increases to infinity as $t \to 0$. Thus, the function has infinitely many maxima and minima and its Fourier series does not exist. If a periodic function $f(t)$ meets the Dirichlet conditions, then its Fourier series converges at every t to a function denoted by

$$\bar{f}(t) = \sum_{m=-\infty}^{\infty} c_m e^{jm\omega_0 t} \tag{6.16}$$

with c_m computed from (6.9). We would expect $\bar{f}(t) = f(t)$ for all t. This is, however, not necessarily the case in general. The reason is simple. Let $f_1(t)$ be a function that differs from $f(t)$ only at a finite number of isolated t. Then, as discussed in Section 3.5.1, we have

$$\int_0^P f_1(t)e^{-jm\omega_0 t}\, dt = \int_0^P f(t)e^{-jm\omega_0 t}\, dt$$

---

[3]Peter Gustaw Lejeune Dirichlet (1805–1857), German mathematician.

Figure 6.7 *Signals that have no Fourier series*

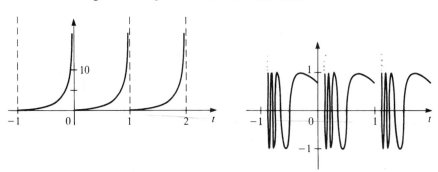

Thus, $f_1(t)$ has the identical Fourier series as $f(t)$. Therefore, even though c_m is computed from $f(t)$, the infinite series in (6.16) may not converge to $f(t)$ at every t. It turns out that $\bar{f}(t)$ equals $f(t)$ at *almost* every t. More specifically, if $f(t)$ is continuous at t_0, then $\bar{f}(t_0) = f(t_0)$. If $f(t)$ is not continuous at t_0, that is, $f(t_0 -) \neq f(t_0 +)$, then

$$\sum_{-\infty}^{\infty} c_m e^{jm\omega_0 t_0} =: \bar{f}(t_0) = \frac{f(t_0 -) + f(t_0 +)}{2}$$

no matter what value $f(t_0)$ assumes. For example, the right-hand side of (6.15) yields $(1 + 0)/2 = 0.5$ at $t = a$ whether $f(a)$ equals 0, 1, or any other value. Because $f(t)$ and $\bar{f}(t)$ have the same properties and the same effect on continuous-time systems, the two functions will be considered the same function in engineering. Thus, we write (6.8) even though it may not hold at some isolated t. From now on, we assume that every periodic function meets the Dirichlet conditions.

Before we discuss the properties and physical meaning of c_m, consider an alternative form of the Fourier series.[4] Using

$$e^{\pm jm\omega_0 t} = \cos m\omega_0 t \pm j \sin m\omega_0 t$$

we can express the complex exponential Fourier series in (6.8) as

$$f(t) = \sum_{m=-\infty}^{-1} c_m e^{jm\omega_0 t} + c_0 + \sum_{m=1}^{\infty} c_m e^{jm\omega_0 t} = \sum_{m=1}^{\infty} c_{-m} e^{-jm\omega_0 t} + c_0 + \sum_{m=1}^{\infty} c_m e^{jm\omega_0 t}$$

$$= c_0 + \sum_{m=1}^{\infty} [(c_m + c_{-m})\cos m\omega_0 t + j(c_m - c_{-m})\sin m\omega_0 t] \tag{6.17}$$

[4]This can be skipped without loss of continuity.

If we define

$$a_0 = c_0 \qquad a_m = c_m + c_{-m} \qquad b_m = j(c_m - c_{-m}) \qquad \text{(6.18)}$$

then (6.17) can be written as

Trig fourier Series

$$f(t) = a_0 + \sum_{m=1}^{\infty} (a_m \cos m\omega_0 t + b_m \sin m\omega_0 t) \qquad \text{(6.19)}$$

Substituting (6.9) into (6.18) yields

• $a_n = 0$ *for odd functions*
• $b_n = 0$ *for even functions*

$$a_0 = \frac{1}{P} \int_{t_0}^{t_0 + P} f(t)\,dt \qquad \text{(6.20a)}$$

• $b_n = 0$ *for n = even for*
odd function (6.20b)
w/ ½ wave symmetry

$$a_m = \frac{2}{P} \int_{t_0}^{t_0 + P} f(t) \cos m\omega_0 t \, dt$$

and

$$f(t) = -f(t - T_0/2)$$

• $b_n = \frac{4}{n\pi}$ *for n = odd*

$$b_m = \frac{2}{P} \int_{t_0}^{t_0 + P} f(t) \sin m\omega_0 t \, dt \qquad \text{(6.20c)}$$

for $m = 1, 2, \dots$. The series in (6.19) with the coefficients in (6.20) is called the *trigonometric Fourier series*. It consists of only positive frequencies. We use mostly complex exponential Fourier series because it can be readily extended to aperiodic functions.

→ *odd square wave:* $f(t) = 1^{st}$ harm + 3^d harm + 5th harm +

6.2.1   Discrete Frequency Spectrum—Distribution of Power

Consider a periodic function $f(t)$ with complex exponential Fourier series

$$f(t) = \sum_{m=-\infty}^{\infty} c_m e^{jm\omega_0 t}$$

where ω_0 is the fundamental frequency and c_m, $m = 0, \pm 1, \pm 2, \dots$, is uniquely determinable from $f(t)$. Conversely, if we disregard some isolated points, the set c_m, $m = 0, \pm 1, \pm 2, \dots$, determines $f(t)$ uniquely. Thus, there is a one-to-one correspondence between $f(t)$ and $\{c_m\}$, and $\{c_m\}$ can be used as an alternative representation of $f(t)$. Because c_m reveals the frequency content of $f(t)$, and because c_m appears only at discrete frequencies $m\omega_0$, $m = 0, \pm 1, \pm 2 \cdots$, the set $\{c_m\}$ is called the *discrete* or *line* frequency spectrum of $f(t)$.

The discrete frequency spectrum is generally a complex-value function of m or $m\omega_0$. Let $c_m = \alpha_m + j\beta_m$. Then, the magnitude and phase of c_m are

$$|c_m| := \sqrt{\alpha_m{}^2 + \beta_m{}^2}$$

and

$$\sphericalangle c_m := \tan^{-1} \frac{\beta_m}{\alpha_m}$$

Thus, c_m can also be expressed in polar form as $c_m = |c_m| e^{j \sphericalangle c_m}$. The complex conjugate of c_m is

$$c_m^* = \alpha_m - j\beta_m = |c_m| e^{-j \sphericalangle c_m}$$

Now we develop some properties of c_m. If $f(t)$ is a real-valued function, that is, $f^*(t) = f(t)$, then we have

$$c_m^* = \left[\frac{1}{P} \int_{-P/2}^{P/2} f(t) e^{-jm\omega_0 t} dt \right]^* = \frac{1}{P} \int_{-P/2}^{P/2} f^*(t) e^{jm\omega_0 t} dt = \frac{1}{P} \int_{-P/2}^{P/2} f(t) e^{jm\omega_0 t} dt = c_{-m}$$

Thus, the discrete frequency spectrum of real $f(t)$ has the property

$$c_m^* = c_{-m}$$

This is called *conjugate symmetry*. If $c_m = \alpha_m + j\beta_m$, then $c_{-m} = \alpha_{-m} + j\beta_{-m} = c_m^* = \alpha_m - j\beta_m$, which implies $\alpha_{-m} = \alpha_m$, $\beta_{-m} = -\beta_m$ and, consequently,

$$|c_m| = |c_{-m}| \qquad \sphericalangle c_m = -\sphericalangle c_{-m}$$

Thus, the plot of $|c_m|$ as a function of m, called the *amplitude spectrum*, is an even function of m, and the plot of $\sphericalangle c_m$, called the *phase spectrum*, is an odd function of m, as Figure 6.8 shows. Because all functions encountered in practice are real valued, their discrete frequency spectra all have these properties.

Even and Odd Real Functions A real or complex function $f(t)$ is even if $f(t) = f(-t)$ and odd if $f(t) = -f(-t)$ for all t. Clearly, $\cos \omega_0 t$ is even, $\sin \omega_0 t$ is odd, and $\cos(\omega_0 t + 0.3)$ is neither even nor odd. If $f(t)$ is a real and even function of t, then c_m is a real and even function of m. If $f(t)$ is a real and odd function of t, then c_m is an imaginary and odd function of m. This can be established from

$$c_m = \frac{1}{P} \int_{-P/2}^{P/2} f(t) e^{-jm\omega_0 t} dt = \frac{1}{P} \int_{-P/2}^{P/2} f(t) [\cos m\omega_0 t - j \sin m\omega_0 t] dt$$

Figure 6.8 *Discrete frequency spectrum*

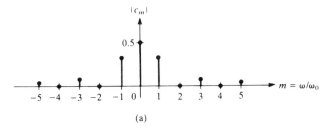

(a)

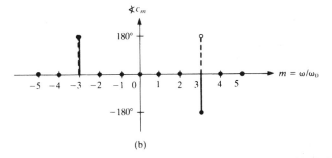

(b)

If $f(t)$ is real, then

$$
Re\ c_m = \frac{1}{P} \int_{-P/2}^{P/2} f(t) \cos m\omega_0 t\, dt
$$

and

$$
Im\ c_m = -\frac{1}{P} \int_{-P/2}^{P/2} f(t) \sin m\omega_0 t\, dt
$$

where *Re* stands for the real part and *Im* stands for the imaginary part. If $f(t)$ is even, then $f(t) \cos(-m\omega_0 t) = f(t) \cos m\omega_0 t$, and we have $Re\ c_m = Re\ c_{-m}$. On the other hand, $f(t) \sin(-m\omega_0 t) = -f(t) \sin m\omega_0 t$, and its integration from $-P/2$ to $P/2$ is zero. Thus, $Im\ c_m = -Im\ c_{-m} = 0$. This establishes that, if $f(t)$ is real and even, so is c_m. The other part can be similarly established. See Problems 6.37 and 6.38.

EXAMPLE 6.2.3

The discrete frequency spectrum of the periodic function $f(t)$ in Figure 6.6(a) for $P = 4$ and $a = 1$ was computed in Example 6.2.2 as

$$c_0 = 0.5 \qquad c_m = \frac{\sin(m\pi/2)}{\pi m}$$

Because $f(t)$ is real and even, so is c_m, as plotted in Figure 6.6(b). Now, let us plot the amplitude and phase spectra. The amplitude plot can be obtained from Figure 6.6(b) simply by changing negative numbers to positive numbers, as shown in Figure 6.8(a). The phase of real c_m is 0 if $c_m \geq 0$ and $\pm 180°$ if $c_m < 0$. If both phases of $c_3 = -1/3\pi$ and $c_{-3} = -1/3\pi$ are taken as $180°$, then the phase spectrum of c_m is even, as indicated by the hollow circles in Figure 6.8(b). This is inconsistent with the claim that the phase spectrum is odd for real $f(t)$. This difficulty can be resolved by choosing $\angle c_3 = -180°$ and $\angle c_{-3} = 180°$, as indicated by the solid circle in Figure 6.8(b). In general, two degrees are considered the same if they differ $360°$ or its multiples.

EXERCISE 6.2.2

Plot the amplitude and phase spectra of the periodic function in Example 6.2.1. ∎

Parseval's Formula and Distribution of Power Every periodic and bounded function $f(t)$ is, as discussed in Section 6.1.2, a power signal. To compute its average power, substitute (6.8) into (6.3) and interchange the order of integration and summation to yield

$$P_{av} := \frac{1}{P}\int_0^P f(t)f^*(t)dt = \frac{1}{P}\int_0^P \left(\sum_{m=-\infty}^{\infty} c_m e^{jm\omega_0 t}\right)f^*(t)dt$$

$$= \sum_{m=-\infty}^{\infty} c_m \left[\frac{1}{P}\int_{t=0}^P f(t)e^{-jm\omega_0 t}dt\right]^*$$

The term inside the brackets is c_m. Thus, we have

$$\boxed{\begin{aligned} P_{av} &= \frac{1}{P}\int_0^P f(t)f^*(t)dt = \frac{1}{P}\int_0^P |f(t)|^2 dt \\ &= \sum_{m=-\infty}^{\infty} c_m c_m^* = \sum_{m=-\infty}^{\infty} |c_m|^2 \end{aligned}}$$

(6.21)

This is called a *Parseval's formula*. It states that the average power equals the sum of $|c_m|^2$. Thus, the discrete frequency spectrum reveals the distribution of power at discrete frequencies $m\omega_0$, $m = 0, \pm 1, \pm 2, \ldots$.

The average power of a real periodic function can also be expressed in terms of a_m and b_m in (6.20). If $f(t)$ is real, then both a_m and b_m in (6.20) are real. Using $a_m = c_m + c_{-m}$ and $b_m = j(c_m - c_{-m})$, we can readily show

$$c_m = \frac{1}{2}(a_m - jb_m) \qquad c_{-m} = \frac{1}{2}(a_m + jb_m)$$

for $m = 1, 2, 3, \ldots$, and $c_0 = a_0$. Thus, we have

$$|c_m|^2 = |c_{-m}|^2 = \frac{1}{4}(a_m^2 + b_m^2)$$

for $m = 1, 2, 3, \ldots$. and (6.21) becomes

$$P_{av} = \frac{1}{P}\int_0^P |f(t)|^2 dt = a_0^2 + \frac{1}{2}\sum_{m=1}^{\infty}(a_m^2 + b_m^2) \qquad \text{(6.22)}$$

In particular, we have

$$\frac{1}{P}\int_0^P (a \cos m\omega_0 t + b \sin m\omega_0 t)^2 dt = \frac{1}{2}(a^2 + b^2)$$

This is consistent with the result in Example 6.1.5.

<hr />

EXAMPLE 6.2.4

Consider the periodic signal $f(t)$ shown in Figure 6.6(a) with $P = 4$ and $a = 1$. Find the percentage of the average power lying inside the frequency range $(-\pi, \pi)$. The total average power of every periodic function can be computed directly from $f(t)$ or indirectly from c_m. For this problem, it is simpler to compute it directly from $f(t)$ as

$$P_{av} = \frac{1}{P}\int_0^P |f(t)|^2 dt = \frac{1}{4}\int_{-1}^1 1 \, dt = \frac{2}{4} = 0.5$$

Because it is not possible to compute the power in the range $(-\pi, \pi)$ directly from $f(t)$, we must compute it from c_m. We see from Figure 6.6(b) that $c_0, c_1,$ and c_{-1} lie inside $(-\pi, \pi)$. Thus the power in the range is

$$(0.5)^2 + \left[\frac{1}{\pi}\right]^2 + \left[\frac{1}{\pi}\right]^2 = 0.25 + 0.10 + 0.10 = 0.45$$

Thus, $0.45/0.5 = 0.9$, or 90% of the power lies inside the frequency range $(-\pi, \pi)$.

$$\frac{P_{-\pi \text{ to } \pi}}{P_{av}} = 0.9$$

EXERCISE 6.2.3

Find the percentages of the average power of

$$f(t) = 3 + 2 \sin 2\pi t - \cos 2.1\pi t$$

at frequencies 0 and 2π rad/s.

[**ANSWERS:** 78.3%, 17.4%.]

do not do

*6.3   SIGNAL APPROXIMATION

This section will develop the Fourier series from a different point of view, showing that it is the solution of a minimization problem. To simplify the discussion, we will assume all functions are real valued.

Consider a real-valued function $f(t)$ defined over the time interval $[t_0, t_1]$. Let $\phi_i(t)$, $i = 1, 2, 3$, be three real-valued functions defined over the same interval. They are called the basis functions. The problem is to find three real constants α_i, $i = 1, 2, 3$, so that the function

$$\overline{f}(t) := \alpha_1 \phi_1(t) + \alpha_2 \phi_2(t) + \alpha_3 \phi_3(t) \qquad (6.23)$$

is as close as possible to $f(t)$ over the time interval $[t_0, t_1]$. In order to discuss closeness, we must develop a way of comparison or a criterion. We define

$$e(t) = f(t) - \overline{f}(t)$$

as the error at time t between $f(t)$ and $\overline{f}(t)$. Because we wish to minimize the error over the entire time interval $[t_0, t_1]$, we must define a total error. One possible definition is

$$E_1 = \int_{t_0}^{t_1} e(t)dt = \int_{t_0}^{t_1} [f(t) - \overline{f}(t)]dt$$

This, however, is not acceptable, because a small E_1 may not imply a small $e(t)$ over $[t_0, t_1]$ due to the possible cancellation of positive and negative errors. To avoid this type of cancellation, we may choose

$$E_2 = \int_{t_0}^{t_1} |e(t)| dt = \int_{t_0}^{t_1} |f(t) - \overline{f}(t)| dt \qquad (6.24)$$

This is the total absolute error. This is an acceptable criterion because a small E_2 implies a small $|e(t)|$. Unfortunately, the solution that minimizes (6.24) cannot be obtained analytically. In fact, the only criterion that can be solved simply and analytically is the quadratic error[5] defined by

$$E = \int_{t_0}^{t_1} e^2(t)dt = \int_{t_0}^{t_1} [f(t) - \bar{f}(t)]^2 dt \tag{6.25}$$

Although it is chosen for mathematical trackability, it turns out to be a good criterion. It penalizes equally positive and negative errors, and it penalizes large errors more heavily than small errors. Thus, a small E will yield a small $|e(t)|$ for all t. The quadratic error is the most often used criterion in system and signal analysis and design.

*6.3.1   Minimization Problem

Now we will restate the problem: Given a real $f(t)$ and three real basis functions ϕ_i, $i = 1, 2, 3$, find real α_i, $i = 1, 2, 3$, in (6.23) to minimize the E in (6.25). The set of α_i that yields the smallest possible E is called the optimal solution. Thus, the problem is called a minimization or an optimization problem.

Consider

$$E = \int_{t_0}^{t_1} [f(t) - \alpha_1 \phi_1(t) - \alpha_2 \phi_2(t) - \alpha_3 \phi_3(t)]^2 dt$$

Necessary conditions for α_i to minimize E are

$$\frac{\partial E}{\partial \alpha_i} = 0 \qquad \text{for } i = 1, 2, 3$$

Because all functions are assumed to be real, we can compute its derivative as

$$\frac{\partial E}{\partial \alpha_i} = \int \frac{\partial}{\partial \alpha_i} [f(t) - \alpha_1 \phi_1(t) - \alpha_2 \phi_2(t) - \alpha_3 \phi_3(t)]^2 dt$$

$$= -\int 2[f(t) - \alpha_1 \phi_1(t) - \alpha_2 \phi_2(t) - \alpha_3 \phi_3(t)]\phi_i(t)dt = 0$$

which implies, for $i = 1, 2, 3$,

$$\alpha_1 \int \phi_1(t)\phi_i(t)dt + \alpha_2 \int \phi_2(t)\phi_i(t)dt + \alpha_3 \int \phi_3(t)\phi_i(t)dt = \int f(t)\phi_i(t)dt$$

---

[5]For fixed and finite t_0 and t_1, the criterion is the same as the *mean square error* (MSE) defined by

$$E = \frac{1}{t_1 - t_0} \int_{t_0}^{t_1} e^2(t)dt$$

Note that if $e(t)$ is complex valued, then we use $|e(t)|^2$ instead of $e^2(t)$.

where all integrations are from t_0 to t_1. These three equations can be expressed in matrix form as

$$
\begin{bmatrix}
\int \phi_1^2(t)dt & \int \phi_2(t)\phi_1(t)dt & \int \phi_3(t)\phi_1(t)dt \\
\int \phi_1(t)\phi_2(t)dt & \int \phi_2^2(t)dt & \int \phi_3(t)\phi_2(t)dt \\
\int \phi_1(t)\phi_3(t)dt & \int \phi_2(t)\phi_3(t)dt & \int \phi_3^2(t)dt
\end{bmatrix}
\begin{bmatrix}
\alpha_1 \\
\alpha_2 \\
\alpha_3
\end{bmatrix}
=
\begin{bmatrix}
\int f(t)\phi_1(t)dt \\
\int f(t)\phi_2(t)dt \\
\int f(t)\phi_3(t)dt
\end{bmatrix}
\quad \textbf{(6.26)}
$$

Because this is a set of linear algebraic equations, solving the minimization problem reduces to solving a set of linear algebraic equations.

Two questions may be raised at this point: Does a solution exist in (6.26)? Because (6.26) is a necessary condition, not a sufficient condition, will its solution be the optimal solution? The answers to these questions are affirmative if $\phi_i(t)$, $i = 1, 2, 3$, is chosen to be linearly independent. The set $\phi_i(t)$, $i = 1, 2, 3$, is said to be linearly independent in $[t_0, t_1]$ if there exist no constant β_i, $i = 1, 2, 3$, not all zero, such that

$$
\beta_1 \phi_1(t) + \beta_2 \phi_2(t) + \beta_3 \phi_3(t) = 0 \qquad \text{for all } t \text{ in } [t_0, t_1] \qquad \textbf{(6.27)}
$$

This concept is an extension of the concept of linear independence of real vectors (see Appendix A) to time functions. It can be shown that the square matrix in (6.26) is nonsingular if and only if ϕ_i, $i = 1, 2, 3$, are linearly independent over the time interval $[t_0, t_1]$.[6] See Reference [4]. Thus, if ϕ_i are linearly independent, a solution exists in (6.26), and it is unique. Furthermore, the quadratic error using the solution will be the smallest possible one. That is, if we use any other α_i, the quadratic error will be larger than the one using the solution of (6.26). Thus, the solution of (6.26) yields the optimal solution of the minimization problem.

EXAMPLE 6.3.1

Consider the real-valued function $f(t)$ defined over $[0, 4]$ and shown in Figure 6.9. The basis functions are chosen as $\phi_1(t) = 1$, $\phi_2(t) = t$, and $\phi_3(t) = t^2$. Find real α_i, $i = 1, 2, 3$, to minimize the quadratic error

$$
E = \int_0^4 [f(t) - \alpha_1 - \alpha_2 t - \alpha_3 t^2]^2 dt
$$

---

[6]If (6.26) is solved using Gaussian elimination with partial pivoting, the nonsingularity of the square matrix in (6.26) will be checked automatically. (See Appendix A.) Therefore, there is no need to separately check the linear independence of the basis functions.

Figure 6.9 *A function defined over [0, 4]*

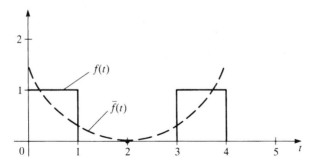

We compute

$$\int_0^4 \phi_1^2(t)dt = \int_0^4 1\,dt = 4$$

$$\int_0^4 \phi_1(t)\phi_2(t)dt = \int_0^4 t\,dt = \left.\frac{1}{2}t^2\right|_0^4 = 8$$

$$\int_0^4 f(t)\phi_1(t)dt = \int_0^1 dt + \int_3^4 dt = 2$$

$$\int_0^4 f(t)\phi_2(t)dt = \int_0^1 t\,dt + \int_3^4 t\,dt = \left.\frac{1}{2}t^2\right|_0^1 + \left.\frac{1}{2}t^2\right|_3^4 = \frac{1}{2} + \frac{7}{2} = 4$$

and so forth. The substitution of these into (6.26) yields

$$\begin{bmatrix} 4 & 8 & 64/3 \\ 8 & 64/3 & 64 \\ 64/3 & 64 & 1024/5 \end{bmatrix} \begin{bmatrix} \alpha_1 \\ \alpha_2 \\ \alpha_3 \end{bmatrix} = \begin{bmatrix} 2 \\ 4 \\ 38/3 \end{bmatrix} \tag{6.28}$$

This is a set of linear algebraic equations, and the methods discussed in Appendix A can be used to solve it. The solution is

$$\alpha_1 = \frac{23}{16} = 1.43 \qquad \alpha_2 = \frac{-45}{32} = -1.4 \qquad \alpha_3 = \frac{45}{128} = 0.35$$

Thus, the optimal solution is

$$\bar{f}(t) = 1.43 - 1.4t + 0.35t^2$$

and is plotted in Figure 6.9. We can see that the error is appreciable. However, if we use any other α_i, the quadratic error in (6.25) will be larger.

The dimension of (6.28) is three, and its solving is relatively simple. If the dimension becomes large, then its solving will become complex. This difficulty can be eliminated if the basis functions are chosen to be orthogonal. The following example demonstrates this.

EXAMPLE **6.3.2**

(Orthogonality of $\phi_i(t)$) Consider the function $f(t)$ in Figure 6.9 and

$$\bar{\phi}_1(t) = 1 \qquad \bar{\phi}_2(t) = \cos\left(\frac{\pi t}{2}\right) \qquad \bar{\phi}_3(t) = \sin\left(\frac{\pi t}{2}\right)$$

Find $\bar{\alpha}_i$ to minimize

$$\bar{E} = \int_0^4 [f(t) - \bar{\alpha}_1\bar{\phi}_1(t) - \bar{\alpha}_2\bar{\phi}_2(t) - \bar{\alpha}_3\bar{\phi}_3(t)]^2 dt$$

It is easy to verify that

$$\int_0^4 \bar{\phi}_i(t)\bar{\phi}_j(t)dt = 0 \qquad \text{for } i \neq j$$

Thus, the functions $\bar{\phi}_i(t)$, $i = 1, 2, 3$, are mutually orthogonal. We compute

$$\int_0^4 \bar{\phi}_1^2(t)dt = \int_0^4 dt = 4$$

$$\int_0^4 \cos^2\left(\frac{\pi t}{2}\right)dt = \int_0^4 \frac{1}{2}[1 + \cos \pi t]dt = \frac{4}{2} = 2$$

and

$$\int_0^4 \sin^2\left(\frac{\pi t}{2}\right)dt = \int_0^4 \frac{1}{2}[1 - \cos \pi t]dt = \frac{4}{2} = 2$$

The substitution of these into (6.26) yields

$$\begin{bmatrix} 4 & 0 & 0 \\ 0 & 2 & 0 \\ 0 & 0 & 2 \end{bmatrix} \begin{bmatrix} \bar{\alpha}_1 \\ \bar{\alpha}_2 \\ \bar{\alpha}_3 \end{bmatrix} = \begin{bmatrix} \int_0^4 f(t)dt \\ \int_0^4 f(t)\cos\left(\frac{\pi t}{2}\right)dt \\ \int_0^4 f(t)\sin\left(\frac{\pi t}{2}\right)dt \end{bmatrix} \qquad \textbf{(6.29)}$$

The square matrix in (6.29) is clearly nonsingular. Because the matrix is diagonal, the matrix equation reduces to three scalar equations and the solution can be readily obtained as

$$\bar{\alpha}_1 = \frac{1}{4} \int_0^4 f(t)dt$$

$$\bar{\alpha}_2 = \frac{1}{2} \int_0^4 f(t) \cos\left(\frac{\pi t}{2}\right)dt \tag{6.30}$$

and

$$\bar{\alpha}_3 = \frac{1}{2} \int_0^4 f(t) \sin\left(\frac{\pi t}{2}\right)dt$$

Thus, if the functions $\bar{\phi}_i(t)$ are orthogonal, the matrix equation can be reduced to scalar equations, and solving the equations becomes very simple. Furthermore, this simplicity holds no matter how large the number of basis functions is.

In (6.30), if we identify $P = 4$, then $\bar{\alpha}_1$, $\bar{\alpha}_2$, and $\bar{\alpha}_3$ equal, respectively, a_0, a_1, and b_1 in (6.20). In other words, the optimal solution of the minimization problem with respect to $\{1, \cos \pi t/2, \sin \pi t/2\}$ is the first three coefficients of the trigonometric Fourier series of $f(t)$ in (6.19). This is true in general, as the next section will establish.

*6.3.2   **From the Minimization Problem to the Fourier Series**

In a minimization problem, if the criterion is chosen as quadratic, and if the basis functions are linearly independent, then the solution can be obtained by solving a set of linear algebraic equations. If the basis functions are, in addition, orthogonal, then the solution can be obtained by solving a set of scalar equations. Using this procedure, we shall develop the Fourier series from a minimization problem.

Consider a real- or complex-valued function $f(t)$ defined over $[0, P]$ and consider the set of complex exponentials

$$\phi_m(t) = \frac{1}{\sqrt{P}} e^{jm\omega_0 t} \qquad \text{for } m = 0, \pm 1, \pm 2, \ldots, \pm N \tag{6.31}$$

for t in $[0, P]$ and $\omega_0 = 2\pi/P$. We shall find a set of complex number α_m so that the function

$$\bar{f}_N(t) = \sum_{m=-N}^{N} \alpha_m \phi_m(t)$$

minimizes the quadratic error

$$E = \int_0^P [f(t) - \bar{f}_N(t)][f(t) - \bar{f}_N(t)]^* dt \tag{6.32}$$

where the asterisk denotes the complex conjugate. This formulation is slightly more general than the one given in Section 6.3.1, where we considered only real functions. Otherwise, the basic idea is the same. Using (6.7), we can readily show

$$\int_0^P \phi_n(t)\phi_m^*(t) = 0 \qquad \text{for } n, m = 0, \pm 1, \cdots, \pm N \text{ and } n \neq m$$

Thus, all $\phi_n(t)$ and $\phi_m(t)$ are mutually orthogonal, and the set $\{\phi_m(t), m = 0, \pm 1, \ldots, \pm N\}$ is called an *orthogonal* set. Because of

$$\int_0^P \phi_m(t)\phi_m^*(t)dt = \frac{1}{P}\int_0^P e^{jm\omega_0 t} e^{-jm\omega_0 t} dt = \frac{P}{P} = 1 \tag{6.33}$$

each $\phi_m(t)$ is said to be normalized. Because the set is orthogonal as well as normalized, the set is called an *orthonormal* set. To simplify the presentation, we shall use the following two notations.

$$(x, y) := \int_0^P x(t)y^*(t)dt = (y, x)^* \tag{6.34}$$

$$|x|^2 := (x, x) = \int_0^P x(t)x^*(t)dt = \int_0^P |x(t)|^2 dt \tag{6.35}$$

(These notations have some special mathematical meanings and properties but, because they are not needed in our development, they will not be discussed.) We expand (6.32) as

$$E = \int_0^P f(t)f^*(t)dt - \sum_{m=-N}^N \alpha_m \int_0^P \phi_m(t)f^*(t)dt - \sum_{m=-N}^N \alpha_m^* \int_0^P f(t)\phi_m^*(t)dt$$
$$+ \sum_{m=-N}^N \sum_{l=-N}^N \alpha_m \alpha_l^* \int_0^P \phi_m(t)\phi_l^*(t)dt \tag{6.36}$$

where one of the indices in the last term has been changed to l to avoid possible confusion. Using the notations of (6.34) and (6.35) and the orthonormal property of $\phi_m(t)$, we can write (6.36) as

$$E = |f|^2 - \sum_{m=-N}^N [\alpha_m(\phi_m, f) + \alpha_m^*(f, \phi_m)] + \sum_{m=-N}^N \alpha_m \alpha_m^*$$

Because α_m is complex, we cannot take the derivative of E directly with respect to α_m as we did in the preceding section. Instead, we will proceed as follows. The subtraction and addition of

$$\sum_{m=-N}^{N} |(f, \phi_m)|^2 = \sum_{m=-N}^{N} (f, \phi_m)(f, \phi_m)^* = \sum_{m=-N}^{N} (f, \phi_m)(\phi_m, f)$$

to E yield

$$\begin{aligned}
E &= |f|^2 - \sum_{m=-N}^{N} |(f, \phi_m)|^2 + \sum_{m=-N}^{N} [(f, \phi_m)(\phi_m, f) - \alpha_m(\phi_m, f) \\
&\quad - \alpha_m^*(f, \phi_m) + \alpha_m \alpha_m^*] \\
&= |f|^2 - \sum_{m=-N}^{N} |(f, \phi_m)|^2 + \sum_{m=-N}^{N} [\alpha_m - (f, \phi_m)][\alpha_m^* - (\phi_m, f)] \\
&= |f|^2 - \sum_{m=-N}^{N} |(f, \phi_m)|^2 + \sum_{m=-N}^{N} |\alpha_m - (f, \phi_m)|^2
\end{aligned}$$

Because the first two terms after the last equality are independent of α_m, and the last term is always nonnegative (zero or positive), E is minimized if and only if

$$|\alpha_m - (f, \phi_m)| = 0$$

or

$$\alpha_m = \int_0^P f(t)\phi^*_m(t)dt = \frac{1}{\sqrt{P}}\int_0^P f(t)e^{-jm\omega_0 t}dt \qquad \textbf{(6.37)}$$

for all m. Consequently the optimal $(2N + 1)$ term approximation of $f(t)$ is

$$\overline{f}_N(t) = \sum_{m=-N}^{N} \alpha_m\phi_m(t) = \sum_{m=-N}^{N} \frac{1}{P}\left[\int_0^P f(t)e^{-jm\omega_0 t}dt\right] e^{jm\omega_0 t} \qquad \textbf{(6.38)}$$

This equation is identical to the first $(2N + 1)$ terms of the complex exponential Fourier series in (6.8) with c_m given in (6.9). Therefore, every finite-term Fourier series of $f(t)$ is an optimal approximation of $f(t)$ in the sense of having the smallest quadratic error. As $N \rightarrow \infty$, $\overline{f}_N(t)$ yields the Fourier series of $f(t)$ and the quadratic error becomes zero, that is,[7]

$$E = \int_0^P (f(t) - \overline{f}_\infty(t))(f(t) - \overline{f}_\infty(t))^* dt = 0 \qquad \textbf{(6.39)}$$

---

[7]$f(t)$ and $\overline{f}_\infty(t)$ may differ at a finite number of isolated t.

This shows that the Fourier series is indeed the optimal solution of a minimization problem.

Gibbs' Phenomenon   In the design of filters, it may be necessary to use a finite number of terms of the Fourier series to approximate a function $f(t)$ over $[0, P]$. Therefore, it is of practical and theoretical interest to see how a truncated Fourier series $\overline{f}_N(t)$ approaches $f(t)$ as $N \to \infty$. Intuitively, one would expect that the error between $\overline{f}_N(t)$ and $f(t)$ at every t in $[0, P]$ would become smaller as N increases. This is indeed the case where $f(t)$ is continuous. However, in the neighborhood of discontinuity points of $f(t)$, an unexpected phenomenon will occur.

Consider the function $f(t)$ in Figure 6.6. Its truncated Fourier series is, as computed in (6.15),

$$\overline{f}_N(t) = 0.5 + \frac{2}{\pi}\cos \omega_0 t - \frac{2}{3\pi}\cos 3\omega_0 t + \frac{2}{5\pi}\cos 5\omega_0 t - \cdots$$

$$+ (-1)^{N-1}\frac{2}{(2N-1)\pi}\cos (2N-1)\omega_0 t \qquad \textbf{(6.40)}$$

for $N = 1, 2, 3, \ldots$. The plots for $N = 3, 15$, and 31 are given in Figure 6.10. From the figure, we can see that the error between $f(t)$ and $\overline{f}_N(t)$, at t other than $t = 1$, does decrease as N increases. However, in the neighborhood of $t = 1$, there is a ripple between $f(t)$ and $\overline{f}_N(t)$. As N increases, the frequency of the ripple increases, and the ripple becomes narrower and moves closer to the discontinuity point. However, the magnitude of the ripple remains roughly the same no matter how large N is. The magnitude is about 9% of the amount of discontinuity. This phenomenon, called Gibbs' phenomenon,[8] occurs at every discontinuity point of a function. This phenomenon is not desirable in filter designs, and ways have been developed to eliminate the ripple. See, for example, Reference [3].

6.4   THE FOURIER TRANSFORM

We have defined the frequency content of periodic signals from the complex exponential Fourier series. Now we shall define the frequency content for aperiodic signals. Consider the signal $f(t)$ shown in Figure 6.6. It is periodic with period P. However, if P increases and becomes infinity, then $f(t)$ becomes aperiodic. Using this fact, we shall develop a transform for aperiodic functions. The complex exponential Fourier series of $f(t)$ in Figure 6.6 is computed in (6.14) as

$$f(t) = \sum_{m=-\infty}^{\infty} c_m e^{jm\omega_0 t}$$

---

[8]This phenomenon was first established by American mathematician Josiah Willard Gibbs (1839-1903).

Figure 6.10 *Truncated Fourier series*

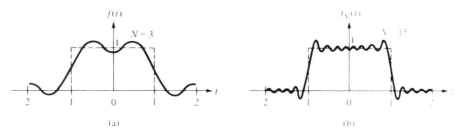

(a) (b)

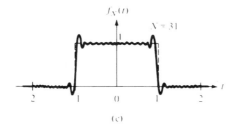

(c)

C_m = discrete/line frequency spectrum

with

pulse w/ period=P, A=1, duration=2a.

$$c_0 = \frac{2a}{P} \quad \text{and} \quad c_m = \frac{2 \sin m\omega_0 a}{m\omega_0 P}$$

✱

The set of c_m is the discrete frequency spectrum of $f(t)$. As P increases, $|c_m|$ becomes smaller and eventually approaches zero. Thus, c_m cannot be used to reveal the frequency content of aperiodic signals. Instead, we consider $Pc_m = 2 \sin m\omega_0 a/m\omega_0$. Its magnitude is independent of P. Figure 6.11 shows the plot of Pc_m for $P = P_1$, $P = 2P_1$, and $P = \infty$. We see that the envelope of Pc_m is the same for all P. Because $\omega_0 = 2\pi/P$, as P increases, the fundamental frequency ω_0 decreases and the discrete frequencies $m\omega_0$ at which Pc_m appears become closer and closer. Eventually, Pc_m appears at every frequency, as shown in Figure 6.11(c).

With this preliminary, we are ready to develop a transform for aperiodic functions. Consider the Fourier series

$$c_m = \frac{1}{P} \int_{-P/2}^{P/2} f(t) e^{-jm\omega_0 t} dt \qquad \textbf{(6.41a)}$$

and

$$f(t) = \frac{1}{P} \sum_{m=-\infty}^{\infty} c_m P e^{jm\omega_0 t} \qquad \textbf{(6.41b)}$$

Figure 6.11 *Trains of pulses with their Fourier coefficients: (a) P = P_1, (b) P = 2P_1, and (c) P = ∞*

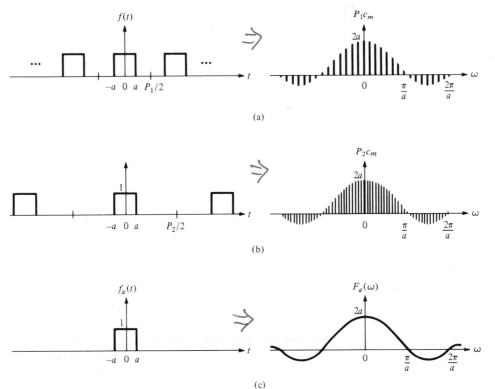

where $\omega_0 = 2\pi/P$. We define $F(m\omega_0) := Pc_m$ and write (6.41) as

$$F(m\omega_0) := c_m P = \int_{-P/2}^{P/2} f(t)e^{-jm\omega_0 t}\,dt \qquad \textbf{(6.42a)}$$

and

$$f(t) = \frac{1}{P}\sum_{m=-\infty}^{\infty} c_m P e^{jm\omega_0 t} = \frac{1}{2\pi}\sum_{m=-\infty}^{\infty} F(m\omega_0)e^{jm\omega_0 t}\omega_0 \qquad \textbf{(6.42b)}$$

Let $\omega := m\omega_0$. As $P \to \infty$, we have $\omega_0 = 2\pi/P \to 0$. In this case, ω becomes a continuum, and ω_0 can be written as $d\omega$. Furthermore, the summation in (6.42b) becomes an integration. Thus, (6.42a) and (6.42b) become, as $P \to \infty$,

Fourier Transform Pair

$$F(\omega) = \int_{-\infty}^{\infty} f(t)e^{-j\omega t}\,dt \qquad \text{for all } \omega \qquad \textbf{(6.43)}$$

$$f(t) = \frac{1}{2\pi}\int_{-\infty}^{\infty} F(\omega)e^{j\omega t}\,d\omega \qquad \text{for all } t \qquad \textbf{(6.44)}$$

This is the *Fourier transform* pair. $F(\omega)$ is called the Fourier transform of $f(t)$, denoted as $F(\omega) = \mathcal{F}[f(t)]$, and $f(t)$ is called the *inverse Fourier transform* of $F(\omega)$, denoted as $f(t) = \mathcal{F}^{-1}[F(\omega)]$. From its derivation, it is justifiable to call $F(\omega)$ the *frequency spectrum* of $f(t)$. More will be said regarding the frequency spectrum in a later section.

The Fourier transform is developed for aperiodic functions. However, not every aperiodic function has a Fourier transform. Sufficient conditions for $f(t)$ to have a Fourier transform are

1. $f(t)$ is absolutely integrable over $(-\infty, \infty)$, that is,

$$\int_{-\infty}^{\infty} |f(t)| \, dt \leq M < \infty$$

for some constant M, and

2. $f(t)$ has a finite number of discontinuities and a finite number of minima and maxima in every finite time interval.

These conditions are similar to the Dirichlet conditions for the Fourier series, and they are also called the Dirichlet conditions. Except for some mathematically contrived pathological cases, such as those in Figure 6.7, most functions satisfy the second condition. Therefore, if $f(t)$ is absolutely integrable, then it is said to be Fourier transformable. Here are some examples of absolutely integrable functions: Consider e^{-2t} for all t shown in Figure 6.4(b). Because the function approaches infinity as $t \to -\infty$, the total area under the function is infinity, and the function is not absolutely integrable. On the other hand, the function $f(t) = e^{-2t}$ for $t \geq 0$ and $f(t) = 0$ for $t < 0$ is absolutely integrable for

$$\int_{-\infty}^{\infty} |f(t)| \, dt = \int_{0}^{\infty} e^{-2t} dt = \left. \frac{1}{-2} e^{-2t} \right|_{t=0}^{\infty} = \frac{1}{-2}[0 - 1] = 0.5$$

The function $e^{-2|t|}$, for all t, shown in Figure 6.4(a) is also absolutely integrable; its integration equals $2 \times 0.5 = 1$. In general, e^{at}, for all t, is not absolutely integrable because any real a; however, $e^{a|t|}$ for all t is absolutely integrable if $a < 0$. Most signals encountered in practice are absolutely integrable because they are bounded (do not go to infinity) and of finite duration (start somewhere and stop somewhere), thus the Fourier transform and the frequency spectrum are widely used in practice. Note that absolute integrability is only a sufficient condition. As we will show later, although periodic functions are not absolutely integrable, their Fourier transforms can be defined by using impulses. Therefore *if $f(t)$ is absolutely integrable or is periodic, it is said to be Fourier transformable.*

As in the Fourier series, if $f(t)$ is not continuous at t_0, then the inverse Fourier transform yields the average of $f(t_0 +)$ and $f(t_0 -)$, or

$$\frac{f(t_0 +) + f(t_0 -)}{2} = \frac{1}{2\pi} \int_{-\infty}^{\infty} F(\omega) e^{j\omega t_0} d\omega$$

no matter what value $f(t_0)$ is. In engineering, two functions that differ only at isolated instants of time have the same effect on continuous-time systems and, therefore, can be considered as the same function. Thus, we may consider that there is a one-to-one relationship between $f(t)$ and its Fourier transform $F(\omega)$.

EXAMPLE **6.4.1**

Consider the function $f(t) = e^{-a|t|}$ for all t, where a is real. If $a < 0$, the function increases exponentially as $t \to \pm\infty$. Hence, it is not absolutely integrable and its Fourier transform does not exist. If $a > 0$, the function is absolutely integrable, and its Fourier transform is

$$F(\omega) = \int_{-\infty}^{\infty} f(t)e^{-j\omega t}dt = \int_{-\infty}^{0} e^{at}e^{-j\omega t}dt + \int_{0}^{\infty} e^{-at}e^{-j\omega t}dt$$

$$= \frac{1}{a - j\omega} + \frac{1}{a + j\omega} = \frac{2a}{a^2 + \omega^2} \tag{6.45}$$

It happens to be real for all ω. Both $f(t) = e^{-a|t|}$, for $a > 0$, and its Fourier transform $F(\omega)$ are plotted in Figure 6.12.

EXERCISE 6.4.1

Show $\mathscr{F}[\delta(t)] = 1$ and $\mathscr{F}[\delta(t - t_0)] = e^{-j\omega t_0}$.   ■

The Fourier transform $F(\omega)$ of real- or complex-valued $f(t)$ is generally a complex-valued function of ω. Therefore, it is often written as

$$F(\omega) = Re\ F(\omega) + j\ Im\ F(\omega) = A(\omega)e^{j\theta(\omega)}$$

with

$$A(\omega) := |F(\omega)| = [(Re\ F(\omega))^2 + (Im\ F(\omega))^2]^{1/2}$$

and

$$\theta(\omega) := \sphericalangle F(\omega) = \tan^{-1}\frac{Im\ F(\omega)}{Re\ F(\omega)}$$

We call $F(\omega)$ the *frequency spectrum*, $A(\omega)$ the *amplitude spectrum*, and $\theta(\omega)$ the *phase spectrum of $f(t)$*. Their physical meanings will be discussed later.

If $f(t)$ is real, or $f(t) = f^*(t)$, where the asterisk denotes the complex conjugate, then

$$F^*(\omega) = (\int_{-\infty}^{\infty} f(t)e^{-j\omega t}\,dt)^* = \int_{-\infty}^{\infty} f(t)e^{j\omega t}\,dt = F(-\omega)$$

This is called conjugate symmetry, which implies

$$A(-\omega)e^{j\theta(-\omega)} = F(-\omega) = F^*(\omega) = Re\ F(\omega) - j\ Im\ F(\omega) = A(\omega)e^{-j\theta(\omega)}$$

Thus, for $f(t)$ real, we have the following important properties

$$\boxed{A(\omega) = A(-\omega)}$$ (6.46a)

and

$$\boxed{\theta(\omega) = -\theta(-\omega)}$$ (6.46b)

In other words, for real $f(t)$, the amplitude spectrum is an even function of ω, and the phase spectrum is an odd function of ω.

A real or complex function $f(t)$ is even if $f(t) = f(-t)$ and odd if $f(t) = -f(-t)$, for all t. If $f(t)$ is a real and even function of t, then $F(\omega)$ is a real and even function of ω. If $f(t)$ is a real and odd function of t, then $F(\omega)$ is an imaginary and odd function of ω. These assertions can be established as in the Fourier series case. For example, the time function in Figure 6.12(a) is real and even; its Fourier transform was computed in Equation (6.45) as

$$F(\omega) = \frac{2a}{a^2 + \omega^2}$$

It is a real-valued and even function of ω as shown in Figure 6.12(b). It can be shown that, if $f(t)$ is absolutely integrable, then $F(\omega)$ is a continuous function of ω. See Reference [36]. For a listing of the symmetric properties of $f(t)$ and $F(\omega)$, see Problem 6.37.

Before proceeding, note that the Fourier transform is a linear operator, that is,

Figure 6.12 (a) f(t) = $e^{-a|t|}$, for a > 0, and (b) its Fourier transform

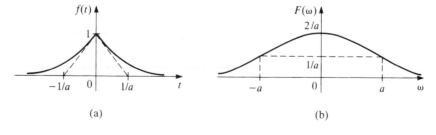

(a) (b)

$$\boxed{\mathscr{F}[\alpha_1 f_1(t) + \alpha_2 f_2(t)] = \alpha_1 \mathscr{F}[f_1(t)] + \alpha_2 \mathscr{F}[f_2(t)]} \tag{6.47}$$

The Fourier transform is defined through an integration. Because the integration is a linear operator, so is the Fourier transform.

6.4.1 From the Laplace Transform to the Fourier Transform— Absolutely Integrable Functions

This section will establish the relationship between the (one-sided) Laplace transform and the Fourier transform. The Fourier transform is defined for a function $f(t)$ defined for all t in $(-\infty, \infty)$. The Laplace transform, however, is defined for $f(t)$ with $t \geq 0$ and completely disregards $f(t)$ for $t < 0$. Therefore, some manipulation is needed before we can establish their relationship.

Consider an absolutely integrable function $f(t)$ defined over $(-\infty, \infty)$. As discussed in (1.13) and (1.16b), we can decompose $f(t)$ into

$$f(t) = f_-(t) + f_+(t) = f(t)q(-t) + f(t)q(t)$$

where $f_-(t) := f(t)q(-t)$ is a negative-time function, $f_+(t) := f(t)q(t)$ is a positive-time function, and $q(t)$ is the unit step function. As discussed in Section 1.3, the negative-time function $f_-(t) = f(t)q(-t)$ becomes a positive-time function if we flip its time as $f_-(-t) = f(-t)q(t)$. Now we show

$$\boxed{\mathscr{F}[f(t)] = \mathscr{F}[f_+(t)] + \mathscr{F}[f_-(t)] = L[f_+(t)]\Big|_{s=j\omega} + L[f_-(-t)]\Big|_{s=-j\omega}} \tag{6.48}$$

In other words, the Fourier transform of (two-sided) absolutely integrable functions can be obtained from the (one-sided) Laplace transform. We will first prove the positive-time part and then prove the negative-time part.

Positive-time Functions The Fourier transform of $f_+(t)$ is

$$F_+(\omega) = \int_{-\infty}^{\infty} f_+(t)e^{-j\omega t} dt = \int_0^{\infty} f_+(t)e^{-j\omega t} dt \tag{6.49}$$

and the Laplace transform of $f_+(t)$ is

$$\overline{F}_+(s) = \int_0^{\infty} f_+(t)e^{-st} dt \tag{6.50}$$

Note that F_+ has been used in (6.49); therefore, we use $\overline{F}_+$ to denote the Laplace transform[9]. If $f_+(t)$ is absolutely integrable, the region of convergence of $\overline{F}_+(s)$ includes the imaginary axis of the s-plane. Thus, $\overline{F}_+(s)$ is defined at $s = j\omega$ for every ω. Substituting $s = j\omega$ into (6.50) yields

---

[9]If we use the same notation, confusion will arise. For example, the Laplace transform of $f(t) = e^{-t}$ for $t \geq 0$ is $1/(s + 1)$. If it is denoted by $F(s)$, then $F(\omega) = 1/(\omega + 1)$, which is not the Fourier transform of $f(t)$.

$$\overline{F}_+(j\omega) = \int_0^\infty f_+(t)e^{-j\omega t}dt$$

which is identical to the Fourier transform in (6.49). Thus, if $f_+(t)$ is a positive-time absolutely integrable function, then

$$F_+(\omega) = \mathscr{F}[f_+(t)] = \left.\underline{L}[f_+(t)]\right|_{s=j\omega} \tag{6.51a}$$

or

$$F_+(\omega) = \mathscr{F}[f(t)q(t)] = \left.\underline{L}[f(t)q(t)]\right|_{s=j\omega} \tag{6.51b}$$

Negative-time Functions The Fourier transform of $f_-(t)$ is

$$F_-(\omega) = \int_{t=-\infty}^\infty f_-(t)e^{-j\omega t}dt = \int_{-\infty}^0 f_-(t)e^{-j\omega t}dt$$

which becomes, after substituting $t = -\tau$,

$$F_-(\omega) = -\int_{\tau=\infty}^0 f_-(-\tau)e^{j\omega \tau}d\tau = \int_0^\infty f_-(-\tau)e^{j\omega \tau}d\tau \tag{6.52}$$

The Laplace transform of the negative-time function $f_-(t)$ is zero, but the Laplace transform of the positive-time function $f_-(-t)$ is

$$\underline{L}[f_-(-t)] = \overline{F}_-(s) = \int_0^\infty f_-(-t)e^{-st}dt$$

which is identical to (6.52) after the substitution of $s = -j\omega$. Thus, we have

$$\mathscr{F}[f_-(t)] = \left.\underline{L}[f_-(-t)]\right|_{s=-j\omega} \quad \text{or} \quad \mathscr{F}[f(t)q(-t)] = \left.\underline{L}[f(-t)q(t)]\right|_{s=-j\omega} \tag{6.53}$$

This completes the proof of (6.48).

EXAMPLE **6.4.2**

Find the Fourier transform of $f(t) = e^{-at}$, for $t \geq 0$, where a is real and positive. The function $f(t)$ is positive-time and absolutely integrable for $a > 0$. Its Laplace transform is $1/(s + a)$. Thus, its Fourier transform is

$$F(\omega) = \mathscr{F}[e^{-at}, t \geq 0] = \left. \frac{1}{s + a} \right|_{s=j\omega} = \frac{1}{j\omega + a}$$

EXAMPLE 6.4.3

Find the Fourier transform of

$$f(t) = \begin{cases} f_+(t) = e^{-2t} & \text{for } t \geq 0 \\ f_-(t) = -e^{2t} & \text{for } t < 0 \end{cases}$$

This function is a real-valued and an odd function of t, as shown in Figure 6.13(a). The Laplace transform of the positive-time function $f_+(t)$ is $1/(s + 2)$. Thus, we have

$$\mathscr{F}[f_+(t)] = \frac{1}{j\omega + 2}$$

The function $f_-(t)$ is negative time. However $f_-(-t) = -e^{-2t} =: \overline{f}(t)$ is positive time. The Laplace transform of $\overline{f}(t)$ is $-1/(s + 2)$. Thus, using (6.53), we have,

$$\mathscr{F}[f_-(t)] = \left. \frac{-1}{s + 2} \right|_{s=-j\omega} = \frac{-1}{2 - j\omega}$$

Figure 6.13 *Amplitude and phase spectra*

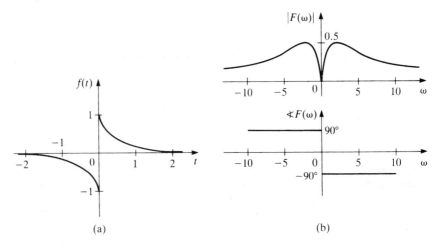

(a)   (b)

and so the Fourier transform of $f(t)$ is

$$F(\omega) = \mathscr{F}[f(t)] = \frac{1}{2 + j\omega} - \frac{1}{2 - j\omega} = \frac{-2j\omega}{4 + \omega^2}$$

It is a pure imaginary function of ω. Because

$$F(-\omega) = \frac{-2j(-\omega)}{4 + (-\omega)^2} = \frac{2j\omega}{4 + \omega^2} = -F(\omega)$$

it also is an odd function of ω. This is consistent with the assertion that if $f(t)$ is real and odd, then $F(\omega)$ is pure imaginary and odd.

The phase of $F(\omega)$ is $-90°$ for all $\omega > 0$ and $90°$ for all $\omega < 0$. The amplitude of $F(\omega)$ is

$$|F(\omega)| = \frac{2|\omega|}{4 + \omega^2}$$

which equals 0 at $\omega = 0$, 0.4 at $\omega = \pm 1$, 0.5 at $\omega = \pm 2$, and so forth. Using these data, the amplitude and phase spectra can be plotted as shown in Figure 6.13(b). Clearly, the amplitude spectrum is even, and the phase spectrum is odd.

Note that, if $f(t)$ is an even function of t, that is, $f_+(t) = f_-(-t)$, then (6.48) reduces to

$$\mathscr{F}[f(t)] = \mathscr{F}[f_+(t)] + \mathscr{F}[f_-(-t)] = \left. \mathcal{L}[f_+(t)] \right|_{s=j\omega} + \left. \mathcal{L}[f_+(t)] \right|_{s=-j\omega}$$

EXERCISE 6.4.2

Find the Laplace and Fourier transforms of e^{-3t}, e^{-t}, e^{2t}, and $e^{-t} \sin 10t$, for $t \geq 0$. Sketch their amplitude and phase spectra.

[**ANSWERS:** $1/(j\omega + 3)$, $1/(j\omega + 1)$, not defined, $10/(-\omega^2 + 2j\omega + 101)$.] ■

EXERCISE 6.4.3

Find the Fourier transforms of $e^{-|t|}$ and $e^{-|t|} \sin 10|t|$, for all t. Are they real and even?

[**ANSWERS:** $2/(1 + \omega^2)$, $20(-\omega^2 + 101)/(\omega^4 - 198\omega^2 + (101)^2)$. Yes.] ■

To conclude this section, we mention that (6.48) holds only if $f(t)$ is absolutely integrable. For example, $f(t) = 1$, for all t, is not absolutely integrable. If we use (6.48) to compute its Fourier transform, then

$$\mathscr{F}[f(t) = 1] = \mathcal{L}[q(t)]\Big|_{s=j\omega} + \mathcal{L}[q(t)]\Big|_{s=-j\omega} = \frac{1}{j\omega} + \frac{1}{-j\omega} = 0$$

This is incorrect, as the next section will disclose.

EXERCISE 6.4.4

Show that, if we use (6.48) to compute the Fourier transform of $f(t) = \sin \omega_0 t$, for all t, the result will be zero. This illustrates that (6.48) is not applicable if $f(t)$ is periodic. ■

6.4.2   Fourier Transform of Periodic Functions

In analysis, we often encounter step functions and sinusoids. These functions and, more generally, periodic functions are not absolutely integrable. However, their Fourier transforms are still defined. Instead of computing them directly, we will discuss first the inverse Fourier transform of the impulse $F(\omega) = \delta(\omega - \omega_0)$. Using (6.44) and the sifting property of the impulse, we have

$$f(t) = \frac{1}{2\pi}\int_{-\infty}^{\infty} F(\omega)e^{j\omega t}d\omega = \frac{1}{2\pi}\int_{-\infty}^{\infty} \delta(\omega - \omega_0)e^{j\omega t}d\omega = \frac{1}{2\pi}e^{j\omega_0 t}$$

for all t. Thus, the inverse Fourier transform of $\delta(\omega - \omega_0)$ is the complex exponential $e^{j\omega_0 t}/2\pi$. Because of the one-to-one relationship between $f(t)$ and $F(\omega)$, we conclude

$$\mathscr{F}[e^{j\omega_0 t}] = 2\pi\delta(\omega - \omega_0) \tag{6.54}$$

and, in particular, by setting $\omega_0 = 0$,

$$\mathscr{F}[f(t) = 1, \text{ for all } t \text{ in } (-\infty, \infty)] = 2\pi\delta(\omega) \tag{6.55}$$

Therefore, the Fourier transform of $f(t) = 1$ is an impulse at $\omega = 0$ with weight 2π, as shown in Figure 6.14(a). Because the Fourier transform is zero for all $\omega \neq 0$, $f(t) = 1$ does not contain any nonzero frequency and is called a DC signal.

Using the linearity property of the Fourier transform and (6.54), we have

$$\mathscr{F}[\sin \omega_0 t] = \mathscr{F}[\frac{1}{2j}(e^{j\omega_0 t} - e^{-j\omega_0 t})] = \frac{\pi}{j}[\delta(\omega - \omega_0) - \delta(\omega + \omega_0)]$$

$$= -\pi j\delta(\omega - \omega_0) + \pi j\delta(\omega + \omega_0) \tag{6.56a}$$

Figure 6.14 *Fourier transform pairs of periodic functions*

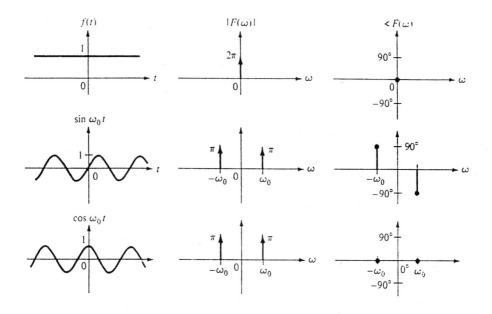

and

$$\mathscr{F}[\cos \omega_0 t] = \mathscr{F}[\frac{1}{2}(e^{-j\omega_0 t} + e^{j\omega_0 t})] = \pi[\delta(\omega - \omega_0) + \delta(\omega + \omega_0)] \quad \textbf{(6.56b)}$$

These are plotted in Figure 6.14(b) and (c). Although the amplitude spectra of sin $\omega_0 t$ and cos $\omega_0 t$ are the same, their phase spectra are different. Their frequency spectra are zero everywhere except at $\pm \omega_0$.

Now let us consider the Fourier transform of a general periodic function. Let the Fourier series of a periodic function $f(t)$ with period P and fundamental frequency $\omega_0 = 2\pi/P$ be

$$f(t) = \sum_{m=-\infty}^{\infty} c_m e^{jm\omega_0 t} \quad \textbf{(6.57)}$$

Then, the Fourier transform of $f(t)$ is, using the linearity property and (6.54),

$$F(\omega) = \mathscr{F}[f(t)] = \sum_{m=-\infty}^{\infty} 2\pi c_m \delta(\omega - m\omega_0) \quad \textbf{(6.58)}$$

It consists of a sequence of impulses with weight $2\pi c_m$ at $\omega = m\omega_0$, $m = 0$, $\pm 1, \pm 2, \dots$. If we replace the coefficient c_m at $\omega = m\omega_0$ of the Fourier series with an impulse with weight $2\pi c_m$ at the same frequency, then we obtain the

Fourier transform. Thus, the Fourier series and Fourier transform of periodic functions are closely related.

EXAMPLE 6.4.4

Consider the sequence of impulses shown in Figure 6.15. It is a periodic function of t with period $P = T$ and can be written as

$$r(t) = \sum_{k=-\infty}^{\infty} \delta(t - kT) \tag{6.59}$$

We express $r(t)$ in the Fourier series as

$$r(t) = \sum_{m=-\infty}^{\infty} c_m e^{jm\omega_0 t}$$

with $\omega_0 = 2\pi/T$ and, using (6.9),

$$c_m = \frac{1}{T} \int_{-T/2}^{T/2} \delta(t) e^{-jm\omega_0 t} dt = \frac{1}{T} e^{-jm\omega_0 t} \bigg|_{t=0} = \frac{1}{T}$$

It is independent of m. Thus, the Fourier series of $r(t)$ in (6.59) is

$$r(t) = \sum_{m=-\infty}^{\infty} \frac{1}{T} e^{jm\omega_0 t} \tag{6.60}$$

Figure 6.15 *A sequence of impulses*

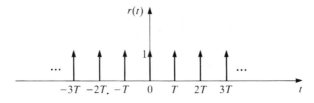

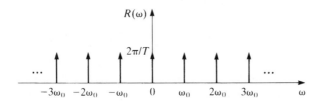

and its Fourier transform is, using (6.58),

$$R(\omega) = \sum_{m=-\infty}^{\infty} \frac{2\pi}{T} \delta(\omega - m\omega_0) \qquad \text{(6.61)}$$

It is also a sequence of impulses, as Figure 6.15 illustrates.

---

*6.4.3   Fourier Transform of Positive-time Periodic Functions[10]

For absolutely integrable positive-time functions, the Fourier transform simply equals the Laplace transform with s replaced by $j\omega$, as discussed in (6.51). Equation (6.51), however, is not applicable to positive-time periodic functions. In this section, we shall modify it for positive-time periodic functions.

Before proceeding, let us consider a property of the Fourier transform. If $F(\omega) = \mathcal{F}[f(t)]$, then

$$\mathcal{F}[f(t)e^{j\omega_0 t}] = F(\omega - \omega_0) \qquad \text{(6.62)}$$

This can be verified from (6.43) as

$$\mathcal{F}[f(t)e^{j\omega_0 t}] = \int_{-\infty}^{\infty} f(t)e^{j\omega_0 t}e^{-j\omega t}dt = \int_{-\infty}^{\infty} f(t)e^{-j(\omega-\omega_0)}dt = F(\omega - \omega_0)$$

This is called the *frequency shifting* property of the Fourier transform.

The Fourier transform of periodic functions was developed from $\mathcal{F}[e^{j\omega_0 t}] = 2\pi\delta(\omega - \omega_0)$. The Fourier transform of positive-time periodic functions will be developed from $\mathcal{F}[e^{j\omega_0 t}q(t)]$ by using (6.62). First, however, we must develop the Fourier transform of the unit step function $q(t)$, which will be developed from the Fourier transform of the signum.

EXAMPLE 6.4.5

(Fourier transform of the signum) Consider the function, called the *signum*, defined by

$$sgn\,(t) = \begin{cases} 1 & \text{if } t > 0 \\ 0 & \text{if } t = 0 \\ -1 & \text{if } t < 0 \end{cases} \qquad \text{(6.63)}$$

---

[10] Although, as the asterisk indicates, this section may be skipped without loss of continuity, it is recommended that the last paragraph of this subsection be read, because why the Fourier transform of positive-time periodic functions is rarely used is explained there.

It is plotted in Figure 6.16. The function is clearly not absolutely integrable. However, if it is approximated by $e^{-a|t|}$ sgn (t), where a is a small positive number, then it is absolutely integrable. We compute, as in (6.45),

$$\mathscr{F}[\text{sgn}\,(t)] = \lim_{a \to 0} \,[-\int_{-\infty}^{0} e^{at} e^{-j\omega t} dt + \int_{0}^{\infty} e^{-at} e^{-j\omega t} dt\,]$$

$$= \lim_{a \to 0} \,[\frac{-1}{a - j\omega} + \frac{1}{a + j\omega}] \tag{6.64}$$

Care must be exercised in setting $a \to 0$ in (6.64). If $\omega \neq 0$, (6.64) is defined as $a \to 0$ and equals $2/j\omega$. If $\omega = 0$, (6.64) becomes

$$\lim_{a \to 0} \,[\frac{-1}{a} + \frac{1}{a}] = 0$$

Thus, we conclude that

$$\mathscr{F}[\text{sgn}\,(t)] = \frac{2}{j\omega} \tag{6.65}$$

except at $\omega = 0$, where it equals 0. This subtlety is often neglected in engineering, and (6.65) is often used without any qualification.

EXAMPLE 6.4.6

(Fourier transform of the step function) Consider the unit step function $q\,(t)$

$$q\,(t) = \begin{cases} 1 & \text{for } t \geq 0 \\ 0 & \text{for } t < 0 \end{cases}$$

Figure 6.16 *The signum*

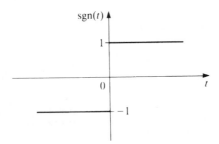

It can be written as

$$q(t) = \frac{1}{2} + \frac{1}{2} \text{ sgn } (t)$$

The Fourier transforms of 1 and sgn (t) are, respectively, $2\pi\delta(\omega)$ and $2/j\omega$; therefore, we have

$$Q(\omega) = \mathcal{F}[q(t)] = \pi\delta(\omega) + \frac{1}{j\omega} \tag{6.66}$$

We see that the Fourier transform of $q(t)$ is different from the Laplace transform of $q(t)$ with s replaced by $j\omega$ because it contains the extra term $\pi\delta(\omega)$.

You may wonder why we used sgn (t) to compute (6.66). Consider $e^{-at}q(t)$. It is absolutely integrable for $a > 0$, and its Fourier transform is $1/(j\omega + a)$. As $a \to 0$, $e^{-at}q(t)$ becomes $q(t)$, and its Fourier transform becomes

$$\overline{Q}(\omega) := \lim_{a \to 0} \frac{1}{j\omega + a}$$

If $\omega \neq 0$, $\overline{Q}$ is the same as (6.66). However, if $\omega = 0$, $\overline{Q}$ becomes indeterminate as $a \to 0$. Therefore, additional manipulation is needed to obtain the term $\pi\delta(\omega)$ in (6.66). This manipulation is not simple, which is the reason we use $\mathcal{F}[\text{sgn } (t)]$ to compute $\mathcal{F}[q(t)]$.

EXAMPLE 6.4.7

Using the shifting property, we have

$$\mathcal{F}[e^{j\omega_0 t}q(t)] = \frac{1}{j(\omega - \omega_0)} + \pi\delta(\omega - \omega_0) \qquad \circledast \tag{6.67}$$

This is a key equation in the following development.

Given these preliminaries, we are ready to modify (6.51) for positive-time periodic functions. Let $f(t)$ be a periodic function defined over $(-\infty, \infty)$. Because $q(t) = 0$, for $t < 0$, the function $f(t)q(t)$ is a positive-time function. Now we claim

$$\mathcal{F}[f(t)q(t)] = \mathcal{L}[f(t)q(t)]\Big|_{s=j\omega} + \frac{1}{2}\mathcal{F}[f(t)] \qquad \circledast \tag{6.68}$$

Indeed, if $f(t)$ is expressed as in (6.57), then the (one-sided) Laplace transform of $f(t)q(t)$ is, using the linearity property and Table 4.1,

$$\mathcal{L}[f(t)q(t)] = \sum_{-\infty}^{\infty} \mathcal{L}[c_m e^{jm\omega_0 t}] = \sum_{-\infty}^{\infty} \frac{c_m}{s - jm\omega_0} \tag{6.69}$$

and the Fourier transform of $f(t)q(t)$ is, using (6.67),

$$\mathcal{F}[f(t)q(t)] = \sum_{-\infty}^{\infty} \mathcal{F}[c_m e^{jm\omega_0 t} q(t)] \tag{6.70}$$

$$= \sum_{-\infty}^{\infty} c_m [\frac{1}{j(\omega - m\omega_0)} + \pi\delta(\omega - m\omega_0)]$$

which contains the terms $\mathcal{L}[f(t)q(t)]$, with s replaced by $j\omega$, and $\pi c_m \delta(\omega - m\omega_0)$, which equals one half of the Fourier transform of $f(t)$, as can be seen from (6.58). This establishes (6.68).

EXAMPLE 6.4.8

Consider $\sin \omega_0 t$, for $t \geq 0$, or $(\sin \omega_0 t)q(t)$, for all t. The Laplace transform of $(\sin \omega_0 t)q(t)$ is, from Table 4.1, $\omega_0/(s^2 + \omega_0^2)$. The Fourier transform of $\sin \omega_0 t$ is computed in (6.56a); thus, we have

$$\mathcal{F}[(\sin \omega_0 t)q(t)] = \frac{\omega_0}{\omega_0^2 - \omega^2} + \frac{\pi j}{2}[\delta(\omega + \omega_0) - \delta(\omega - \omega_0)] \tag{6.71}$$

EXERCISE 6.4.5

Verify

$$\mathcal{F}[(\cos \omega_0 t)q(t)] = \frac{j\omega}{\omega_0^2 - \omega^2} + \frac{\pi}{2}[\delta(\omega + \omega_0) + \delta(\omega - \omega_0)] \tag{6.72}$$

For easy reference, some of the preceding Fourier transform pairs are listed in Table 6.1. The Fourier transform of positive-time periodic functions is rarely used in analysis. One reason for this is that its frequency spectrum is not intuitively appealing. For example, a step function is often considered as a DC signal. As such, it should not contain any nonzero frequency component. However, the spectrum of a step function is nonzero for all ω, as shown in Table 6.1. Now, if we extend the step function to negative time, then its

Table 6.1 *Some Fourier Transform Pairs*

Waveform	$f(t)$		$F(\omega)$	$|F(\omega)|$
	$\delta(t)$	$\longleftrightarrow$	1	
	1	$\longleftrightarrow$	$2\pi\delta(\omega)$	
	$u(t)$	$\longleftrightarrow$	$\dfrac{1}{j\omega} + \pi\delta(\omega)$	
	$\sin \omega_0 t$	$\longleftrightarrow$	$-j\pi\delta(\omega - \omega_0) + j\pi\delta(\omega + \omega_0)$	
	$(\sin \omega_0 t)\,u(t)$	$\longleftrightarrow$	$\dfrac{\omega_0}{\omega_0^2 - \omega^2} + \dfrac{\pi j}{2}\left[\delta(\omega + \omega_0) - \delta(\omega - \omega_0)\right]$	

spectrum is zero everywhere except at $\omega = 0$. This is consistent with the usual notion of DC signals. Similar remarks apply to positive-time sine and cosine functions. Thus, in signal analysis, periodic functions are usually defined for all t in $(-\infty, \infty)$ or are two-sided. Yet another reason for not using the Fourier transform of positive-time periodic functions is that its use is considerably more complicated than the use of the Laplace transform in system analysis, as will be shown in Section 6.5.1.

6.4.4 **Frequency Spectrum—Spectral Density—Distribution of Energy**

The set of complex exponential Fourier coefficients $\{c_m\}$ reveals the frequency content of periodic $f(t)$ at discrete frequencies $m\,2\pi/P$, $m = 0, \pm 1, \pm 2, \ldots$, and is called the discrete or line frequency spectrum. The Fourier transform $F(\omega)$ of aperiodic functions is defined as Pc_m with $P \to \infty$. Therefore, $F(\omega)$ also reveals the frequency content of $f(t)$ and will be called the frequency spectrum of $f(t)$. In this section, we will discuss its physical meaning.

Consider an absolutely integrable function $f(t)$. Let $F(\omega)$ be its Fourier transform or frequency spectrum. We can express the total energy of $f(t)$ in terms of the spectrum. As was discussed in Section 6.1.2, the total energy of $f(t)$ is

$$\text{total energy} := E = \int_{-\infty}^{\infty} |f(t)|^2 = \int_{-\infty}^{\infty} f(t)f^*(t)dt \tag{6.73}$$

Substituting (6.44) into (6.73) and interchanging the order of integrations yield

$$E = \int_{-\infty}^{\infty} |f(t)|^2 dt = \int_{-\infty}^{\infty} f(t) \left[\frac{1}{2\pi}\int_{-\infty}^{\infty} F(\omega)e^{j\omega t} d\omega \right]^* dt$$

$$= \frac{1}{2\pi}\int_{-\infty}^{\infty} F^*(\omega) \left[\int_{-\infty}^{\infty} f(t)e^{-j\omega t} dt \right] d\omega \tag{6.74}$$

which becomes, after substituting (6.43),

$$\boxed{E = \int_{-\infty}^{\infty} |f(t)|^2 dt = \frac{1}{2\pi}\int_{-\infty}^{\infty} F^*(\omega)F(\omega)d\omega = \frac{1}{2\pi}\int_{-\infty}^{\infty} |F(\omega)|^2 d\omega} \tag{6.75}$$

This is the counterpart of (6.21) for aperiodic functions and is also called a Parseval's formula.

Unlike the discrete frequency spectrum c_m that is defined only at discrete frequencies, the Fourier transform $F(\omega)$ is defined at every ω in the entire frequency range from $-\infty$ to ∞. The integration of $|F(\omega)|^2/2\pi$ over the entire frequency range yields the total energy of the signal. In the discrete frequency spectrum, the power contained at frequency $m\omega_0$ is $|c_m|^2$. In the present case, it is meaningless to talk about energy at a discrete frequency, because

$$\int_{\omega_0-}^{\omega_0+} |F(\omega)|^2 dt = 0$$

unless $F(\omega)$ contains an impulse at ω_0. Thus, for a spectrum defined over the continuum of frequency, energy must be defined over a band of frequency. For example, the energy contained in (ω_0, ω_1), with $\omega_1 > \omega_0$, is

$$\frac{1}{2\pi} \int_{\omega_0}^{\omega_1} |F(\omega)|^2 d\omega \tag{6.76}$$

Thus, $F(\omega)$ reveals the distribution of the energy of $f(t)$ over the entire frequency range and is called, more informatively, the spectral density of $f(t)$.

The integration in (6.75) is carried out over positive and negative frequencies. If $f(t)$ is real, then $|F(\omega)| = |F(-\omega)|$ and (6.75) can be reduced to

$$E = \frac{1}{2\pi} \int_{-\infty}^{\infty} |F(\omega)|^2 d\omega = \frac{1}{\pi} \int_{0}^{\infty} |F(\omega)|^2 d\omega$$

Now the integration is carried over only positive frequencies. The quantity $|F(\omega)|^2/\pi$ is sometimes called the *energy spectral density*.

EXAMPLE 6.4.9

Consider the positive-time function $f(t) = e^{-t}$, for $t \geq 0$. Its total energy is

$$E = \int_{-\infty}^{\infty} |f(t)|^2 dt = \int_{0}^{\infty} e^{-2t} dt = \frac{1}{-2} e^{-2t} \bigg|_{t=0}^{\infty} = \frac{1}{2}$$

and its frequency spectrum is

$$F(\omega) = \frac{1}{s+1} \bigg|_{s=j\omega} = \frac{1}{1+j\omega} = \frac{1}{\sqrt{1+\omega^2}\, e^{j\tan^{-1}\omega}} = \frac{1}{\sqrt{1+\omega^2}} e^{-j\tan^{-1}\omega}$$

Its amplitude and phase spectra are plotted in Figure 6.17(a). Now we find the frequency interval $(-\omega_0, \omega_0)$ that contains half of the total energy of $f(t)$. We compute, using the evenness of $|F(\omega)|$ or $|F(-\omega)| = |F(\omega)|$,

$$\frac{1}{2\pi} \int_{-\omega_0}^{\omega_0} |F(\omega)|^2 d\omega = \frac{1}{\pi} \int_{0}^{\omega_0} \frac{1}{1+\omega^2} d\omega = \frac{1}{\pi} \tan^{-1} \omega_0$$

If this is required to equal half of the total energy of $f(t)$, or $1/4$, then

$$\tan^{-1} \omega_0 = \frac{\pi}{4}$$

which implies $\omega_0 = 1$ rad/s. Thus, half of the total energy of $f(t)$ is contained in the frequency range $(-1, 1)$, and we may say that $f(t) = e^{-t}$ is a low-frequency signal.

EXAMPLE 6.4.10

Find the frequency spectrum of the positive-time function $f_1(t) = e^{-t} \cos 100t$, for $t \geq 0$. Because $|e^{-t} \cos 100t| \leq |e^{-t}|$ and because e^{-t} is absolutely integrable, $f_1(t)$ is absolutely integrable. Its Laplace transform is, using Table 4.1,

$$\overline{F}_1(s) = \frac{s + 1}{(s + 1)^2 + 100^2}$$

Thus, the frequency spectrum of $f_1(t)$ is

$$F_1(\omega) = \overline{F}_1(j\omega) = \frac{j\omega + 1}{(j\omega + 1)^2 + 10000} = \frac{j\omega + 1}{(-\omega^2 + 10001) + j2\omega}$$

Clearly, we have

$$F_1(0) = \frac{1}{10001} \approx 0 \qquad F_1(\infty) = \frac{j\omega}{-\omega^2}\Big|_{\omega=\infty} = 0 \cdot e^{-j90°}$$

At $\omega = \sqrt{10001} \approx 100$, we have

$$F_1(100) \approx \frac{j100}{j2 \times 100} = \frac{1}{2} = 0.5$$

If we compute more points, then the amplitude and phase spectra can be plotted as shown in Figure 6.17(b). We see that most of the energy is concentrated around $\omega = \pm 100$. We also see that the spectrum of $e^{-t} \cos 100t$ is the shifting of the spectrum of e^{-t} to ± 100 rad/s. This is, in fact, the frequency-shifting property or frequency modulation, which will be discussed later.

EXERCISE 6.4.6

What is the frequency spectrum of $e^{-t} \sin 100t$. Roughly plot its amplitude and phase spectra. At what frequencies is most of its energy concentrated?

[ANSWERS: $F(\omega) = 100/(10001 - \omega^2 + 2j\omega)$, ± 100 rad/s.]   ∎

Figure 6.17 *Amplitude and phase spectra*

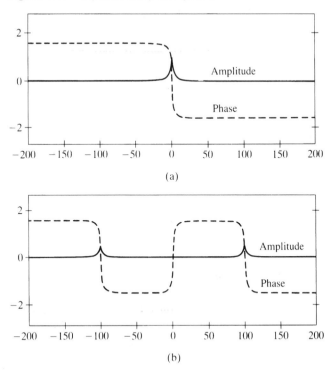

Table 6.2 lists the discrete spectrum and spectral density. Because the spectral density is applicable to both periodic and aperiodic functions, we shall use spectral density exclusively in the remainder of this text. Furthermore, the Fourier transform, spectral density, and frequency spectrum will be used synonymously.

Duration and Bandwidth   To conclude this section, let us discuss an important linking between a time function and its Fourier transform. Consider the time function $f(t)$ and its spectrum $F(\omega)$ shown in Figure 6.12. We shall use the concepts of *duration* for time functions and *bandwidth* for Fourier transforms. Instead of defining them rigorously (see Problem 6.33 for a definition), we shall proceed intuitively. As Figure 6.12(a) shows, the tangent of $f(t)$ at $t = 0 +$ intersects the time axis at $1/a$ and the tangent of $f(t)$ at $t = 0 -$ intersects the time axis at $-1/a$ (see Problem 6.18). Therefore, the width or duration of $f(t)$ is roughly proportional to $2/a$. Consider now the frequency spectrum in Figure 6.12(b). The amplitude of $F(\omega)$ decreases to half of $F(0)$ at $\omega = \pm a$. Therefore, the width or bandwidth of $F(\omega)$ can be considered to be proportional to $2a$. We see that the larger the duration $2/a$ of the time function, the smaller

Table 6.2 *Frequency Spectrum of Analog Signals*

Frequency Spectrum $(-\infty < \omega < \infty)$	Periodic Signal $f(t) = f(t + P)$ $\omega_0 = 2\pi/P$	Absolutely Integrable Signal $f(t)$
Fourier series	$c_m = \dfrac{1}{P}\displaystyle\int_0^P f(t)e^{-jm\omega_0 t}\,dt$	$F(m\omega_0) = Pc_m$
(Discrete frequency spectrum)	$f(t) = \displaystyle\sum_{m=-\infty}^{\infty} c_m e^{jm\omega_0 t}$	$P \to \infty$
Fourier transform	$F(\omega) = \displaystyle\sum_{m=-\infty}^{\infty} 2\pi c_m \delta(\omega - m\omega_0)$	$F(\omega) = \displaystyle\int_{-\infty}^{\infty} f(t)e^{-j\omega t}\,dt$
(Frequency spectrum)		$f(t) = \dfrac{1}{2\pi}\displaystyle\int_{-\infty}^{\infty} F(\omega)e^{j\omega t}\,d\omega$
Remarks	Power signal if $f(t)$ is bounded: $\dfrac{1}{P}\displaystyle\int_0^P f(t)f^*(t)\,dt = \sum_{m=-\infty}^{\infty} c_m c^*_m$	Energy signal if $f(t)$ is bounded: $\displaystyle\int_{-\infty}^{\infty} f(t)f^*(t)\,dt$ $= \dfrac{1}{2\pi}\displaystyle\int_{-\infty}^{\infty} F(\omega)F^*(\omega)\,d\omega$

the bandwidth $2a$ of its spectrum. Conversely, the smaller the duration of the time function, the larger the bandwidth of its spectrum. For easy reference, we call this property the *inverse relationship* between time and frequency or between (time) duration and (frequency) bandwidth. More examples will be given later to support this property.

EXERCISE 6.4.7

Verify the time-frequency inverse relationship for $f(t) = \delta(t)$ and $f(t) = 1$, for all t. ∎

6.5   CONVOLUTION

The zero-state response of every linear time-invariant continuous-time system can be described, as discussed in Section 3.5.2, by the convolution

$$y(t) = \int_0^\infty h(t - \tau)u(\tau)\,d\tau \tag{6.77a}$$

if the input is applied from $t = 0$, or by the convolution

$$y(t) = \int_{-\infty}^{\infty} h(t - \tau) u(\tau) d\tau \tag{6.77b}$$

if the input is applied from $t = -\infty$, where $h(t)$ is the impulse response. Let us apply the Fourier transform to (6.77b). By definition, we have

$$Y(\omega) = \int_{-\infty}^{\infty} y(t) e^{-j\omega t} dt = \int_{-\infty}^{\infty} [\int_{-\infty}^{\infty} h(t - \tau) u(\tau) d\tau] e^{-j\omega(t-\tau)} e^{-j\omega\tau} dt$$

which, by interchanging the order of integrations, can be written as

$$Y(\omega) = \int_{-\infty}^{\infty} [\int_{-\infty}^{\infty} h(t - \tau) e^{-j\omega(t-\tau)} dt] u(\tau) e^{-j\omega\tau} d\tau \tag{6.78}$$

Let $t' := t - \tau$. The integration inside the brackets becomes

$$\int_{-\infty}^{\infty} h(t') e^{-j\omega t'} dt' = H(\omega)$$

which is independent of τ and can be moved outside the integration in (6.78). Thus, we have

$$Y(\omega) = H(\omega) \int_{-\infty}^{\infty} u(\tau) e^{-j\omega\tau} d\tau = H(\omega) U(\omega)$$

or

$$\boxed{Y(\omega) = \mathscr{F}[\int_{-\infty}^{\infty} h(t - \tau) u(\tau) d\tau] = H(\omega) U(\omega)} \tag{6.79}$$

where $H(\omega)$ and $U(\omega)$ are the Fourier transforms of $h(t)$ and $u(t)$. In this derivation, it is implicitly assumed that $y(t)$, $h(t)$, and $u(t)$ are all Fourier transformable; otherwise, (6.79) is meaningless.[11]

Like (4.11), (6.79) is an algebraic equation and can be used in system analysis. However, the employment of (6.79) is much more complicated than employing the Laplace transform, as the following example demonstrates.

---

[11]Recall that a function is Fourier transformable if it is absolutely integrable or periodic. If $h(t)$ is periodic, the output $y(t)$ may not be Fourier transformable for some Fourier transformable $u(t)$. If $h(t)$ is absolutely integrable, then the output $y(t)$ is Fourier transformable for every Fourier transformable $u(t)$. Thus, strictly speaking, (6.79) is applicable only if $h(t)$ is absolutely integrable or the system is BIBO stable, which will be discussed in Chapter 8.

*EXAMPLE 6.5.1

Consider an LTIL system with transfer function $\overline{H}(s) = 1/(s + 1)$. To compute the unit step response of the system, we first use the Laplace transform and then use the Fourier transform. The Laplace transform of $u(t) = q(t)$ is $\overline{U}(s) = 1/s$. Thus, we have

$$\overline{Y}(s) = \overline{H}(s)\overline{U}(s) = \frac{1}{s + 1} \cdot \frac{1}{s} = \frac{-1}{s + 1} + \frac{1}{s}$$

Its inverse Laplace transform is

$$y(t) = 1 - e^{-t} \qquad \text{for } t \geq 0 \tag{6.80}$$

This is the response of the system.

The impulse response of the system is $h(t) = e^{-t}$, for $t \geq 0$. Both $h(t)$ and $u(t)$ are Fourier transformable, so (6.79) can be used to compute the response. The Fourier transform of $h(t)$ is $1/(j\omega + 1)$. The Fourier transform of $u(t) = q(t)$, as computed in (6.66), is

$$U(\omega) = \frac{1}{j\omega} + \pi\delta(\omega)$$

Thus, the output is

$$Y(\omega) = H(\omega)U(\omega) = \frac{1}{j\omega + 1} \cdot [\frac{1}{j\omega} + \pi\delta(\omega)]$$

$$= \frac{1}{j\omega + 1} \cdot \frac{1}{j\omega} + \frac{\pi\delta(\omega)}{j\omega + 1} \tag{6.81}$$

The first term after the equality can be expanded as

$$\frac{1}{j\omega + 1} \cdot \frac{1}{j\omega} = \frac{-1}{j\omega + 1} + \frac{1}{j\omega}$$

Because $\delta(\omega) = 0$ for all ω except $\omega = 0$, the second term can be written as,

$$\frac{\pi\delta(\omega)}{j\omega + 1} = \frac{\pi\delta(\omega)}{0 + 1} = \pi\delta(\omega)$$

Thus, $Y(\omega)$ in (6.81) becomes

$$Y(\omega) = \frac{-1}{j\omega + 1} + [\frac{1}{j\omega} + \pi\delta(\omega)]$$

The inverse Fourier transform of the term inside the bracket is $q(t)$. The inverse Fourier transform of $1/(j\omega + 1)$ is e^{-t}, for $t \geq 0$. Thus, we have

$$y(t) = 1 - e^{-t} \quad \text{for } t \geq 0$$

which, as expected, equals (6.80).

This example shows that the Fourier transform is more complicated to use in system analysis than is the Laplace transform. If we replaced $u(t) = 1$, for $t \geq 0$, with $u(t) = \sin \omega_0 t$, for $t \geq 0$, in the example, the complexity of using (6.79) would be even more announced. Because the Laplace transform is not only easier to use, it is also more general—it does not require both $h(t)$ and $u(t)$ to be Fourier transformable—the Fourier transform is rarely used in system analysis. The Fourier transform and (6.79) are, however, very important in signal analysis: The concept of frequency spectrum is developed from sinusoids, the Fourier series of periodic functions, and the Fourier transform of aperiodic functions, so a study of the Fourier transform leads naturally to the concept of spectrum. And, as will be discussed in Chapter 8, Equation (6.79) gives the relationship between the spectra of the input and output under the assumptions of stability and steady state.

6.6 PROPERTIES OF THE FOURIER TRANSFORM

Because of the close relationship between the Laplace transform and the Fourier transform, most properties of the former are applicable, with minor modification, to the latter. In this section, we will discuss only those properties that are useful in signal analysis, and derive a number of useful Fourier-transform pairs.

Time-shifting Property Consider the time function $f(t - t_0)$ that equals $f(t)$ shifted to t_0. If $F(\omega) = \mathscr{F}[f(t)]$, then

$$\boxed{\mathscr{F}[f(t - t_0)] = e^{-j\omega t_0} F(\omega)}$$

This can be established from

$$\mathscr{F}[f(t - t_0)] = \int_{-\infty}^{\infty} f(t - t_0) e^{-j\omega t} dt$$

which, after introducing $t' := t - t_0$, becomes

$$\mathscr{F}[f(t - t_0)] = \int_{-\infty}^{\infty} f(t') e^{-j\omega(t' + t_0)} dt' = e^{-j\omega t_0} \int_{-\infty}^{\infty} f(t') e^{-j\omega t'} dt'$$

$$= e^{-j\omega t_0} F(\omega)$$

This property has a very important implication. Because of

$$|e^{-j\omega t_0} F(\omega)| = |e^{-j\omega t_0}| \, |F(\omega)| = |F(\omega)| \qquad \text{\Large ✳}$$

and

$$\measuredangle \left[e^{-j\omega t_0} F(\omega) \right] = \measuredangle e^{-j\omega_0 t} + \measuredangle F(\omega) = -\omega t_0 + \measuredangle F(\omega) \qquad \text{\Large ✳}$$

 shifting a signal will not change its amplitude spectrum; it merely introduces a linear phase into the phase spectrum. Therefore, if we are interested in only the amplitude spectrum, the time function can be shifted to any position. Note the duality of this time shifting with the frequency shifting discussed in (6.62).

EXAMPLE 6.6.1

Consider the rectangular function $f_a(t)$ shown in Figure 6.11(c), that is,

$$f_a(t) = \begin{cases} 1 & \text{for } |t| \le a \\ 0 & \text{for } |t| > a \end{cases} \tag{6.82}$$

Its Fourier transform is

$$F_a(\omega) = \int_{-\infty}^{\infty} f_a(t) e^{-j\omega t} dt = \int_{-a}^{a} e^{-j\omega t} dt = \frac{-1}{j\omega} [e^{-j\omega a} - e^{j\omega a}]$$

$$= \frac{2 \sin a\omega}{\omega} \tag{6.83}$$

Because $f_a(t)$ is a real and even function of t, $F(\omega)$ is a real and even function of ω, as plotted in Figure 6.11(c).

EXAMPLE 6.6.2

Consider the rectangular function defined as

$$f_b(t) = \begin{cases} 1 & \text{for } 0 \le t \le 2a \\ 0 & \text{for } t < 0 \text{ and } t > 2a \end{cases} \tag{6.84}$$

It is the shifting of $f_a(t)$ in (6.82) to the right by a, that is,

$$f_b(t) = f_a(t - a)$$

Thus, we have

$$F_b(\omega) = e^{-j\omega a} F_a(\omega) = \frac{2e^{-j\omega a} \sin a\omega}{\omega} \tag{6.85}$$

We can compute (6.85) once again using a different method. The function in (6.84) is positive time and absolutely integrable; thus, we have

$$\mathscr{F}[f_b(t)] = \mathscr{L}[f_b(t)]\Big|_{s=j\omega}$$

Because $f_b(t)$ equals $q(t) - q(t - 2a)$, where $q(t)$ is the unit step function, we have, as in (4.93),

$$\mathscr{L}[f_b(t)] = \frac{1}{s} - \frac{1}{s}e^{-2as}$$

Thus, we have

$$\mathscr{F}[f_b(t)] = \frac{1 - e^{-2aj\omega}}{j\omega} = \frac{e^{-ja\omega}(e^{ja\omega} - e^{-ja\omega})}{j\omega}$$
$$= \frac{2e^{-ja\omega}\sin a\omega}{\omega}$$

which is the same as (6.85).

EXERCISE 6.6.1

Find the Fourier transform of $\cos \omega_0 t$ from that of $\sin \omega_0 t$ using the identity $\cos \theta = -\sin (\theta - \pi/2)$.

[**ANSWER:** $j\pi e^{-j\pi\omega/2\omega_0} \delta(\omega - \omega_0) - j\pi e^{-j\pi\omega/2\omega_0} \delta(\omega + \omega_0) = j\pi e^{-j\pi\omega_0/2\omega_0}$
$\delta(\omega - \omega_0) - j\pi e^{-j\pi\omega_0/2\omega_0} \delta(\omega + \omega_0) = j\pi(-j)\delta(\omega - \omega_0) - j\pi(-j)\delta(\omega + \omega_0) = \pi\delta(\omega - \omega_0) - \pi\delta(\omega + \omega_0)$] ■

EXERCISE 6.6.2

Find the spectrum of

$$f(t) = \begin{cases} 2e^{-t} & \text{for } t \geq 2 \\ 0 & \text{for } t < 2 \end{cases}$$

[**ANSWER:** Define

$$g(t) = \begin{cases} 2e^{-2}e^{-t} & \text{for } t \geq 0 \\ 0 & \text{for } t < 0 \end{cases}$$

Then $f(t) = g(t-2)$ and $F(\omega) = e^{-j2\omega}2e^{-2}/(j\omega + 1).$] ∎

Frequency-shifting Property If $F(\omega) = \mathscr{F}[f(t)]$, then

$$\boxed{\mathscr{F}[f(t)e^{j\omega_0 t}] = F(\omega - \omega_0)}$$

This was established in (6.62) and used in (6.67). It is an important formula.

Duality Property If

$$F(\omega) = \mathscr{F}[f(t)]$$

then

$$f(-\omega) = \frac{1}{2\pi}\mathscr{F}[F(t)]$$

This follows directly from (6.43) and (6.44).

EXAMPLE 6.6.3

Let $f(t) = \delta(t)$. Then

$$\Delta(\omega) := \mathscr{F}[f(t)] = \int_{-\infty}^{\infty} \delta(t)e^{-j\omega t}dt = e^{j\omega t}\Big|_{t=0} = 1$$

for all ω. Using the duality property, we have

$$\mathscr{F}[\Delta(t) = 1, \text{ for all } t] = 2\pi\delta(-\omega) = 2\pi\delta(\omega)$$

This is the same as (6.55).

Example 6.6.3 also shows the inverse relationship between the duration of a time function and the bandwidth of its frequency spectrum. The duration of $\delta(t)$ is zero and the bandwidth of its spectrum $\Delta(\omega) = 1$ is infinity. The duration of $\Delta(t) = 1$ is infinity and the bandwidth of its spectrum $2\pi\delta(\omega)$ is zero. This is the extreme case of the inverse relationship.

Sinc Function A special function defined by

$$\boxed{\text{sinc } x := \frac{\sin \pi x}{\pi x}} \tag{6.86}$$

is called the *sinc function*. It is plotted in Figure 6.18. It is an even function
because sinc $(-x) = \sin(-\pi x)/(-\pi x) = (-\sin \pi x)/(-\pi x) = \sin \pi x / \pi x = $ sinc x. Its
value at $x = 0$, using l'Hopital's rule, is

$$\cos\emptyset = 1$$

$$\text{sinc } 0 = \left.\frac{\sin \pi x}{\pi x}\right|_{x=0} = \left.\frac{\pi \cos \pi x}{\pi}\right|_{x=0} = 1$$

Because $\sin \pi x = 0$ at $x = \pm 1, \pm 2, \dots$, the value of sinc x is zero at
$x = \pm 1, \pm 2, \dots$. Clearly, sinc x approaches zero as $x \to \pm \infty$. This function
often arises in signal analysis and filter design.

EXAMPLE 6.6.4

Consider the Fourier transform pair in (6.82) and (6.83). We can rewrite it,
using (6.86), as

$$\mathscr{F}[f_a(t)] = \frac{2 \sin a\omega}{\omega} = \frac{2a \sin (\pi \cdot \frac{a\omega}{\pi})}{\pi \cdot \frac{a\omega}{\pi}} = 2a \text{ sinc } (\frac{a\omega}{\pi}) =: F_a(\omega)$$

Using the duality property, we have

$$\frac{1}{2\pi}\mathscr{F}[F_a(t)] = \mathscr{F}[\frac{a}{\pi} \text{ sinc } (\frac{at}{\pi})] = f_a(-\omega) = f_a(\omega)$$

The transforms are plotted in Figure 6.19. The Fourier transform of the pulse
$f_a(t)$ is a sinc function and the Fourier transform of a sinc function is a pulse.
Note that both are real and even.

Figure 6.18 *Sinc function*

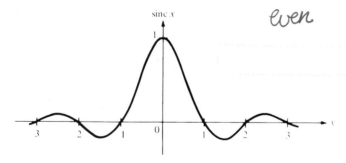

even

Figure 6.19 *Sinc function and its inverse Fourier tranform*

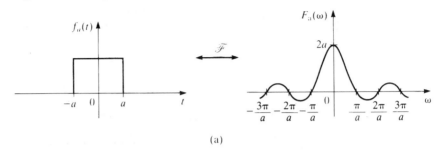

(a)

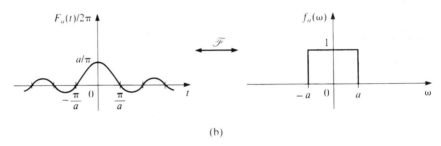

(b)

The plot of $f_a(\omega)$ in Figure 6.19(b), as will be discussed in Chapter 8, characterizes an ideal lowpass filter. Its inverse Fourier transform is

$$\mathscr{F}^{-1}[f_a(\omega)] = \frac{a}{\pi} \operatorname{sinc}\left(\frac{at}{\pi}\right) \tag{6.87}$$

which is the impulse response of the ideal lowpass filter. <u>A necessary and sufficient condition for a system to be causal is that its impulse response be zero for $t < 0$.</u> The impulse response of the ideal lowpass filter is not zero for $t < 0$, as shown in Figure 6.19(b). Thus, an ideal lowpass filter is not causal and, therefore, cannot be built in the real world. A major task of filter design is to find causal or physically realizable filters to approximate the ideal filter as closely as possible. See Reference [3].

Example 6.6.4 also shows the inverse relationship between time duration and frequency bandwidth. The duration of $f_a(t)$ is $2a$. The width of the main lobe of $F_a(\omega)$ is, as can be seen from Figure 6.19, $2\pi/a$. They are indeed inversely proportional.

Time-scaling Property If $F(\omega) = \mathscr{F}[f(t)]$, then

$$\mathscr{F}[f(\alpha t)] = \frac{1}{|\alpha|} F\left(\frac{\omega}{\alpha}\right) \tag{6.88}$$

where α is a nonzero real constant.

By definition, we have

$$\mathscr{F}[f(\alpha t)] = \int_{-\infty}^{\infty} f(\alpha t)e^{-j\omega t}\,dt = \int_{-\infty}^{\infty} f(\alpha t)e^{-j(\omega/\alpha)\alpha t}\,dt$$

If $\alpha > 0$, the substitution of $t' = \alpha t$ yields

$$\mathscr{F}[f(\alpha t)] = \frac{1}{\alpha}\int_{-\infty}^{\infty} f(t')e^{-j(\omega/\alpha)t'}\,dt' = \frac{1}{\alpha}F(\frac{\omega}{\alpha})$$

If $\alpha < 0$, in the substitution of $t' = \alpha t$ and $dt' = \alpha dt$, the lower and upper limits of the integration must be reversed as

$$\mathscr{F}[f(\alpha t)] = \frac{1}{\alpha}\int_{\infty}^{-\infty} f(t')e^{-j(\omega/\alpha)t'}\,dt'$$

which can be written as

$$\mathscr{F}[f(\alpha t)] = \frac{-1}{\alpha}\int_{-\infty}^{\infty} f(t')e^{-j(\omega/\alpha)t'}\,dt' = \frac{1}{|\alpha|}F(\frac{\omega}{\alpha})$$

This establishes the time-scaling property.

Consider $f(t)$ and its spectrum shown in Figure 6.20(a). Define $f_1(t) := f(\alpha t)$. Figure 6.20(b) and (c) plot $f(\alpha t)$ with $\alpha = 2$ and $\alpha = 0.5$. We

Figure 6.20 *Stretching and compressing of time function*

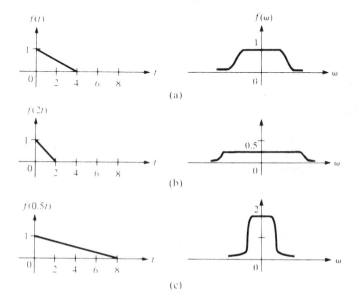

see that, if $|\alpha| > 1$, the function $f(\alpha t)$ is being speeded up or its duration is compressed. If $|\alpha| < 1$, the function $f(\alpha t)$ is being slowed down or its duration is stretched. The corresponding spectra are also shown in Figure 6.20. We see that, aside from the factor in magnitude, *time compression causes frequency expansion, and time expansion causes frequency compression*. This supports once again the inverse relationship between time and frequency.

The time scaling can be used to explain the change of pitch of an audio tape when its speed is changed. When we increase the speed of the tape, its spectrum spreads out and, therefore, contains higher frequency components. Thus, the pitch is higher. When we decrease the speed, the spectrum is compressed and contains lower frequency components. Thus, the pitch is lower.

If $\alpha = -1$ in (6.88), the time scaling becomes time reversal or time flipping. In particular, a negative-time signal becomes a positive-time signal. Let $f_-(t)$ be a negative-time function and let $F_-(\omega)$ be its Fourier transform. Then (6.88) implies, with $\alpha = -1$,

$$\mathscr{F}[f_-(-t)] = F_-(-\omega)$$

Because $f_-(-t)$ is positive time, (6.51) implies

$$\mathscr{F}[f_-(-t)] = \left. \mathcal{L}[f_-(-t)] \right|_{s=j\omega}$$

Thus, we have $F_-(-\omega) = \left. \mathcal{L}[f_-(-t)] \right|_{s=j\omega}$, or

$$F_-(\omega) = \mathscr{F}[f_-(t)] = \left. \mathcal{L}[f_-(-t)] \right|_{s=-j\omega} \tag{6.89}$$

This establishes (6.53) once again.

To conclude this section, some properties of the Fourier transform that are useful in signal analysis are listed in Table 6.3.

*6.7 MODULATION

Every device is designed to operate in only a limited frequency range. For example, motors are classified as DC and AC motors. A DC motor can be driven by DC signals or signals with frequency spectra centered at $\omega = 0$. A 60 Hz AC motor cannot be driven by a DC signal; it will operate most efficiently if the frequency spectrum of the applied signal is centered around 60 Hz. Now suppose we have a signal $u(t)$ with frequency spectrum $|U(\omega)|$, as shown in Figure 6.21. Because the frequency spectrum of $u(t)$ does not overlap with the operational frequency range of the motor, the application of $u(t)$ will not drive the motor. This problem can be resolved if the spectrum of $u(t)$ is shifted to center at 60 Hz, which can be achieved by modulation. Thus, the purpose of modulation is to match up the frequency spectrum of a signal with the opera-

Table 6.3 *Properties of Fourier transform useful in signal analysis*

Property	Time function	Fourier transform
Linearity	$x_1 f_1(t) + x_2 f_2(t)$	$x_1 F_1(\omega) + x_2 F_2(\omega)$
Frequency shifting	$f(t)e^{j\omega_0 t}$ $f(t)e^{-j\omega_0 t}$	$F(\omega - \omega_0)$ $F(\omega + \omega_0)$
Time shifting	$f(t - t_0)$	$e^{-j\omega t_0} F(\omega)$
Time scaling	$f(\alpha t)$ $f(t/\alpha)$	$\dfrac{1}{\|\alpha\|} F\left(\dfrac{\omega}{\alpha}\right)$ $\alpha F(\alpha \omega)$
Time reversal	$f(-t)$ $f(t)$	$F(-\omega)$ $F(\omega)$
Duality	$v(t)$ $F(\omega) = \mathcal{F}[f(t)]$ $v(t)$	$v(f)$ $f(-\omega) = \dfrac{1}{2\pi}\mathcal{F}[F(t)]$ $v(f)$

• Multiplication,
$$v_1(t)v_2(t)$$
$$\Leftrightarrow \int_{-\infty}^{\infty} V(\rho)V(f-\rho)d\rho$$

• Differentiation : $\dfrac{d^n v(t)}{dt^n} \Leftrightarrow (j2\pi f)^n V(t)$

• Integration : $\int_{-\infty}^{t} v(\tau)d\tau \Leftrightarrow \dfrac{V(0)}{2}\delta(f) + \dfrac{V(f)}{j2\pi f}$

• Convolution : $\int_{-\infty}^{\infty} v_1(\tau)v_2(t-\tau)d\tau \Leftrightarrow V_1(f)V_2(f)$

tional frequency range of a device. This type of modulation is widely used in control systems.

Modulation is also important in radio communication. The frequency spectrum of human voices lies generally between 200 Hz and 4 kHz. The frequency range audible by human beings, however, is usually wider, between 20 Hz and 200 kHz. This is called the audio frequency range. Signals in audio frequency are not transmitted directly for three reasons: First, the wavelength of audio signals is very long. The speed of light is 3×10^8 meters per second. Therefore, the wavelength of a 200 Hz signal is $(3 \times 10^8)/200 = 1.5 \times 10^6$ meters. In order to transmit such signals efficiently, the size of an antenna must be at least one-tenth of its wavelength, or 1.5×10^5 meters. This is impractical. Second, audio signals attenuate rapidly in atmosphere. And third, interference will occur if two or more audio signals are transmitted simultaneously. For these reasons, audio signals are modulated before transmission.

There are many modulation schemes. We will discuss only the suppressed-carrier amplitude modulation and amplitude modulation here.

Figure 6.21 *Mismatch of frequency spectra*

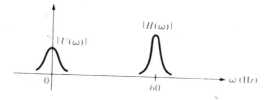

*6.7.1 Suppressed-carrier Amplitude Modulation

Consider a signal $u(t)$ with frequency spectrum $U(\omega)$ centered at $\omega = 0$, as shown in Figure 6.21. If we wish to shift the spectrum $U(\omega)$ to center at ω_c, we can multiply $u(t)$ with $\cos \omega_c t$ to yield

$$u_m(t) := u(t) \cos \omega_c t \qquad (6.90)$$

The function $u_m(t)$ is obtained by multiplying $u(t)$ by $\cos \omega_c t$ point by point, as shown in Figure 6.22. The phase of $u_m(t)$ is the same as $\cos \omega_c t$ if $u(t) > 0$. They differ by 180° if $u(t) < 0$. We see that the envelope of $u_m(t)$ has the same waveform as $u(t)$. The signal $\cos \omega_c t$ is called the *carrier signal*, and ω_c is called the *carrier frequency*. The amplitude of the carrier signal is modified or modulated by $u(t)$; therefore, $u(t)$ is called the *modulating signal* and $u_m(t)$ is called the *modulated signal*.

Let us compute the frequency spectrum of $u_m(t)$. Using $\cos \omega_c t = (e^{j\omega_c t} + e^{-j\omega_c t})/2$ and the frequency shifting property in (6.62), we have

$$U_m(\omega) := \mathcal{F}[u_m(t)] = \mathcal{F}[\frac{1}{2} u(t)(e^{j\omega_c t} + e^{-j\omega_c t})] \qquad (6.91)$$

$$= \frac{1}{2}[U(\omega - \omega_c) + U(\omega + \omega_c)]$$

Figure 6.22 *Suppressed-carrier amplitude modulation*

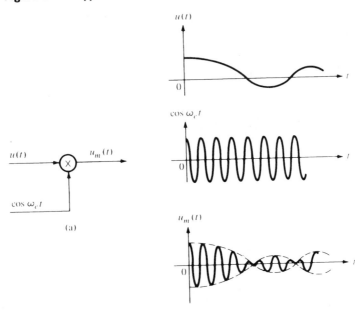

Its amplitude spectrum is plotted in Figure 6.23. We see that the spectrum is shifted to $\pm\omega_c$ and the amplitude is reduced by half. Thus, if ω_c is chosen properly, the spectrum of an input signal can be matched with the operational frequency range of a device. This type of modulation is often used in control systems. The spectrum of the carrier signal $\cos \omega_c t$ is two impulses at $\pm\omega_c$. They do not appear in the spectrum of $u_m(t)$. Thus, the modulation is called the *suppressed-carrier amplitude modulation (SCAM)*.

*6.7.2   Amplitude Modulation

The modulation in (6.90) is just one of many possible modulation schemes. A different modulation that is widely used in communication, especially in radio transmission, is discussed here. Consider the setup shown in Figure 6.24. It is a modification of the one in Figure 6.22(a): $A \cos \omega_c t$ has been added to the output. Thus, we have

$$\bar{u}_m(t) = [A + u(t)] \cos \omega_c t \qquad \qquad \text{(6.92)}$$

where $\cos \omega_c t$ and ω_c are, respectively, the carrier signal and the carrier frequency. If $u(t)$ and $\cos \omega_c t$ are as shown in Figure 6.22(b) and if $A \geq |u(t)|$, for all t, then the modulated signal $\bar{u}_m(t)$ is as shown in Figure 6.25(a). The Fourier transform of $\bar{u}_m(t)$, using (6.56b) and (6.91), is

$$\mathcal{T}[\bar{u}_m(t)] = \mathcal{T}[A \cos \omega_c t] + \mathcal{T}[u(t) \cos \omega_c t]$$
$$= A\pi[\delta(\omega - \omega_c) + \delta(\omega + \omega_c)] \qquad \text{(6.93)}$$
$$+ \frac{1}{2}[U(\omega - \omega_c) + U(\omega + \omega_0)]$$

Figure 6.23 *Frequency spectra of modulated signals*

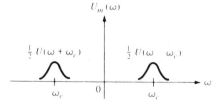

Figure 6.24 *Amplitude modulation*

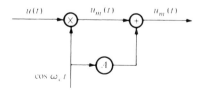

Figure 6.25 *Amplitude modulation*

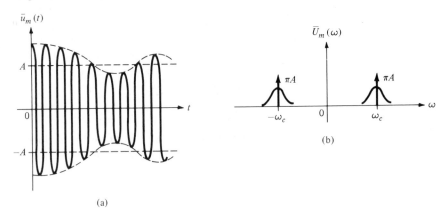

(a)

(b)

It is plotted in Figure 6.25(b), where we see that the spectrum of $u(t)$ is shifted to $\pm \omega_c$. The spectrum of $\bar{u}_m(t)$ in (6.93) is the sum of the spectra of $u_m(t)$ in (6.90) and the carrier signal $\cos \omega_c t$. Because the spectrum of $\cos \omega_c t$ does not appear in Figure 6.23, the modulation in (6.90) is called the suppressed-carrier amplitude modulation (SCAM) and the one in (6.92) is called the *amplitude modulation (AM)*. The amplitude modulation is more complicated than the suppressed-carrier modulation. However, its demodulation, namely, recovering $u(t)$ from $\bar{u}_m(t)$, is much simpler (see Problems 6.32 and 8.14). Thus, AM— rather than SCAM—is used in radio transmission.

In radio transmission, each radio station is assigned a carrier frequency inside the radio frequency band between 540 and 1600 kHz. Because the frequency spectra of human voice and music are limited to 4 kHz, if the carrier frequencies of different stations are 10 kHz apart, no interference will occur, as Figure 6.26 illustrates. Therefore, a number of ratio stations can transmit simul-

Figure 6.26 *Frequency-division multiplexing*

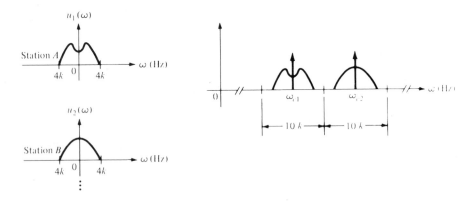

taneously through the radio-frequency band. This is called the _frequency-division multiplexing_.

Modulation and multiplexing are fundamental in communication. The interested reader is referred to, for example, Reference [12].

*6.7.3 Complex Convolution

Modulation is achieved by the multiplication of a signal $u(t)$ and the cosine function $\cos \omega_c t$. Now we shall study the general case, namely, the multiplication of two arbitrary time functions. Let $f_i(t)$, $i = 1, 2$, be two Fourier transformable functions defined over $(-\infty, \infty)$, and let $F_i(\omega)$ be their Fourier transforms. The Fourier transform of the product of $f_1(t)$ and $f_2(t)$ is

$$\mathscr{F}[f_1(t)f_2(t)] = \int_{-\infty}^{\infty} f_1(t)f_2(t)e^{-j\omega t}dt$$

$$= \int_{-\infty}^{\infty} f_1(t)\left[\frac{1}{2\pi}\int_{-\infty}^{\infty} F_2(\beta)e^{j\beta t}d\beta\right] e^{-j\omega t}dt$$

where we have used β as a dummy variable inside the brackets to avoid confusion with the original variable ω. Interchanging the order of integrations yields

$$\mathscr{F}[f_1(t)f_2(t)] = \frac{1}{2\pi}\int_{-\infty}^{\infty}\left[\int_{-\infty}^{\infty} f_1(t)e^{-j(\omega-\beta)t}dt\right] F_2(\beta)d\beta$$

which, because the integration inside the brackets equals $F_1(\omega - \beta)$, becomes

$$\boxed{\begin{aligned}\mathscr{F}[f_1(t)f_2(t)] &= \frac{1}{2\pi}\int_{-\infty}^{\infty} F_1(\omega - \beta)F_2(\beta)d\beta \\ &= \frac{1}{2\pi}\int_{-\infty}^{\infty} F_1(\beta)F_2(\omega - \beta)d\beta\end{aligned}} \tag{6.94}$$

The last equality can be verified by changing the variable as in (3.8) and (3.56). The convolution in (6.94) is called the _complex convolution_. Its graphical computation is identical to the one discussed in Chapter 3 and will not be repeated. Note that if either F_i is an impulse, then the convolution achieves shifting (see Problems 3.23 and 3.24). It is important to compare the complex convolution with

$$\mathscr{F}\left[\int_{-\infty}^{\infty} f_1(t - \tau)f_2(\tau)d\tau\right] = F_1(\omega)F_2(\omega) \tag{6.95}$$

which was developed in (6.79). Thus, convolution in the time domain corresponds to multiplication in the frequency domain. Conversely, multiplication in the time domain corresponds to convolution in the frequency domain.

EXAMPLE 6.7.1

We can use (6.94) to reestablish (6.91). Since $\mathscr{F}[u(t)] = U(\omega)$ and $\mathscr{F}[\cos \omega_c t]$ $= \pi[\delta(\omega - \omega_c) + \delta(\omega + \omega_c)]$, we have

$$\mathscr{F}[u(t) \cos \omega_c t] = \frac{1}{2\pi} \int_{-\infty}^{\infty} U(\omega - \beta)\pi[\delta(\beta - \omega_c) + \delta(\beta + \omega_c)]d\beta$$

which, using the shifting property of the impulse, becomes

$$\mathscr{F}[u(t) \cos \omega_0 t] = \frac{1}{2} [U(\omega - \beta)\big|_{\beta = \omega_c} + U(\omega - \beta)\big|_{\beta = -\omega_c}]$$

$$= \frac{1}{2}[U(\omega - \omega_c) + U(\omega + \omega_c)]$$

This is the same as (6.91). Note that $U(\omega)$ convolves with the impulses $\delta(\omega \pm \omega_c)$. Thus, $U(\omega)$ is being shifted to $\pm \omega_c$.

EXERCISE 6.7.1

Use (6.94) to establish (6.71).   ■

6.8   **SUMMARY**

1. A periodic signal with period P and fundamental frequency $\omega_0 = 2\pi/P$ can be expressed in various forms of the Fourier series. The set of the coefficients c_m of the complex exponential Fourier series at $m\omega_0$, $m = 0, \pm 1, \pm 2, \ldots$, is called the discrete or line frequency spectrum.

2. The Fourier transform is developed from the Fourier series $(F(\omega) = Pc_m)$ by extending the period P to infinity. It is applicable only to absolutely integrable and periodic functions.

3. If $f(t)$ is absolutely integrable and is written as the summation of a positive-time function and a negative-time function, or $f_+(t) + f_-(t)$, then

$$\mathscr{F}[f(t)] = \mathscr{L}[f_+(t)]\big|_{s=j\omega} + \mathscr{L}[f_-(-t)]\big|_{s=-j\omega}$$

If $f(t)$ is periodic with fundamental frequency ω_0 over $(-\infty, \infty)$ and its Fourier series is

$$f(t) = \sum_{m=-\infty}^{\infty} c_m e^{jm\omega_0 t}$$

then the Fourier transform of $f(t)$ is

$$F(\omega) = \mathcal{F}[f(t)] = \sum_{m=-\infty}^{\infty} 2\pi c_m \delta(\omega - m\omega_0)$$

Pg 345/346

If $f(t)$ is periodic only for positive time, see (6.68) and (6.70).

4. The Fourier transform of a signal is called the frequency spectrum or spectral density. If the signal is absolutely integrable, then the spectrum is a continuous function of ω and will not contain any impulses. If the spectrum contains impulses, the signal contains a periodic part.

5. In signal analysis, it is convenient to consider signals to be defined from $-\infty$ to ∞. By so doing, the frequency spectrum of a constant will consist of an impulse at $\omega = 0$, and the frequency spectrum of $\sin \omega_0 t$ or $\cos \omega_0 t$ will consist of two impulses at $\omega = \pm \omega_0$.

6. In general, the duration of a time function is inversely proportional to the bandwidth of its frequency spectrum. Thus, time compression (speeding up) will cause frequency expansion (higher pitch), and time expansion (slowing down) will cause frequency compression (lower pitch). This is called the time-frequency inverse relationship.

7. The Fourier transform is not as convenient to use nor is it as general as the Laplace transform in system analysis. It is, however, useful in steady-state analysis under the assumption of stability, as will be discussed in Chapter 8.

PROBLEMS

6.1 Which of the following signals are periodic? For those that are periodic, find their fundamental periods and fundamental frequencies.

(a) $\sin 10t + 10 \cos 10t + \sin 4t$

(b) $\sin 10t + 10 \cos 10\pi t$

(c) $\sin 3t + \sin(9t - \pi/4)$

(d) $\sin 3t - \cos 4t + \sin 5t$

6.2 Which of the following are energy signals?

(a) $e^{-t} \sin 2t$, for all t *Pg. 307*

(b) $(e^{-t} \sin 2t) q(t)$ for all t

(c) $e^{-|t|} \sin 2t$, for all t

6.3 Consider a sequence of rectangular pulses all separated by one unit of time. The width of the pulses increase by one, as shown in Figure P6.3. Is it periodic? Is it a power signal?

Figure P6.3

6.4 (a) Show that $1/t$ is not absolutely integrable in $[0,1]$ and that $1/\sqrt{t}$ is absolutely integrable in $[0,1]$. Is it true that, if $f(t)$ is absolutely integrable, then $|f(t)|^2$ is absolutely integrable?

(b) Show that, if $f(t)$ is absolutely integrable in $(-\infty, \infty)$ and is bounded, then $f(t)$ is an energy signal, or

$$\int_{-\infty}^{\infty} |f(t)|^2 dt < \infty$$

6.5 Find the Fourier series in complex exponential form for the following signals.

(a) $f_1(t) = \sin 4t + \cos 6t$ *Pg. 312 Example*

(b) $f_2(t)$ is periodic with period 1 and $f_2(t) = t$ for $0 \le t \le 1$, as shown in Figure 6.3(b).   *Pg. 317 formulae*

(c) $f_3(t) = |\sin \pi t|$

6.6 Compute the average powers of the signals in Problem 6.5.

6.7 Plot the discrete frequency spectra of the signals in Problem 6.5.

6.8 Show that, if $f(t)$ is even, then all coefficients associated with the sine functions in (6.19) are zero. If $f(t)$ is odd, then all coefficients associated with the cosine functions in (6.19) are zero.

6.9 Find the Fourier series of the periodic function shown in Figure P6.9.

Pg. 317 formulae

Figure P6.9

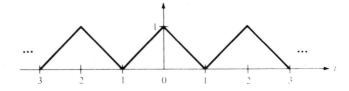

6.10 A periodic function with period P is called half-wave odd symmetry if it has the property

$$f(t) = -f(t + \frac{P}{2})$$

Show that its Fourier series has only odd harmonics, that is, $c_m = 0$ for m even.

6.11 A periodic function $f(t)$ with period P is also periodic with period $2P$. What would happen in computing the Fourier series of $f(t)$ if we used $\omega_0 = 2\pi/2P$ rather than $\omega_0 = 2\pi/P$ as the fundamental frequency?

6.12 (a) If we apply $u(t) = \sin \pi t$ to a device, called a half-wave rectifier, then the output consists of the part of $u(t)$ with $u(t) > 0$, that is,

$$y(t) = \begin{cases} \sin \pi t & \text{for } 2k \leq t < 2k + 1 \\ 0 & \text{for } 2k + 1 \leq t < 2(k + 1) \end{cases}$$

for $k = 0, \pm 1, \pm 2, \ldots$. Find the Fourier series of $y(t)$. What are the average powers of $u(t)$ and $y(t)$? What percentage of the input power is transmitted to the output?

(b) If we apply $u(t) = \sin \pi t$ to a device, called a full-wave rectifier, then the output is

$$y(t) = |\sin \pi t|$$

What is the fundamental period of $y(t)$? What are the average powers of $u(t)$ and $y(t)$? What percentage of the input power is transmitted to the output?

6.13 Show that the only constants β_i, $i = 1, 2, 3$, meeting

$$\beta_1 \cdot 1 + \beta_2 t + \beta_3 t^2 = 0 \qquad \text{for all } t \text{ in } [0, 4]$$

are $\beta_i = 0$, for $i = 1, 2$, and 3. Thus, $\{1, t, t^2\}$ is linearly independent in the time interval $[0, 4]$.

6.14 Show that $\phi_1(t) = 1$ and $\phi_2(t) = \sqrt{3}(1 - 2t)$ are orthonormal in $[0, 1]$. Find α_1 and α_2 so that the function

$$\alpha_1 \phi_1(t) + \alpha_2 \phi_2(t)$$

is the best approximation of $\sin t$ in $[0, 1]$ in the sense of having the smallest quadratic error

$$\int_0^1 [\sin t - \alpha_1 \phi_1(t) - \alpha_2 \phi_2(t)]^2 dt$$

6.15 The functions $\phi_i(t)$, $i = 1, 2, 3, 4$, defined in Figure P6.15, are called the *Walsh functions*. Show that they are orthonormal in $[0, 1]$.

6.16 Find a set of α_i, $i = 1, 2, 3, 4$, so that the function

$$\bar{f}(t) = \alpha_1 \phi_1(t) + \alpha_2 \phi_2(t) + \alpha_3 \phi_3(t) + \alpha_4 \phi_4(t)$$

where $\phi_i(t)$ are as defined in Figure P6.15, minimizes

$$\int_0^1 [\sin t - \bar{f}(t)]^2 dt$$

6.17 Find, if they exist, the Fourier transforms of the following functions.

(a) e^{-2t}, for all t in $(-\infty, \infty)$

(b) e^{-2t}, for $t \geq 0$ and zero for $t < 0$

(c) e^{2t}, for $t \geq 0$ and zero for $t < 0$

(d) $e^{-2|t|}$, for all t

(e) e^{-3t}, for $t \geq 0$, and $2e^{4t}$, for $t < 0$

Figure P6.15

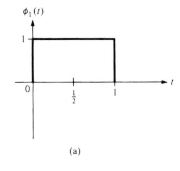

(a)

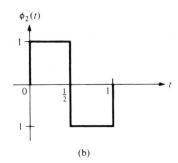

(b)

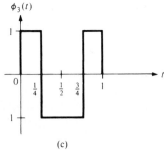

(c)

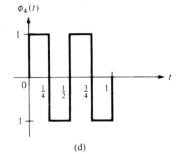

(d)

6.18 Consider $f(t) = e^{-a|t|}$ with $a > 0$, as shown in Figure 6.12. Show that the slope of $f(t)$ at $t = 0+$ intersects the time axis at $1/a$.

6.19 Consider the function $u(t) = e^{(\alpha+j\omega)t}$, for all t in $(-\infty, \infty)$. Find the condition on α and ω for $u(t)$ to be absolutely integrable. Find the condition for $u(t)$ to be Fourier transformable.

[**ANSWERS**: Not absolutely integrable for any real α. Fourier transformable for $\alpha = 0$.]

6.20 Let $H(\omega) = \mathscr{F}[h(t)]$. Show

$$H(0) = \int_{-\infty}^{\infty} h(t)dt$$

and

$$h(0) = \frac{1}{2\pi}\int_{-\infty}^{\infty} H(\omega)d\omega = \int_{-\infty}^{\infty} H(2\pi f)df$$

where $\omega = 2\pi f$. That is, the net area under the function $h(t)$ equals the value of $H(\omega)$ at $\omega = 0$, and the net area under $H(f)$ equals the value of $h(t)$ at $t = 0$.

6.21 (a) Show that, if $f(t)$ is even (real or complex), then $F(\omega)$ is even. If $f(t)$ is odd, then $F(\omega)$ is odd.

(b) Show that, if $f(t)$ is real and even, then its Fourier transform $F(\omega)$ is real and even. If $f(t)$ is real and odd, then $F(\omega)$ is imaginary and odd.

6.22 (a) Show that every function $f(t)$ can be decomposed uniquely as

$$f(t) = f_e(t) + f_o(t)$$

where $f_e(t)$ is even and $f_o(t)$ is odd. Express $f_e(t)$ and $f_o(t)$ in terms of $f(t)$.

(b) Show

$$\int_{-\infty}^{\infty} |f(t)|^2dt = \int_{-\infty}^{\infty} |f_e(t)|^2dt + \int_{-\infty}^{\infty} |f_o(t)|^2dt$$

6.23 (a) Find the Fourier transforms of the periodic functions in Problem 6.5.

(b) Find the frequency spectrum of

$$f(t) = \begin{cases} 3\sin 2\pi t + e^{-2t} & \text{for } t \geq 0 \\ 3\sin 2\pi t & \text{for } t < 0 \end{cases}$$

6.24 (a) Find the frequency spectrum of the positive-time function $f_1(t) = e^{-2t}$, for $t \geq 0$. Find the smallest frequency interval that contains half of the total energy of $f_1(t)$.

(b) Repeat (a) for the two-sided function $f_2(t) = e^{-2|t|}$, for all t.

(c) Repeat (a) for the positive-time function $f_3(t) = e^{-2t} \sin 100t$, for $t \geq 0$.

6.25 (a) Consider the function

$$f(t) = \begin{cases} t^{-2/3} & \text{for } 0 \leq t \leq 1 \\ 0 & \text{otherwise} \end{cases}$$

Show

$$\int_{-\infty}^{\infty} |f(t)| \, dt < \infty \quad \text{and} \quad \int_{-\infty}^{\infty} f^2(t) \, dt = \infty$$

Is an absolutely integrable signal always an energy signal?

(b) Consider the function

$$f(t) = \begin{cases} 1/(t+1) & \text{for } t \geq 0 \\ 0 & \text{for } t < 0 \end{cases}$$

Show

$$\int_{-\infty}^{\infty} |f(t)| \, dt = \infty \quad \text{and} \quad \int_{-\infty}^{\infty} f^2(t) \, dt < \infty$$

Is an energy signal always an absolutely integrable signal?

6.26 (a) Show that, if $F(\omega) = \mathcal{F}[f(t)]$, then

$$\mathcal{F}[\frac{d}{dt} f(t)] = j\omega F(\omega)$$

Can you obtain this formula from the Laplace transform if $f(t)$ is positive time?

(b) In (a), if $f(t) = 1$, for all t, will you encounter any difficulty?

6.27 Find the Fourier transform of the three pulses shown in Figure P6.27. Use the linearity and time-shifting property of the Fourier transform.

6.28 Find the Fourier transforms of the functions shown in Figure P6.28.

6.29 Suppose a signal is measured as shown in Figure P6.29. Compute its Fourier transform by using the dotted straight-line approximation.

Figure P6.27

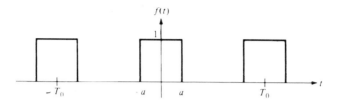

Figure P6.28

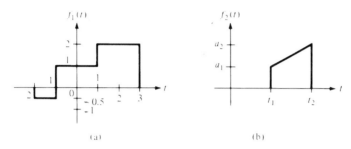

Figure P6.29

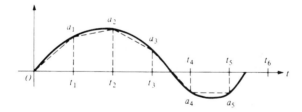

6.30 Consider the pulse signal shown in Figure P6.30(a). It can be written as

$$h(t) := Af_a(t) \cos \omega_c t$$

where $f_a(t)$ is defined as in Figure 6.11(c). It is called a radio-frequency (RF) pulse when ω_c falls in the radio-frequency band. Show that its amplitude spectrum is as shown in Figure P6.30(b).

6.31 Consider a signal $u(t)$ with the frequency spectrum shown in Figure P6.31. Because $U(\omega) = 0$ for $|\omega| > W$, $u(t)$ is said to be band-limited to W. Plot the frequency spectra of $u(t) \cos \omega_c t$ with $\omega_c = 0.8W$ and $\omega_c = 5W$. Which one has the same form, except for a factor, as the spectrum of $u(t)$? Can you recover $u(t)$ from $u(t) \cos \omega_c t$ if $\omega_c < W$?

Figure P6.30

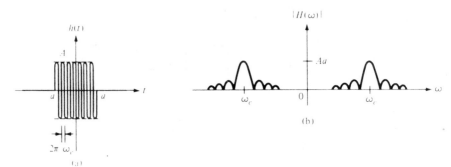

(a)

(b)

Figure P6.31

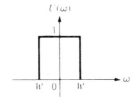

6.32 (a) Consider the signal $u(t)$ shown in Figure 6.22, in which $u(0) > |u(t)|$ for all t, and the amplitude of the largest negative value of $u(t)$ is larger than $0.5u(0)$. Plot $\bar{u}_m(t) = (A + u(t)) \cos 10\pi t$ with $A = 0.5u(0)$ and $A = 2u(0)$.

(b) If $\bar{u}_m(t)$ is applied to the circuit shown in Figure P6.32, where the diode is assumed to be ideal, what are the outputs for $A = 0.5u(0)$ and $A = 2u(0)$. The envelope of which output will resemble $u(t)$? Can you give a reason for requiring $A \geq |u(t)|$, for all t, in the amplitude modulation in (6.92)? Note that the envelope can be extracted by connecting a capacitor in parallel with the resistor R. Thus, $u(t)$ can be easily recovered from $\bar{u}_m(t)$ in amplitude modulation.

Figure P6.32

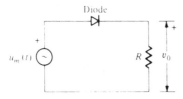

6.33 In this chapter we referred to the duration of a signal and the bandwidth of its frequency spectrum without giving any definitions. Many sets of definitions are available in the literature. For a real energy signal $f(t)$ with $F(0) \geq |F(\omega)|$ for all ω, we may define the duration as

$$D = \frac{(\int_{-\infty}^{\infty} |f(t)| dt)^2}{\int_{-\infty}^{\infty} |f(t)|^2 dt}$$

and the bandwidth as

$$B = \frac{\int_{-\infty}^{\infty} |F(\omega)|^2 d\omega}{2|F(0)|^2}$$

Show $DB \geq \pi$, thereby showing the inverse relationship between D and B.

6.34 Let $f(t)$ be a periodic function with period P and fundamental frequency ω_0. Define

$$\bar{f}(t) = \begin{cases} f(t) & \text{for } -P/2 \leq t \leq P/2 \\ 0 & \text{otherwise} \end{cases}$$

That is, $\bar{f}(t)$ is one period of $f(t)$. Show

$$Pc_m = \bar{F}(m\omega_0)$$

where c_m is the Fourier series coefficient of $f(t)$ and $\bar{F}(\omega)$ is the Fourier transform of $\bar{f}(t)$.

6.35 Let $F_i(\omega)$ be the Fourier transform of $f_i(t)$. Show

$$\int_{-\infty}^{\infty} f^*_1(t) f_2(t) dt = \frac{1}{2\pi} \int_{-\infty}^{\infty} F^*_1(\omega) F_2(\omega) d\omega$$

This is a more general form of the Parseval's formula given in (6.75).

6.36 (a) A real- or complex-valued function $f(t)$ is called symmetric or even if $f(t) = f(-t)$, conjugate symmetric if $f(t) = f^*(-t)$, antisymmetric or odd if $f(t) = -f(-t)$, and conjugate antisymmetric if $f(t) = -f^*(-t)$. Classify the following.

$$f_1(t) = t^2 + jt^2$$
$$f_2(t) = t^2 + jt$$
$$f_3(t) = t + jt$$
$$f_4(t) = t + jt^2$$

(b) Let $f(t)$ be expressed as $f(t) = f_r(t) + jf_i(t)$, where f_r and f_i are real functions denoting, respectively, the real and imaginary parts of $f(t)$. Show the following.

$f(t)$ is even $\longleftrightarrow f_r(t)$ and $f_i(t)$ are both even

$f(t)$ is conjugate symmetric $\longleftrightarrow f_r(t)$ is even and $f_i(t)$ is odd

$f(t)$ is odd $\longleftrightarrow f_r(t)$ and $f_i(t)$ are both odd

$f(t)$ is conjugate antisymmetric $\longleftrightarrow f_r(t)$ is odd and $f_i(t)$ is even

6.37 Establish the properties in Table 6.4.

6.38 Develop a table of symmetric properties (like Table 6.4) for discrete frequency spectra. Is there any difference?

Table 6.4 *Symmetric Properties*

$f(t)$	$F(\omega) = \int_{-\infty}^{\infty} f(t)e^{-j\omega t}\,dt$
Even $[f(t) = f(-t)]$	Even $[F(\omega) = F(-\omega)]$
Conjugate symmetric $[f(t) = f^*(-t)]$	Real $[F(\omega) = F^*(\omega)]$
Odd $[f(t) = -f(-t)]$	Odd $[F(\omega) = -F(-\omega)]$
Conjugate antisymmetric $[f(t) = -f^*(-t)]$	Imaginary $[F(\omega) = -F^*(\omega)]$
Real $[f(t) = f^*(t)]$	Conjugate symmetric $[F(\omega) = F^*(-\omega)]$
Imaginary $[f(t) = -f^*(t)]$	Conjugate antisymmetric $[F(\omega) = -F^*(-\omega)]$
Real and even $[f(t) = f^*(t) = f(-t)]$	Real and even $[F(\omega) = F^*(\omega) = F(-\omega)]$
Real and odd $[f(t) = f^*(t) = -f(-t)]$	Imaginary and odd $[F(\omega) = -F^*(\omega) = -F(-\omega)]$
Imaginary and even $[f(t) = -f^*(t) = f(-t)]$	Imaginary and even $[F(\omega) = -F^*(\omega) = F(-\omega)]$
Imaginary and odd $[f(t) = -f^*(t) = -f(-t)]$	Real and odd $[F(\omega) = F^*(\omega) = -F(-\omega)]$

USING **MATLAB**

MATLAB will be used to compute frequency spectra here. Consider the frequency spectrum in Example 6.4.10. The following commands

```
w = -200:1:200;
F = (j*w+1)./((-w.^2+10001)+2*j*w);
plot(w,abs(F),w,angle(F))
```

yield the plot in Figure 6.17(b). Note that we have used ./ and .^ for element by element operations (they are typed without a period for matrix operations). The frequency spectrum of this example is obtained from the rational function

$$\bar{F}(s) = \frac{s + 1}{(s + 1)^2 + 100^2} = \frac{s + 1}{s^2 + 2s + 10001}$$

by the substitution $s = j\omega$. In this case, the frequency spectrum can also be obtained by using the command $freqs$, as follows.

```
w = logspace(-1, 3, 200);        (generates two hundred equally
                                  spaced points in [10^-1, 10^3])
n = [1 1]; d = [1 2 10001];
F = freqs(n,d,w);
plot(w,abs(F),w,angle(F))
```

This yields the plot in Figure M6.1. If we use $logspace(a,b)$ without the third argument, then the default is 50. In other words, $logspace(a,b)$ generates fifty equally spaced points in $[10^a, 10^b]$. Because the amplitude spectrum is even and the phase spectrum is odd, the frequency spectrum for negative ω can easily be obtained from positive ω.

The frequency spectrum of the time function $e^{-t} \cos 100t$ of this example also can be obtained directly from the time function. We type

```
t = 0:0.01:10;
f = exp(-t).* cos(100 *t);
F = 0.01 * fft(f,512);
w = 2 * π * (0:511)/(512 * 0.01);
plot(w,abs(F),w,angle(F))
```

which yields the plot in Figure M6.2. The preceding statements will be developed in detail in the next chapter. We will discuss their meanings only briefly here. First, we sample f from $t = 0$ to 10 with sampling period $T = 0.01$. Then, we use a 512-point fft (fast Fourier transform) to compute the frequency spectrum of the sampled sequence. Note that the number of sampled

Figure M6.1

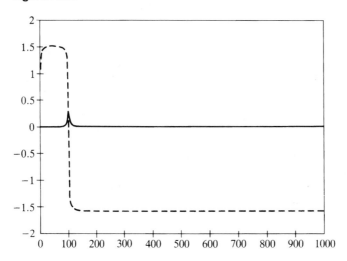

points should be larger than 512; otherwise trailing zeros will be added to the sampled sequence. The frequency spectrum of f equals that of its sampled sequence multiplied by a factor of T. Thus, the multiplication of 0.01 and `fft` yields the spectrum of f. The computed frequency spectrum lies in the frequency range $[0, 2\pi/T)$ or $[0, 628)$ in rad/s. Command `w` generates 512 equally spaced frequencies in $[0, 628)$. If we extend the spectrum periodically to negative frequencies, then the spectrum in $(-314, 314]$ is the spectrum of f, and the spectrum of f is zero for $|\omega| > 314$. If the sampling period T is chosen to be sufficiently small, and aliasing in frequency is small, then the computed spectrum is close to the actual spectrum. This is indeed the case for the amplitude spectrum shown. Because the phase spectrum does not approach zero as $\omega \to \pm\infty$, the computed phase spectrum is not close to the actual one.

MATLAB PROBLEMS

Use MATLAB to compute the following.

M6.1 Compute the frequency spectrum of the function in Example 6.4.9. Compute it directly and use command `freqs`.

M6.2 Use a 512-point `fft` to repeat Problem M6.1. Try $T = 1, 0.1$, and 0.01.

M6.3 Compute the frequency spectrum of the positive-time signal

$$f(t) = e^{-0.1t}(2 + 3 \sin 5t - \cos 2.5t)$$

for $t \geq 0$.

Figure M6.2

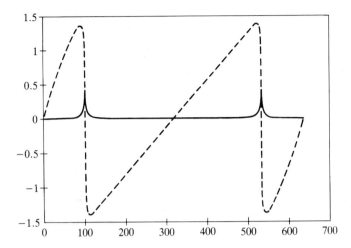

Discrete-Time Signal Analysis and Computation of Spectra

INTRODUCTION

In this chapter, we will study discrete-time signal analysis, the counterpart of our study of continuous-time signals in Chapter 6. As in the analog case, signals in this chapter will be extended to the negative time and be defined for all integer k in $(-\infty, \infty)$; they can be real- or complex-valued sequences. The frequency spectrum of continuous-time signals was developed from the complex exponentials $e^{j\omega_0 t}$; the frequency spectrum of discrete-time signals will be developed from the complex exponential sequences $e^{j\omega_0 kT}$. We will begin by discussing some properties of $e^{j\omega_0 kT}$.

A discrete-time signal $f[k]$ is periodic with period N, where N is a positive integer, if

$$ f[k] = f[k + N] $$

for all integers k in $(-\infty, \infty)$. The smallest such N is called the *fundamental period*. As discussed in Section 1.6, the complex exponential sequence

$$ f[k] = e^{j\omega_0 kT} $$

is periodic if and only if $\omega_0 T/\pi$ is a rational number. Its frequency, however, is defined for all $\omega_0 T$ whether the sequence is periodic or not. As in (1.30), the frequency of $e^{j\omega_0 kT}$ is defined as the frequency of the analog exponential $e^{j\omega t}$ with

$$ -\pi < \omega T \le \pi \quad \text{or} \quad -\frac{\pi}{T} < \omega \le \frac{\pi}{T} \tag{7.1} $$

such that the sample of $e^{j\omega t}$, with sampling period T, equals $\sin \omega_0 kT$; that is,

$$e^{j\omega t}\Big|_{t=kT} = e^{j\omega_0 kT} \tag{7.2}$$

Because

$$e^{j2\pi nk} = 1 \tag{7.3}$$

for all integers n and k and consequently, as in (1.33),

$$e^{j\omega_1 kT} = e^{j\omega_2 kT} \qquad \text{if } \omega_1 = \omega_2 \ (\text{modulo } \frac{2\pi}{T})$$

two frequencies are considered the same if they differ by $2n\pi/T$ for every integer n (negative, zero, or positive). In order to eliminate this nonuniqueness, the frequency of discrete-time exponential sequences will be restricted to a range of $2\pi/T$, such as

$$\frac{-\pi}{T} < \omega \le \frac{\pi}{T}, \qquad 0 \le \omega < \frac{2\pi}{T} \qquad \text{or} \qquad \frac{\pi}{T} \le \omega < \frac{3\pi}{T} \tag{7.4}$$

If ω is restricted to the range $(-\pi/T, \pi/T]$, then the frequency of $e^{j\omega kT}$ is simply ω.

To summarize, the frequency range of continuous-time $e^{j\omega t}$ is $(-\infty, \infty)$, but the frequency range of discrete-time $e^{j\omega kT}$ is $(-\pi/T, \pi/T]$. This is the most important difference between the two cases. Other than this, most of the concepts discussed in Chapter 6 can be applied directly to the discrete-time case. Therefore, the discussion here will be brief.

7.1.1 Orthogonality of complex exponential sequences

To simplify discussion, we will assume from now on, unless stated otherwise, that the sampling period T equals 1. Thus, the frequency range becomes $(-\pi, \pi]$ in rad/s or $(-0.5, 0.5]$ in Hz. Consider the complex exponential sequence $e^{j\omega_0 k}$ with

$$\omega_0 = \frac{2\pi}{N}$$

Because

$$e^{\frac{j2\pi}{N}(k+N)} = e^{\frac{j2\pi k}{N}} e^{j2\pi} = e^{\frac{j2\pi k}{N}}$$

the sequence $e^{j2\pi k/N}$ is periodic with fundamental period N. Consider

$$\phi_m[k] := e^{jm(\frac{2\pi}{N})k} = e^{jm\omega_0 k} \qquad m = 0, \pm 1, \pm 2, \pm 3, \ldots \qquad (7.5)$$

Note that k denotes time instants and m denotes different exponentials. Thus the set consists of infinitely many complex exponentials. However, there are only N distinct complex exponentials, and all of them are periodic with period N. Indeed, we have

(*i*) $\qquad \phi_m[k + N] = e^{jm(2\pi/N)[k + N]} = e^{jm(2\pi/N)k} e^{jm2\pi} = e^{jm(2\pi/N)k} = \phi_m[k]$

and

(*ii*) $\qquad \phi_{m+N}[k] = e^{j(m+N)(2\pi/N)k} = e^{jm(2\pi/N)k} = \phi_m[k]$

Thus, $\phi_m[k]$ is periodic in k with (not necessarily fundamental) period N and is also periodic in m. Therefore, there are only N distinct periodic sequences $\phi_m[k]$ with period N.

Which N sequences of $\phi_m[k]$ shall we use in analysis? If

$$m = -\frac{N}{2} + 1, -\frac{N}{2} + 2, \ldots, -1, 0, 1, \ldots, \frac{N}{2} + 1 \qquad (7.6a)$$

for N even, or

$$m = -\frac{N-1}{2}, -\frac{N-1}{2} + 1, \ldots, -1, 0, 1, \ldots, \frac{N-1}{2} \qquad (7.6b)$$

for N odd, then the frequency of

$$\phi_m[k] := e^{jm(\frac{2\pi}{N})k} = e^{jm\omega_0 k} = \cos m(2\pi/N)k + j \sin m(2\pi/N)k$$

is simply $m\omega_0$. However, writing (7.6) is complicated; it is much simpler to write $m = 0, 1, \ldots, N - 1$. Therefore, instead of using $\phi_m[k]$ with m in (7.6), we will use $\phi_m[k]$, with $m = 0, 1, \ldots, N - 1$, in analysis.

As in the continuous-time complex exponentials shown in Figure 1.8(a), it is difficult to plot $\phi_m[k]$ with k as a coordinate. It is simpler to plot $\phi_m[k]$ on the unit circle as shown in Figure 7.1, because $|\phi_m[k]| = 1$ for all integers m and k. For example, if $N = 3$, then we have, for $k = -3, -2, -1, 0, 1, 2, 3, 4, 5,$

$$\phi_0[k] = e^{j0(\frac{2\pi}{3})k} : \quad \{ \ldots, a, a, a, a, a, a, a, a, a, \ldots \}$$

$$\phi_1[k] = e^{j1(\frac{2\pi}{3})k} : \quad \{ \ldots, a, b, c, a, b, c, a, b, c, \ldots \}$$

$$\phi_2[k] = e^{j2(\frac{2\pi}{3})k} : \quad \{ \ldots, a, c, b, a, c, b, a, c, b, \ldots \}$$

Figure 7.1 *(a) Complex exponential sequences for* N = 3 *and (b) complex exponential sequences for* N = 4

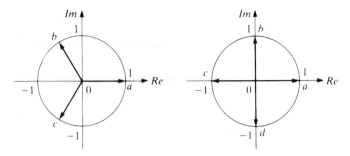

where a, b, and c are the three vectors shown in Figure 7.1(a) or

$$a = e^{j0} = 1 \qquad b = e^{j\frac{2\pi}{3}} = e^{j\,120°} = \cos 120° + j \sin 120°$$

$$= -0.5 + j\,0.866 \qquad c = e^{j\frac{4\pi}{3}} = -0.5 - j\,0.866$$

The sequence $\phi_0[k]$ has fundamental period 1 and frequency 0. The sequence $\phi_1[k]$ has fundamental period 3 and frequency $2\pi/3$ rad/s. The sequence $\phi_2[k] = \phi_{-1}[k]$ has fundamental period 3; its frequency is $-2\pi/3$ which is obtained as $2\omega_0 - 2\pi = 2(2\pi/3) - 2\pi = -2\pi/3$.

EXERCISE 7.1.1

Consider the complex exponential sequence $\phi_m[k]$ with $N = 4$. Compute the four vectors in Figure 7.1(b) and then express $\phi_m[k]$ explicitly for $m = 0, 1, 2, 3$, and $k = 0, 1, 2, 3, 4, 5, 6$. What are their fundamental periods and frequencies.

[**ANSWERS:** $a = 1$, $b = j$, $c = -1$ and $d = -j$; a,a,a,a,a,a,a, 1, 0; a,b,c,d,a,b,c,d, 4, $\pi/2$; a,c,a,c,a,c,a,c, 2, π; a,d,c,b,a,d,c,b, 4, $-\pi/2$] ∎

Now let us discuss some general properties of $\phi_m[k]$. As in (6.6) for the continuous-time case, we have

$$\sum_{k=0}^{N-1} \phi_m[k] = \sum_{k=0}^{N-1} e^{jm(2\pi/N)k} = \begin{cases} N & \text{for } m = 0 \text{ (modulo } N) \\ 0 & \text{otherwise} \end{cases} \qquad (7.7)$$

Two integers equal each other's modulo N if their difference equals N or its multiple. Thus $m = 0$ (modulo N) means that $m = 0, \pm N, \pm 2N, \pm 3N$, and so forth. If $m = 0$, then every term in the summation equals 1 and there are N terms. Thus, the sum equals N. This sum also equals N if $m = N$ or, more generally, $m = 0$ (modulo N), because $\phi_m[k]$ is periodic in m with period N. We use

$$\sum_{k=0}^{N-1} r^k = 1 + r + r^2 + \cdots + r^{N-1} = \frac{1 - r^N}{1 - r} \tag{7.8}$$

to write

$$\sum_{k=0}^{N-1} e^{jm(2\pi/N)k} = \frac{1 - e^{jm(2\pi/N)N}}{1 - e^{jm(2\pi/N)}} \tag{7.9}$$

If $m \neq 0$ (modulo N), then $e^{jm(2\pi/N)} \neq 1$, $e^{jm(2\pi/N)N} = e^{jm2\pi} = 1$, and (7.9) becomes 0. This establishes (7.7). Note that (7.7) can also be verified directly using the vectors in Figure 7.1.

EXAMPLE 7.1.1

We can verify (7.7) for $N = 3$ directly from the plot in Figure 7.1(a). If $m = 0$, (7.7) becomes $a + a + a = 1 + 1 + 1 = 3$. If $m = 1$ or 2, (7.7) becomes $a + b + c = 1 + (-0.5 + j0.866) + (-0.5 - j0.866) = 0$.

EXERCISE 7.1.2

Verify (7.7) for $N = 4$ from Figure 7.1(b). ∎

Letting the asterisk denote the complex conjugate; (7.7) then implies

$$\sum_{k=0}^{N-1} \phi_m[k]\phi_n^*[k] = \sum_{k=0}^{N-1} e^{j(m-n)(2\pi/N)k} = \begin{cases} N & \text{for } m = n \text{ (modulo } N) \\ 0 & \text{for } m \neq n \text{ (modulo } N) \end{cases} \tag{7.10}$$

This is called the *orthogonality property* of the set $\phi_m[k]$ and is the discrete counterpart of (6.7). Note that the summation from $k = 0$ to $N - 1$ in (7.7) and (7.10) can be replaced by $k = i$ to $i + N - 1$ for any integer i.

7.2   DISCRETE-TIME FOURIER SERIES

Every analog periodic $f(t)$ with period P that meets the Dirichlet conditions can be expressed in the (continuous-time) Fourier series as

$$f(t) = \sum_{m=-\infty}^{\infty} c_m e^{jm\omega_0 t}$$

where $\omega_0 = 2\pi/P$. It is an infinite complex-exponential series and consists of infinitely many discrete frequencies. In this section, we develop its discrete counterpart.

Let $f[k]$ be a periodic sequence with period N. Define

$$\omega_0 = \frac{2\pi}{N}$$

Then $f[k]$ can be expressed in the following *discrete-time Fourier series*

$$f[k] = \sum_{m=0}^{N-1} c_m e^{jm\omega_0 k} = \sum_{m=0}^{N-1} c_m e^{jm(2\pi/N)k} \qquad (7.11)$$

with

$$c_m = \frac{1}{N} \sum_{k=0}^{N-1} f[k] e^{-jm(2\pi/N)k} \qquad (7.12)$$

for $m = 0, 1, 2, \ldots, N-1$. Strictly speaking, we should use the range in (7.6) for the summation in (7.11); however, we use the range from $m = 0$ to $N-1$ to simplify presentation and computation. In fact, we can use any consecutive N numbers in the summations in (7.11) and (7.12).

To establish (7.12), we multiply $e^{-jn(2\pi/N)k}$ on both sides of (7.11) and then sum them up over N to yield

$$\sum_{k=0}^{N-1} f[k] e^{-jn(2\pi/N)k} = \sum_{k=0}^{N-1} \left[\sum_{m=0}^{N-1} c_m e^{jm(2\pi/N)k} \right] e^{-jn(2\pi/N)k}$$

$$= \sum_{m=0}^{N-1} c_m \left[\sum_{k=0}^{N-1} e^{j(m-n)(2\pi/N)k} \right] \qquad (7.13)$$

where we have changed the order of summations. The summation in the brackets equals 0 if $m \neq n$ and N if $m = n$, thus, (7.13) reduces to

$$\sum_{k=0}^{N-1} f[k] e^{-jn(2\pi/N)k} = c_n N$$

which is (7.12) after replacing n with m. This establishes the discrete-time Fourier series.

Although (7.12) is stated for m in the range from 0 to $N-1$, the equation is actually applicable to any integer m. However, the set of c_m, for all integer m, is periodic with period N. For example, we have

$$c_{m+N} = \frac{1}{N} \sum_{k=0}^{N-1} f[k]e^{-j(m+N)(2\pi/N)k}$$

$$= \frac{1}{N} \sum_{k=0}^{N-1} f[k]e^{-jm(2\pi/N)k} e^{-jN(2\pi/N)k} \qquad (7.14)$$

Because $e^{-j2\pi k} = 1$, for all integers k, (7.14) reduces to

$$c_{m+N} = \frac{1}{N} \sum_{k=0}^{N-1} f[k]e^{-jm(2\pi/N)k} = c_m \qquad (7.15)$$

This shows the periodicity of c_m. When discussing symmetric properties, we must extend c_m periodically to all m. However, we need only N consecutive c_m in the discrete-time Fourier series.

The discrete-time Fourier series expresses $f[k]$ as a linear combination of the complex-exponential sequences with frequencies $m(2\pi/N)$, $m = 0, 1, \ldots,$ $N-1$, and weight c_m. Thus the set of c_m is called the *discrete* or *line* frequency spectrum. As in the analog case, c_m is, in general, complex valued. Its amplitude $|c_m|$ is called the discrete amplitude spectrum, and its phase $\sphericalangle c_m$ is called the discrete phase spectrum. If $f[k]$ is a real-valued sequence, that is, $f^*[k] = f[k]$, where the asterisk denotes the complex conjugate, then we have

$$c_m^* = \left[\sum_{k=0}^{N-1} f[k]e^{-jm(2\pi/N)k} \right]^* = \sum_{k=0}^{N-1} f[k]e^{jm(2\pi/N)k} = c_{-m}$$

which, as in the analog case, implies

$$|c_m| = |c_{-m}| \qquad \text{and} \qquad \sphericalangle c_m = -\sphericalangle c_{-m}$$

Thus, if $f[k]$ is real, its discrete frequency spectrum c_m is conjugate symmetric, its amplitude spectrum is even, and its phase spectrum is odd.

The frequency range of discrete-time signals with $T = 1$ is $(-\pi, \pi]$. Thus we are interested in c_m with $m\omega_0 = m(2\pi/N)$ in $(-\pi, \pi]$. The c_m computed in (7.12) has frequencies at $m\omega_0 = m(2\pi/N)$, for $m = 0, 1, \ldots, N-1$, or inside the frequency range $[0, 2\pi)$ (not including 2π). However, because of the periodicity, the c_m in $(-\pi, \pi]$ can be readily obtained from those in $[0, 2\pi)$. In conclusion, the frequency range of discrete-time signals is $(-\pi, \pi]$. However, for convenience, we compute the frequency spectrum in $[0, 2\pi)$. By periodic extension, we can then obtain the frequency spectrum in $(-\pi, \pi]$.

EXAMPLE **7.2.1**

Consider how to find the discrete-time Fourier series of the periodic sequence with period $N = 4$ and defined by

$$f[0] = f[1] = 1 \qquad f[2] = f[3] = 0 \tag{7.16}$$

From (7.12), we have

$w/_{m=0}:$
$$c_0 = \frac{1}{4} \sum_{k=0}^{3} f[k] e^{-j \cdot 0 \cdot \frac{2\pi}{4} k} = \frac{1}{4}[1 \cdot 1 + 1 \cdot 1 + 0 \cdot 1 + 0 \cdot 1]$$
$$= 0.5 = 0.5 e^{j0°} \tag{7.17a}$$

$w/_{m=1}:$
$$c_1 = \frac{1}{4} \sum_{k=0}^{3} f[k] e^{-j \frac{2\pi}{4} k} = \frac{1}{4}[1 \cdot 1 + 1 \cdot e^{-j\frac{\pi}{2}} + 0 \cdot e^{-j\pi} + 0 \cdot e^{-j\frac{3\pi}{2}}]$$
$$= \frac{1}{4}(1 - j) = 0.35 e^{-j45°} \tag{7.17b}$$

$w/_{m=2}:$
$$c_2 = \frac{1}{4} \sum_{k=0}^{3} f[k] e^{-j\pi k} = \frac{1}{4}[1 \cdot 1 + 1 \cdot (-1) + 0 \cdot 1 + 0 \cdot (-1)]$$
$$= \frac{1}{4}(1 - 1) = 0 \tag{7.17c}$$

and

$w/_{m=3}:$
$$c_3 = \frac{1}{4} \sum_{k=0}^{3} f[k] e^{-j\frac{3\pi}{2} k} = \frac{1}{4}[1 \cdot 1 + 1 \cdot j]$$
$$= \frac{1}{4}(1 + j) = 0.35 e^{j45°} \tag{7.17d}$$

Thus, the discrete-time Fourier series of $f[k]$ in (7.16) is

$$f[k] = \sum_{m=0}^{3} c_m e^{jm \frac{2\pi}{4} k} = 0.5 + 0.25(1 - j) \cdot e^{j\frac{\pi}{2} k} + 0 \cdot e^{j\pi k}$$
$$+ 0.25(1 + j) \cdot e^{j\frac{3\pi}{2} k} \tag{7.18a}$$
$$= 0.5 + 0.35 e^{j(\frac{\pi}{2} k - \frac{\pi}{4})} + 0.35 e^{j(\frac{3\pi}{2} k + \frac{\pi}{4})} \tag{7.18b}$$

Figure 7.2 shows the discrete amplitude and phase spectra of the periodic sequence. They are plotted using the solid dots for m in $[0, 3]$ or, equivalently, for frequency in $[0, 2\pi)$. They are extended periodically as shown. The periodic sequence in (7.16) consists of three frequencies at 0, $\pi/2$, and $3\pi/2$ or $-\pi/2$.

7.2.1 Discrete frequency spectrum—distribution of power

What is the physical meaning of the discrete frequency spectrum? As in the analog case, we may define the average power of a periodic sequence of period N as

Figure 7.2   *Discrete frequency spectrum of the sequence in (7.16)*

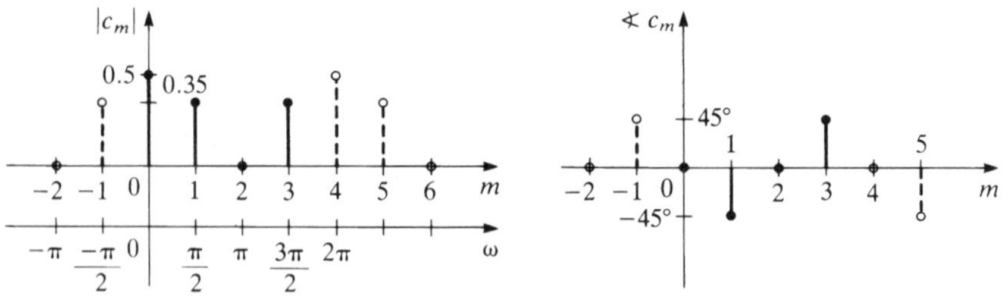

[Two stem plots. Left plot: vertical axis labeled $|c_m|$, with values 0.5 and 0.35 marked. Horizontal axis labeled m with marks at $-2, -1, 0, 1, 2, 3, 4, 5, 6$ and a second axis labeled ω with marks at $-\pi, -\frac{\pi}{2}, 0, \frac{\pi}{2}, \pi, \frac{3\pi}{2}, 2\pi$. Right plot: vertical axis labeled $\angle c_m$, with $45°$ and $-45°$ marked, value 1 and 5 indicated. Horizontal axis labeled m with marks at $-2, -1, 0, 2, 3, 4$.]

average
Power

$$P_{av} = \lim_{K \to \infty} \frac{1}{2K} \sum_{k=-K}^{K} f[k]f^*[k] = \frac{1}{N} \sum_{k=0}^{N-1} f[k]f^*[k] = \frac{1}{N} \sum_{k=0}^{N-1} |f[k]|^2$$

Substituting (7.11) and interchanging the order of summations yield

$$\frac{1}{N} \sum_{k=0}^{N-1} f[k]f^*[k] = \frac{1}{N} \sum_{k=0}^{N-1} \left[\sum_{m=0}^{N-1} c_m e^{jm(2\pi/N)k} \right] f^*[k]$$

$$= \frac{1}{N} \sum_{m=0}^{N-1} c_m \left[\sum_{k=0}^{N-1} f^*[k]e^{jm(2\pi/N)k} \right] = \sum_{m=0}^{N-1} c_m \left[\frac{1}{N} \sum_{k=0}^{N-1} f[k]e^{-jm(2\pi/N)k} \right]^*$$

which becomes, after substituting (7.12),

$$P_{av} = \frac{1}{N} \sum_{k=0}^{N-1} |f[k]|^2 = \sum_{m=0}^{N-1} c_m c_m^* = \sum_{m=0}^{N-1} |c_m|^2$$

This is the counterpart of (6.21). It states that the average power of $f[k]$ equals the sum of $|c_m|^2$. Thus, the discrete frequency spectrum reveals the distribution of power at discrete frequencies. Note that the total energy of periodic sequences is infinity; thus, we discuss only the average power.

EXAMPLE 7.2.2 $f[0] = f[+1] = 1$; $f[2] = f[3] = 0$

Consider the periodic sequence $f[k]$ in (7.16) with period $N = 4$. Its average power can be computed directly from $f[k]$ as

$$P_{av} = \frac{1}{4}[1^2 + 1^2] = \frac{2}{4} = 0.5$$

It can also be computed from the frequency spectrum as

$$P_{av} = \sum_{m=0}^{3} c_m c_m^* = \frac{1}{2} \cdot \frac{1}{2} + \frac{1-j}{4} \cdot \frac{1+j}{4} + 0 + \frac{1+j}{4} \cdot \frac{1-j}{4}$$

$$= \frac{4}{16} + \frac{2}{16} + \frac{2}{16} = \frac{1}{2} = 0.5$$

which yields the same result, as expected.

7.3   DISCRETE-TIME FOURIER TRANSFORM

In the continuous-time case, the Fourier transform is obtained from the Fourier series by extending the period to infinity. Using the identical procedure, we can develop the discrete-time Fourier transform. Consider the periodic sequence shown in Figure 7.3. It is periodic with period $N = N_1 + N_2 + 1$. Using (7.11) and (7.12), we can express its discrete-time Fourier series as

$$f[k] = \sum_{m=0}^{N-1} c_m e^{jm(2\pi/N)k} \tag{7.19a}$$

and

$$c_m = \frac{1}{N} \sum_{k=-N_1}^{N_2} f[k] e^{-jm(2\pi/N)k} \tag{7.19b}$$

The summation in (7.19b) is from $k = -N_1$ to N_2 rather than from 0 to $N - 1$. This is permitted because of the periodicities of $f[k]$ and the complex exponentials. Defining

$$\boxed{\omega := m\frac{2\pi}{N}} \quad \text{and} \quad \boxed{F(\omega) := Nc_m}$$

Figure 7.3 *A periodic sequence*

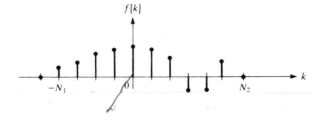

we rewrite (7.19) as

$$F(\omega) = Nc_m = \sum_{k=-N_1}^{N_2} f[k]e^{-j\omega k} \tag{7.20a}$$

and

$$f[k] = \frac{1}{2\pi} \sum_{m=0}^{N-1} Nc_m e^{jm(2\pi/N)k} \cdot \frac{2\pi}{N} = \frac{1}{2\pi} \sum_{m=0}^{N-1} F(\omega)e^{j\omega k} \cdot \frac{2\pi}{N} \tag{7.20b}$$

As N approaches infinity, the sequence becomes aperiodic and $2\pi/N \to 0$. In this case, $\omega = m2\pi/N$ becomes a continuum and $2\pi/N$ can be written as $d\omega$. Furthermore, the summation in (7.20b) becomes an integration. Thus, as $N \to \infty$, (7.20) becomes

$$F(\omega) = \sum_{k=-\infty}^{\infty} f[k]e^{-jk\omega} =: \mathscr{T}_d[f[k]] \tag{7.21a}$$

and

$$f[k] = \frac{1}{2\pi} \int_{\omega=0}^{2\pi} F(\omega)e^{jk\omega}d\omega = \frac{1}{2\pi} \int_{<2\pi>} F(\omega)e^{jk\omega}d\omega =: \mathscr{T}_d^{-1}[F(\omega)] \tag{7.21b}$$

This pair is called the *discrete-time Fourier transform* and is the counterpart of (6.43) and (6.44). Because $F(\omega)$ is defined as Nc_m with $N \to \infty$, and because c_m denotes the weight associated with frequency $m(2\pi/N)$, it is justifiable to call $F(\omega)$ the *frequency spectrum* of $f[k]$. From now on, the terms *frequency spectrum* and *discrete-time Fourier transform* will be used synonymously.

The discrete-time Fourier transform is developed for aperiodic sequences. However, not every aperiodic sequence has a discrete-time Fourier transform. A sufficient condition for $f[k]$ to have a discrete-time Fourier transform is that $f[k]$ is *absolutely summable* or

$$\sum_{-\infty}^{\infty} |f[k]| \le M < \infty \tag{7.22}$$

for some constant M. For example, the sequences $f[k] = b^k$ or $b^k \sin \omega_0 k$, for $k \ge 0$ and 0 for $k < 0$, are absolutely summable if $|b| < 1$ because

$$\sum_{k=-\infty}^{\infty} |f[k]| = \sum_{k=0}^{\infty} |b^k \sin \omega_0 k| \le \sum_{k=0}^{\infty} |b|^k = \frac{1}{1-|b|}$$

The sequence b^k for all k, however, is not absolutely summable for any b (why?). If a sequence is bounded and has a finite number of nonzero entries, then it is absolutely summable.

If $f[k]$ is absolutely summable, then (7.21a) implies

$$|F(\omega)| \le \sum_{k=-\infty}^{\infty} |f[k]e^{-j\omega k}| \le \sum_{k=-\infty}^{\infty} |f[k]| < \infty$$

for all ω. It can also be shown that $F(\omega)$ is a continuous function of ω under the same condition. See Reference [36]. Thus, *if $f[k]$ is absolutely summable, then its frequency spectrum $F(\omega)$ is a bounded and continuous function of ω.* ✳

Now let us discuss some properties of $F(\omega)$. Because, using (7.3),

$$e^{-jk(\omega+2n\pi)} = e^{-jk\omega}e^{-j2\pi nk} = e^{-jk\omega}$$

for every integer k and n, the complex exponential $e^{-jk\omega}$ is a periodic function of ω with period 2π. Thus, $F(\omega)$ in (7.21a) is also a periodic function of ω with period 2π. In general, the frequency spectrum $F(\omega)$ is a complex-valued function of ω. If we write

$$F(\omega) = Re\ F(\omega) + j\ Im\ F(\omega) =: A(\omega)e^{j\theta(\omega)} \qquad (7.23)$$

where Re and Im stand for the real part and the imaginary part, then

$$A(\omega) := |F(\omega)| = [(Re\ F(\omega))^2 + (Im\ F(\omega))^2]^{1/2}$$

and

$$\theta(\omega) := \sphericalangle F(\omega) = \tan^{-1}\frac{Im\ F(\omega)}{Re\ F(\omega)}$$

$A(\omega)$ is called the *amplitude spectrum* and $\theta(\omega)$ is called the *phase spectrum*. ✳

If $f[k]$ is a real-valued sequence, then we have

$$F(-\omega) = F^*(\omega) \qquad (7.24a)$$

where the asterisk denotes the complex conjugate, and

✳ $$A(\omega) = A(-\omega) \quad\text{and}\quad \theta(\omega) = -\theta(-\omega)$$ ✳ $\qquad (7.24b)$

Thus, if $f[k]$ is real, then its frequency spectrum is conjugate symmetric, its amplitude spectrum is even, and its phase spectrum is odd. Here, we implicitly assume that the spectrum has been extended periodically to all ω. A sequence $f[k]$ is even if $f[k] = f[-k]$ and odd if $f[k] = -f[-k]$ for all k. If $f[k]$ is a real and even function of k, then its discrete-time Fourier transform $F(\omega)$ is a real and even function of ω. If $f[k]$ is real and odd, then $F(\omega)$ is imaginary and odd. These statements can be established as in the analog case, and their proofs are left as exercises.

EXAMPLE 7.3.1

Consider the aperiodic sequence shown in Figure 7.4(a). It equals 1 for $k = -2, -1, 0, 1, 2$ and equals 0 elsewhere. Its discrete-time Fourier transform is

$$F(\omega) = \sum_{k=-2}^{2} e^{-jk\omega} = e^{j2\omega} + e^{j\omega} + 1 + e^{-j\omega} + e^{-j2\omega} = 1 + 2\cos\omega + 2\cos 2\omega$$

and is plotted in Figure 7.4(b). It is real and even because $f[k]$ is real and even. Figures 7.4(c) and (d) show the amplitude and phase spectra of $f[k]$. Note that if $F(\omega)$ is positive, then its phase is 0. If $F(\omega)$ is negative, then its phase is $180°$ or $-180°$. Although it is all right to choose the phase to be $180°$ for both positive and negative ω, we prefer to choose the phase to be $180°$ (or $-180°$) for positive ω and $-180°$ (or $180°$) for negative ω so that the phase of $F(\omega)$ is an odd function of ω, as shown in Figure 7.4(d).

EXERCISE 7.3.1

Consider the aperiodic sequence shown in Figure 7.4(e). Verify that its spectrum is as shown in Figure 7.4(f). Which sequence, the one in Figure 7.4(a) or the one in Figure 7.4(e), has higher frequency components? ■

7.3.1   Frequency spectrum—distribution of energy

In this section, we will discuss the physical meaning of $F(\omega)$. As in the analog case, the total energy of $f[k]$ can be defined as

$$E := \sum_{k=-\infty}^{\infty} f[k]f^*[k] = \sum_{k=-\infty}^{\infty} |f[k]|^2 \tag{7.25}$$

If $f[k]$ is absolutely summable, then its total energy is finite. Indeed, if

$$\sum_{k=-\infty}^{\infty} |f[k]| < \infty$$

then $f[k]$ must be bounded. That is, there exists a constant M such that

$$|f[k]| \le M \qquad \text{for all } k$$

Thus, we have

$$E = \sum_{k=-\infty}^{\infty} |f[k]|^2 = \sum_{k=-\infty}^{\infty} |f[k]||f[k]| \le M \sum_{k=-\infty}^{\infty} |f[k]|$$

Figure 7.4 *Two sequences and their spectra*

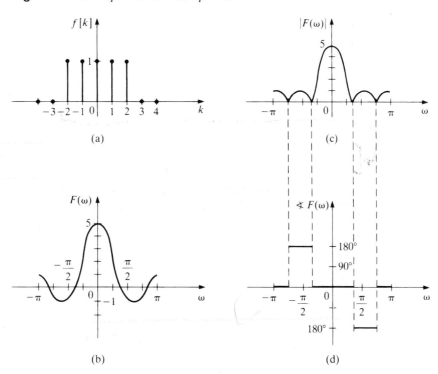

(a)

(c)

(b)

(d)

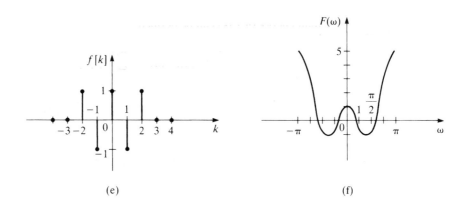

(e)

(f)

This shows that, if $f[k]$ is absolutely summable, then $f[k]$ has a finite total energy. In order to express the total energy of $f[k]$ in terms of its frequency spectrum, we substitute (7.21b) into (7.25) and interchange the order of summation and integration to yield

$$E = \sum_{k=-\infty}^{\infty} \left[\frac{1}{2\pi} \int_{\omega=0}^{2\pi} F(\omega)e^{jk\omega}d\omega \right] f^*[k] = \frac{1}{2\pi} \int_{\omega=0}^{2\pi} F(\omega) \left[\sum_{k=-\infty}^{\infty} f[k]e^{-jk\omega} \right]^* d\omega$$

Thus, we have

$$E = \sum_{k=-\infty}^{\infty} |f[k]|^2 = \frac{1}{2\pi} \int_{\omega=0}^{2\pi} F(\omega)F^*(\omega)d\omega = \frac{1}{2\pi} \int_{<2\pi>} |F(\omega)|^2 d\omega \qquad (7.26)$$

This is the counterpart of (6.75) and is also called a Parseval's formula. Because of the periodicity of $F(\omega)$, the integration can be carried out from $-\pi$ to π or over any interval of 2π, denoted by $<2\pi>$. Unlike the frequency spectrum of periodic sequences, which is defined at discrete frequencies, the frequency spectrum of aperiodic sequences is defined at every ω. Thus, the energy at a discrete frequency is infinitesimal, and it is only meaningful to discuss energy over a nonzero frequency interval.[1] This situation is identical to the continuous-time case. Table 7.1 is the discrete counterpart of Table 6.2. The discrete-time Fourier transform of periodic sequences in the table will be developed in a later section.

Table 7.1 *Frequency spectrum of discrete-time sequences*

Frequency Spectrum ($-\pi < \omega \le \pi$)	Periodic Sequence $f[k] =$ $f[k+N]$ $\omega_0 = 2\pi/N$	Absolutely Summable Sequence $f[k]$
Discrete-time Fourier series (Discrete frequency spectrum)	$c_m = \dfrac{1}{N}\sum_{0}^{N-1} f[k]e^{-jm\omega_0 k} = c_{m+N}$ $f[k] = \sum_{m=0}^{N-1} c_m e^{jm\omega_0 k}$	$F(m\omega_0) = Nc_m$ $N \to \infty$
Discrete-time Fourier transform (Frequency spectrum)	$F(\omega) = \sum_{m=0}^{N-1} 2\pi c_m \delta(\omega - m\omega_0)$ $F(\omega) = F(\omega + 2\pi)$	$F(\omega) = \sum_{-\infty}^{\infty} f[k]e^{-j\omega k} = F(\omega + 2\pi)$ $f[k] = \dfrac{1}{2\pi}\int_{0}^{2\pi} F(\omega)e^{j\omega k}d\omega$
Remarks	Average power= $\dfrac{1}{N}\sum_{0}^{N} f[k]f^*[k] = \sum_{m=0}^{N-1} c_m c^*_m$	Total energy= $\sum_{-\infty}^{\infty} f[k]f^*[k] = \dfrac{1}{2\pi}\int_{0}^{2\pi} F(\omega)F^*(\omega)d\omega$

[1] Because of this, it is immaterial whether the frequency spectrum at $\omega = 2\pi$ is included or not. That is, there is no difference between using the frequency range $[0, 2\pi)$ or $[0, 2\pi]$. This is, however, not the case for discrete frequency spectra.

7.3.2   **From the Fourier transform to the discrete-time Fourier transform**

In the preceding sections, we developed the frequency spectrum of discrete-time signals using first the discrete-time Fourier series and then the discrete-time Fourier transform. In this section, we shall show that it can also be obtained directly from the (continuous-time) Fourier transform. Consider a discrete-time sequence $f[k]$, $k = 0, \pm1, \pm2, \ldots.$ If we apply the (continuous-time) Fourier transform to $f[k]$, then $\mathcal{T}[f[k]] = 0$ and the result is useless. Now we shall modify the sequence as

$$\mathcal{T} f_s(t) := \sum_{-\infty}^{\infty} f[k]\delta(t - k) \tag{7.27}$$

See (5.11). It is a sequence of impulses with weight $f[k]$ at $t = k$. It is defined for all t but assumes nonzero values only at discrete time instants k. Therefore, $f_s(t)$ can be considered as a continuous-time representation of the discrete-time sequence $f[k]$. The application of the Fourier transform to (7.27) yields, using the linearity property,

$$\mathcal{T}[f_s(t)] = \sum_{k=-\infty}^{\infty} f[k].\mathcal{T}[\delta(t - k)]$$

Using the sifting property of impulses, we have

$$\mathcal{T}[\delta(t - k)] = \int_{-\infty}^{\infty} \delta(t - k)e^{-j\omega t}dt = e^{-j\omega t}\Big|_{t=k} = e^{-j\omega k}$$

Hence, the Fourier transform or frequency spectrum of $f_s(t)$ is

$$\mathcal{T}[f_s(t)] = \sum_{k=-\infty}^{\infty} f[k]e^{-j\omega k} \tag{7.28}$$

which is identical to (7.21a). Thus, we have established

$$\mathcal{T}[f_s(t)] = \mathcal{T}_d[f[k]] \tag{7.29a}$$

or

$$\text{continuous-time Fourier transform of } \sum_{-\infty}^{\infty} f[k]\delta(t - k)$$
$$= \text{discrete-time Fourier transform of } f[k] \tag{7.29b}$$

Therefore, the relationship between the continuous-time and discrete-time Fourier transforms is similar to the one between the Laplace transform and the z-transform, which was discussed in Section 5.2.2.

7.3.3   **From the z-transform to the discrete-time Fourier transform—absolutely summable sequences**

Consider an absolutely summable sequence $f[k]$ defined over $(-\infty,\infty)$. We can decompose it into two parts as

$$f[k] = f_-[k] + f_+[k] \tag{7.30}$$

where $f_-[k]$ is a negative-time sequence—that is, $f_-[k] = 0$, for $k = 1$, $2, 3, \ldots$—and $f_+[k]$ is a positive-time sequence—that is, $f_+[k] = 0$, for $k = -1, -2, -3, \ldots$. Note that the decomposition in (7.30) is unique except for the value at $k = 0$. It is immaterial where $f[0]$ is assigned; $f[0]$ can be assigned entirely in $f_+[k]$, entirely in $f_-[k]$, or partially in $f_+[k]$ and partially in $f_-[k]$. Note that the negative-time sequence $f_-[k]$ becomes a positive-time sequence if we reverse its time as $f_-[-k]$.

Now we show

$$
\begin{aligned}
\text{frequency spectrum of } f[k] &= \mathscr{F}_d[f[k]]\\
&= \mathscr{Z}[f_+[k]]\Big|_{z=e^{j\omega}} + \mathscr{Z}[f_-[-k]]\Big|_{z=e^{-j\omega}}
\end{aligned}
\tag{7.31}
$$

which is the discrete counterpart of (6.48). We will discuss only the negative-time sequence here. By definition, we have

$$
\mathscr{F}_d[f_-[k]] = \sum_{k=-\infty}^{\infty} f_-[k]e^{-jk\omega} = \sum_{k=-\infty}^{0} f_-[k]e^{-jk\omega}
$$

$$
= \sum_{k'=0}^{\infty} f_-[-k']e^{jk'\omega} \tag{7.32}
$$

where we have changed the variable $k' := -k$ in the last step. Note that $f_-[-k]$ is positive time and its z-transform is

$$
\mathscr{Z}[f_-[-k]] = \sum_{k=0}^{\infty} f_-[-k]z^{-k} \tag{7.33}
$$

The sequence $f_-[k]$ is absolutely summable; thus, the infinity summation in (7.33) converges on the unit circle of the z-plane. The substitution of $z = e^{-j\omega}$ into (7.33) yields

$$
\mathscr{Z}[f_-[-k]]\Big|_{z=e^{-j\omega}} = \sum_{k=0}^{\infty} f_-[-k]e^{jk\omega} \tag{7.34}
$$

which is identical to (7.32). This establishes the negative-time part of (7.31). The positive-time part can be similarly proved.

Before proceeding, note that, if the sampling period is included in $f(kT)$, then (7.31) must be modified as frequency spectrum of $f\,(kT) = \mathscr{F}_d\,[f\,(kT)]$

$$= \left. \mathscr{Z}[f_+(kT)]\right|_{z=e^{j\omega T}} + \left. \mathscr{Z}[f_-(-kT)]\right|_{z=e^{-j\omega T}} \qquad (7.35)$$

In conclusion, if a sequence is absolutely summable, then its frequency spectrum can be obtained from its z-transform by the substitution of $z = e^{\pm j\omega T}$. This is illustrated in the following examples. Because F has been used to denote the spectrum of $f\,[k]$, the z-transform of $f\,[k]$ will be denoted by $\bar{F}$. ✳

EXAMPLE **7.3.2**

Consider the positive-time sequence $f\,[k] = (0.9)^k$, b^k $k = 0, 1, 2, \ldots$. It is absolutely summable. Its z-transform is, using Table 5.1,

'w_2 $b < 1$ $$\bar{F}(z) = \frac{1}{1 - 0.9z^{-1}} = \frac{z}{z - 0.9}$$

Thus, the frequency spectrum of $f\,[k]$ is

$$F(\omega) = \bar{F}(e^{j\omega}) = \frac{1}{1 - 0.9e^{-j\omega}}$$

We compute $F(0) = 1/(1 - 0.9 \cdot 1) = 1/0.1 = 10$, $F(\pi) = 1/(1 - 0.9(-1)) = 1/1.9$ $= 0.53$,

$$F(\frac{\pi}{4}) = \frac{1}{1 - 0.9(0.7 - j0.7)} = \frac{1}{0.37 + j0.63} = \frac{1}{0.73e^{j60°}} = 1.37e^{-j60°}$$

and so forth. Using these data, we can plot the amplitude and phase spectra of $f\,[k]$ as shown in Figure 7.5.

Figure 7.5 *Frequency spectrum*

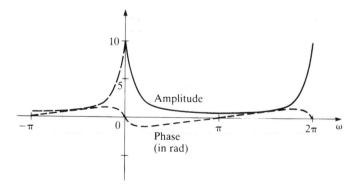

EXAMPLE 7.3.3

Consider how to find the frequency spectrum of $f[k] = (0.9)^k \cos \pi k$, $k = 0, 1, 2, \ldots$. The sequence is absolutely summable and its z-transform is, using Table 5.1,

$$\bar{F}(z) = \frac{(z - 0.9(\cos \pi))z}{z^2 - 2 \cdot 0.9(\cos \pi)z + (0.9)^2} = \frac{1 + 0.9z^{-1}}{1 + 1.8z^{-1} + 0.81z^{-2}}$$

Thus, the frequency spectrum of $f[k]$ is

$$F(\omega) = \bar{F}(e^{j\omega}) = \frac{1 + 0.9e^{-j\omega}}{1 + 1.8e^{-j\omega} + 0.81e^{-j2\omega}}$$

which is plotted in Figure 7.6. Most of the energy of this sequence centers around $\omega = \pi$, the highest frequency of digital signals. Thus, this is a high-frequency signal. The sequence in Example 7.3.2 is a low-frequency signal.

EXAMPLE 7.3.4

 b^k

Consider how to find the frequency spectrum of $f[k] = 2^k$, $k = 0, 1, 2, \ldots$. Its z-transform is $\bar{F}(z) = z/(z - 2)$. However,

$$F(\omega) = \bar{F}(e^{j\omega}) = \frac{e^{j\omega}}{e^{j\omega} - 2}$$

is meaningless because $f[k]$ is not absolutely summable. Thus, its frequency spectrum is not defined.

$b > 1$

Figure 7.6 *Frequency spectrum*

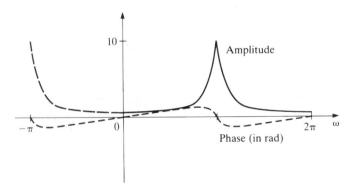

EXERCISE 7.3.2

Find the frequency spectra of $(0.8)^k$ and $(0.8)^k \cos 2k$, for $k = 0, 1, 2, \ldots$.

[**ANSWERS:** $1/(1 - 0.8e^{-j\omega})$; $(1 + 0.33e^{-j\omega})/(1 + 0.66e^{-j\omega} + 0.64e^{j2\omega})$.] ■

EXAMPLE **7.3.5**

To find the frequency spectrum of $f[k] = (0.5)^{|k|}$, $k = 0, \pm1, \pm2, \ldots$, define

$$f_+[k] := (0.5)^k \qquad \text{for } k = 0, 1, 2, \ldots$$

and

$$f_-[k] := (0.5)^{-k} \qquad \text{for } k = 0, -1, -2, \ldots$$

Because $f[0] = 1$ appears in both $f_+[0]$ and $f_-[0]$, we have[2]

$$\boxed{f[k] = f_+[k] + f_-[k] - 1 \cdot \delta[k]} \Rightarrow \text{for discrete-time case}$$

where $\delta[k]$ is the impulse sequence. The z-transform of $f_+[k]$ is $z/(z - 0.5) = 1/(1 - 0.5z^{-1})$. Thus, we have

$$\text{spectrum of } [f_+[k] - \delta[k]] = \left[\frac{1}{1 - 0.5z^{-1}} - 1\right]\Bigg|_{z=e^{j\omega}}$$

The z-transform of $f_-[-k]$ is $z/(z - 0.5) = 1/(1 - 0.5z^{-1})$ and

$$\text{spectrum of } [f_-[-k]] = \frac{1}{1 - 0.5z^{-1}}\Bigg|_{z=e^{-j\omega}}$$

Thus, the frequency spectrum of $f[k]$ is

$$F(\omega) = \frac{1}{1 - 0.5e^{-j\omega}} - 1 + \frac{1}{1 - 0.5e^{j\omega}}$$

$$= \frac{1 - 0.5e^{j\omega} - (1 - 0.5e^{-j\omega})(1 - 0.5e^{j\omega}) + 1 - 0.5e^{-j\omega}}{(1 - 0.5e^{-j\omega})(1 - 0.5e^{j\omega})}$$

---

[2]In the continuous-time case, the value of $f(t)$ at $t = 0$, an isolated point, is immaterial. Therefore, it is unnecessary to subtract $f(0)$ from $f(t) = f_+(t) + f_-(t)$, even though $f_+(t)$ and $f_-(t)$ both contain $f(0)$. ✱

which can be simplified as

$$F(\omega) = \frac{0.75}{1.25 - \cos \omega}$$

It is real and even, and it is plotted in Figure 7.7.

EXERCISE 7.3.3

Find the frequency spectra of the following.

  (a) $f[k] = (0.8)^k$, $k = 0, \pm 1, \pm 2, \ldots$

  (b) $f[k] = (0.8)^{|k|}$, $k = 0, \pm 1, \pm 2, \ldots$

  (c) $f[k] = \begin{cases} (0.8)^k & \text{for } k = 0, 1, 2, \ldots \\ (0.5)^{-k} & \text{for } k = -1, -2, -3, \ldots \end{cases}$

[**ANSWERS:** (a) Not defined; (b) $0.36/(1.64 - 1.6 \cos \omega)$; (c) $[1/(1 - 0.8e^{-j\omega})]$ $- 1 + [1/(1 - 0.5e^{j\omega})]$.] ∎

7.3.4   Discrete-time Fourier transform of periodic sequences

Just like the continuous-time case, the frequency spectrum of discrete-time periodic signals contains impulses. Instead of directly computing the frequency spectrum of periodic sequences, it is simpler to compute the time sequence of the spectrum $F(\omega) = \delta(\omega - \omega_0)$, an impulse located at $\omega = \omega_0$. Using (7.21b) and the sifting property of the impulse, we have

$$f[k] = \frac{1}{2\pi} \int_{-\pi}^{\pi} \delta(\omega - \omega_0)e^{jk\omega}d\omega = \frac{1}{2\pi}e^{jk\omega_0} \qquad (7.36)$$

Figure 7.7 *Frequency spectrum of a two-sided sequence*

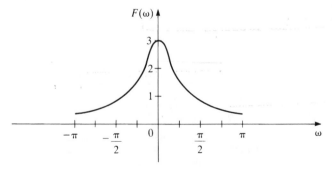

for all k. In this equation, we have assumed implicitly that ω_0 lies inside the range $(-\pi, \pi]$. Thus, we have

$$\mathscr{F}_d[e^{jk\omega_0}] = 2\pi\delta(\omega - \omega_0)$$ (7.37)

In particular, we have

$$\mathscr{F}_d[1] = 2\pi\delta(\omega)$$

$$\mathscr{F}_d[\sin k\omega_0] = \frac{\pi}{j}[\delta(\omega - \omega_0) - \delta(\omega + \omega_0)]$$

and

$$\mathscr{F}_d[\cos k\omega_0] = \pi[\delta(\omega - \omega_0) + \delta(\omega + \omega_0)]$$

These are identical to (6.54), (6.55), and (6.56). In general, if the discrete-time Fourier series of a periodic sequence of period N is

$$f[k] = \sum_{m=0}^{N-1} c_m e^{jm(2\pi/N)k}$$

then the discrete-time Fourier transform of the sequence is

$$\text{frequency spectrum of } f[k] = \mathscr{F}_d[f[k]] = \sum_{m=0}^{N-1} 2\pi c_m \delta(\omega - m\frac{2\pi}{N})$$ (7.38)

This is similar to (6.57) and (6.58).

7.4   PROPERTIES OF THE DISCRETE-TIME FOURIER TRANSFORM

In this section, we will discuss some properties of the discrete-time Fourier transform. Because of the close relationship between the continuous-time and discrete-time Fourier transforms, most properties of the former also hold for the latter.

Time-shifting property   Let $F(\omega) = \mathscr{F}_d[f[k]]$ and let k_0 be an integer. Then, we have

$$\mathscr{F}_d[f[k - k_0]] = e^{-j\omega k_0} F(\omega)$$ (7.39)

By definition, we have

$$\mathscr{F}_d[f[k - k_0]] = \sum_{k=-\infty}^{\infty} f[k - k_0]e^{-jk\omega} = e^{-jk_0\omega} \sum_{k=-\infty}^{\infty} f[k - k_0]e^{-j(k-k_0)\omega}$$

which becomes (7.39) after setting $k' := k - k_0$. Equation (7.39) implies

Time shifting

$$\left| e^{-jk_0\omega} F(\omega) \right| = \left| e^{-jk_0\omega} \right| \left| F(\omega) \right| = \left| F(\omega) \right|$$

and

$$\sphericalangle [e^{-jk_0\omega} F(\omega)] = \sphericalangle e^{-jk_0\omega} + \sphericalangle F(\omega) = -k_0\omega + \sphericalangle F(\omega)$$

Thus, time shifting does not affect the amplitude spectrum of a sequence; it merely introduces a linear phase into the phase spectrum. This property can be used to facilitate the computation of frequency spectra.

EXERCISE 7.4.1

Plot the amplitude and phase spectra of

$$f[k] = 1 \quad \text{for } k = 0, 1, 2, 3, 4 \qquad f[k] = 0 \quad \text{otherwise}$$

What are their relationships with the ones in Figure 7.4(c) and (d)?

[**ANSWERS:** The amplitude spectrum is the same, and the phase spectrum equals the one in Figure 7.4(d) shifted by -2ω.] ■

Frequency-shifting property If $F(\omega) = \mathscr{F}_d[f[k]]$, then

$$\mathscr{F}_d[e^{j\omega_0 k} f[k]] = F(\omega - \omega_0) \tag{7.40}$$

This can easily be proved using (7.21a). Thus, the multiplication of a sequence by a complex exponential with frequency ω_0 merely shifts the spectrum to ω_0. Because both $F(\omega)$ and $e^{j\omega k}$ are periodic with period 2π, we may assume, without loss of generality, that ω_0 lies inside $(-\pi, \pi]$ or $[0, 2\pi)$. Because

$$\cos \omega_0 k = \frac{e^{j\omega_0 k} + e^{-j\omega_0 k}}{2}$$

we have

$$\mathscr{F}_d[f[k] \cos \omega_0 k] = \frac{1}{2}[F(\omega - \omega_0) + F(\omega + \omega_0)]$$

This is the discrete-time counterpart of (6.91) and may be referred to as the modulation property.

EXAMPLE 7.4.1

$Ex\ 7.3.2 \rightarrow$ Consider the sequence $f[k] = (0.9)^k$ for $k \geq 0$. Its spectrum $F(\omega)$ is plotted in Figure 7.5. The frequency spectrum of $(0.9)^k \cos \pi k$ equals, using the frequency-shifting property, $0.5F(\omega - \pi) + 0.5F(\omega + \pi)$. The functions $F(\omega \pm \pi)/2$ are the shifting of $F(\omega)/2$ to $\pm \pi$. Their sum yields the spectrum of $(0.9)^k \cos \pi k$, as shown in Figure 7.6.

$\curvearrowleft Ex.\ 7.3.3$

7.4.1 | Discrete convolution

The zero-state response of every linear time-invariant discrete-time system can be described by the discrete convolution

$$y[k] = \sum_{i=0}^{\infty} h[k-i]u[i] \tag{7.41a}$$

if the input is applied from $k = 0$, or

$$y[k] = \sum_{i=-\infty}^{\infty} h[k-i]u[i] \tag{7.41b}$$

if the input is applied from $k = -\infty$, where $h[k]$ is the impulse response of the system and $h[k] = 0$ for $k < 0$ if the system is causal. The application of the discrete-time Fourier transform to (7.41) yields

$$Y(\omega) = \sum_{k=-\infty}^{\infty} y[k]e^{-jk\omega} = \sum_{k=-\infty}^{\infty} \left[\sum_{i=-\infty}^{\infty} h[k-i]u[i] \right] e^{-jk\omega}$$

which, by interchanging the order of summations, can be written as

$$Y(\omega) = \sum_{i=-\infty}^{\infty} \left[\sum_{k=-\infty}^{\infty} h[k-i]e^{-j(k-i)\omega} \right] u[i]e^{-ji\omega}$$

Let $k' = k - i$. Then, the term inside the brackets becomes $H(\omega)$ and the remainder equals $U(\omega)$. Thus, we have

$$\boxed{Y(\omega) = H(\omega)U(\omega)} \tag{7.42}$$

In other words, the discrete-time Fourier transform transforms a convolution into a multiplication. It is important to mention that (7.42) holds only if $y[k]$, $h[k]$, and $u[k]$ are discrete-time Fourier transformable.

If $y[k]$, $h[k]$, and $u[k]$ are positive-time sequences, then (7.42) can also be obtained from the z-transform. Let $\bar{Y}(z)$, $\bar{H}(z)$, and $\bar{U}(z)$ be the z-transforms of $y[k]$, $h[k]$, and $u[k]$, respectively. Then we have, as developed in Chapter 5,

$$\bar{Y}(z) = \bar{H}(z)\bar{U}(z) \tag{7.43}$$

which becomes, by substituting $z = e^{j\omega}$,

$$\bar{Y}(e^{j\omega}) = \bar{H}(e^{j\omega})\bar{U}(e^{j\omega})$$

which reduces to (7.42) because $Y(\omega) = \bar{Y}(e^{j\omega})$, and so forth. Equation (7.43) is more general than (7.42) because it does not require $y[k]$, $h[k]$, and $u[k]$ to be absolutely summable or periodic. Its manipulation is also simpler. Therefore, even though the discrete-time Fourier transform *can* be used in system analysis, because it is neither as simple nor as general as the z-transform, it is *not* used. It is, however, important in signal analysis and in computer computation, which will be discussed in a later section.

*7.5   THE DISCRETE FOURIER TRANSFORM

This section introduces the discrete Fourier transform, or DFT for short. Unlike the discrete-time Fourier transform, which is defined for sequences with finite or infinite length, the DFT is defined only for sequences with finite length. Before giving a formal definition, we shall develop the DFT from the discrete-time Fourier transform with computer computation in mind.

Consider the following sequence of length N

$$f[k] \quad \text{for } k = 0, 1, 2, \cdots, N-1 \tag{7.44}$$

Its discrete-time Fourier transform is

$$F(\omega) = \sum_{k=-\infty}^{\infty} f[k]e^{-j\omega k} = \sum_{k=0}^{N-1} f[k]e^{-j\omega k} \tag{7.45}$$

It is periodic with period 2π. The frequency interval of interest is $(-\pi, \pi]$. There are infinitely many points in the interval. If we use a digital computer to compute (7.45), then we can compute only a finite number of ω in $F(\omega)$. If $f[k]$ has N points, we compute N equally spaced ω in $(-\pi, \pi]$. These N points are

$$\omega_m = m\frac{2\pi}{N} \tag{7.46}$$

with m ranging as shown in (7.6). This range is very complicated; it is much simpler to use the range $m = 0, 1, 2, \ldots, N - 1$. Thus, in computing, we compute $F(\omega)$ at $\omega = m(2\pi)/N$, for $m = 0, 1, \ldots, N - 1$, and the frequency range becomes $[0, 2\pi)$ (not including 2π). Once $F(\omega)$ with ω in $[0, 2\pi)$ is computed, $F(\omega)$ with ω in $(-\pi, \pi]$ can be readily obtained by periodic extension.

We define $F[m] := F(\omega_m) = F(2\pi m/N)$, for $m = 0, 1, 2, \ldots, N - 1$, and substitute (7.46) into (7.45) to yield

$$\boxed{F[m] = F(m\frac{2\pi}{N}) = \sum_{k=0}^{N-1} f[k]e^{-jmk\,2\pi/N}} \qquad \text{DFT}$$

(7.47)

Note that there are two indices, k and m, both ranging from 0 to $N - 1$. The index k denotes time instant, and m denotes discrete frequency. The two indices should not be confused. Equation (7.47) is the discrete Fourier transform of $f[k]$.

Now let us consider the inverse discrete Fourier transform. The time sequence $f[k]$ can be computed from its frequency spectrum by using (7.21b). If the values of $F(\omega)$ are available only at $\omega_m = 2\pi m/N$, $m = 0, 1, 2, \ldots, N - 1$, it is reasonable to approximate (7.21b) as

$$f[k] = \frac{1}{2\pi} \int_0^{2\pi} F(\omega)e^{jk\omega}d\omega \approx \frac{1}{2\pi} \sum_{m=0}^{N-1} F(\omega_m)e^{jkm\,2\pi/N} \cdot (\frac{2\pi}{N})$$

where $2\pi/N$ is the frequency interval between subsequent ω_m. Using $F[m]$, we can write the equation as

$$f[k] \approx \frac{1}{N} \sum_{m=0}^{N-1} F[m]e^{jkm\,2\pi/N} \qquad \text{IDFT}$$

(7.48)

for $k = 0, 1, 2, \ldots, N - 1$. Equation (7.48) is, in fact, the inverse discrete Fourier transform. We see that computer computation of discrete-time Fourier transform leads naturally to the DFT.

We have developed the DFT and its inverse in (7.47) and (7.48) from the computational point of view. Now we shall formally establish them as a transform pair. Consider a sequence of N real or complex numbers $f[k]$, $k = 0, 1, \ldots, N - 1$, where we implicitly assume $f[k] = 0$ for $k < 0$ and $k > N - 1$. We define

$$W = e^{-j2\pi/N}$$

(7.49)

Then, the DFT of $f[k]$ is

$$\boxed{F[m] = \sum_{k=0}^{N-1} f[k]e^{-j2\pi km/N} = \sum_{k=0}^{N-1} f[k]W^{km} =: \mathscr{D}[f[k]]}$$

(7.50)

$\rightarrow$ requires N multiplications

for $m = 0, 1, \ldots, N - 1$, and the inverse DFT of $F[m]$ is

$$f[k] = \frac{1}{N} \sum_{m=0}^{N-1} F[m] e^{j2\pi mk/N} = \frac{1}{N} \sum_{m=0}^{N-1} F[m] W^{-mk} =: \mathscr{D}^{-1}[F[m]] \quad (7.51)$$

for $k = 0, 1, \ldots, N - 1$. Note that we have used $F(\omega)$ to denote the discrete-time Fourier transform and $F[m]$ to denote the DFT. To show that they are indeed a transform pair, we substitute (7.50), after changing the summation index from k to n, into (7.51) to yield

$$\mathscr{D}^{-1}[F[m]] = \frac{1}{N} \sum_{m=0}^{N-1} \left[\sum_{n=0}^{N-1} f[n] e^{-j2\pi nm/N} \right] e^{j2\pi mk/N}$$

which becomes, after changing the order of summations and using (7.10),

$$\mathscr{D}^{-1}[F[m]] = \frac{1}{N} \sum_{n=0}^{N-1} f[n] \left[\sum_{m=0}^{N-1} e^{j2\pi(k-n)m/N} \right] = f[k]$$

This establishes the DFT pair. It is rather surprising that, although (7.48) is obtained by approximation, it turns out to be an equality.

We discuss some properties of the DFT pair. Because $W^{knN} = e^{-jkn2\pi} = 1$, for every integer k and n, we have

$$W^{k(m+nN)} = W^{km} W^{knN} = W^{km} \cdot 1 = W^{km}$$

Thus, W is periodic with period N. This property implies

$$\boxed{F[m] = F[m \pm N] = F[m \pm 2N] = \cdots}$$

Thus, $F[m]$ is also periodic with period N. Recall that m is actually a frequency at $m2\pi/N$. Thus, the period of $F[m]$ is $N2\pi/N = 2\pi$ in rad/s, same as the period of $F(\omega)$, and we have

$$F[m] = \mathscr{D}[f[k]] = \mathscr{F}_d[f[k]] \Big|_{\omega=m(2\pi/N)} = F(\omega) \Big|_{\omega=m(2\pi/N)} = F\left(m\frac{2\pi}{N}\right)$$

In other words, the DFT of $f[k]$ computes the samples of the discrete-time Fourier transform of $f[k]$.

The inverse DFT in (7.51) actually applies to every integer k. Because of the periodicities of W^{mk} and $F[m]$, the inverse yields a periodic $\bar{f}[k]$ with period N which is the periodic extension of $f(t)$, or

$$\mathscr{D}^{-1}[F[m]] = \bar{f}[k] := \begin{cases} f[k] & \text{for } 0 \leq k \leq N - 1 \\ \text{periodic extension of } f[k] \end{cases} \quad (7.52)$$

In contrast, the inverse discrete-time Fourier transform of $F(\omega)$ yields

$$\mathscr{F}_d^{-1}[F(\omega)] = \begin{cases} f[k] & \text{for } 0 \le k \le N-1 \\ 0 & \text{otherwise} \end{cases} \tag{7.53}$$

This is the major difference between the inverse DFT and inverse discrete-time Fourier transform. To stress the difference, we can write the DFT more informatively as

$$F[m] = \mathscr{D}[f[k]] = \mathscr{D}[\bar{f}[k]] \tag{7.54a}$$

and

$$\bar{f}[k] = \mathscr{D}^{-1}[F[m]] \tag{7.54b}$$

for all integers m and k, where $\bar{f}[k]$ is the periodic extension of $f[k]$. Therefore, even though the DFT is defined for finite sequences, it can also be considered to be defined for periodic sequences of infinity length.

To conclude this section, let us consider the relationship between the DFT and the discrete-time Fourier series. By comparing (7.12) and (7.50), we have

$$F[m] = Nc_m$$

Substituting this into (7.51) yields (7.11). Thus, *the DFT is essentially the same as the discrete-time Fourier series*; they differ only by a constant.

*7.5.1   Properties of the DFT

Because the DFT and the discrete-time Fourier series are essentially the same, all properties of the latter apply to the former. However, it is important to point out that we must consider the DFT to be defined for periodic sequences of period N. For example, while the finite sequence $f[k]$, for $k = 0, 1, 2, \ldots, N-1$, can never be even or odd, its periodic extension $\bar{f}[k]$ can have either property.

A number of properties of the DFT are listed here, but their proofs are omitted.

Linearity property The DFT is a linear operator. That is,

$$\mathscr{D}[a_1 f_1[k] + a_2 f_2[k]] = a_1 \mathscr{D}[f_1[k]] + a_2 \mathscr{D}[f_2[k]]$$

Symmetric properties If $\bar{f}[k]$ is real, then $F[m]$ is conjugate symmetric, or $F[-m] = F^*[m]$. If $\bar{f}[k]$ is real and even, then $F[m]$ is real and even. If $\bar{f}[k]$ is real and odd, then $F[m]$ is imaginary and odd. If fact, all symmetric properties listed in Table 6.4 apply to the DFT.

Duality property Let $F[m]$ and $\bar{f}[k]$ be the DFT pair. That is,

$$F[m] = \mathscr{D}[\bar{f}[k]] \quad \text{and} \quad \bar{f}[k] = \mathscr{D}^{-1}[F[m]]$$

Then, we have

$$\bar{f}[m] = \left[\mathscr{D}\left[\frac{F^*[k]}{N}\right]\right]^*$$

where the asterisk denotes the complex conjugate. This follows directly from (7.50) and (7.51). Thus, the DFT can be used to compute the inverse DFT.

Time-shifting property If $F[m] = \mathscr{D}[\bar{f}[k]]$, then, for any integer k_0,

$$\mathscr{D}[\bar{f}(k - k_0)] = e^{-jmk_0} F[m]$$

Thus, time shifting does not affect the amplitude of the DFT; it merely introduces a linear phase into the original DFT.

Frequency-shifting property If $F[m] = \mathscr{D}[\bar{f}[k]]$, then

$$\mathscr{D}[\bar{f}[k]e^{jkm_0}] = F[m - m_0]$$

where m_0 is an integer. Note that m is associated with frequency. Thus, the multiplication of $\bar{f}[k]$ by e^{jkm_0} introduces frequency shifting.

*7.5.2   Periodic convolution

Consider two sequences $h[k]$ and $u[k]$ for $k = 0, 1, \ldots, N - 1$. Both are of length N and are implicitly assumed to be 0 for $k < 0$ and $k > N - 1$. Consider

$$y[k] = \sum_{i=0}^{k} h[k - i]u[i] = \sum_{i=0}^{k} h[i]u[k - i] \tag{7.55}$$

This is the (discrete) convolution studied in Section 3.2.1. As discussed in Figure 3.4, we have $y[k] = 0$ for $k < 0$ and $k > 2N - 1$. Now we extend $h[k]$ and $u[k]$ periodically with period N and denote them by $\bar{h}[k]$ and $\bar{u}[k]$. Then the equation

$$\bar{y}[k] := \sum_{i=0}^{N-1} \bar{h}[k - i]\bar{u}[i] = \sum_{i=0}^{N-1} \bar{h}[i]\bar{u}[k - i] \tag{7.56}$$

is called the *periodic* or *circular convolution*. The summation in (7.55) is from 0 to k; the summation in (7.56) is over a complete cycle, that is, from 0 to $N - 1$. Because $h[k]$ and $\bar{u}[k]$ are periodic with period N, so is $\bar{y}[k]$. The graphical procedure for computing (7.55) can be applied directly to compute (7.56). An example illustrates.

EXAMPLE 7.5.1

Consider the two sequences of length 3 shown in the left-hand column of Figure 7.8(a) and (b). That is, $h[0] = 2$, $h[1] = 1$, $h[2] = 1$, $u[0] = 1$, $u[1] = 0$, and $u[2] = -1$. Figure 7.8(c) and (d) plot the computation of $y[3]$. First we flip $u[i]$, and then we shift it to $k = 3$, as shown in Figure 7.8(c). The multiplication of $h[i]$ and $u[3 - i]$ yields the plot in Figure 7.8(d). The algebraic sum of all entries in part (d) yields $y[3] = -1$, as shown in Figure 7.8(e), where $y[k]$ for all k is also plotted. Note that, for $k < 0$ and $k > 2N - 1 = 5$, $h[i]$ and $u[k - i]$ do not overlap and their product is zero. Thus, $y[k] = 0$ for $k < 0$ and $k > 2N - 1 = 5$.

To compute the periodic convolution in (7.56) for $k = 3$, we first extend $h[i]$ and $u[i]$ periodically, as shown in the right-hand column of Figure 7.8(a) and (b). We then flip $\bar{u}[i]$ and shift it to $k = 3$ to yield $\bar{u}[3 - i]$, as shown in Figure 7.8(c). The product of $\bar{h}[i]$ and $\bar{u}[3 - i]$ is shown in Figure 7.8(d). The alge-

Figure 7.8 *Convolution and periodic convolution*

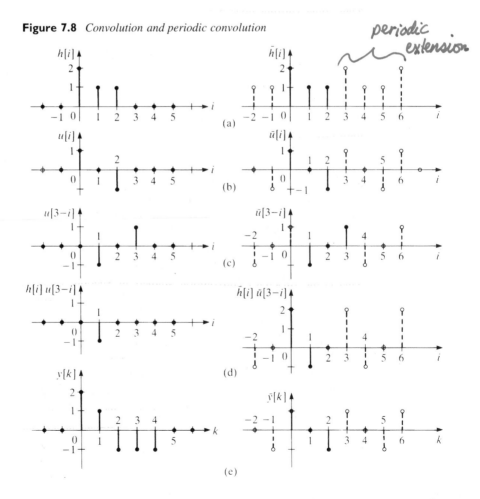

periodic extension

braic sum of the entries at $i = 0$, 1, and 2 yields $\bar{y}[3] = 1$, as shown in Figure 7.8(e), where $\bar{y}[k]$ for all k is also plotted. We see that $\bar{y}[k]$ is periodic with period $N = 3$. We also see that the convolution and the periodic convolution yield entirely different results.

We give a different interpretation of the periodic convolution. First, we express (7.56) explicitly, for $N = 3$ and $k = 0$, as

$$\bar{y}[0] = \sum_{i=0}^{2} \bar{h}[i]\bar{u}[0 - i] = \bar{h}[0]\bar{u}[0] + \bar{h}[1]\bar{u}[-1] + \bar{h}[2]\bar{u}[-2]$$

Because $\bar{u}[k]$ is periodic with period $N = 3$, we have $\bar{u}[-1] = \bar{u}[-1 + 3] = \bar{u}[2]$ $= u[2]$ and $\bar{u}[-2] = u[1]$. Thus, the equation can be written as

$$\bar{y}[0] = h[0]u[0] + h[1]u[2] + h[2]u[1] \tag{7.57a}$$

Similarly, we have

$$\bar{y}[1] = \sum_{i=0}^{2} \bar{h}[i]\bar{u}[1 - i] = h[0]u[1] + h[1]u[0] + h[2]u[2] \tag{7.57b}$$

and

$$\bar{y}[2] = \sum_{i=0}^{2} \bar{h}[i]\bar{u}[2 - i] = h[0]u[2] + h[1]u[1] + h[2]u[0] \tag{7.57c}$$

Now we plot $h[k]$ and $u[k]$ on circles, as shown in Figure 7.9. Because one of them must be flipped, one is arranged in the counterclockwise direction and the other in the clockwise direction. If we line up $h[0]$ and $u[0]$, as in Figure 7.9(a), then the sum of the product of the corresponding u and h yields $y[0]$, as in (7.57a). If we rotate u one step in the counterclockwise direction, as in Figure 7.9(b), then the result yields $y[1]$. Rotating u one more step yields $y[2]$, as in Figure 7.9(c). Unlike the (discrete) convolution, which flips and shifts linearly, the periodic convolution flips and shifts circularly. Thus, it is also called the *circular convolution*. In contrast, the convolution in (7.55) is also called the *linear convolution*.

The application of the discrete-time Fourier transform to the convolution in (7.55) yields $Y(\omega) = H(\omega)U(\omega)$. Although the transform can also be applied to the periodic convolution in (7.56), the result will not be simple and will not be discussed. Now let us apply the DFT to (7.56). By definition, we have

$$Y[m] = \mathscr{D}[\bar{y}[k]] = \sum_{k=0}^{N-1} \bar{y}[k]W^{-km} = \sum_{k=0}^{N-1} [\sum_{i=0}^{N-1} \bar{h}[i]\bar{u}[k - i]]W^{-[k-i]m}W^{-im}$$

Figure 7.9 *Periodic or circular convolution*

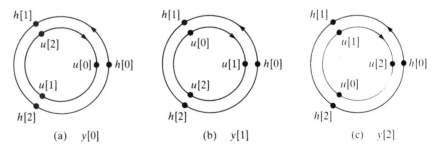

(a) $y[0]$ (b) $y[1]$ (c) $y[2]$

which becomes, after changing the order of summations and changing the index $p = k - i$,

$$Y[m] = \sum_{i=0}^{N-1} \overline{h}[i]W^{-im}\left[\sum_{k=0}^{N-1}\overline{u}[k-i]W^{-[k-i]m}\right] = \sum_{i=0}^{N-1}\overline{h}[i]W^{-im}\left[\sum_{p=-i}^{N-1-i}\overline{u}[p]W^{-pm}\right]$$

$$= \sum_{i=0}^{N-1}\overline{h}[i]W^{-im}\left[\sum_{p=0}^{N-1}\overline{u}[p]W^{-pm}\right] = H[m]U[m]$$

where we have used the periodicities of $\overline{u}$ and W. Thus, the DFT transforms the periodic convolution into a multiplication or

See class notes for more

$$\mathscr{D}[\sum_{i=0}^{N-1}\overline{h}[i]\overline{u}[k-i]] = \mathscr{D}[\overline{h}[k]]\mathscr{D}[\overline{u}[k]] = \mathscr{D}[h[k]]\mathscr{D}[u[k]] \quad \textbf{(7.58)}$$

This is an important result and will be used in computer computation of convolutions and system responses.

It would be possible to list some other properties of the DFT. However, they are less useful in computer computation and so will not be discussed here.

*7.5.3 Fast Fourier transform

This section focuses on the digital computation of the DFT. If we compute (7.50) directly, each $F[m]$ requires N multiplications. Thus, to compute N of $F[m]$ requires N^2 multiplications. This number increases rapidly as N increases, as shown in Table 7.2. Thus, the DFT, for N large, is not computed directly on digital computers.

Efficient algorithms, called the fast Fourier transform (FFT), have been developed to compute the DFT. The basic idea of the FFT is as follows: We divide the original N sequence into two ($N/2$)-point sequences. To compute the DFT of each ($N/2$)-point sequence requires $(N/2)^2$ multiplications. Because of the structure of the DFT, it is possible to obtain the DFT of the original N-point sequence from the DFTs of the two ($N/2$)-point sequences using N multiplications. Thus, this method of computing the DFT of N-point sequences requires

Table 7.2 *Numbers of multiplications in computing the DFT*

N	Direct DFT (N^2)	FFT ($0.5N\log_2 N$)
2	4	1
8	64	12
32	1024	80
64	4096	192
$2^{10} = 1024$	1048576	5120
2^{20}	10^{12}	10^7

$$N + 2 \left[\frac{N}{2} \right]^2 = N + \frac{N^2}{2}$$

multiplications, about half of N^2 for N large. If we divide $(N/2)$-point sequences into half, then the number of multiplications can be cut further by half. Proceeding forward, the number of multiplications in computing the DFT of N-point sequences can be reduced to $0.5N \log_2 N$. Table 7.2 lists this number for various N. The saving in the FFT is very significant for N large, as the table shows.

A large number of FFT subroutines are available and are discussed mostly in texts on digital signal processing. See, for example, Reference [3]. Some FFT subroutines require N to be a power of 2; some do not. Because of the duality of the DFT pair, the same FFT can be used to compute both DFT and inverse DFT. In MATLAB, the command fft computes DFT and the command ifft computes inverse DFT. In the remainder of this chapter, we will discuss some applications of the DFT or, equivalently, the FFT.

*7.6 COMPUTING FREQUENCY SPECTRA OF FINITE SEQUENCES

This section discusses the computation of the frequency spectra of sequences of finite length. Here, sequences of finite length will be called finite-length sequences or, simply, finite sequences. Consider the finite sequence $h[k]$, for $k = 0, 1, 2, \ldots, N-1$. Its frequency spectrum $H(\omega)$ is, by definition, the discrete-time Fourier transform of $h[k]$ or

$$H(\omega) = \sum_{k=-\infty}^{\infty} h[k]e^{-jk\omega} = \sum_{k=0}^{N-1} h[k]e^{-jk\omega} \tag{7.59}$$

with frequency interval of interest $(-\pi, \pi]$. The DFT of $h[k]$ is, by definition,

$$H[m] = \sum_{k=0}^{N-1} h[k]e^{-j2\pi km/N} \tag{7.60}$$

for $m = 0, 1, 2, \ldots, N - 1$. A comparison of (7.59) and (7.60) yields

$$\boxed{H[m] = H(m\frac{2\pi}{N})} \tag{7.61}$$

for $m = 0, 1, 2, \ldots, N - 1$. Thus, the DFT or FFT computes N equally spaced
samples of $H(\omega)$ in $[0, 2\pi)$ in rad/s or $[0, 1)$ in Hz. Once $H(\omega)$ with ω in $[0, 2\pi)$
is computed, $H(\omega)$ with ω in $(-\pi, \pi]$ can be readily obtained by periodic
extension. An example illustrates this.

EXAMPLE 7.6.1

Consider the sequence

$$h[k] = \begin{cases} 1 & \text{for } k = 0, 1, 2 \\ 0 & \text{otherwise} \end{cases} \tag{7.62}$$

It is a sequence of length 3. The frequency spectrum of $h[k]$ is

$$H(\omega) = \sum_{k=-\infty}^{\infty} h[k]e^{-jk\omega} = \sum_{k=0}^{2} h[k]e^{-jk\omega} = 1 + e^{-j\omega} + e^{-j2\omega}$$

$$= e^{-j\omega}(e^{j\omega} + 1 + e^{-j\omega}) = e^{-j\omega}(1 + 2\cos\omega)$$

Figure 7.10 shows the amplitude spectrum of $H(\omega)$; it is defined for every ω.
To compute the DFT of $h[k]$, we first compute

$$W = e^{-j2\pi/N} = e^{-j2\pi/3} = e^{-j120°} = \cos 120° - j\sin 120° = -0.5 - j0.866$$

and

$$W^0 = 1 \qquad W^2 = e^{-j4\pi/3} = e^{-j240°} = e^{j120°} = \cos 120° + j\sin 120°$$
$$= -0.5 + j0.866$$

Note that W is periodic with period $N = 3$. Thus, we have $W^3 = W^0 = 1$ and
$W^4 = W$. The DFT of $h[k]$ is, then, using (7.50) for $m = 0, 1, 2$,

$$H[0] = W^0 + W^0 + W^0 = 1 + 1 + 1 = 3$$
$$H[1] = W^0 + W + W^2 = 1 + (-0.5 - j0.866) + (-0.5 + j0.866) = 0$$

Figure 7.10 *Amplitude spectrum of* h[k] *in (7.62)*

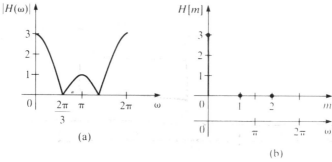

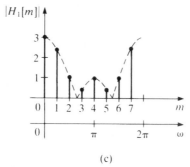

(c)

and

$$H[2] = W^0 + W^2 + W^4 = 1 + W^2 + W^1 = 0$$

If we use the command fft in MATLAB, then the commands

```
h = [1;1;1];
H = fft(h)
```

yield

```
ans = 3.0000
     -0.0000 - 0.0000i
      0.0000 - 0.0000i
```

This result is the same as the one computed by hand.

Let us briefly discuss the notation used in MATLAB. Data are arranged in a matrix row by row separated by semicolons. Data in each row are separated by space. Thus, $h = [1\ 1\ 1]$ is a 1×3 row vector and $h = [1;1;1]$ is a 3×1 column vector. When using the command fft, if h is expressed as a column, then the answer will appear as a column. If h is expressed as a row, such as $h = [1\ 1\ 1]$, then the answer will appear as a row. Figure 7.10(b) plots $H[m]$. It is indeed the samples of $H(\omega)$ in Figure 7.10(a).

Padding with trailing zeros The number of points computed in the DFT or the FFT equals the number of data points. The number of data in the preceding example is 3, and the FFT yields 3 samples of $H(\omega)$. From the three samples in Figure 7.10(b), it is not possible to visualize or to extrapolate the spectrum in Figure 7.10(a). In a computation, if the number of points computed is small or the interval between computed points is large, then the *resolution* is said to be poor. Thus, the resolution in Figure 7.10(b) is very poor.

In order to increase the resolution, we must compute more points. Suppose we wish to compute 8 points, then we modify $h[k]$ to make it a sequence of length 8. This can be easily achieved by padding zeros to the end of $h[k]$ to yield

$$h_1[k] = \begin{cases} h[k] & \text{for } k = 0, 1, 2 \\ 0 & \text{for } k = 3, 4, 5, 6, 7 \end{cases}$$

It is clear that the frequency spectrum of $h_1[k]$ equals that of $h[k]$. Thus, $h_1[k]$ can be used to compute the spectrum of $h[k]$. If $N = 8$, then $W = e^{-j2\pi/8} = e^{-j45°}$ and (7.50) becomes

$$H_1[0] = 1 \cdot W^0 + 1 \cdot W^0 + 1 \cdot W^0 + 0 \cdot W^0 + 0 \cdot W^0 + 0 \cdot W^0 + 0 \cdot W^0 + 0 \cdot W^0$$
$$H_1[1] = 1 \cdot W^0 + 1 \cdot W^1 + 1 \cdot W^2 + 0 \cdot W^3 + 0 \cdot W^4 + 0 \cdot W^5 + 0 \cdot W^6 + 0 \cdot W^7$$
$$H_1[2] = 1 \cdot W^0 + 1 \cdot W^2 + 1 \cdot W^4 + 0 \cdot W^6 + 0 \cdot W^8 + 0 \cdot W^{10} + 0 \cdot W^{12}$$
$$+ 0 \cdot W^{14}$$

and so forth. Even for this simple problem, hand computation of the DFT is complicated. If we type

```
h1 = [1;1;1;0;0;0;0;0];
H1 = fft (h1);
[abs(H1),angle(F1).*180./pi]    (changes angles from radians to
                                 degrees)
```

then MATLAB will yield

```
ans = 3.0000        0
      2.4142      -45.0000
      1.0000      -90.0000
      0.4142       45.0000
      1.0000        0
      0.4142      -45.0000
      1.0000       90.0000
      2.4142       45.0000
```

The amplitude is plotted in Figure 7.10(c). The resolution is better than the one in Figure 7.10(b).

Some `fft` commands will automatically add zeros to a sequence. For example, if we use MATLAB and type

```
h = [1 1 1];
H = fft (h,128); (uses 128 point fft; the program will add 125
                 trailing zeros to the end of h)
f = 2*pi*(0:127)/128; (generates 128 equally spaced frequencies
                       in [0, 2π) in rad)
plot(f,abs(H))
```

then a plot identical to the one in Figure 7.10(a) will appear on the screen. Recall that the frequency range of digital signals with $T = 1$ is $(-\pi, \pi]$ in rad/s or $(-0.5, 0.5]$ in Hz. In order for frequencies to lie in this range, the index m must assume the range in (7.6). This is complicated. Thus, we use $m = 0, 1, \ldots, N - 1$, and the frequency range becomes $[0, 2\pi)$ in rad/s or $[0, 1)$ in Hz. The command `f = (0:N - 1)/N` in MATLAB generates N equally spaced frequencies in $[0, 1)$. Once the spectrum in $[0, 1)$ is obtained, the spectrum in $(-0.5, 0.5]$ can be readily obtained by periodic extension.

*7.6.1   Computing frequency spectra of infinite sequences

The frequency spectrum of an infinite sequence is defined only if the sequence can be decomposed into two parts: one part must be absolutely summable and the other must be periodic. If a sequence is neither absolutely summable nor periodic, then its frequency spectrum is not defined. This section will discuss first the computation of the spectrum of absolutely summable sequences and then of periodic sequences. To simplify discussion, we will assume all sequences start from k = 0. If a sequence does not start from k = 0, we may shift it to k = 0 and compute its spectrum. The frequency spectrum of the original sequence can then be obtained using the time-shifting property. See Problem 7.10.

Consider an infinite sequence $h[k]$ defined for all $k \geq 0$. No digital computer can handle infinitely many data; therefore, the sequence must be truncated before computation. If the sequence is absolutely summable, then $|h[k]|$ approaches zero as k approaches infinity. Therefore, intuitively, all we have to do is to choose an N such that $|h[k]|$ is very small for all $k \geq N$ and the frequency spectrum of $h[k]$ will then approximately equal that of $h[k]$ for k in $[0, N - 1]$. The only problem is: How to choose N? An example discusses this problem.

EXAMPLE **7.6.2**

Consider

$$h[k] = 0.9^k \cos \pi k \tag{7.63}$$

for $k \geq 0$. Its frequency spectrum is computed analytically in Example 7.3.3 and plotted in Figure 7.6. Now we shall compute it numerically, using $N = 4$, 16, 32, 64, and 128. The following MATLAB commands will generate the amplitude spectra shown in Figure 7.11.

```
k = 0:1:1000;
h = (0.9.^k).*cos (pi.*k);
H1 = fft(h,4);
f1 = (0:3)/4;      (generates 4 equally spaced frequencies in [0, 1) in
                   Hz)
H2 = fft(h,16);
f2 = (0:15)/16;
H3 = fft(h,32);
f3 = (0:31)/32;
H4 = fft(h,64);
f4 = (0:63)/64;
H5 = fft(h,128);
f5 = (0:127)/128;
plot(f1,abs(H1),f2,abs(H2),f3,abs(H3),f4,abs(H4),
  f5,abs(H5))
```

Note that, if $N = 4$, the `fft` command will compute 4 points in [0, 1) at 0, 0.25, 0.5, and 0.75 Hz; the results are connected by straight lines, as in the

Figure 7.11 *Amplitude spectrum of* h[k] *in (7.63)*

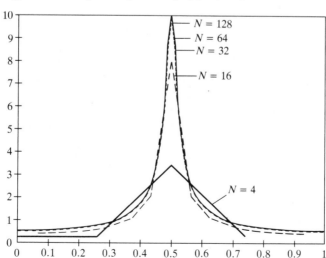

figure. We see that the spectra change drastically as N increases from 4 to 16 and from 16 to 32. As N increases from 32 to 64 and from 64 to 128, the amplitude spectra hardly change, except the maximum heights increase slightly. The amplitude spectrum for $N = 256$ is indistinguishable from the one for $N = 128$. Thus, the spectrum computed for $N = 128$ should be close to the spectrum of the infinite sequence $h[k]$. Indeed, it is very close to the one in Figure 7.6.

From this example, we can now develop a procedure for choosing N. First, we arbitrarily choose an N, preferably to be a power of 2 to take full advantage of the FFT. No matter how large N is, there is no way to tell how close the computed spectrum is to the actual spectrum. (If we knew the actual spectrum, then there would be no need for this computation.) We then choose a larger N, say $N_1 = 2N$, and repeat the computation. If the result is very close to the one computed by using N, we may stop the computation. Otherwise, we choose an N_2 that is larger than N_1 or $N_2 = 2N_1 = 4N$ and compute the spectrum. We repeat the process until the spectra of two subsequent computations are very close. Thus, in actual computation, N is chosen by trial and error.

Periodic sequences   The frequency spectrum of every periodic sequence consists of impulses. For absolutely summable sequences, because $|h[k]| \rightarrow 0$ as $k \rightarrow \infty$, the effect of truncation diminishes as N increases. Periodic sequences do not approach zero as k increases; thus, the effect of truncation will always remain. This is illustrated in the next example.

EXAMPLE 7.6.3

Consider $h[k] = \sin 0.6\pi k$. It is a sinusoid with frequency 0.6π rad/s or $0.6\pi/2\pi = 0.3$ Hz. Its frequency spectrum, as was discussed in Section 7.3.4, consists of two impulses at ± 0.3 Hz in the frequency range $(-0.5, 0.5]$ in Hz or at 0.3 and 0.7 Hz in the frequency range $[0, 1)$ in Hz. Recall that the spectrum of $h[k]$ is periodic with period 2π in rad/s or 1 in Hz.

Now we compute the spectra of $h[k]$ using $N = 64, 128, 256,$ and 512. The following MATLAB commands yield the plot shown in Figure 7.12.

```
k  = 0:1:1000;
h  = sin (0.6.*pi.*k);
H1 = fft(h,64);
f1 = (0:63)/64;
H2 = fft(h,128);
f2 = (0:127)/128;
```

Figure 7.12 *Effects of truncation*

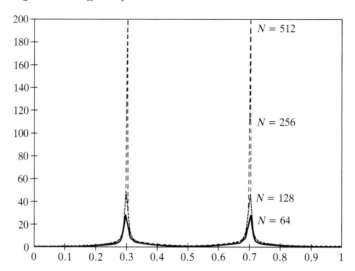

```
H3 = fft(h,256);
f3 = (0:255)/256;
H4 = fft(h,512);
f4 = (0:511)/512;
plot(f1,abs(H1),f2,abs(H2),f3,abs(H3),f4,abs(H4))
```

We see that the amplitude spectra spread around 0.3 and 0.7. As N increases, the maximum heights increase from about 30 to 50 to 120 to 190, the spreadings also become less. Unlike the plot in Figure 7.11, the maximum height will not cease to increase as N increases. Here, as N continues to increase, the spreading becomes less and the maximum height continues to increase. When this occurs, we can stop the computation and conclude the existence of a periodic sequence with frequency at 0.3Hz.

The frequency spectrum of $\sin \omega_0 k$ is an impulse at ω_0. The frequency spectrum of truncated $\sin \omega_0 k$ will not be an impulse; it will spread out around ω_0. This spreading is called *leakage*. The larger the N, the smaller the leakage. Although it is not apparent from the plot in Figure 7.12, truncation will also introduce ripples into the spectra, as shown in Figure 6.10. See also, for example, Reference [3]. Thus, truncation will introduce both spreading or leakage and ripples into frequency spectra. The following example illustrates the effect of spreading.

EXAMPLE **7.6.4**

Consider

$$h[k] = \sin(0.6\pi k) + 0.8\sin(0.64\pi k + 1.6) \qquad (7.64)$$

which consists of two sinusoids, one with frequency 0.3 Hz and the other with frequency 0.32 Hz. They have a phase difference of 1.6 rad/s. The sequence is plotted in Figure 7.13 from $k = 0$ to $k = 100$. Although the plot clearly shows the existence of at least two periodic subsequences, it is difficult to determine the frequencies, amplitudes, and phases of the subsequences from the plot.

To compute the frequency spectrum of 16-point $h[k]$, enter the following MATLAB commands.

```
k = 0:1:1000;
h = sin (0.6.*pi.*k)+0.8.* sin (0.64.*pi.*k+1.6);
H1 = fft(h,16);
f1 = (0:15)/16;
plot(f1,abs(H1),f1,angle(H1))
```

These will yield the amplitude spectrum (solid lines) and phase spectrum (dashed lines), in radians, in Figure 7.14. From this plot, we cannot detect the existence of the two sinusoids, which means that the resolution is not good enough to distinguish the two sinusoids. The reason for the poor resolution is simple: the spreadings of the spectra, for $N = 16$, of the two sinusoids overlap

Figure 7.13   *The sequence in (7.64)*

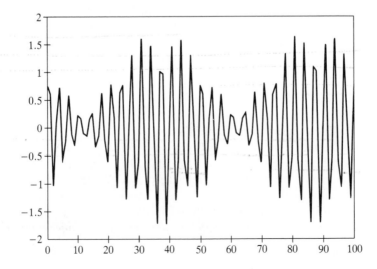

Figure 7.14 *Amplitude and phase spectra of 16-point* h[k] *in (7.64)*

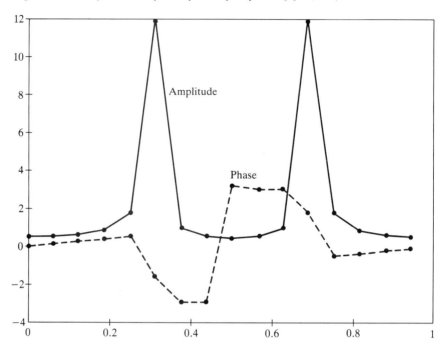

each other and blur the individual spectrum. Note that the program generates 16 points which are then connected by straight lines.

Next we compute the frequency spectrum of 512-point $h[k]$. The commands `H2 = fft(h,512); f2 = (0:511)/512;plot(f2, abs(H2), f2, angle(H2).*180/pi)` yield the amplitude spectrum (solid line) and phase spectrum (dotted line) in Figure 7.15. The amplitude spectrum shows two sinusoids with frequencies at 0.3Hz and 0.32Hz. Their largest magnitudes are roughly 200 and 180—not exactly the relative amplitudes of sin $(0.6\pi k)$ and 0.8 sin $(0.64\pi k + 1.6)$—but their relative magnitudes are indicated. It is difficult to tell the relative phase of the two sinusoids from the phase plot, which is expressed in degrees. If we are interested in the average power, then the phase plot is not needed. We need only the amplitude plot.

With the preceding examples in mind, let us now discuss the spectrum of general infinite sequences. Given an infinite sequence, we compute its N- and $2N$-point spectra. If the two spectra differ greatly, we know that we have not used enough points and we try again, this time computing a $4N$-point spectrum. We continue until two subsequent spectra are roughly the same, excepting possible heights of very narrow spikes. If there are no narrow spikes, the infinite

Figure 7.15 *Amplitude and phase spectra of 512-point h[k] in (7.64)*

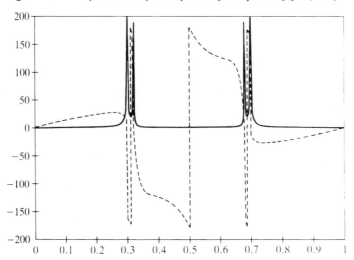

sequence has no periodic part. If there are narrow spikes with heights increasing greatly whenever the number of points used is doubled, then the sequence contains sinusoids; their frequencies are given by the locations of the spikes. In conclusion, the DFT or the FFT can be used to compute the spectrum of finite or infinite sequences.

*7.7   ALIASING IN TIME

In this section we will discuss the relationship between the inverse discrete-time Fourier transform of $H(\omega)$ and the inverse DFT of the N equally spaced samples of $H(\omega)$, or of

$$H[m] = H(\omega)\Big|_{\omega = m\frac{2\pi}{N}} = H(m\frac{2\pi}{N})$$

for $m = 0, 1, 2, \ldots, N - 1$. We will use $h[k]$ to denote the inverse discrete-time Fourier transform and $\bar{h}[k]$ to denote the inverse DFT. Now we shall establish the relationship between $h[k]$ and $\bar{h}[k]$. By definition, we have

$$H(\omega) = \sum_{k=-\infty}^{\infty} h[k]e^{-jk\omega} \tag{7.65}$$

and

$$H[m] = H(m\frac{2\pi}{N}) = \sum_{k=0}^{N-1} \bar{h}[k]e^{-j2\pi km/N} \tag{7.66}$$

Substituting $\omega = 2\pi m/N$ into (7.65) and dividing $(-\infty, \infty)$ into sections of length N yield

$$H(m\frac{2\pi}{N}) = \sum_{k=-\infty}^{\infty} h[k]e^{-j2\pi km/N} = \sum_{p=-\infty}^{\infty} \sum_{k=pN}^{pN+N-1} h[k]e^{-j2\pi km/N}$$

which, after introducing the new index $\bar{k} = k - pN$, using $e^{-j2\pi pm} = 1$ for every integer p and m, and interchanging the order of summations, becomes

$$H(m\frac{2\pi}{N}) = \sum_{p=-\infty}^{\infty} \sum_{\bar{k}=0}^{N-1} h[\bar{k} + pN]e^{-j2\pi(\bar{k}+pN)m/N} = \sum_{\bar{k}=0}^{N-1} \left[\sum_{p=-\infty}^{\infty} h[\bar{k} + pN] \right] e^{-j2\pi\bar{k}m/N}$$

A comparison of this with (7.66) immediately yields

$$\bar{h}[k] = \sum_{p=-\infty}^{\infty} h[k + pN] \qquad (7.67)$$

This is an important result. Let us discuss its physical meaning. Let $h[k]$ be a finite sequence of length 4, as shown in Figure 7.16(a), and let $N = 6$. Then, $h[k + N]$ is the shifting of $h[k]$ to the left and $h[k - N]$ is the shifting of $h[k]$ to the right six units of time, as shown in Figure 7.16(b). Thus, $\bar{h}[k]$ is the sum of $h[k]$ and its repetitive shifting to the right and left every N units of time. If N is equal to or larger than the length of $h[k]$, then $h[k \pm pN]$ do not overlap each other, as in Figure 7.16(b), and $\bar{h}[k]$ is simply the periodic extension of $h[k]$ with period N. This is exactly what we discussed in (7.52) and (7.53). In this case, we have $h[k] = \bar{h}[k]$ for k in $[0, N - 1]$. If $N = 3$, smaller than the length of $h[k]$, then $h[k]$ and its repetitions at $\pm pN$ overlap each other, which is called *aliasing in time* and is shown in Figure 7.16(c). The sum of these yields $\bar{h}[k]$, as shown in Figure 7.16(d). We see that $h[k]$ is different from $\bar{h}[k]$ for k in

Figure 7.16 *A sequence, its repetition, and aliasing in time*

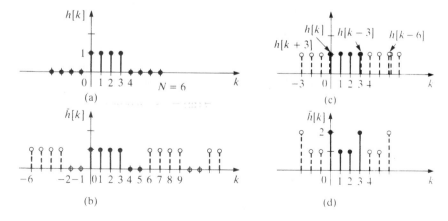

[0, $N - 1$]. Thus, if N is smaller than the length of a sequence, aliasing in time occurs and we do not have $h[k] = \bar{h}[k]$ for k in [0, $N - 1$]. On the other hand, if N is equal to or larger than the length of a sequence, then

$$h[k] = \mathscr{F}_d^{-1}[H(\omega)] = \bar{h}[k] = \mathscr{D}^{-1}[H(m\frac{2\pi}{N})]$$

for $k = [0, N - 1]$. Thus, the inverse DFT can be used to compute the inverse discrete-time Fourier transform.

If $h[k]$ is an infinite sequence, no matter how large N is, repetitions of $h[k]$ will always overlap, and the resulting $\bar{h}[k]$ will be different from $h[k]$ in [0, $N - 1$]. However, if $h[k]$ is absolutely summable, then $|h[k]| \rightarrow 0$ as k approaches infinity. In this case, if N is chosen to be sufficiently large, then aliasing will be small and we have $\bar{h}[k] \approx h[k]$ for k in [0, $N - 1$]. Based on this property, the inverse DFT can be used to compute time sequences from frequency spectra, as will be discussed next.

*7.7.1   Computing time sequences from frequency spectra

The inverse discrete-time Fourier transform, $h[k]$, of a given frequency spectrum, $H(\omega)$, can be obtained from the formular in (7.21b). Computing (7.21b) analytically is complicated and now rarely done in engineering. It is much simpler to use the inverse DFT or the command `ifft` to compute $h[k]$. Note that the inverse DFT yields $\bar{h}[k]$, which is related to $h[k]$ by (7.67). However, if $h[k]$ is a finite sequence and if N is chosen to be sufficiently large, then we can obtain an exact $h[k]$ from $\bar{h}[k]$. If $h[k]$ is an infinite sequence, then the best we can obtain is only an approximate $h[k]$.

EXAMPLE 7.7.1

Given the frequency spectrum

$$H(\omega) = e^{-j\omega}(1 + 2\cos\omega) \tag{7.68}$$

which is, as was studied in Example 7.6.1, the frequency spectrum of [1, 1, 1], a sequence of length 3. The three equally spaced samples of $H(\omega)$ in [0, 2π) is [3, 0, 0], which can also be generated by the command `fft([1;1;1])`, as discussed in Example 7.6.1. We compute the time sequence of [3, 0, 0] by typing

```
H = [3 0 0];
ifft(h)
```

The answer is [1 1 1]. This is the inverse discrete-time Fourier transform of $H(\omega)$ in (7.68).

In reality, when we are given $H(\omega)$, there is no way to tell *a priori* whether or not its time sequence is finite and, if it is, what its length is. However, we can simply arbitrarily choose an N, preferably to be a power of 2 to take full advantage of the FFT, and proceed to compute its inverse DFT by calling `ifft`. For example, for $H(\omega)$ in (7.68), we choose $N = 16$ and type

```
w = 2*pi*(0:15)/16; (generates 16 equally spaced frequencies in
                     [0, 2π) in rad)
H = exp (-i.*w).*(1 + 2.*cos (w));
ifft(H)
```

Then the answer is

```
1.0000-0.0000i  1.0000-0.0000i  1.0000+0.0000i
0.0000+0.0000i  0.0000-0.0000i  ...  0.0000+0.0000i
```

The first three entries are 1, and the remaining thirteen entries are 0. Thus, the time sequence of the spectrum is again [1 1 1].

The inverse discrete-time Fourier transform of the spectrum in the preceding example yields a time sequence starting from $k = 0$. This may not be so for other $H(\omega)$. Suppose time sequences of given spectra are as shown in Figure 7.17(a). First, we choose an N, say $N = N_1$, with N larger than the lengths of the sequences. We then compute N_1-point `ifft` to yield the sequences in Figure 7.17(b). Note that the inverse DFT yields a periodic sequence with period N_1. From this sequence, we cannot determine the exact location of the

Figure 7.17 *Determination of locations of time sequences*

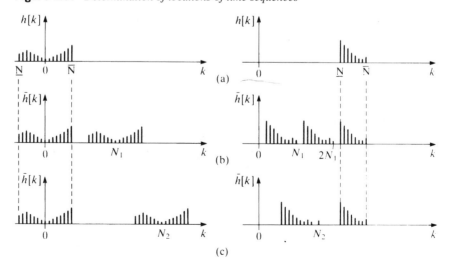

(a)

(b)

(c)

sequences. Next we choose $N_2 > N_1$ and compute N_2-point `ifft` to yield the sequences in Figure 7.17(c). They are periodic sequences with period N_2. The sections where they coincide (as shown in Figure 7.17) yield the time sequences of the given frequency spectra. Thus, if the N_1-point `ifft` and the N_2-point `ifft` of a given frequency spectrum have the same number of nonzero entries and the corresponding nonzero entries are identical, then the inverse discrete-time Fourier transform of the spectrum is a finite sequence. The exact location of the sequence is to be determined by periodic extensions, as shown in Figure 7.17. For example, the 3-point and 16-point `ifft` of the spectrum in Example 7.7.1 have three nonzero entries; the corresponding entries are the same. They coincide at time instants $k = 0$ to $k = 2$. Thus, the time sequence of the spectrum in Example 7.7.1 is the finite sequence $h[0] = h[1] = h[2] = 1$ and $h[k] = 0$ otherwise.

Infinite sequences   Consider a frequency spectrum $H(\omega)$. It is assumed that $H(\omega)$ does not contain impulses and its inverse discrete-time Fourier transform is an infinite sequence. If we use the inverse DFT or the command `ifft` to compute its time sequence, then aliasing always occurs, no matter what N we choose. However, if N is sufficiently large, then the aliasing will be small, and we can obtain a time sequence that is very close to the actual one. The procedure of choosing N is as follows: First, we compute N_1-point `ifft` and then compute N_2-point `ifft` with $N_2 > N_1$. If the N_2-point sequence has N_1 points very close to the N_1-point sequence and the remaining $N_2 - N_1$ points very close to 0, then we may stop the computation. Otherwise, we continue to increase N until two subsequent computations have those properties. From the periodic extensions of the two sequences, we can then use the idea in Figure 7.17 to determine the exact location of the sequence. An example illustrates this.

EXAMPLE 7.7.2

Consider the frequency spectrum shown in Figure 7.18(a), that is,

$$H(\omega) = \begin{cases} 1 & \text{for } |\omega| \leq 1 \\ 0 & \text{for } 1 < |\omega| < \pi \end{cases} \tag{7.69}$$

This actually characterizes a digital ideal lowpass filter with cutoff frequency $\omega_c = 1$ rad/s. Recall that $H(\omega)$ is periodic with period 2π and that, in computer computation, we use $H(\omega)$ in the frequency range $[0, 2\pi)$, as shown in Figure 7.18(b). First we compute the 16-point `ifft` of $H(\omega)$. For $N = 16$, we have

$$\omega_m = m\frac{2\pi}{16}$$

If $m = 0$, 1, and 2, then $\omega_m < 1$. Thus, 16 equally spaced samples of $H(\omega)$ in Figure 7.18(a) are

Figure 7.18 *Ideal lowpass digital filter*

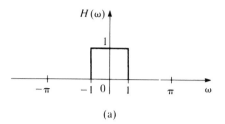

(a)

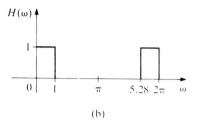

(b)

$$H[0] = 1, \quad H(1) = 1, \quad H[2] = 1, \quad H(m) = 0 \quad \text{for } m = 3, 4, \cdots, 13,$$
$$H[14] = 1, \quad H[15] = 1$$

Note that $H[16] = 1$ at $\omega_{16} = 2\pi$ is the same as $H[0]$ and is not included in the samples. The following commands in MATLAB

```
H1 = [1 1 1 0 0 0 0 0 0 0 0 0 0 0 1 1];
h1 = ifft(H1)
```

will yield a 16-point time sequence. The typing of the preceding H1 is complicated. It can be replaced by

```
h1 = [ones(1,3) zeros(1,11) ones(1,2)]
```

where `ones(1,3)` generates three 1 and `zeros(1,11)`, eleven 0. Using this notation, we compute the 128-point and 256-point inverse FFTs of $H(\omega)$. For $N = 128$, we have

$$\omega_m = m\frac{2\pi}{N} = m\frac{2\pi}{128}$$

The inequality $\omega_m \leq 1$ implies $\omega \leq 128/2\pi = 20.3$. Thus, the largest integer m to meet $\omega_m \leq 1$ is $m = 20$. Therefore, the first 21 samples of $H(\omega)$ in Figure 7.18(b) are 1, the last 20 samples are also 1, and the remaining $128 - 21 - 20 = 87$ samples are 0. Similarly, if $N = 256$, the first 41 and last 40 samples in Figure 7.18(b) are 1, and the remaining $256 - 41 - 40 = 175$ samples are 0. The following commands in MATLAB yield the plot in Figure 7.19.

```
H2 = [ones(1,21) zeros(1,87) ones(1,20)];
h2 = ifft(H2);
k2 = 0:127;
H3 = [ones(1,41) zeros(1,175) ones(1,40)];
h3 = ifft(H3);
k3 = 0:255;
plot(k2,h2,k3,h3)
```

Figure 7.19 *The impulse response of an ideal lowpass filter*

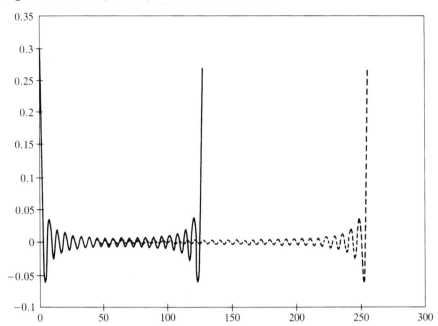

The first $128/2 = 64$ points of the 256-point sequence are very close to the first half of the 128-point sequence, and the last 64 points of the 256-point sequence are very close to the second half of the 128-point sequence. The remaining central portion of the 256-point sequence is very small. Thus, we stop the compuation. Because h2 and h3, after they are extended periodically, are real and even, and because they coincide for all $|k| \leq 128/2 = 64$, we conclude that the inverse discrete-time Fourier transform $h[k]$ of $H(\omega)$ in (7.69) is a two-sided sequence and is real and even. Because the central portion of the 256-point sequence is not identically zero, $h[k]$ is not a finite sequence. In fact, $h[k]$ is an infinite sequence. Thus, the larger N is, the more accurately $h[k]$ is computed. If $N = 256$, then

$$h[k] = \begin{cases} \text{h3}[k] & \text{for } |k| \leq 128 \\ 0 & \text{for } |k| > 128 \end{cases} \tag{7.70}$$

Note that $h[k]$ in (7.70) is the impulse response sequence of the ideal lowpass filter in (7.69). Because $h[k]$ is not zero for $k < 0$, the ideal filter is not causal and cannot be built in the real world. How to design a causal filter that approximates the ideal filter is beyond scope of this text. See, for example, Reference [3].

COMPUTING CONVOLUTIONS

This section focuses on the use of the DFT or FFT to compute discrete convolutions. Recall that every linear time-invariant system can be described by a convolution of the form

$$y[k] = \sum_{i=0}^{k} h[k-i]u[i] \tag{7.71}$$

Thus, this discussion can be applied to the computation of responses of discrete-time LTI systems. If the z-transform of $h[k]$ is a rational function of z, then the response can be more easily computed using the state-variable equations that will be discussed in Chapter 10 or using the command `filter` in MATLAB (see the Using MATLAB section in Chapter 5). If the z-transform of $h[k]$ is not a rational function of z or if $h[k]$ is given graphically, then we must use the method discussed here.

Digital computers can handle only finite sequences; thus, we assume $h[k]$ to be a finite sequence of length P and $u[k]$ to be a finite sequence of length Q. We select an integer N such that

$$N \geq P + Q - 1$$

Note that N can also be chosen to be a power of 2. Now we pad trailing zeros to $h[k]$ and $u[k]$ to make them both of length N and call them $h_N[k]$ and $u_N[k]$, that is,

$$h_N[k] = \begin{cases} h[k] & \text{for } k = 0, 1, \cdots, P-1 \\ 0 & \text{for } k = P, P+1, \cdots, N-1 \end{cases} \tag{7.72}$$

and

$$u_N[k] = \begin{cases} u[k] & \text{for } k = 0, 1, \cdots, Q-1 \\ 0 & \text{for } k = Q, Q+1, \cdots, N-1 \end{cases} \tag{7.73}$$

Let $\bar{h}_N[k]$ and $\bar{u}_N[k]$ be periodic extensions of $h_N[k]$ and $u_N[k]$ with period N. For example, if $h[k]$ and $u[k]$ are as shown in the left column of Figure 7.20(a) and (b), then $\bar{h}_N[k]$ and $\bar{u}_N[k]$ are as shown in the right column of Figure 7.20(a) and (b). The convolution

$$\bar{y}_N[k] = \sum_{i=0}^{N-1} \bar{h}_N[k-i]\bar{u}_N[i] \tag{7.74}$$

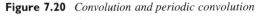

Figure 7.20   *Convolution and periodic convolution*

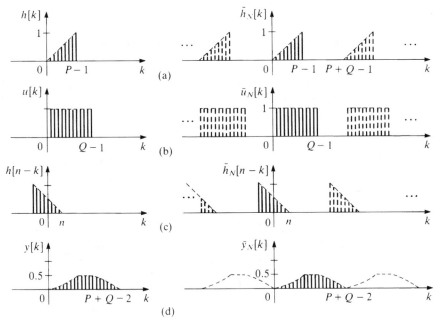

is called, as defined in (7.56), the periodic convolution. Convolutions and periodic convolutions generally yield, as shown in Figure 7.8, different results because of possible overlapping. However, if we pad trailing zeros to $h[k]$ and $u[k]$ as in (7.72) and (7.73), then the periodic convolution in (7.74) yields the same result as the convolution in (7.71) for k in $[0, N-1]$. To see this, look at the plots of $h[n-k]$ and $\bar{h}_N[n-k]$ in Figure 7.20(c). The product of $h[n-k]$ and $u[k]$ is the same as the product of $\bar{h}_N[n-k]$ and $\bar{u}_N[k]$ in $[0, N-1]$. Thus, we have $y[k] = \bar{y}_N[k]$ for all k in $[0, N-1]$, as shown in Figure 7.20(d).

The application of the DFT to (7.74), as derived in (7.58), yields

$$\mathscr{D}[\bar{y}_N[k]] = \mathscr{D}[\bar{h}_N[k]]\mathscr{D}[\bar{u}_N[k]]$$

which implies

$$\bar{y}_N[k] = \mathscr{D}^{-1}\left[\mathscr{D}[h_N[k]]\mathscr{D}[u_N[k]]\right] \tag{7.75}$$

Thus, $\bar{y}_N[k]$ and, consequently, $y[k]$ in (7.71) can be computed using the commands `fft` twice and `ifft` once.

Even though using the FFT to compute convolutions appears to be complicated, it is actually much more efficient than the direct computation for N large. To simplify discussion, we assume $P = Q = N/2$ and consider only the number of multiplications. To compute $y[k]$ in (7.71) directly requires $k + 1$ multiplications. Thus, to compute $y[k]$, for $k = 0, 1, \ldots, N - 1$, requires

$$1 + 2 + \cdots + N \approx \frac{N^2}{2}$$

multiplications. Because $h[k]$ and $u[k]$ have only $N/2$ nonzero entries, the preceding number can again be cut in half. Thus, to compute the convolution in (7.71) directly requires roughly $N^2/4$ multiplications or 262144 if $N = 1024$ (see Table 7.1). To compute $y[k]$ using (7.75) requires two `fft` commands, N multiplications, and then one `ifft` command. The numbers of multiplications in `fft` and `ifft` are the same (see the duality property of the DFT and inverse DFT). Thus, to compute $y[k]$ using (7.75) requires, using Table 7.1,

$$3(0.5N\log_2 N) + N$$

or 16384 if $N = 1024$. This is one-sixteenth of the multiplications needed by direct computation.[3] Thus, for N large, it is much more efficient to use the FFT to compute convolutions.

To conclude this section, we mention that the FFT can also be used to compute the convolution of a finite sequence and an infinite sequence. For example, consider the sequences shown in Figure 7.21(a) and (b). First, we divide the infinite sequence $u[k]$ into finite sequences $u_1[k]$, $u_2[k]$, $\ldots$, as shown. We then use the FFT command to compute the convolution of $h[k]$ and $u_i[k]$, for $i = 1, 2, \ldots$, as shown in Figure 7.21(c), (d), and (e). Because of linearity, the result of the convolution of the finite sequence and the infinite sequence is simply the sum of the sequences in Figure 7.21(c), (d), (e), $\ldots$.

7.9 SPECTRA OF ANALOG SIGNALS AND THEIR SAMPLED SIGNALS—ALIASING IN FREQUENCY

The frequency spectrum of continuous-time signals will be computed from the frequency spectra of their sampled sequences. Thus we discuss first their relationship. Consider a continuous-time signal $f(t)$ with frequency spectrum

$$F(\omega) = \int_{t=-\infty}^{\infty} f(t)e^{-j\omega t}dt \qquad (7.76)$$

Let $f(t)$ be applied to the input of the system shown in Figure 7.22(a). The system consists of only a switch and is called an ideal sampler. The switch

---

[3]If $h[k]$ and $u[k]$ are real, then direct computation involves only real multiplications. The computations using the `fft` and `ifft` commands generally involve complex multiplications. Each complex multiplication requires four real multiplications. Thus, the saving in using the FFT will be reduced by a factor 4.

Figure 7.21   *Convolution of a finite and an infinite sequence*

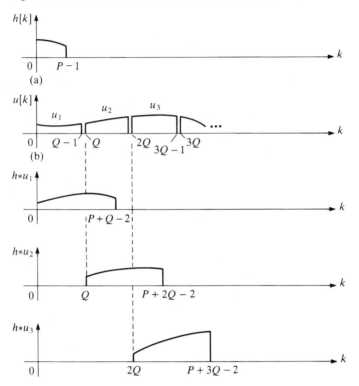

closes every T seconds and opens immediately thereafter. Therefore, if the input is $f(t)$, the output is $f(kT)$, for $k = 0, \pm 1, \pm 2, \ldots$. We call $f(kT)$ the sampled sequence of $f(t)$, T the sampling period, and $\omega_s := 2\pi/T$ in rad/s or $f_s = 1/T$ in Hz the sampling frequency.

Before proceeding, let us discuss the frequency spectrum of $f(kT)$. In this sequence, the sampling period is generally different from 1, and the discrete-time Fourier transform pair in (7.21) must be modified as

$$F_s(\omega) = \sum_{k=-\infty}^{\infty} f(kT)e^{-jk\omega T} \tag{7.77a}$$

and

$$f(kT) = \frac{T}{2\pi} \int_{\omega=0}^{2\pi/T} F_s(\omega)e^{jk\omega T}\, d\omega \tag{7.77b}$$

Figure 7.22 *Sampling of continuous-time signals*

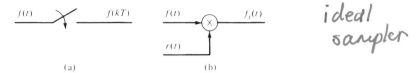

(a)   (b)

Because $F(\omega)$ has been used to denote the frequency spectrum of $f(t)$, we use $F_s(\omega)$ to denote the frequency spectrum of $f(kT)$. The frequency spectrum of $f(t)$ is defined for all ω in $(-\infty, \infty)$. The frequency spectrum of $f(kT)$ is periodic with period $2\pi/T$ and only the spectrum in the frequency range $(-\pi/T, \pi/T]$ is of interest. Note that, as the sampling period T approaches 0, the frequency range $(-\pi/T, \pi/T]$ approaches $(-\infty, \infty)$.

In order to establish the relationship between $F(\omega)$ and $F_s(\omega)$, we express the sampled sequence $f(kT)$ as

$$f_s(t) = \sum_{k=-\infty}^{\infty} f(kT)\delta(t - kT) \tag{7.78}$$

This means that every sample $f(kT)$ is replaced by an impulse at kT with weight $f(kT)$, as in (7.27). Because

$$f(kT)\delta(t - kT) = f(t)\delta(t - kT)$$

for every k, we can write (7.78) as

$$f_s(t) = \sum_{k=-\infty}^{\infty} f(t)\delta(t - kT) = f(t) \sum_{k=-\infty}^{\infty} \delta(t - kT) =: f(t)r(t) \tag{7.79}$$

where

$$r(t) = \sum_{-\infty}^{\infty} \delta(t - kT) \tag{7.80}$$

Equation (7.79) can be represented as shown in Figure 7.22(b), which is the same as Figure 6.22(a). Thus, sampling can be considered to be a modulation process, as discussed in Section 6.7.

The function $r(t)$ in (7.80) consists of a train of impulses, as shown in Figure 6.15(a); it is periodic with period T and has the following Fourier series representation, as derived in Example 6.4.4,

$$r(t) = \sum_{-\infty}^{\infty} \delta(t - kT) = \frac{1}{T} \sum_{k=-\infty}^{\infty} e^{jk\omega_s t}$$

with $\omega_s = 2\pi/T$. Substituting this into (7.79) yields

$$f_s(t) = f(t)\frac{1}{T}\sum_{k=-\infty}^{\infty} e^{jk\omega_s t} = \frac{1}{T}\sum_{k=-\infty}^{\infty} f(t)e^{jk\omega_s t}$$

As developed in (7.29), the frequency spectrum of $f(kT)$ equals the (continuous-time) Fourier transform of $f_s(t)$ or

$$F_s(\omega) = \mathscr{F}[f_s(t)] = \mathscr{F}\left[\frac{1}{T}\sum_{k=-\infty}^{\infty} f(t)e^{jk\omega_s t}\right] = \frac{1}{T}\sum_{k=-\infty}^{\infty} \mathscr{F}[f(t)e^{jk\omega_s t}]$$

which becomes, using the frequency-shifting property of the Fourier transform,

$$\boxed{F_s(\omega) = \frac{1}{T}\sum_{k=-\infty}^{\infty} F(\omega - k\omega_s) = \frac{1}{T}\sum_{k=-\infty}^{\infty} F(\omega - \frac{2k\pi}{T})} \tag{7.81}$$

This equation relates the spectrum of $f(t)$ and that of its sampled sequence $f(kT)$ and is of fundamental importance.

Let us discuss the implication of (7.81). For easy plotting, we assume $F(\omega)$ to be real valued and of the form shown in Figure 7.23. It is band limited to W in the sense

$$F(\omega) = 0 \quad \text{for } |\omega| > W \tag{7.82}$$

Thus, its highest frequency component is W. The function $F(\omega - 2k\pi/T)$ is the shifting of $F(\omega)$ from $\omega = 0$ to $\omega = 2k\pi/T$. Hence, $F_s(\omega)$ is the sum of all repetitions of $F(\omega)/T$ centering at $2k\pi/T$, $k = 0, \pm 1, \pm 2, \ldots$. Figure 7.24 plots the frequency spectra $F_s(\omega)$ for four different T. We see that, if $\pi/T > W$ or if the sampling period is smaller than π/W, as in Figure 7.24(a), then the frequency spectrum of $F_s(\omega)$ in the frequency range $(-\pi/T_1, \pi/T_1]$ is identical to that of $F(\omega)$, except for the factor $1/T$. On the other hand, if $\pi/T \leq W$ or if the sampling period is larger than or equal to π/W, as in Figure 7.24(b), (c), and (d),

Figure 7.23 *The frequency spectrum of an analog signal*

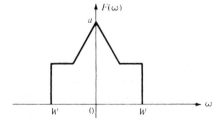

Figure 7.24 *Spectra of the sampled sequences of the signal in Figure 7.23*

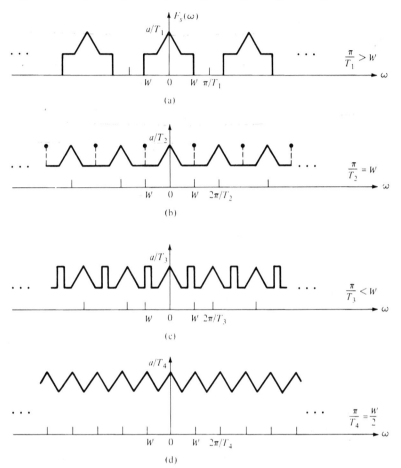

then the repetitions of $F(\omega)$ will overlap and the resulting $F_s(\omega)$ in $[-W, W]$ will not resemble $F(\omega)$. This type of overlapping is called *aliasing in frequency*. Whenever there is aliasing, $F_s(\omega)$ will differ from $F(\omega)$ in the frequency range $(-\pi/T, \pi/T]$.

To conclude this section, we state the sampling theorem. Because there is a one-to-one correspondence between the spectrum and its time function, we have essentially established the following sampling theorem.

Sampling Theorem Let $f(t)$ be a continuous-time signal whose frequency spectrum is band limited to W (in radians per second). Then, $f(t)$ can be recovered from its sampled sequence $f(kT)$ if the sampling period T is smaller than π/W or the sampling frequency $\omega_s = 2\pi/T$ is larger than $2W$.

This theorem is a generalization of the sampling thoerem discussed in Section 1.6.1 for sinusoids. We restate the theroem using Hz rather than rad/s as follows: If $f(t)$ is band limited to f_{max} (in Hz), then $f(t)$ can be recovered from its sampled sequence if the sampling frequency f_s is at least twice of f_{max}. Note that we need ideal lowpass filters to recover $f(t)$ from $f(kT)$. Because ideal lowpass filters cannot be built, we must use nonideal filters in the actual recovery of $f(t)$. To compensate for this, we usually sample $f(t)$ at a higher sampling frequency than $2f_{max}$. This is called *oversampling*.

*7.9.1   Computing frequency spectra of analog signals

When using a digital computer to compute the frequency spectrum of an analog signal, we must first sample the signal to yield a discrete-time signal. If the signal is band limited, it is possible to obtain the exact frequency spectrum of $f(t)$ from its sampled sequence $f(kT)$. To be more specific, if $f(t)$ is band limited to W, and if the sampling period T is chosen to be smaller than π/W, then we have, inside the frequency interval $(-\pi/T, \pi/T]$,

$$\begin{aligned}&\text{Fourier transform of } f(t) \\ &= T \times \text{discrete-time Fourier transform of } f(kT)\end{aligned} \tag{7.83}$$

or, moving T inside the transform,

$$\text{frequency spectrum of } f(t) = \text{frequency spectrum of } Tf(kT) \tag{7.84}$$

The need for the factor T can be seen from (7.81) or Figure 7.24. If T is small, (7.83) may yield a more accurate result than (7.84). Thus, if a continuous-time signal is band limited, its spectrum can be obtained from the spectrum of its sampled sequence.

If a signal is not band limited, then aliasing in frequency will always occur, no matter how small the chosen sampling period is. For example, consider the continuous-time signal $f(t)$ with spectrum $F(\omega)$ shown in Figure 7.25(a). It is not band limited. Even so, if we choose a sampling period T such that

$$|F(\omega)| \approx 0 \quad \text{or} \quad |F(\omega)| < \varepsilon \quad \text{for all } |\omega| > \pi/T \tag{7.85}$$

where ε is a very small positive number, then the aliasing would be small, as shown in Figure 7.25(b). Thus, we have

$$\text{frequency spectrum of } f(t) \approx \text{frequency spectrum of } Tf(kT) \tag{7.86}$$

for $|\omega| \leq \pi/T$. Therefore, the spectrum of $f(t)$, band limited or not, can be computed from its sampled sequence if the sampling period is properly chosen.

In actual application, we may *not* have any prior knowledge of the spectrum of a given function $f(t)$. Thus, (7.85) cannot be used to choose T. Instead, we proceed by first choosing an arbitrary sampling period T_1 and computing

Figure 7.25 *Aliasing in frequency*

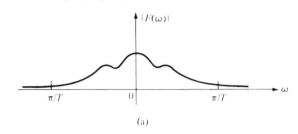

(a)

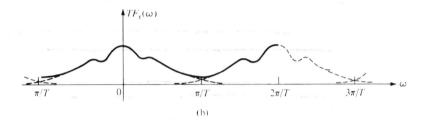

(b)

the spectrum of $T_1 f(kT_1)$. At this point, there is no way to know whether or not the computed spectrum is close to the actual one. Next, we choose a $T_2 < T_1$, say $T_2 = T_1/2$, and repeat the computation. If the spectrum of $T_2 f(kT_2)$ is very close to the one of $T_1 f(kT_1)$ in the frequency range $0 \le \omega < \pi/T_1$, and if the spectrum of $T_2 f(kT_2)$ approximately equals zero for $\pi/T_1 \le \omega < \pi/T_2$, as shown in Figure 7.26, we stop the computation and

$$\text{spectrum of } f(t) \approx \begin{cases} \text{spectrum of } T_2 f(kT_2) & \text{for } |\omega| \le \pi/T_2 \\ 0 & \text{for } |\omega| > \pi/T_2 \end{cases}$$

If the spectrum of $T_2 f(kT_2)$ is quite different from that of $T_1 f(kT_1)$, then we may choose $T_3 = T_2/2$ and repeat the computation. We repeat the process until the results of two subsequent computations are very close. Thus, in practice, the sampling period is chosen by trial and error. Once the sampling period T is chosen, the question of how many data points to use arises. Answering this question may again require some trial and error. Nevertheless, by combining the processes of choosing N and T, a fairly accurate frequency spectrum of any continuous-time signal can be obtained.

7.10   SUMMARY

1. The complex exponential sequences $\phi_m[k] = e^{jm\omega_0 k}$ with $\omega_0 = 2\pi/N$, for all integers m, are periodic in m with period N. Thus, there are only N distinct complex exponential sequences in $\phi_m[k]$. In the analog case, there are infinitely many distinct complex exponentials in $e^{jm\omega_0 t}$, for all integers m.

Figure 7.26 *Computation of spectra*

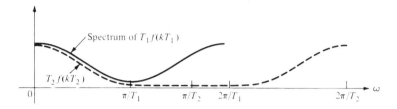

2. A periodic sequence with period N can be expressed in the discrete-time Fourier series, which, unlike the continuous-time case, consists of only N terms. The discrete-time Fourier transform is developed from the discrete-time Fourier series by extending the period N to infinity. The discrete-time Fourier transform can also be developed from the continuous-time Fourier transform. It is applicable only to absolutely summable sequences, periodic sequences, and their linear combinations.

3. The discrete-time Fourier transform pair is defined as

$$F(\omega) = \sum_{k=-\infty}^{\infty} f(kT)e^{-jk\omega T} \quad \text{, peridic w/period}$$
$$N = \frac{2\pi}{T}$$

$$f(kT) = \frac{T}{2\pi} \int_{\omega=0}^{2\pi/T} F(\omega)e^{jk\omega T}d\omega$$

If $T = 1$, then the pair reduces to (7.21). The transform $F(\omega)$ reveals the distribution of energy of $f(kT)$ in the frequency range $(-\pi/T, \pi/T]$ and is called the frequency spectrum. It is periodic with period $2\pi/T$.

4. If $f(kT)$ is absolutely summable, and if $f(kT)$ is written as a summation of a positive-time and a negative-time sequence, such as $f(kT) = f_+(kT) + f_-(kT)$, then the spectrum of $f(kT)$ is

$$\text{frequency spectrum of } f(kT) = \mathcal{F}_d[f(kT)]$$
$$= \mathcal{Z}[f_+(kT)]\Big|_{z=e^{j\omega T}} + \mathcal{Z}[f_-(-kT)]\Big|_{z=e^{-j\omega T}}$$

If $f(kT)$ is periodic, the preceding formula cannot be used. See (7.38) and (7.39).

5. In digital computer computation of the spectrum of discrete-time signals, we can use only a finite number of data points and compute only a finite number of frequencies. This leads to the discrete Fourier transform (DFT). The DFT is developed to compute the discrete-time Fourier transform. The fast Fourier transform (FFT) is an efficient way of computing the DFT.

6. Physically, if $T = 1$, the frequency range of digital signals is $(-\pi, \pi]$ in rad/s or $(-0.5, 0.5]$ in Hz. However, for convenience of indexing, we use

the frequency range $[0, 2\pi)$ in rad/s or $[0, 1)$ in Hz in numerical computation.

7. Let $h[k]$ be the inverse discrete-time Fourier transform of $H(\omega)$ and let $\bar{h}[k]$ be the inverse discrete Fourier transform (DFT) of N samples of $H(\omega)$. Then we have

$$\bar{h}[k] = \sum_{p=-\infty}^{\infty} h[k + pN]$$

If repetitions of $h[k]$ overlap with each other, aliasing in time occurs in forming $\bar{h}[k]$.

8. If $h[k]$ is a finite sequence with frequency spectrum $H(\omega)$, and if N is equal to or larger than the length of the sequence, then $h[k]$ can be computed, using the FFT, from the N samples of $H(\omega)$. If $h[k]$ is an infinite sequence, then we can obtain, using the FFT, only an approximate $h[k]$ from the samples of $H(\omega)$.

9. Let $F(\omega)$ be the (continuous-time) Fourier transform of $f(t)$ and let $F_s(\omega)$ be the discrete-time Fourier transform of the sample of $f(t)$ with sampling period T. Then we have

$$F_s(\omega) = \frac{1}{T} \sum_{k=-\infty}^{\infty} F(\omega - \frac{2\pi k}{T})$$

If repetitions of $F(\omega)$ overlap with each other, aliasing in frequency occurs in forming $F_s(\omega)$.

10. If a continuous-time signal $f(t)$ is band limited to W, then we have

$$\text{frequency spectrum of } f(t) = \text{frequency spectrum of } Tf(kT)$$

if the sampling period T is chosen to be smaller than π/W (see the sampling theorem). If $f(t)$ is not band limited, then aliasing in frequency will occur. However, if the chosen T is sufficiently small, the spectrum of $f(t)$ will be roughly equal to the spectrum of $Tf(kT)$ in $(-\pi/T, \pi/T]$.

11. In MATLAB, the command `fft` computes the DFT and the command `ifft` computes the inverse DFT.

PROBLEMS

7.1 Consider the set of complex exponentials of k

$$\phi_m[k] := e^{jm\frac{3\pi}{7}k}$$

for all integers m. How many different exponentials does the set have?

7.2 Consider the periodic sequence of period 3 defined by

$$f[0] = 0 \qquad f[1] = 1 \qquad \text{and} \qquad f[2] = 2$$

What is its discrete-time Fourier series?

7.3 Show that, if a periodic sequence is real, then the coefficient c_k of its discrete-time Fourier series has the property $c_k^* = c_{-k}$ (called the conjugate symmetric property), where the asterisk stands for the complex conjugate.

7.4 Express the following periodic sequences in discrete-time Fourier series. Plot their line spectra, also.

(a) $\sin 0.2\pi k$

(b) $\sin (7\pi k + 0.524) + \cos 2\pi k$

(c) $\sin (2\pi k/3) \cos (2\pi k/3)$

(d) $\sin (2\pi k/3) \cos (\pi k/4)$

(e) the two sequences shown in Figure P7.4

7.5 Compute the average powers of the sequences in Problem 7.4.

7.6 What are the time sequences with the following discrete-time Fourier series?

(a) $c_0 = 0$, $c_1 = 0.5 - j0.2$, and $c_2 = 0.5 + j0.2$

(b) $c_0 = 0$, $c_1 = 1$, and $c_2 = 2$

Are they both real sequences? If not, tell why.

7.7 Find the discrete-time Fourier transforms, if they exist, of the following discrete-time signals.

(a) $f_1[k] = \begin{cases} 1 & \text{for } k = 0, 1, 2, 3, 4 \\ 0 & \text{for } k > 4 \end{cases}$

(b) $f_2[k] = e^{-0.2k}$ for $k = 0, 1, 2, \ldots$

(c) $f_3[k] = 2^{-2k}$ for all k

(d) $f_4[k] = 2^{-2|k|}$ for all k

(e) $f_5[k] = \begin{cases} -e^{-2k} & \text{for } k = 0, 1, 2, \ldots \\ e^{3k} & \text{for } k = -1, -2, -3, \ldots \end{cases}$

7.8 Compute the total energies of the sequences in Problem 7.7.

7.9 (a) Compute the frequency spectrum $F(\omega)$ of

$$f[0] = 0 \qquad f[1] = 1 \qquad f[2] = 2 \qquad \text{and} \qquad f[k] = 0 \qquad \text{otherwise}$$

Compare $F(\omega)$ with the Fourier series computed in Problem 7.2.

(b) Let $\overline{f}[k]$ be a periodic sequence of period N. Define

$$f[k] = \begin{cases} \overline{f}[k] & \text{for } 0 \le k \le N - 1 \\ 0 & \text{otherwise} \end{cases}$$

Figure P7.4

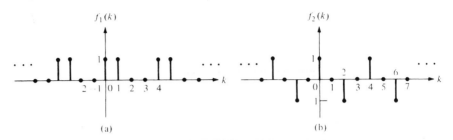

(a)   (b)

That is, $f[k]$ is the first period of $\overline{f}[k]$. Show

$$Nc_m = F(m\frac{2\pi}{N})$$

where c_m is the discrete-time Fourier series coefficients of $\overline{f}[k]$ and $F(\omega)$ is the discrete-time Fourier transform of $f[k]$.

7.10 (a) Consider a sequence $f[k]$ and its shifted sequence defined by

$$h[k] = f[k - k_0]$$

Let $F(\omega)$ and $H(\omega)$ be their frequency spectra. Show

$$|H(\omega)| = |F(\omega)| \quad \text{and} \quad \sphericalangle H(\omega) = \sphericalangle F(\omega) - k_0\omega$$

(b) Consider $f[k]$, for $k = k_0, k_0 + 1, \ldots$, that is, $f[k]$ starts from k_0. Define $h[k] = f[k + k_0]$. Show that $h[k]$ starts from $k = 0$ and

$$|H(\omega)| = |F(\omega)| \quad \text{and} \quad \sphericalangle F(\omega) = \sphericalangle H(\omega) - k_0\omega$$

7.11 Find the positive-time sequences with the following frequency spectra.

(a) $\dfrac{e^{j\omega}}{e^{j2\omega} - 0.01}$

(b) $\dfrac{e^{j\omega}}{e^{j2\omega} + 0.1e^{j\omega} - 0.02}$

7.12 Find a time sequence $f[k]$ that starts at $k = -5$ and has an amplitude spectrum equal to the one in Problem 7.11(a).

7.13 Compute the frequency spectra of the sequences in Problem 7.4.

7.14 Consider the finite sequence of length 5 studied in Example 7.3.1, that is, the sequence $f[k] = 1$ for $k = 0, \pm 1$, and ± 2 with frequency spectrum $F(\omega) = 1 + 2 \cos \omega + 2 \cos 2\omega$. What are the inverse DFTs of N samples of $F(\omega)$ for $N = 3, 4, 5, 6$, and 10?

Figure P7.15

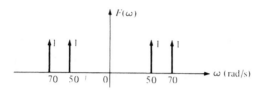

7.15 Consider the continuous-time signal $f(t)$ with the frequency spectrum shown in Figure P7.15. Find the frequency spectra of its sampled sequences if

(a) the sampling frequency is $\omega_s = 40$ rad/s,

(b) $\omega_s = 90$ rad/s, and

(c) $\omega_s = 200$ rad/s.

7.16 Consider the continuous-time signal $f(t)$ with the frequency spectrum shown in Figure P7.16. Find the frequency spectra of its sampled sequences if the sampling frequency is

(a) $\omega_s = 20$ rad/s,

(b) $\omega_s = 30$ rad/s, and

(c) $\omega_s = 50$ rad/s.

Figure P7.16

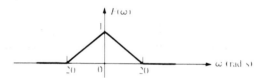

USING **MATLAB**

MATLAB can be used to compute frequency spectra. Consider the frequency spectrum in Example 7.3.3. The following commands yield the plot in Figure 7.6 in [0, 2π].

```
w = 2*pi*(0:99)/100; (generates 100 points equally spaced in
                                    [0, 2π])
F=(1+0.9.*exp(-j*w))./(1+1.8.*exp(-j*w)+0.81.*
  exp(-j*2*w));
plot(w,abs(F),w,angle(F))
```

The frequency spectrum of this example is obtained from the rational function

$$\bar{F}(z) = \frac{1 + 0.9z^{-1}}{1 + 1.8z^{-1} + 0.81z^{-2}}$$

by the substitution $z = e^{j\omega}$. The frequency spectrum can also be obtained by using the command freqz as follows.

```
n = [1 0.9];d = [1 1.8 0.81];
[F,w]= freqz(n,d,100,'whole');
plot(w,abs(F),w,angle(F))
```

This also yields the plot in Figure 7.6. Note that freqz(n,d,100) computes 100 points between 0 and π; freqz(n,d,100,'whole') computes 100 points between 0 and 2π.

Because the use of the command fft to compute frequency spectra and convolutions is explained in the chapter text, we will not discuss it here.

MATLAB PROBLEMS

Use MATLAB to compute the following.

M7.1 Use the preceding two methods to compute the frequency spectra of the two sequences in Exercise 7.3.2.

M7.2 Use the command fft to compute the spectra of the sequences in Problem M7.1.

M7.3 Use a 6-point fft to compute the frequency spectrum of $h[k] = 2$, for $k = 0, 1, \ldots, 5$, and $h[k] = 0$, for $k < 0$ and $k > 5$. Repeat the computation for $N = 64, 128, 256,$ and 512.

M7.4 Compute the frequency spectrum of

$$h[k] = (0.9)^k \sin 0.5\pi k + (0.5)^k \cos 0.8\pi k$$

M7.5 Compute the frequency spectrum of

$$h[k] = (0.9)^k \sin 0.5\pi k + \cos 0.8\pi k$$

Is there a spike whose height increases with N?

M7.6 Find the sequence with the frequency spectrum

$$H(\omega) = e^{-j2\omega}(5 + 2\cos\omega - 3\sin 2\omega)$$

Is it a finite sequence?

M7.7 Use the command $\mathtt{ifft}$ to solve Problem 7.12.

M7.8 Compute the frequency spectrum of the positive-time signal

$$f(t) = e^{-t}\sin 2t + 2^{-0.2t}\cos 3t.$$

M7.9 (a) Find the inverse Laplace transform of

$$H(s) = \frac{s-2}{(s+1)^2}$$

and plot it.

(b) Use the command $\mathtt{ifft}$ to repeat (a). Compare the results.

8

BIBO Stability and Frequency Response

30

8.1 INTRODUCTION

We focused on system analysis in Chapters 4 and 5 and signal analysis in Chapters 6 and 7. Here, we shall consider them together as we discuss the processing of signals by systems. First, some examples will illustrate the practical application of signal processing.

The voltage supplied by household electric outlets is

$$v(t) = 120\sqrt{2}\,\sin 2\pi \cdot 60t$$

as shown in Figure 8.1(a). It is a pure sinusoidal function with amplitude $120\sqrt{2}$ volts and frequency 60 Hz (cycles per second) or $60 \cdot 2\pi$ radians per second. However, if we use a television and a vacuum cleaner at the same time, the television picture will be snowed or distorted because the voltage fed into the television will be corrupted by noise, as shown in Figure 8.1(b). This interference can be eliminated if we design a system called a filter to eliminate the noise. Filters in radios and televisions pick up signals from a desired station and reject signals from other stations. They are examples of simple signal processing. A much more difficult application is eliminating ghosts in television pictures. Unlike when the power supply is corrupted by noise, ghosts occur when the signal that carries the picture is reflected by tall buildings or geographical obstacles. Eliminating ghosts is a difficult problem in signal processing.

When transmitting a telephone conversation over a long distance, the sound level may become inaudible. A system called an amplifier will boost the sound to its original level. Filters that eliminate crosstalk or noise in the transmission may be required as well. Amplifiers are also needed in radios and televisions. In addition, the EKG waves in Figure 1.1(a), which are in the order of mV (10^{-3}

Figure 8.1   *(a) Sinusoidal signal and (b) sinusoidal signal corrupted by noise*

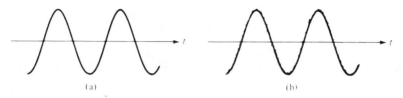

(a)   (b)

volt), and the brain waves in Figure 1.1(b), which are in the order of μV (10^{-6} volt), must be amplified in order to be visible.

Signal processing—in particular, picture processing—has found many applications in recent years. By freezing, enlarging, sharpening, or enhancing the pictures taken prior to the explosion of the space shuttle Challenger, it was possible to reconstruct the exact sequence, to the hundredth of a second, of the disastrous flight and identify possible causes of the blowup. Signal processing can be used to find tumors from body scans, compute the size of a heart or detect tooth cavities from x-rays, and analyze electrocardiograms. Indeed, the practical applications of signal processing are many.

Every signal processing is achieved by designing a system, which can be an electronic circuit, a computer program, or an optical system. Figure 1.4 showed an example of picture processing using an optical system. The picture in Figure 1.4(a) was blurred due to linear motion. It was reconstructed in Figure 1.4(b) using a holographic image deblurring filter. See Reference [33].

Every system designed to process signals must be stable. If an electrical system is not stable, it will likely burn out or saturate after the application of an input, no matter how small. If a mechanical system or a structure is not stable, it will generally blow up or disintegrate, as shown in Figure 8.2. If a computer program is unstable, it will overflow. Thus, every system must be designed to be stable in order to function properly. If a system is built with linear time-invariant passive RLC elements, then the system is automatically stable and the study of stability becomes unnecessary. However, if a system is built with operational amplifier circuits, then there is no guarantee that the system will be stable. A computer program, too, can be stable or unstable. The study of stability is essential in such systems.

The response of every linear time-invariant system can be decomposed into the zero-state response and zero-input response. It is customary to study the stability of these two responses separately. In this chapter, we will study only the stability of the zero-state response. The stability of the zero-input response will be studied in the next chapter.

 The zero-state response of every linear time-invariant causal (lumped or distributed) system can be described by the convolution integral[1]

---

[1]If $h(t)$ contains an impulse at $t = 0$, then the lower limit of the integrations should be replaced by 0−. See Section 3.5.2.

Figure 8.2 *Collapse of a bridge*

$$y(t) = \int_0^t h(t-\tau)u(\tau)d\tau = \int_0^t h(t)u(t-\tau)d\tau \qquad\qquad \textbf{(8.1)}$$

where u, y, and h are respectively the input, output, and impulse response of the system. If an LTI system is lumped, then it can also be described by

$$Y(s) = H(s)U(s) \qquad\qquad \textbf{(8.2)}$$

where

$$H(s) := \frac{N(s)}{D(s)} := \frac{b_m s^m + b_{m-1}s^{m-1} + \cdots + b_1 s + b_0}{a_n s^n + a_{n-1}s^{n-1} + \cdots + a_1 s + a_0} \qquad\qquad \textbf{(8.3)}$$

and $Y(s)$, $U(s)$, and $H(s)$ are the Laplace transforms of $y(t)$, $u(t)$, and $h(t)$. $H(s)$ is called the transfer function of the system. It is a rational function of s or a ratio of two polynomials $N(s)$ and $D(s)$. The transfer function describes only the zero-state response. The poles of $H(s)$ are, as discussed in Chapter 4, the roots of $D(s)$ after canceling all common factors between $N(s)$ and $D(s)$. To simplify the discussion in this chapter, we will assume that $N(s)$ and $D(s)$ have no common factors. Under this assumption, all roots of $D(s)$ are the poles of $H(s)$.

8.2   STABILITY OF ZERO-STATE RESPONSES— BIBO STABILITY

The zero-state response of every LTIL system is describable by (8.1) and (8.2). Since they describe only the relationship between the input and output signals, the stability study can only question whether the output will have the same certain property that the input has. For example, if an input signal has a finite energy, will the excited output also have a finite energy?

In stability studies, the property most often queried is boundedness. A positive-time signal $u(t)$ is said to be bounded if it does not go to infinity or, equivalently, if there exists a constant N such that

$$|u(t)| \le N < \infty$$

for all t in $[0, \infty)$.

Definition 8.1   A system is said to be BIBO (bounded-input bounded-output) stable or, more precisely, the zero-state response of a system is said to be BIBO stable if *every* bounded input excites a bounded output. Otherwise, the system is said to be not BIBO stable.[2]

EXAMPLE 8.2.1

Consider the network shown in Figure 8.3(a). The input u is a current source; the output y is the voltage across the capacitor. Using the equivalent Laplace-transform circuit in Figure 8.3(b), we can readily obtain

$$Y(s) = \frac{s \cdot \dfrac{1}{s}}{s + \dfrac{1}{s}} U(s) = \frac{s}{s^2 + 1} U(s)$$

If we apply the bounded input $u(t) = 1$ for $t \ge 0$, then the output is

$$Y(s) = \frac{s}{s^2 + 1} \cdot \frac{1}{s} = \frac{1}{s^2 + 1}$$

or

$$y(t) = \sin t$$

---

[2]Our definition is different from the following: A system is BIBO stable if a bounded input excites a bounded output. While the definitions differ in only one word (*every* is replaced by *a*), their meanings are entirely different.

Figure 8.3 *Unstable system*

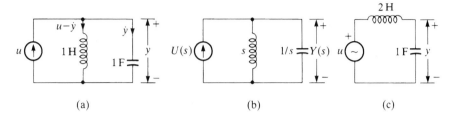

| (a) | (b) | (c) |

It is clearly bounded. If we apply the bounded input $u(t) = \sin at$, for $t \geq 0$, where a is a positive real constant and $a \neq 1$, then the output is

$$Y(s) = \frac{s}{s^2 + 1} \cdot \frac{a}{s^2 + a^2} = \frac{as[(s^2 + a^2) - (s^2 + 1)]}{(a^2 - 1)(s^2 + 1)(s^2 + a^2)}$$

$$= \frac{a}{a^2 - 1} \cdot \frac{s}{s^2 + 1} - \frac{a}{a^2 - 1} \cdot \frac{s}{s^2 + a^2}$$

(handwritten annotation: $= \sin at$)

which implies, using Table 4.1,

$$y(t) = \frac{a}{a^2 - 1}[\cos t - \cos at]$$

The output is bounded for any $a \neq 1$. Thus, the outputs excited by the bounded inputs $u(t) = 1$ and $\sin at$ with $a \neq 1$ are all bounded. Even so, we cannot conclude the BIBO stability of the network because we have not yet checked *every* possible bounded input. Consider the application of $u(t) = \sin t$, which yields

$$Y(s) = \frac{s}{s^2 + 1} \cdot \frac{1}{s^2 + 1} = \frac{s}{(s^2 + 1)^2}$$

This implies, using Table 4.1,

$$y(t) = \frac{1}{2}t \sin t$$

Clearly, $y(t)$ approaches positive or negative infinity as $t \to \infty$. Thus, the bounded input $u(t) = \sin t$ excites an unbounded output, and the system in Figure 8.3(a) is *not* BIBO stable.

EXERCISE 8.2.1

Show that, if $u(t) = \sin at$, for any integer a, the output $y(t)$ of the network in Figure 8.3(c) is bounded. Is the system BIBO stable? If not, find a bounded input that excites an unbounded output.

[**ANSWERS:** No; $\sin \sqrt{2}\,t$.] ∎

EXERCISE 8.2.2

Consider a system with transfer function $1/s$. It is called an integrator. What, if any, bounded input would excite an unbounded output? Is the system BIBO stable?

[**ANSWERS:** Step function; no.] ∎

When using Definition 8.1, if we find a bounded input that excites an unbounded output, we can conclude that the system is not BIBO stable. However, it is difficult, if not impossible, to conclude that a system is BIBO stable using Definition 8.1, because there are infinitely many bounded inputs to be checked. Fortunately, we have the following very powerful theorem.

Theorem 8.1 An LTI system with impulse response $h(t)$ is BIBO stable if and only if $h(t)$ is absolutely integrable in $[0, \infty)$; that is,

$$\int_0^\infty |\,h(t)\,|\;dt \leq M < \infty$$

for some constant M.

PROOF We will show first that, if $h(t)$ is absolutely integrable in $[0, \infty)$, then the system is BIBO stable. Indeed, if $u(t)$ is bounded or $|\,u(t)\,| \leq M_1$ for all t in $[0, \infty)$, then (8.1) implies

$$|y(t)| = \left|\int_0^t h(\tau)u(t-\tau)d\tau\right| \leq \int_0^t |h(\tau)|\,|u(t-\tau)|d\tau$$
$$\leq M_1 \int_0^t |h(\tau)|d\tau \leq M_1 M$$

for all t in $[0, \infty)$. Thus, the output is bounded.

Now we will show that, if $h(t)$ is not absolutely integrable, then the system is not BIBO stable. A rigorous proof of this part will overwhelm the basic idea; thus, we will proceed intuitively. If $h(t)$ is not absolutely integrable, then there exists a t_1 such that

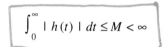

$$\int_0^{t_1} |h(\tau)|d\tau = \infty \tag{8.4}$$

Let $u(t)$ be the bounded input defined by

$$u(t_1 - \tau) = \begin{cases} 1 & \text{if } h(\tau) \geq 0 \\ -1 & \text{if } h(\tau) < 0 \end{cases}$$

Then, we have

$$y(t_1) = \int_0^{t_1} h(\tau)u(t_1 - \tau)d\tau = \int_0^{t_1} |h(\tau)|d\tau = \infty$$

which is not bounded. This establishes the theorem.

EXAMPLE 8.2.2

Consider the positive feedback system shown in Figure 8.4(a). Its impulse response is computed in Example 3.5.3 as

$$h(t) = \sum_{i=1}^{\infty} a^i \delta(t - i)$$

We compute

$$\int_0^{\infty} |h(t)|dt \leq \sum_{i=1}^{\infty} |a|^i \int_0^{\infty} |\delta(t - i)|dt = \sum_1^{\infty} |a|^i$$

where we have assumed $\delta(t) \geq 0$ and $|\delta(t)| = \delta(t)$. The summation is infinity if $|a| \geq 1$. Thus, the positive feedback system is not BIBO stable if the magnitude of a is equal to or larger than 1. If $|a| < 1$, then

$$\sum_{i=1}^{\infty} |a|^i = |a| \sum_0^{\infty} |a|^i = |a| \cdot \frac{1}{1 - |a|} < \infty$$

and the positive feedback system is BIBO stable.

Figure 8.4 *(a) Positive feedback system and (b) negative feedback system*

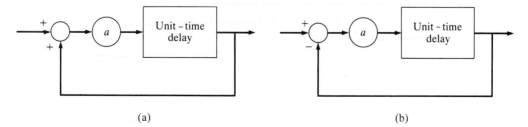

(a)   (b)

EXERCISE 8.2.3

Show that the negative feedback system in Figure 8.4(b) is BIBO stable if and only if $|a| < 1$. Is it true that every negative feedback system is always BIBO stable and every positive feedback system is always not BIBO stable?

[**ANSWER:** False.]   ■

EXERCISE 8.2.4

The unit step response of the positive feedback system in Figure 8.4(a) with $a = 1$ approaches infinity, as discussed in Example 4.12.3. Will the unit step response of the negative feedback system in Figure 8.4(b) with $a = 1$ approach infinity? If not, find a bounded input that will excite an unbounded output.

[**ANSWERS:** No, see Figure 4.25. $u(t) = 1$ for $2k \le t < 2k + 1$ and $u(t) = -1$ for $2k + 1 \le t < 2k + 2$, $k = 0, 1, 2 \ldots$.]   ■

EXERCISE 8.2.5

Is a unit-time delay system $[y(t) = u(t - 1)]$ BIBO stable? Answer the question by using Definition 8.1 and Theorem 8.1.

[**ANSWER:** Yes. $\displaystyle\int_0^\infty |\delta(t - 1)|\, dt = 1 < \infty.$]   ■

The transfer function is the Laplace transform of the impulse response. Therefore, it is possible to express the stability condition in terms of the transfer function. The transfer function of LTI *distributed* systems is *not* a rational function of s; its stability condition is, in general, not transparent and is rarely used. The transfer function of LTI *lumped* systems is a rational function of s; its stability condition is very simple and is widely used.

Theorem 8.2   An LTIL system with proper rational transfer function $H(s)$ is BIBO stable if and only if every pole of $H(s)$ has a negative real part or, equivalently, lies inside the open left half–s-plane.

PROOF   Open left half–s-plane refers to the left half–s-plane excluding the $j\omega$-axis. This theorem implies that a system is BIBO stable if, and only if, the system has no pole with zero or positive real part. We will give an intuitive argument to justify the theorem. If $H(s)$ has poles on the $j\omega$-axis or inside the right half–s-plane, such as

$$\frac{1}{s}, \qquad \frac{1}{s^2 + 1}, \qquad -\frac{1}{s - 2}$$

then their inverses are, for $t \geq 0$,

$$1, \quad \sin t, \quad e^{2t}$$

Clearly, they are not absolutely integrable in $[0, \infty)$. Therefore, if $H(s)$ has one or more poles on the $j\omega$-axis or inside the right half–s-plane, the zero-state response of the system is not BIBO stable. This shows the necessity of the theorem.

To show the sufficiency, let a_i, $i = 1, 2, \ldots, m$, be distinct poles of $H(s)$. They may be simple or repeated. Using partial fraction expansion, we expand $H(s)$ as

$$H(s) = k_0 + \sum_{i=1}^{m} \left[\frac{k_{i1}}{(s - a_i)} + \frac{k_{i2}}{(s - a_i)^2} + \cdots \right]$$

Its inverse Laplace transform is, using Table 4.1,

$$h(t) = k_0 \delta(t) + \sum_{i=1}^{m} (k_{i1} e^{a_i t} + k_{i2} t e^{a_i t} + \cdots)$$

Now we will show that, if $Re \; a_i < 0$, then every term of $h(t)$ is absolutely integrable. The integration of $\delta(t)$ is 1. Thus, the first term is absolutely integrable. Let $a_i = \alpha + j\beta$. If all poles of $H(s)$ have negative real parts, then $\alpha = Re \; a_i < 0$. We compute

$$\int_0^\infty |e^{a_i t}| \, dt = \int_0^\infty |e^{(\alpha + j\beta)t}| \, dt = \int_0^\infty e^{\alpha t} dt = \frac{1}{\alpha} e^{\alpha t} \Big|_{t=0}^{\infty} = -\frac{1}{\alpha} < \infty \qquad \textbf{(8.5)}$$

where we have used $|e^{j\beta t}| = 1$, for all t, and $e^{\alpha t} = 0$ at $t = \infty$ under the assumption $\alpha < 0$. Thus, $e^{a_i t}$ is absolutely integrable in $[0, \infty)$. Next, using integration by parts, we compute

$$\int_0^\infty |te^{a_i t}| \, dt = \int_0^\infty te^{\alpha t} dt = \frac{1}{\alpha} \int_0^\infty t \frac{de^{\alpha t}}{dt}$$

$$= \frac{1}{\alpha} \left[te^{\alpha t} \Big|_{t=0}^{\infty} - \int_0^\infty e^{\alpha t} \, dt \right] \qquad \textbf{(8.6)}$$

Clearly, we have $te^{\alpha t} = 0$ at $t = 0$. At $t = \infty$, $te^{\alpha t}$ is the product of an infinity and a zero. Therefore, we must use l'Hopital's rule to compute its value. We write $te^{\alpha t} = t/e^{-\alpha t}$. It is a ratio of infinity over infinity at $t = \infty$. We differentiate it to yield, at $t = \infty$,

$$te^{\alpha t} = \frac{t}{e^{-\alpha t}} = \frac{1}{-\alpha e^{-\alpha t}} \rightarrow \frac{1}{-\infty} = 0$$

Therefore, (8.6) reduces, after the substitution of (8.5), to

$$\int_0^t |te^{a_i t}| \, dt = -\frac{1}{\alpha} \int_0^\infty e^{\alpha t} \, dt = \frac{1}{\alpha^2}$$

Thus, $te^{a_i t}$ is also absolutely integrable in $[0, \infty)$. Proceeding forward, we can show that every term of $h(t)$ is absolutely integrable. Since $h(t)$ consists of only a finite number of such terms, we conclude that $h(t)$ is absolutely integrable and the system is BIBO stable. This establishes Theorem 8.2.

 Note that a system's BIBO stability depends only on the poles of $H(s)$ and does not depend on the zeros. If all poles of $H(s)$ lie inside the open left half–s-plane, the system is BIBO stable, no matter where the zeros are. The zeros can lie inside the right or left half–s-plane or on the $j\omega$-axis. For easy reference, we call a pole a *stable* pole if it lies in the open left half-plane and an *unstable* pole if it lies in the closed right half-plane (the right half-plane including the $j\omega$-axis).

Let us use Theorem 8.2 to study the network in Figure 8.3(a). Its transfer function is

$$H(s) = \frac{s}{s^2 + 1}$$

Its poles are $\pm j$. They have zero real parts. Thus, the network is not BIBO stable.

EXAMPLE 8.2.3

No structure is built to be completely rigid. The twin towers at the World Trade Center in New York City will swing on windy days; the wings of every Boeing 747 will vibrate fiercely when it flies into a storm. Even a concrete elevated highway will oscillate if it is hit by a force. The collapse of the elevated highway in Oakland, California, in the 1989 earthquake can be explained by modeling the highway as shown in Figure 8.5(b), or

$$H(s) = \frac{Y(s)}{U(s)} = \frac{1}{s^2 + \omega_0^2}$$

where the input is a horizontal force and the output is the displacement at the point where the input is applied or at some other point. We call ω_0 the structural resonance frequency. Suppose an earthquake generates the seismic force $\sin at$. If the seismic frequency a equals ω_0, then the output is

Figure 8.5 *(a) Collapse of an elevated highway, (b) model, (c) more realistic model, (d) seismic force, and (e) responses*

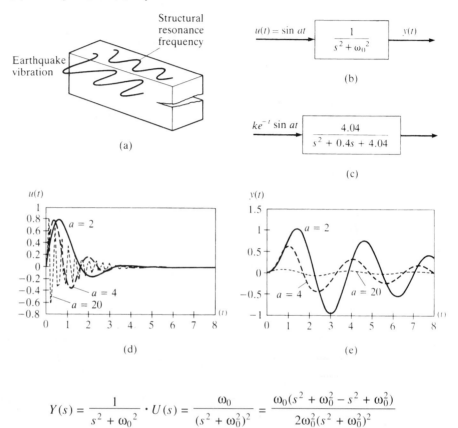

$$Y(s) = \frac{1}{s^2 + \omega_0^2} \cdot U(s) = \frac{\omega_0}{(s^2 + \omega_0^2)^2} = \frac{\omega_0(s^2 + \omega_0^2 - s^2 + \omega_0^2)}{2\omega_0^2(s^2 + \omega_0^2)^2}$$

$$= \frac{1}{2\omega_0(s^2 + \omega_0^2)} - \frac{s^2 - \omega_0^2}{2\omega_0(s^2 + \omega_0^2)^2}$$

which implies, using Table 4.1,

$$y(t) = \frac{1}{2\omega_0^2} \sin \omega_0 t - \frac{1}{2\omega_0} t \cos \omega_0 t$$

Thus, the displacement increases with time, and the highway eventually collapses. The same reason can be used to explain the collapse of the suspension bridge shown in Figure 8.2, which occurred in 1940 in Washington state. When the frequency of gust coincided with the structural resonance frequency of the bridge, the amplitude of the oscillation increased and the bridge collapsed.

The highway is modeled as an unstable system in Figure 8.5(b). Although this gives a simple explanation of the collapse of the highway, the model is unrealistic. It means that, if the highway is hit by a force, it will generate an oscillation that will never diminish. In reality, the oscillation will decay to zero. Therefore, a more realistic model is

$$H(s) = \frac{4.04}{(s + 0.2)^2 + 2^2} = \frac{4.04}{s^2 + 0.4s + 4.04}$$

with structural resonance frequency 2 rad/s and time constant $1/0.2 = 5$ s, as shown in Figure 8.5(c). Its impulse response is $2.02e^{-0.2t} \sin 2t$ and will decay to zero in about $5 \times 5 = 25$ seconds. Now, the seismic wave is modeled as

$$u(t) = ke^{-t} \sin at$$

Because its time constant is 1, the earthquake lasts about 5 seconds. We apply $u(t)$ to the bridge with $a = 2$, 4, and 20 and $k = 15.5$, 1.2, and 1. These k are chosen so that the maximum values of $u(t)$ for the three different a are roughly the same as shown in Figure 8.5(d). The response of the system is

$$Y(s) = \frac{4.04}{s^2 + 0.4s + 4.04} \cdot \frac{ka}{(s + 1)^2 + a^2}$$

and is plotted in Figure 8.5(e). We see that the largest magnitudes of the responses are, respectively, about 0.15, 0.7, and 1.05 for $a = 20$, 4, and 2. Suppose the elevated highway will collapse when the displacement is larger than 1. Then, the highway will survive the earthquake when the seismic frequency is 20 or 4, but it will collapse when the seismic frequency equals the structure resonance frequency. This explanation is more realistic than the one based on the unstable model in Figure 8.5(b). (This problem will be discussed further in Section 8.6.)

EXERCISE 8.2.6

Which of the following systems are BIBO stable? For any that are not, find a bounded input that excites an unbounded output.

(a) $\dfrac{1}{s + 10}$ (b) $\dfrac{1}{(s - 1)}$

(c) $\dfrac{s - 1}{s(s + 10)}$ (d) $\dfrac{s^2 - 2s + 1}{s^2 - 1}$

(e) $\dfrac{s^2 - 1}{4s^2 + 1}$ (f) $\dfrac{s - 10}{s^2 + 2s + 2}$

[**ANSWERS:** (a) Yes, (b) no (step input), (c) no (step input), (d) yes, (e) no (sin 0.5t), (f) yes.]

If a system is not BIBO stable, such as

$$H(s) = \frac{1}{s-2} \quad \text{or} \quad H(s) = \frac{1}{(s^2+1)^2}$$

then its output approaches infinity for almost any applied input. The exception is the input Laplace transform that has unstable poles of $H(s)$ as its zeros, such as

$$U(s) = \frac{s-2}{(s+1)(s+2)(s+3)} \quad \text{or} \quad U(s) = \frac{(s^2+1)^2}{(s+1)(s+2)(s+3)}$$

In this case, the output becomes

$$Y(s) = \frac{1}{s-2} \cdot \frac{s-2}{(s+1)(s+2)(s+3)} = \frac{1}{(s+1)(s+2)(s+3)}$$

and is bounded. Thus, if this type of pole-zero cancellations occurs, the output will not approach infinity. However, this type of exact cancellation is impossible to achieve in practice for two reasons: First, exact values of physical components often are not known. For example, a 1K resistor may have a resistance anywhere between 0.9K and 1.1K; the capacitance of a 1 μF capacitor may deviate more than 20%. Therefore, the transfer function of a system is often obtained by approximation. Furthermore, the transfer function may change due to aging or due to changes of load or environment. Thus, the poles of a physical system may drift, and there is no way to cancel them. Second, we have no control over the applied input and, therefore, cannot introduce zeros to cancel poles.

In conclusion, the output of a system with open right half-plane poles or repeated imaginary poles will grow without bound no matter what input is applied. If a system has simple $j\omega$-axis poles, such as

$$H(s) = \frac{1}{(s+2)(s+j5)(s-j5)} = \frac{1}{(s+2)(s^2+25)}$$

then the output generally contains the term sin $5t$, no matter what input is applied. Because of this sustained oscillation, the system cannot be used to process signals. Thus, except for oscillators that are designed to generate sustained oscillations (see Section 9.8), every practical system is designed to be BIBO stable.

8.3 THE ROUTH TEST

Consider an LTIL system with proper transfer function $N(s)/D(s)$. It is assumed that $N(s)$ and $D(s)$ have no common factors. To employ Theorem 8.2, we must compute the roots of $D(s)$. If the degree of $D(s)$ is 3 or higher, hand computation of the roots will be very complicated. Furthermore, we do not need to know the exact locations of the roots to determine stability. Therefore, it is desirable to have a method for determining stability that does not require solving for the roots. Such a method is introduced here. The method is called the Routh test or the Routh-Hurwitz test, because Hurwitz developed a method that is equivalent to the Routh test.[3]

Definition 8.2 A polynomial $D(s)$ with real coefficients is called a Hurwitz polynomial if every root of $D(s)$ has a negative real part.

A polynomial $D(s)$ is not a Hurwitz polynomial if it has one or more roots with zero or positive real parts. The Routh test is a method for checking whether or not a polynomial is Hurwitz without solving for the roots. The most important application of the test is in checking the stability of LTIL systems. Consider the polynomial with a positive leading coefficient

$$D(s) = a_0 s^n + a_1 s^{n-1} + \cdots + a_{n-1} s + a_n \quad \text{with } a_0 > 0 \quad \textbf{(8.7)}$$

where a_i are real constants. If the leading coefficient a_0 is negative, we may simply multiply -1 to $D(s)$ to yield a positive a_0. Note that $D(s)$ and $-D(s)$ have the same set of roots; therefore, $a_0 > 0$ does not impose any restriction on $D(s)$.

Necessary condition for a polynomial to be Hurwitz If $D(s)$ is Hurwitz, then all coefficients must be positive. In other words, if $D(s)$ has a missing term (a zero coefficient) or a negative coefficient, then $D(s)$ is not Hurwitz. We use an example to establish this condition. Assume that $D(s)$ has two real roots and a pair of complex conjugate roots and is factored as

$$\begin{aligned} D(s) &= a_0(s + \alpha_1)(s + \alpha_2)(s + \beta_1 + j\gamma_1)(s + \beta_1 - j\gamma_1) \\ &= a_0(s + \alpha_1)(s + \alpha_2)(s^2 + 2\beta_1 s + \beta_1^2 + \gamma_1^2) \end{aligned} \quad \textbf{(8.8)}$$

The roots of $D(s)$ are $-\alpha_1$, $-\alpha_2$, and $-\beta_1 \pm j\gamma_1$. If $D(s)$ is Hurwitz, then $\alpha_1 > 0$, $\alpha_2 > 0$, and $\beta_1 > 0$. Hence, all coefficients in the factors are positive. Clearly, all coefficients remain positive after multiplying out the factors. This shows that, if $D(s)$ is Hurwitz, then its coefficients must be all positive. This condition is necessary but not sufficient. For example, the polynomial

$$s^3 + 2s^2 + 9s + 68 = (s + 4)(s - 1 + 4j)(s - 1 - 4j)$$

has all positive coefficients, but is not Hurwitz.

For a polynomial of degree 1 or 2 with a positive leading coefficient, the preceding necessary condition is sufficient as well. That is, a polynomial of

---

[3]In honor of Edward John Routh (1831−1907), who was born in Canada and spent most of his life in England, and Adolph Hurwitz (1859−1919), a Swiss mathematician.

degree 1 or 2 with a positive leading coefficient is Hurwitz if, and only if, all coefficients are positive. This statement can be proved directly or be reduced from Theorem 8.3 which will be established next.

Necessary and sufficient condition For easy presentation, consider the following polynomial of degree 6:

$$D(s) = a_0 s^6 + a_1 s^5 + a_2 s^4 + a_3 s^3 + a_4 s^2 + a_5 s + a_6 \qquad \text{with } a_0 > 0$$

We form Table 8.1.[4] The first two rows are formed from the coefficients of $D(s)$. The first coefficient (in the descending power of s) is put in the (1, 1) position, the second in the (2, 1) position, the third in the (1, 2) position, the fourth in the (2, 2) position, and so forth. Next, we compute $k_1 := a_0/a_1$. This is the ratio of the first elements of the first two rows. The third row is obtained as follows: the first row subtracts the product of the second row and k_1, that is,

$$b_0 = a_0 - k_1 a_1, \qquad b_1 = a_2 - k_1 a_3, \qquad b_2 = a_4 - k_1 a_5, \qquad b_3 = a_6 - k_1 \cdot 0$$

The result is placed at the right-hand side of the second row. The first element $b_0 = a_0 - k_1 a_1 = a_0 - a_0 = 0$ is always zero and is discarded. We then use b_i, $i = 1, 2, 3$, to form the third row, as shown. The fourth row is obtained in the same manner from its two previous rows. That is, we compute $k_2 = a_1/b_1$ and subtract the product of the third row and k_2 from the second row:

$$c_0 = a_1 - k_2 b_1, \qquad c_1 = a_3 - k_2 b_2, \qquad c_2 = a_5 - k_2 b_3$$

Table 8.1 *The Routh table*

	s^6	a_0	a_2	a_4	a_6	
$k_1 = \frac{a_0}{a_1}$	s^5	a_1	a_3	a_5		(1st row) $- k_1$(2nd row) $= [b_0 \ b_1 \ b_2 \ b_3]$
$k_2 = \frac{a_1}{b_1}$	s^4	b_1	b_2	b_3		(2nd row) $- k_2$(3rd row) $= [c_0 \ c_1 \ c_2]$
$k_3 = \frac{b_1}{c_1}$	s^3	c_1	c_2			(3rd row) $- k_3$(4th row) $= [d_0 \ d_1 \ d_2]$
$k_4 = \frac{c_1}{d_1}$	s^2	d_1	d_2			(4th row) $- k_4$(5th row) $= [e_0 \ e_1]$
$k_5 = \frac{d_1}{e_1}$	s^1	e_1				(5th row) $- k_5$(6th row) $= [f_0 \ f_1]$
	s^0	f_1				

[4] The presentation is slightly different from the method of cross-product. This presentation requires less computation and is easier to understand.

We then drop the first element, which is zero, and arrange c_1 and c_2 as shown. We repeat the process until the row corresponding to $s^0 = 1$. If the degree of $D(s)$ is n, there are a total of $(n + 1)$ rows. The table is called the Routh table.

We discuss the size of the table. If $n = \deg D(s)$ is even, the first row has one more entry than the second row. If n is odd, the first two rows have the same number of entries. In both cases, the number of entries decreases by one at odd powers of s. For example, the numbers of entries in the rows of s^5, s^3, and s decrease by one from their previous rows. Note also that the right-most entries of the rows corresponding to even powers of s are the same. For example, in Table 8.1, we have $a_6 = b_3 = d_2 = f_1$. Thus, the last entry of the table equals the constant term of $D(s)$.

Theorem 8.3 (**The Routh test**) A polynomial with a positive leading coefficient is a Hurwitz polynomial if, and only if, every entry in the Routh table is positive or if, and only if, every entry in the first column, namely, $a_1, b_1, c_1, d_1, \cdots$, is positive.

It is clear that the first condition implies the second condition. It is rather surprising that the second condition implies the first condition. Either condition can be used. A proof of this theorem is beyond the scope of this text and can be found in Reference [4]. This theorem implies that, if a zero or a negative number appears in the table, the polynomial is not Hurwitz. In this case, it is unnecessary to complete the table. However, to conclude that a polynomial *is* Hurwitz, we must complete the table.

EXAMPLE 8.3.1

Consider $2s^4 + s^3 + 5s^2 + 3s + 4$. We form

$$
k_1 = \frac{2}{1} \quad
\begin{array}{c|ccc}
s^4 & 2 & 5 & 4 \\
s^3 & 1 & 3 & \\
s^2 & -1 & 4 &
\end{array}
\qquad [0 \ -1 \ 4] = (\text{1st row}) - k_1(\text{2nd row})
$$

Clearly, we have $k_1 = 2/1$, the ratio of the first entries of the first two rows. The result of the first row subtracting the product of the second row and k_1 is listed on the right-hand side of the s^3 row. We drop the first zero and put the rest in the s^2 row. A negative number appears in the table; therefore, the polynomial is not Hurwitz.

EXAMPLE 8.3.2

Consider $2s^5 + s^4 + 7s^3 + 3s^2 + 4s + 2$. We form

$$
k_1 = \frac{2}{1} \quad
\begin{array}{c|ccc}
s^5 & 2 & 7 & 4 \\
s^4 & 1 & 3 & 2 \\
s^3 & 1 & 0 \\
\end{array}
\qquad [0 \ 1 \ 0] = (\text{1st row}) - k_1(\text{2nd row})
$$

A zero appears in the table. Thus, the polynomial is not Hurwitz.

EXAMPLE 8.3.3

Consider $2s^5 + s^4 + 7s^3 + 3s^2 + 4s + 1.5$. We form

$$
\begin{array}{c|ccc}
 & s^5 & 2 & 7 & 4 \\
k_1 = \dfrac{2}{1} & s^4 & 1 & 3 & 1.5 \\
k_2 = \dfrac{1}{1} & s^3 & 1 & 1 \\
k_3 = \dfrac{1}{2} & s^2 & 2 & 1.5 \\
k_4 = \dfrac{2}{0.25} & s^1 & 0.25 \\
 & s^0 & 1.5 \\
\end{array}
$$

$[0 \ 1 \ 1] = (\text{1st row}) - k_1(\text{2nd row})$

$[0 \ 2 \ 1.5] = (\text{2nd row}) - k_2(\text{3rd row})$

$[0 \ 0.25] = (\text{3rd row}) - k_3(\text{4th row})$

$[0 \ 1.5] = (\text{4th row}) - k_4(\text{5th row})$

All entries in the table are positive, hence the polynomial is a Hurwitz polynomial.

EXERCISE 8.3.1

Which of the following polynomials are Hurwitz?

(a) $s^4 + 3s^2 + 3s + 1$

(b) $s^4 + s^3 + s^2 + s + 2$

(c) $2s^4 + 5s^3 + 5s^2 + 2s + 1$

(d) $-s^5 - 3s^4 - 10s^3 - 12s^2 - 7s - 3$

[**ANSWERS:** (a) No, (b) no, (c) yes, (d) yes.] ■

EXERCISE 8.3.2

Show that a polynomial of degree 1 or 2 is Hurwitz if and only if all coefficients are of the same sign. ∎

EXERCISE 8.3.3

Show that $a_0 s^3 + a_1 s^2 + a_2 s + a_3$ with $a_i > 0$, $i = 0, 1, 2,$ and 3, is Hurwitz if and only if $a_1 a_2 - a_0 a_3 > 0$. ∎

It is possible to obtain the distribution of roots from the Routh table. Generally, the number of changes of signs in the first column of the table equals the number of roots in the open right half–s–plane. This information is rarely used in engineering and, therefore, will not be discussed here. If we are interested in the distribution, we may employ a digital computer subroutine to compute all roots of a polynomial. This gives not only the distribution but also the exact location of roots.

8.4   STEADY-STATE RESPONSE OF BIBO STABLE SYSTEMS

In this section, we will discuss an implication of BIBO stability. If a system is not BIBO stable, its output excited by a sinusoidal input generally approaches infinity as $t \to \infty$. On the other hand, if a system is BIBO stable, its output will approach a sinusoidal function with the same frequency as the input. The amplitude and phase, however, will be modified by the transfer function of the system.

Theorem 8.4 Consider an LTIL system with proper transfer function $H(s)$. If $H(s)$ is BIBO stable and if we apply the input $u(t) = \sin \omega_0 t$ or $u(t) = \cos \omega_0 t$, then the output $y(t)$ will approach

$$y_{ss}(t) := \lim_{t \to \infty} y(t) = A(\omega_0) \sin(\omega_0 t + \theta(\omega_0)) \qquad (8.9a)$$

or

$$y_{ss}(t) := \lim_{t \to \infty} y(t) = A(\omega_0) \cos(\omega_0 t + \theta(\omega_0)) \qquad (8.9b)$$

where

$$A(\omega_0) := |H(j\omega_0)| := [(Re\ H(j\omega_0))^2 + (Im\ H(j\omega_0))^2]^{1/2}$$

and

$$\theta(\omega_0) := \sphericalangle H(j\omega_0) = \tan^{-1}[Im\ H(j\omega_0)/Re\ H(j\omega_0)]$$

The response $y_{ss}(t)$ is called the *steady-state response*, and $H(j\omega)$ is called the *frequency response*.

PROOF   The Laplace transform of $u(t) = \sin \omega_0 t$ is $\omega_0/(s^2 + \omega_0^2)$. Because $H(s)$ has no poles on the $j\omega$-axis, $s = \pm j\omega_0$ are simple poles of $Y(s) = H(s)U(s)$. Using partial fraction expansion, we expand

$$Y(s) = H(s)U(s) = H(s)\frac{\omega_0}{s^2 + \omega_0^2} = H(s)\frac{\omega_0}{(s - j\omega_0)(s + j\omega_0)}$$

$$= \frac{k_1}{s - j\omega_0} + \frac{k_1^*}{s + j\omega_0} + \text{(terms due to the poles of } H(s)) \qquad \textbf{(8.10)}$$

where

$$k_1 = H(s) \cdot \frac{\omega_0}{s^2 + \omega_0^2}(s - j\omega_0)\Big|_{s=j\omega_0} = H(j\omega_0)\frac{\omega_0}{2j\omega_0} = \frac{A(\omega_0)e^{j\theta(\omega_0)}}{2j}$$

and k_1^* is the complex conjugate of k_1 or

$$k_1^* = \frac{A(\omega_0)e^{-j\theta(\omega_0)}}{-2j}$$

The inverse Laplace transform of (8.10) is

$$y(t) = \frac{A(\omega_0)e^{j\theta(\omega_0)}}{2j}e^{j\omega_0 t} + \frac{A(\omega_0)e^{-j\theta(\omega_0)}}{-2j}e^{-j\omega_0 t}$$

$$+ \text{(terms due to the poles of } H(s))$$

Because all poles of $H(s)$ are assumed to have negative real parts, all terms inside the parentheses approach zero as $t \to \infty$. Thus, we have

$$y_{ss}(t) := \lim_{t\to\infty} y(t) = \frac{A(\omega_0)}{2j}[e^{j(\omega_0 t + \theta(\omega_0))} - e^{-j(\omega_0 t + \theta(\omega_0))}]$$

$$= A(\omega_0)\sin(\omega_0 t + \theta(\omega_0))$$

This establishes (8.9a). Equation (8.9b) can be similarly proved. The steady-state response of $H(s)$ excited by $\sin \omega_0 t$ depends only on the value of $H(s)$ at $s = j\omega_0$. Thus, $H(j\omega)$ is called the *frequency response*. This is why the Laplace-transform domain is also called the frequency domain.

In the proof we compute only the zero-state response. If a system is completely characterized by its transfer function (see Section 4.10), then the theorem holds even if initial conditions are nonzero. Most practical systems meet the condition of complete characterization; thus, Theorem 8.4 holds for zero-state and zero-input responses. An example illustrates.

EXAMPLE 8.4.1

Consider an LTIL system described by

$$\dot{y}(t) + 0.4y(t) = 3u(t) \tag{8.11}$$

We compute its response due to $y(0-) = 0.5$ and $u(t) = \sin 2t$ for $t \geq 0$. The application of the Laplace transform to (8.11) yields

$$sY(s) - y(0-) + 0.4Y(s) = 3U(s)$$

which implies

$$Y(s) = \underbrace{\frac{y(0-)}{s+0.4}}_{\substack{\text{zero-input} \\ \text{response}}} + \underbrace{\frac{3}{s+0.4}U(s)}_{\substack{\text{zero-state} \\ \text{response}}} = \underbrace{\frac{0.5}{s+0.4}}_{\substack{\text{zero-input} \\ \text{response}}} + \underbrace{\left[\frac{3}{s+0.4} \cdot \frac{2}{s^2+4}\right]}_{\substack{\text{zero-state} \\ \text{response}}}$$

$$= \underbrace{\frac{0.5}{s+0.4} + \left[\frac{1.44}{s+0.4}\right.}_{\substack{\text{transient} \\ \text{response}}} + \underbrace{\left.\frac{1.47e^{-j1.37}}{2j(s-2j)} - \frac{1.47e^{j1.37}}{2j(s+2j)}\right]}_{\substack{\text{steady-state} \\ \text{response}}}$$

Its inverse Laplace transform is

$$y(t) = 1.94e^{-0.4t} + \frac{1.47}{2j}(e^{j(2t-1.37)} - e^{-j(2t-1.37)})$$

$$= \underbrace{1.94e^{-0.4t}}_{\substack{\text{transient} \\ \text{response}}} + \underbrace{1.47 \sin(2t - 1.37)}_{\substack{\text{steady-state} \\ \text{response}}} \tag{8.12}$$

Clearly, as $t \to \infty$, we have

$$y_{ss}(t) = \lim_{t \to \infty} y(t) = 1.47 \sin(2t - 1.37)$$

The response $y(t)$ is plotted in Figure 8.6 with the solid line.

The response of an LTIL system can be decomposed into the zero-state response and zero-input response. Now we shall give a different decomposi-

Figure 8.6 *The complete response of (8.12)*

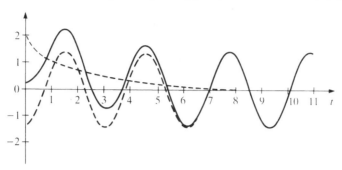

tion. The part of the response as $t \to \infty$ is called the *steady-state response* and the remainder is called the *transient response*. From the example, we see that the transient response is excited partly by nonzero initial conditions and partly by the input. The steady-state response, however, is excited entirely by the applied input. In other words, whereas initial conditions will excite only transient responses, inputs will generally excite both transient and steady-state responses. Every transient response is a linear combination of responses of all poles of the system. Thus, if a system is BIBO stable, its transient response will die out as $t \to \infty$.

In the example, the steady-state response of (8.12) excited by $u(t) = \sin 2t$ is obtained by computing the total response and then letting $t \to \infty$. The computation yields both the transient and steady-state responses. If we are interested in only the steady-state response, we may employ Theorem 8.4. The transfer function of (8.11) is

$$H(s) = \frac{3}{s + 0.4}$$

We compute

$$H(j2) = \frac{3}{2j + 0.4} = \frac{3}{2.04e^{j1.37}} = 1.47e^{-j1.37}$$

Thus, the steady-state response of (8.11) excited by $u(t) = \sin 2t$ is, using Theorem 8.4,

$$y_{ss}(t) = 1.47 \sin (2t - 1.37)$$

This computation is much simpler than the one in the example, but it yields less information; no transient response is computed.

EXERCISE 8.4.1

Use Theorem 8.4 to compute the steady-state response of

$$H(s) = \frac{2}{s^2 + 2s + 2}$$

excited by $u(t) = 3 \sin \sqrt{2}\, t$.

[**ANSWER:** $y_{ss}(t) = 2.1 \sin (\sqrt{2}\, t - \pi/2)$]

Theorem 8.4 is very important because it is the basis for signal processing. It also provides a method for measuring the transfer function, as will be discussed later.

Theorem 8.5 Consider a system with proper transfer function $H(s)$. The unit step response of the system approaches a constant as $t \rightarrow \infty$ if and only if the system is BIBO stable and the constant equals

$$y_{ss}(t) = \lim_{t \to \infty} y(t) = H(0) \tag{8.13}$$

PROOF We expand

$$Y(s) = H(s)U(s) = H(s)\frac{1}{s} = \frac{k_1}{s} + (\text{terms due to poles of } H(s))$$

with

$$k_1 = \left. \frac{H(s)}{s} \cdot s \right|_{s=0} = H(0)$$

If $H(s)$ is stable, every pole lies inside the open left half-plane and its time response approaches zero as $t \rightarrow \infty$. Thus, we have (8.13). If the system is not BIBO stable, then $H(s)$ has at least one closed right half-plane pole. If the pole is in the open right half-plane, the unit step response will approach infinity. If the pole is on the imaginary axis, the response will contain a sustained oscillation. If the pole is at the origin, then the response will contain the factor t that approaches infinity as $t \rightarrow \infty$. Thus we conclude that the unit step response approaches a constant if and only if the system is BIBO stable. This establishes the theorem.

We mention that (8.13) also can be established using the final-value theorem. Because all poles of $sY(s) = H(s)$ lie inside the open left half-plane, the final-value theorem can be used. Thus, we have

$$\lim_{t \to \infty} y(t) = \lim_{s \to 0} sY(s) = H(0)$$

EXERCISE 8.4.2

What are the steady-state responses of

$$H_1(s) = \frac{2}{s-2} \qquad H_2(s) = \frac{s+1}{s^2 + 2s + 10} \qquad H_3(s) = \frac{s-1}{s^2 + 2s + 10}$$

excited by a unit step input?

[**ANSWERS:** ∞, $1/10$, $-1/10$.]

8.4.1   Infinite time

The response of a BIBO stable system excited by a step or sinusoidal input consists of the transient response and steady-state response. The response reaches steady state only after the transient response dies out completely. Mathematically speaking, the transient will die out only as $t \to \infty$ and the response will not reach steady state in a finite time. However, in engineering, the transient is often considered to have died out when its value decreases to less then 1% of its initial value. Using this convention, the response of a BIBO stable system can reach steady state in a finite time, as will be discussed in the following.

Consider a system with BIBO stable transfer function $H(s)$. To simplify discussion, we assume that $H(s)$ has a simple pole at λ_1 and a repeated pole with multiplicity 2 at λ_2. Then the response of the system excited by a unit step input or the sinusoidal input $\sin \omega t$ is of the form

$$y(t) = k_1 e^{\lambda_1 t} + k_2 e^{\lambda_2 t} + k_3 t e^{\lambda_2 t} + steady\ state\ [H(0)\ or$$
$$|H(j\omega)| \sin(\omega t + \sphericalangle H(j\omega))]$$

This is obtained by combining the discussion in Section 4.7.1 and Theorems 8.4 and 8.5. We see that the form of the transient response is determined by the poles of $H(s)$. For example, the transient response of the system in (8.11) with transfer function $3/(s + 0.4)$ is $1.94 e^{-0.4t}$ which is the response of the pole at -0.4. We defined in Section 1.2.1 the *time constant* of $e^{-0.4t}$ as $1/0.4$. This time constant can also be defined from the real pole -0.4; it is simply the inverse of the magnitude of the pole or $1/0.4 = 2.5$. As was discussed in Section 1.2.1, the response $e^{-0.4t}$ decreases to less than 1% of its original value in five time constants or $5 \times 2.5 = 12.5$. Thus the system in (8.11) will be considered to have reached steady state in 12.5 seconds. This is a very short infinity indeed!

We can extend the concept of time constants to complex poles and stable transfer functions. The time constant of a stable complex pole is defined as the inverse of the magnitude of the real part. The time constant of a stable $H(s)$ is defined, then, as the largest time constant of all poles of $H(s)$. For example, if $H(s)$ has poles -1, -3, $-0.1 \pm j2$, the three time constants are 1, $1/3 = 0.33$, and $1/0.1 = 10$. Thus, the time constant of $H(s)$ is 10 seconds. If all poles of $H(s)$ are plotted as shown in Figure 8.7, then we have

$$\text{time constant} = \frac{1}{\text{the shortest distance of all poles from the imaginary axis}}$$

In engineering, the response of $H(s)$ excited by a step or sinusoid input is considered to have reached steady state in five time constants of $H(s)$. If $H(s)$ is not BIBO stable, its time constant is not defined.

EXERCISE 8.4.3

What are the time constants of the following transfer functions?

(a) $\dfrac{1}{s - 2}$

(b) $\dfrac{s - 5}{(s + 2)(s + 4)}$

(c) $\dfrac{s + 4}{(s + 0.1)(s^2 + 4s + 4)}$

(d) a system with the poles and zeros shown in Figure 8.7.

[**ANSWERS:** (a) Not defined, (b) 0.5, (c) 10, (d) 1.]

The time constant of a BIBO stable transfer function $H(s)$ as defined is open to argument. It is possible to find a transfer function whose response

Figure 8.7 *Pole-zero pattern*

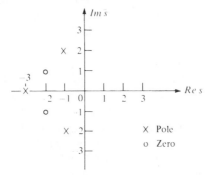

excited by a step input will not reach steady state in five time constants. The following example illustrates.

EXAMPLE 8.4.2

Consider $H(s) = 1/(s + 1)^3$. It has a repeated pole with multiplicity 3 at $s = -1$. The time constant of $H(s)$ is 1 second. The unit step response of $H(s)$ is

$$Y(s) = \frac{1}{(s + 1)^3} \cdot \frac{1}{s} = \frac{1}{s} + \frac{-1}{(s + 1)^3} + \frac{-1}{(s + 1)^2} + \frac{-1}{(s + 1)}$$

or

$$y(t) = 1 - 0.5t^2 e^{-t} - te^{-t} - e^{-t}$$

Figure 8.8 plots the transient response $-(0.5t^2 + t + 1)e^{-t}$ (dashed line) and total response (solid line). The transient response decreases to -0.126 or 13% of its initial value at $t = 5$ and to -0.007 or 0.7% at $t = 9$. Therefore, the response of this system can be considered to have reached the steady state in nine time constants, not in five time constants.

This example shows that, if a transfer function has repeated poles or, more generally, a cluster of poles in a small region close to the imaginary axis of the s-plane, then the rule of five time constants is not applicable. In fact, the situation is even more complicated. The zeros of a transfer function also affect the transient response (see Example 4.7.7 and Figure 4.6), but they are not considered when defining the time constant. Thus, it is extremely difficult to state precisely in how many time constants the response will reach steady state. The rule of five time constants is, however, useful in pointing out that infinity in engineering does not necessarily mean the mathematical infinity.

Figure 8.8 *Unit step response*

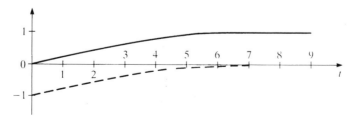

FUNDAMENTAL CONCEPT OF FILTERING— FREQUENCY RESPONSE

Consider the signal

$$u(t) = \sin t + 0.1 \sin 100t \qquad (8.14)$$

which is shown in Figure 8.9(a). The first component in (8.14) is assumed to carry information, and second component is noise. The ratio of the amplitudes of the information-bearing signal and noise, called the signal-to-noise (S/N) ratio, is 10. We process the signal using the network shown in Figure 8.9(b). The transfer function of the network is

$$H(s) = \frac{1}{1 + 0.1s} = \frac{10}{s + 10} \qquad (8.15)$$

Its time constant is 1/10, or 0.1 second. Therefore, the output will reach the steady state after 0.5 second. We compute

$$H(j1) = \frac{1}{1 + j0.1} = \frac{1}{1.005e^{j0.1}} = 0.995e^{-j0.1}$$

and

$$H(j100) = \frac{1}{1 + j10} \approx \frac{1}{10e^{j1.47}} = 0.1e^{-j1.47}$$

Thus, the steady-state response is

$$y_{ss}(t) = 0.995 \sin (t - 0.1) + 0.01 \sin (100t - 1.47)$$

as shown in Figure 8.9(c). We see that the S/N ratio at the output is now 0.995/0.01 = 99.5, and the noise is essentially eliminated or filtered out. The price we pay to achieve the filtering is a small attenuation of the amplitude of the information and a small time delay.

Figure 8.9 *Effect of filtering*

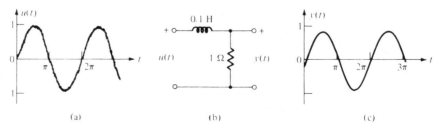

The steady-state response of $H(s)$ excited by $\sin \omega_0 t$ depends only on the value of $H(s)$ at $s = j\omega_0$. Thus, $H(j\omega)$ is called the *frequency response*. We compute the frequency response of the $H(s)$ in (8.15) as follows

$$H(j\omega) = \frac{10}{j\omega + 10} = \begin{cases} 1e^{j0°} & \text{at } \omega = 0 \\ 0.995e^{-j6°} & \text{at } \omega = 1 \\ 0.707e^{-j45°} & \text{at } \omega = 10 \quad \leftarrow \quad \omega_{3dB} \\ 0.45e^{-j63.5°} & \text{at } \omega = 20 \end{cases}$$

and plot it in Figure 8.10. Because $H(j\omega)$ is generally complex, the plot consists of two parts: amplitude and phase. These are called the *amplitude* and *phase characteristics*. If the frequency response plot is available, the steady-state response excited by any sinusoid input can be read out directly from that plot.

EXERCISE 8.5.1

Find the steady-state responses of the system in (8.15) due to the following inputs. Read them directly from Figure 8.10.

(a) $u(t) = 1 + \sin 5t$

(b) $u(t) = \sin 20t + 3 \cos (20t - 0.2)$

[**ANSWERS:** (a) $1 + 0.9 \sin (5t - 0.46)$, (b) $0.45 \sin (20t - 1.1) + 1.35 \cos (20t - 1.3)$.]

Figure 8.10 *The amplitude and phase characteristics of (8.15)*

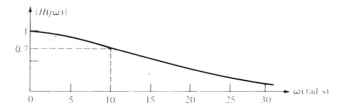

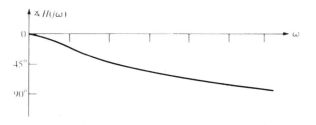

EXERCISE 8.5.2

Plot the amplitude and phase characteristics of

$$H(s) = \frac{1}{0.1s - 1} \qquad (8.16)$$

What is its steady-state response excited by the signal in (8.14)?

[**ANSWERS:** The amplitude plot is identical to the one in Figure 8.10, and the phase plot is the same if negative degrees in the coordinate are changed to positive. The steady-state response is infinity, because the system is not BIBO stable.] ∎

From the amplitude characteristic in Figure 8.10, we see that the network in Figure 8.9(b) will pass sinusoidal signals with low frequencies and will stop sinusoidal signals with high frequencies. Therefore, the network is called a *lowpass* filter. The amplitude characteristic of an *ideal* lowpass filter is shown in Figure 8.11(a). Also plotted in Figures 8.11(b) and (c) are the amplitude plots of bandpass and highpass filters. The filter design problem is finding linear time-invariant lumped and causal systems that have amplitude plots as close as possible to those in Figure 8.11. This design problem is outside the scope of this text. However, we must stress two important points: *stability* and *steady state*. If a system is not BIBO stable, it cannot be used as a filter, even if it has the desired amplitude characteristic as Exercise 8.5.2 illustrated. In filter design, we often consider only the steady-state response; the transient response is often disregarded. A good filter should, however, have an acceptable transient response. For example, the unit step response of a filter should not take too long to reach the steady state and should not introduce too much oscillation.

There is one serious deficiency in our discussion up to this point. The signals we encounter may not all be pure sinusoids, as in (8.14). This limitation will be removed in the next section.

Figure 8.11 *The amplitude characteristics of ideal lowpass, bandpass, and highpass filters*

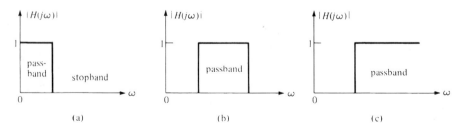

8.5.1   Response of stable systems—periodic signals

Consider an LTIL system with stable proper transfer function $H(s)$. We show that the output excited by any periodic input will approach a periodic function with the same period as $t \to \infty$. This will extend Theorem 8.4 from sinusoidal signals to periodic signals.

First we show that if $H(s)$ is BIBO stable, its steady-state response excited by $u(t) = e^{j\omega_0 t}$, for $t \geq 0$, is

$$y_{ss}(t) = \lim_{t \to \infty} y(t) = H(j\omega_0)e^{j\omega_0 t} \tag{8.17}$$

Let $A(\omega_0)$ and $\theta(\omega_0)$ be the amplitude and phase of $H(j\omega_0)$, that is,

$$H(j\omega_0) = A(\omega_0)e^{j\theta(\omega_0)}$$

Then, its steady-state response excited by $\cos \omega_0 t + j \sin \omega_0 t$ is, using Theorem 8.4,

$$y_{ss}(t) = A(\omega_0)\cos(\omega_0 t + \theta(\omega_0)) + jA(\omega_0)\sin(\omega_0 t + \theta(\omega_0))$$
$$= A(\omega_0)e^{j(\omega_0 t + \theta(\omega_0))} = A(\omega_0)e^{j\theta(\omega_0)}e^{j\omega_0 t} = H(j\omega_0)e^{j\omega_0 t}$$

This establishes (8.17). In (8.17), the input is applied for $t \geq 0$; therefore, the response $y(t)$ contains a transient response and we do *not* have

$$y(t) = H(j\omega_0)e^{j\omega_0 t} \qquad \text{for all } t \geq 0$$

For mathematical convenience, we shall now assume that the input $e^{j\omega_0 t}$ is applied from $t = -\infty$. In this case, we may assume that the transient has died out at $t = -\infty$ and the response has reached the steady state for all t. Therefore, if we apply $u(t) = e^{j\omega_0 t}$ for all t, we have

$$y(t) = H(j\omega_0)e^{j\omega_0 t} \tag{8.18}$$

for all t. This is one reason for extending positive-time signals to minus infinity. By so doing, we may disregard transient responses and consider only steady-state responses. It is important to note that $H(s)$ must be BIBO stable, otherwise (8.18) does not hold.

Now let us study the general case. Let $u(t)$ be a periodic function with period P and fundamental frequency $\omega_0 = 2\pi/P$ expressed as

$$u(t) = \sum_{k=-\infty}^{\infty} u_k e^{jk\omega_0 t} \qquad \text{for all } t \tag{8.19}$$

If $H(s)$ is BIBO stable, then its output $y(t)$ is, using the linearity property,

$$y(t) = \sum_{k=-\infty}^{\infty} u_k H(jk\omega_0) e^{jk\omega_0 t} \qquad \text{for all } t \qquad \textbf{(8.20)}$$

This is the Fourier series of the output. It is a periodic function with the same period as $u(t)$. If we use the trigonometric form

$$u(t) = a_0 + \sum_{k=1}^{\infty} (a_k \cos k\omega_0 t + b_k \sin k\omega_0 t)$$

then the output is

$$y(t) = a_0 H(0) + \sum_{k=1}^{\infty} A(k\omega_0)[a_k \cos (k\omega_0 t + \theta(k\omega_0))$$
$$+ b_k \sin (k\omega_0 t + \theta(k\omega_0))] \qquad \textbf{(8.21)}$$

Once again, (8.20) and (8.21) hold only if the periodic input is applied from $-\infty$. If the input is applied from $t = 0$, then (8.20) and (8.21) hold only after the transient dies out.

EXAMPLE 8.5.1

Consider the input signal and system in (8.14) and (8.15), which were shown in Figure 8.9. Let us now study the average power at the input and output. In this study, we will consider only the steady-state response. For the input signal

$$u(t) = \sin t + 0.1 \sin 100t \qquad \text{for all } t$$

the average power is, using (6.22),

$$P_{av}(u) = \frac{1}{2}[1 + (0.1)^2] = 0.505$$

The transfer function of the system is

$$H(s) = \frac{10}{s + 10}$$

and $H(j1) = 0.995e^{-j0.1}$, $H(j100) = 0.1e^{-j1.47}$. Thus, we have

$$y(t) = 0.995 \sin (t - 0.1) + 0.01 \sin (100t - 1.47)$$

for all t. The average power of $y(t)$ is, using (6.22),

$$P_{av}(y) = \frac{1}{2}[(0.995)^2 + (0.01)^2] = 0.495$$

The average power of y is smaller than the one of u because some of the power is dissipated in the resistor of the system. Note that $(0.995)^2/1 = 99\%$ of the power of sin t is transmitted to the output while only $(0.01/0.1)^2 = 1\%$ of the power of sin $100t$ is transmitted to the output. Thus, the network in Figure 8.9(b) would pass sin t but would stop sin $100t$; it is a lowpass filter.

EXAMPLE 8.5.2

Consider the periodic input $u(t)$ shown in Figure 6.6 with $P = 4$ and $a = 1$. The function is computed in (6.15) as

$$u(t) = 0.5 + \frac{1}{\pi}(e^{j\pi t/2} + e^{-j\pi t/2}) - \frac{1}{3\pi}(e^{3\pi t/2} + e^{-j3\pi t/2}) + \cdots$$

Now suppose we have an ideal lowpass filter with the amplitude characteristic shown in Figure 8.12(b), that is,

$$H(j\omega) = \begin{cases} 1 & \text{if } |\omega| \le \pi \\ 0 & \text{if } |\omega| > \pi \end{cases}$$

The filter will pass sinusoids with frequencies smaller than or equal to π and stop sinusoids with frequencies larger than π. Thus, the output of the filter is

$$y(t) = 0.5 + \frac{1}{\pi}(e^{j\pi t/2} + e^{-j\pi t/2})$$

as shown in Figure 8.12(c). The spectrum of $y(t)$ is simply the product of $H(j\omega)$ in Figure 8.12(b) and the spectrum of $u(t)$ in Figure 8.12(a).

The average power of $u(t)$ is

$$P_u = \frac{1}{4}\int_{-2}^{2} u^2(t)dt = \frac{1}{4}\int_{-1}^{1} dt = \frac{2}{4} = 0.5$$

The average power of $y(t)$ is, using (6.21),

$$P_y = (0.5)^2 + (\frac{1}{\pi})^2 + (\frac{1}{\pi})^2 = 0.45$$

Thus, $P_y/P_u = 0.45/0.5 = 0.9$, or 90% of the input power is transmitted to the output.

Figure 8.12   *Response of an ideal lowpass filter*

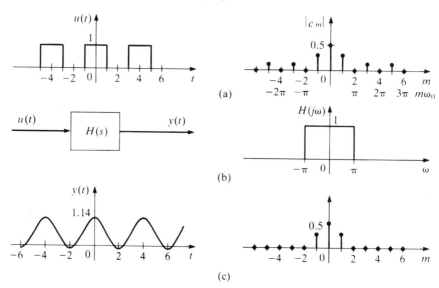

(a)

(b)

(c)

EXERCISE 8.5.3

Find the response of the signal in Figure 8.12 if $H(s)$ is replaced by an ideal bandpass filter with $H(j\omega) = 1$, for ω in $[\pi, 2\pi]$. What percentage of the power is transmitted from the input to the output?

[**ANSWERS:** $y(t) = -(1/(3\pi))(e^{3\pi t/2} + e^{-j3\pi t/2})$, 4%.]

8.6   RESPONSE OF STABLE SYSTEMS—GENERAL SIGNALS

In this section, we will study the output frequency spectrum of stable systems excited by aperiodic signals. Before proceeding, let us discuss the difference between the frequency spectrum and the frequency response. The former is defined for signals; the latter is defined for systems. The frequency spectrum of a signal is defined only if the signal is absolutely integrable or periodic. The frequency response of a system is defined only if the system is BIBO stable or, equivalently, its impulse response is absolutely integrable. Thus, these two conditions are quite similar. The frequency spectrum of $f(t)$ is the Fourier transform of $f(t)$. The frequency response of a system with transfer function $H(s)$ is $H(j\omega)$. Let $\overline{H}(\omega)$ be the Fourier transform of the impulse response $h(t)$. Note that we have used H without an overbar to denote the Laplace transform, so we will use a different notation (H with an overbar) to denote the Fourier transform. If the system is stable and causal, then $h(t)$ is absolutely integrable and positive time. Thus, we have, as discussed in (6.51),

$$\overline{H}(\omega) = \mathcal{F}[h(t)] = \mathcal{L}[h(t)]\Big|_{s=j\omega} =: H(j\omega)$$

Thus the frequency spectrum and frequency response are essentially the same. The former is defined for absolutely integrable or periodic signals; the latter is defined for stable systems.

The response of every LTI system with impulse response $h(t)$ can be described by

$$y(t) = \int_{-\infty}^{\infty} h(\tau)u(t-\tau)d\tau$$

if the input is applied from $-\infty$. Now we show that, if the system is BIBO stable, and if the input is absolutely integrable, so is $y(t)$. Indeed, we have

$$|y(t)| \le \int_{-\infty}^{\infty} |h(\tau)| \, |u(t-\tau)|d\tau$$

and

$$\int_{-\infty}^{\infty} |y(t)| \, dt \le \int_{t=-\infty}^{\infty} \int_{\tau=-\infty}^{\infty} |h(\tau)| \, |u(t-\tau)| d\tau dt$$

$$= \int_{\tau=-\infty}^{\infty} |h(\tau)| \int_{t=-\infty}^{\infty} |u(t-\tau)| dt d\tau = \int_{\tau=-\infty}^{\infty} |h(\tau)| d\tau \int_{t'=-\infty}^{\infty} |u(t')| dt'$$

where we have changed the order of integrations and the variable $t' = t - \tau$. The preceding inequality implies that if $u(t)$ is absolutely integrable, so is $y(t)$. As shown in the preceding section, if the input is periodic, so is the output if the system is BIBO stable. Thus, we conclude that, under the BIBO stability condition, if the input is Fourier transformable, so is the output. If a system is not BIBO stable, its frequency response is not defined—see (8.16)—and, in general, its output frequency spectrum is not defined, even if the input frequency spectrum is defined. Thus, in the remainder of this section we assume that every system is BIBO stable.

Let $\overline{Y}(\omega)$, $\overline{U}(\omega)$, and $\overline{H}(\omega)$ be, respectively, the Fourier transforms of $y(t)$, $u(t)$, and $h(t)$.[5] Then, as established in (6.79), we have

$$\overline{Y}(\omega) = \overline{H}(\omega)\overline{U}(\omega) \tag{8.22a}$$

---

[5]In Chapter 6, we used $Y(\omega)$ to denote the Fourier transform and $\overline{Y}(s)$ to denote the Laplace transform because the Fourier transform was the main topic there. In this chapter, we use mostly the Laplace transform. Thus, the notations are reversed here.

If $y(t)$, $h(t)$, and $u(t)$ are positive time, then the Laplace transform of the convolution is $Y(s) = H(s)U(s)$. This equation holds whether the system is stable or not and whether the input and output are Fourier transformable or not. However, if the system is stable and $u(t)$ is absolutely integrable, then we have

$$Y(j\omega) = H(j\omega)U(j\omega) \tag{8.22b}$$

which is the same as (8.22a), that is, $H(j\omega) = \overline{H}(\omega)$ is the frequency response and $U(j\omega) = \overline{U}(\omega)$ and $Y(j\omega) = \overline{Y}(\omega)$ are the frequency spectra.

Now we discuss the implication of (8.22). It states that the output frequency spectrum is simply the product of the frequency response of the system and the frequency spectrum of the input signal. For example, if the frequency spectrum of $u(t)$ and the frequency response of $H(s)$ are as shown in Figures 8.13(a) and (b), then the frequency spectrum of $y(t)$ is as shown in Figure 8.13(c). It is obtained by multiplying $\overline{U}(\omega)$ and $\overline{H}(\omega)$, point by point, for all ω. Because $\overline{H}(\omega) = 0$ for $|-\omega| > \omega_c$, so is $\overline{Y}(\omega)$.

EXAMPLE 8.6.1

Consider a signal with the frequency spectrum shown in Figure 8.14(a). If the signal is applied to the ideal lowpass filter shown in Figure 8.14(b), what percentage of its energy passes through the filter?

The frequency spectrum of the output is shown in Figure 8.14(c). It is obtained by multiplying $\overline{U}(\omega)$ and $\overline{H}(\omega)$. The total energy of the input is, using (6.75),

$$E_u = \frac{1}{2\pi} \int_{-\infty}^{\infty} |\overline{U}(\omega)|^2 d\omega = \frac{1}{2\pi} [1 \times 2 + 4 \times 2 + 1 \times 2] = \frac{12}{2\pi}$$

The total energy of the output is

$$E_y = \frac{1}{2\pi} [4 \times 2] = \frac{8}{2\pi}$$

Thus, $E_y/E_u = 8/12 = 67\%$ of the input energy passes through the ideal filter.

Figure 8.13 *Effect on the spectrum*

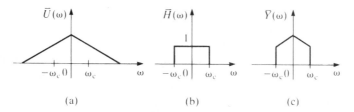

(a)   (b)   (c)

Figure 8.14 *Lowpass filtering*

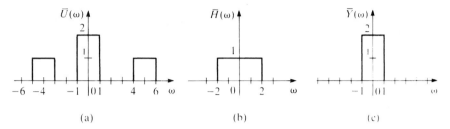

(a)   (b)   (c)

EXAMPLE 8.6.2

Consider a system with the frequency response shown in Figure 8.15(b). If inputs with the frequency spectra shown in Figure 8.15(a) are applied to the system, the output frequency spectra are as shown in Figure 8.15(c). The frequency spectrum of the first input does not overlap with the system's frequency response. Thus, the output frequency spectrum is identically zero and the input does not excite any output. The frequency spectrum of the second input overlaps with the system's frequency response. Thus, the second output has the frequency spectrum shown and its total energy is

$$\text{output energy} = \frac{1}{2\pi}[2 \times 2^2 + 2 \times 2^2] = \frac{16}{2\pi} = \frac{8}{\pi}$$

The idea in Example 8.6.2 can be used to explain the collapse of the elevated highway discussed in Example 8.2.3. We first plot the seismic fre-

Figure 8.15 *Effects on the spectra*

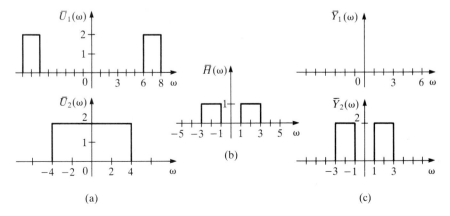

(a)   (c)

Figure 8.16 *Energy interpretation of earthquake*

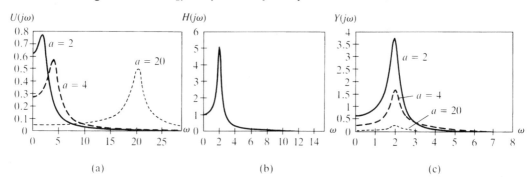

(a)   (b)   (c)

quency spectrum of $u(t) = ke^{-t} \sin at$ for the three cases in Figure 8.5(d). Because $u(t)$ is positive time and absolutely integrable, its spectrum is

$$U(s)\Big|_{s=j\omega} = \frac{ka}{(s+1)^2 + a^2}\Bigg|_{s=j\omega} = \frac{ka}{(j\omega+1)^2 + a^2}$$

Figure 8.16(a) shows the three plots for $k = 15.5$, $a = 2$, for $k = 1.2$, $a = 4$, and for $k = 1$, $a = 20$. Note that they can also be obtained using the FFT, which was discussed in Chapter 7. The transfer function of the highway is modeled as $H(s) = 4.04/(s^2 + 0.4s + 4.04)$, and its frequency response is plotted in Figure 8.16(b). The product of the frequency spectra in Figure 8.16(a) and the frequency response in Figure 8.16(b) yields the output frequency spectra in Figure 8.16(c). We see from Figure 8.16(b) that the amplitude of the frequency response is largest in the neighborhood of $\omega = 2$ rad/s. This neighborhood can be called the *structural resonance frequency range*. The seismic frequency spectrum for $a = 20$ hardly overlaps with the structural resonance frequency range of the highway. Thus, the output frequency spectrum is very small and not much vibration is excited by the seismic wave. However, most of the seismic frequency spectrum for $a = 2$ overlaps with the resonance frequency range. Thus, the output frequency spectrum is large and, consequently, the output energy will be large, which means that a large vibration will be excited and the highway will collapse.

The seismic wave is modeled as $u(t) = ke^{-t} \sin at$ in Example 8.2.3. This model may differ appreciably from actual seismic waves. Actually, there is no need to find a model for seismic waves. We can compute their spectra from actual measured data, as was discussed in Chapter 7. The frequency response of a structure can also be obtained by measurement, as will be discussed later. When designing a structure, we ensure that the structural resonance frequency range does not overlap with the seismic frequency spectrum.

EXAMPLE 8.6.3

Consider the signal $u(t) = 1 + 0.1 \sin (1000t)$ shown in Figure 8.17(a). Suppose the high-frequency sinusoid is noise to be eliminated. We consider two filters with transfer functions

$$H_1(s) = \frac{0.5}{s + 0.5} \quad \text{and} \quad H_2(s) = \frac{5}{s + 5}$$

Clearly, we have $H_1(0) = H_2(0) = 1$, $H_1(j\,1000) \approx 0$, and $H_2(j\,1000) \approx 0$. Thus, the two filters will pass the DC part without attenuation and will eliminate the noise, as shown in Figure 8.17(b). However, the time constants of $H_1(s)$ and $H_2(s)$ are, respectively, $1/0.5 = 2$ and $1/5 = 0.2$. Thus, while the first filter takes 10 seconds to reach the steady-state response, the second filter takes only 1 second and is a better filter.

It is important to reemphasize that a filter must be stable; otherwise, it cannot be used in practice and its frequency response is meaningless. In the actual design of filters, the speed of response also must be considered.

8.6.1 Bandwidth

This section will introduce the concept of bandwidth. The *bandwidth* of a BIBO system with transfer function $H(s)$ is defined as the positive frequency range in which the amplitude characteristic meets the condition

$$|H(j\omega)| \geq 0.707\,|H(j\omega)|_{\max}$$

where $|H(j\omega)|_{\max}$ denotes the maximum value of $|H(j\omega)|$. The frequency ω_c with the property $|H(j\omega_c)| = 0.707\,|H(j\omega)|_{\max}$ is called the *cutoff frequency*.

Figure 8.17 *Filtering of signals*

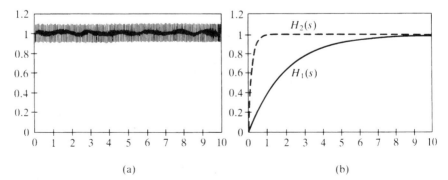

(a)   (b)

For the lowpass filters shown in Figures 8.10 and 8.12(b), because $|H(j\omega)|_{max} = |H(j0)|$, the bandwidth is the positive frequency range meeting

$$|H(j\omega)| \geq 0.707 |H(j0)|$$

and the cutoff frequency is given by

$$|H(j\omega_c)| = 0.707 |H(j0)|$$

For $H(s)$ in (8.15), the cutoff frequency is 10 rad/s, and the bandwidth is also 10 rad/s.

Now let us discuss the physical meaning of the bandwidth. If we apply the input $u(t) = e^{j\omega t}$ for all t, then the output is, as derived in (8.18), $y(t) = H(j\omega)e^{j\omega t}$, for all t. The average power of $u(t)$, using (6.21), is 1, and the average power of $y(t)$ is $|H(j\omega)|^2$. Thus, if $|H(j\omega)| \geq 0.707 |H(j\omega)|_{max}$, then

$$|H(j\omega)|^2 \geq (0.707)^2 |H(j\omega)|^2_{max} = 0.5 |H(j\omega)|^2_{max}$$

This implies that, for ω inside the bandwidth, the transferred power is at least 50% of the maximum power that can be transferred by the system. Thus, the bandwidth is called the *half-power bandwidth* and ω_c is called the *half-power point*.

The half-power point can also be defined using decibel (dB). The dB is defined as the unit of the logarithmic magnitude $20 \log_{10} |H(j\omega)| = a$, that is,

$$20 \log_{10} |H(j\omega)| = a \text{ dB} \qquad (8.23)$$

If $|H(j\omega)|_{max} = 1$, the amplitude at ω_c is

$$20 \log_{10} 0.707 = -3 \text{ dB}$$

Thus, ω_c is also called the -3 dB point. Note that, if a system is not BIBO stable, its frequency response is not defined and, consequently, its bandwidth is meaningless.

*8.6.2   Measuring Frequency Responses

Two methods for measuring frequency responses will be presented here. The frequency response of a system is defined only if the system is BIBO stable. Thus, we assume every system is BIBO stable in this section.

Method I   As discussed in (8.22), the input and output frequency spectra are related by

$$Y(j\omega) = H(j\omega)U(j\omega)$$

If we can generate an impulse as an input, then $U(j\omega) = 1$ for all ω in $(-\infty, \infty)$ and $Y(j\omega) = H(j\omega)$, as shown in Figure 8.18(a). Thus the frequency response can be obtained by computing the frequency spectrum of the output, which was discussed in Chapter 7. In practice, it is not possible to generate an impulse. Fortunately, in application, it is unnecessary to generate an impulse to measure the frequency response.

The frequency responses of most systems are band limited in the sense that $H(j\omega) \approx 0$ for ω larger than a certain number, for example, larger than W in Figure 8.18(a). If we can find an input whose frequency spectrum roughly equals 1 for $\omega < W$, as shown in Figure 8.18(b), then $Y(\omega) \approx H(\omega)$ and the output frequency spectrum roughly equals the frequency response of the system. Consider the pulse shown in Figure 8.18(c) with width a and height $1/a$. Its Laplace transform can be obtained from (4.93) as $U(s) = (1 - e^{-as})/as$. Thus, its frequency spectrum is

$$U(j\omega) = \left.\frac{(1 - e^{-as})}{as}\right|_{s=j\omega} = \frac{1 - e^{-ja\omega}}{ja\omega} = \frac{e^{-j0.5a\omega}(e^{j0.5a\omega} - e^{-j0.5a\omega})}{ja\omega}$$

$$= \frac{e^{-j0.5a\omega}2j \sin 0.5a\omega}{ja\omega} = \frac{e^{-j0.5a\omega} \sin 0.5a\omega}{0.5a\omega}$$

Figure 8.18 *Measurement of frequency responses*

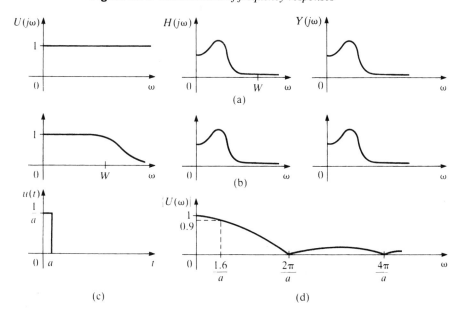

(a)

(b)

(c)   (d)

Because $|e^{-j0.5a\omega}| = 1$ for all ω, we have

$$|U(\omega)| = \left| \frac{\sin 0.5a\omega}{0.5a\omega} \right|$$

Using the sinc function in Figure 6.18, we can readily plot $|U(\omega)|$, as shown in Figure 8.18(d). The value of $|U(j\omega)|$ is 1 at $\omega = 0$ and decreases to 0.9 at $|\omega| = 1.6/a$. Thus, $U(j\omega)$ can be considered roughly constant in the frequency range $[0, 1.6/a]$. In conclusion, when the width of the pulse in Figure 8.18(c) is chosen properly, the output frequency spectrum of the system excited by the pulse yields approximately the frequency response of the system.

EXAMPLE 8.6.4

The measurement of frequency responses or amplitude characteristics is useful in manufacturing. For example, the amplitude characteristic of a loudspeaker must meet a certain specification; it must be fairly constant over a frequency range, as shown in Figure 8.19(a). If the measured amplitude characteristic is as shown in Figure 8.19(b) or (c), then the loudspeaker must be rejected. The response of the speaker can be measured using the arrangement shown in Figure 8.19(d). If the highest frequency of the frequency response is W, the width of the input pulse can be chosen as

$$\frac{1.6}{a} = W \quad \text{or} \quad a = \frac{1.6}{W}$$

Figure 8.19 *Testing of speakers*

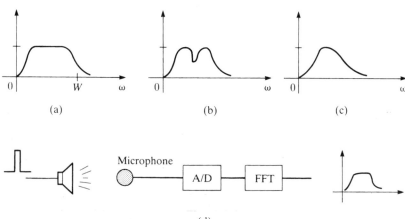

(a)   (b)   (c)

(d)

Then, the frequency spectrum of the output roughly equals the frequency response of the loudspeaker. The output is sampled using an A/D converter with sampling period a, and then the FFT is used to compute the spectrum. This yields the frequency response of the loudspeaker. This method of checking speakers is more efficient and more reliable than inspection by eye. In this measurement, what is important is the flatness of the input frequency spectrum in the range $[0, W]$, which is determined by the width of the pulse; the magnitude of the frequency spectrum is not important. Thus, the height of the pulse need not be exactly $1/a$; it can be smaller as long as the pulse generates enough power to drive the speaker.

Method 2 The second method uses Theorem 8.4 to measure the frequency response. By sweeping ω_0 in $u(t) = \sin \omega_0 t$ over a range such as $\omega_0 = k\Delta\omega$, $k = 1, 2, 3, \ldots$, $|H(jk\Delta\omega)|$ and $\sphericalangle H(jk\Delta\omega)$ can be obtained from the steady-state response of $u(t)$. Commercial devices such as a network analyzer, spectrum analyzer, or Hewlett Packard's HP3563A Control system analyzer are available to carry out these measurements. This method is more accurate than the first method, because in the first method the input frequency spectrum is not exactly constant over the frequency range of $H(j\omega)$. This second method, however, takes much longer to complete. For each ω, we must wait until the response reaches the steady state. If we measure 20 points of $H(j\omega)$, then we need 20 runs. The first method requires only one run.

The frequency response can be plotted as shown in Figure 8.10 using linear scales. It is possible to use different scales to plot the same plot. For example, the plot in Figure 8.10 can be plotted as shown in Figure 8.20, which is called the Bode plot. The frequency in the Bode plot is in logarithmic scale. Thus, $\omega = 1$ becomes $\log_{10} 1 = 0$ and $\omega = 0$ becomes $-\infty$. The amplitude is expressed in dB (decibels), as defined in (8.23), and the phase is expressed in degrees. The advantage of using the Bode plot is that the amplitude plot can be approximated by two straight lines, as shown in Figure 8.20. Furthermore, the intersection of the two straight lines, called the corner frequency, yields the pole location. This topic is discussed extensively in every control text—see, for example, Reference [8]—and will not be discussed further here.

*8.7 DISCRETE-TIME CASE

In this section, we will study the BIBO stability of LTI discrete-time systems. Most of the concepts developed for continuous-time systems are directly applicable here; therefore, the discussion will be brief. The zero-state response of every LTI discrete-time system can be described by

$$y[k] = \sum_{i=0}^{k} h[k-i]u[i] = \sum_{i=0}^{k} h[i]u[k-i] \tag{8.24}$$

Figure 8.20   *The Bode plot of Figure 8.10*

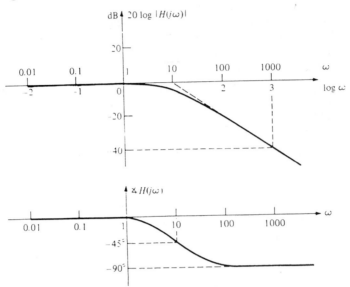

where u, y, and h are, respectively, the input, output, and impulse response of the system. If the system is also lumped, then it can be described by

$$Y(z) = \frac{N(z)}{D(z)} U(z) =: H(z)U(z) \tag{8.25}$$

with

$$H(z) = \frac{N(z)}{D(z)} = \frac{b_m z^m + b_{m-1} z^{m-1} + \cdots + b_0}{z^n + a_{n-1} z^{n-1} + \cdots + a_1 z + a_0} \tag{8.26}$$

where $Y(z)$, $U(z)$, and $H(z)$ are, respectively, the z-transforms of $y[k]$, $u[k]$, and $h[k]$. Because we study only causal systems in this text, we assume $H(z)$ to be proper. As in the continuous-time case, we assume that $N(z)$ and $D(z)$ have no common factors.

A sequence $u[k]$ is bounded if it does not go to infinity or, equivalently, there exists a constant M_1 such that

$$|u[k]| \leq M_1 < \infty$$

for all $k \geq 0$. An LTI discrete-time system is BIBO stable if, and only if, every bounded input sequence excites a bounded output sequence.

Theorem 8.6 A system with impulse response $h[k]$, $k = 0, 1, 2, \ldots$, is BIBO stable if and only if $h[k]$ is absolutely summable, that is,

$$\sum_{k=0}^{\infty} |h[k]| \leq M < \infty \tag{8.27}$$

PROOF The proof is identical to the proof of Theorem 8.1, except that the integration is replaced by a summation. Indeed, if $u[k]$ is bounded, that is, $|u[k]| \leq M_1$ for $k \geq 0$, then

$$|y[k]| = \left| \sum_{i=0}^{k} h[i]u[k-i] \right| \leq \sum_{i=0}^{k} |h[i]| \, |u[k-i]| \leq M_1 \sum_{i=0}^{k} |h[i]| \leq M_1 M$$

This shows the sufficiency. Conversely, if

$$\sum_{i=0}^{k'} |h[i]| = \infty$$

for some k', then for the bounded sequence

$$u[k'-i] = \begin{cases} 1 & \text{if } h[i] \geq 0 \\ -1 & \text{if } h[i] < 0 \end{cases}$$

the output is

$$y[k'] = \sum_{i=0}^{k'} h[i]u[k'-i] = \sum_{i=0}^{k'} |h[i]| = \infty$$

which is not bounded. This shows the necessity of the theorem.

EXAMPLE 8.7.1

Consider an LTI discrete-time system with impulse response $h[k] = 1/k$, for $k = 1, 2, \ldots$, and $h(0) = 0$. We compute

$$S := \sum_{k=0}^{\infty} |h[k]| = \sum_{k=1}^{\infty} \frac{1}{k} = 1 + \frac{1}{2} + \frac{1}{3} + \frac{1}{4} + \frac{1}{5} + \frac{1}{6} + \frac{1}{7} + \cdots$$

$$= (1) + \left(\frac{1}{2}\right) + \left(\frac{1}{3} + \frac{1}{4}\right) + \left(\frac{1}{5} + \frac{1}{6} + \frac{1}{7} + \frac{1}{8}\right) + \left(\frac{1}{9} + \cdots + \frac{1}{16}\right) + \cdots$$

We see that the term in every pair of parentheses is larger than 1/2. Thus, we have

$$S \geq \frac{1}{2} + \frac{1}{2} + \frac{1}{2} + \cdots = \infty$$

and the system is not BIBO stable. Note that the condition $|h[k]| \to 0$, as $k \to \infty$, does not imply the BIBO stability of systems.

EXERCISE 8.7.1

Show that the LTI system with impulse response $h[k] = 1/k$, for $k = 1, 2, \ldots,$ 1000, and $h[k] = 0$, for $k = 0$ and $k > 1000$, is BIBO stable. ■

If an LTI discrete-time system is distributed, its transfer function $H(z)$ is not a rational function of z. For example, the transfer function of the system in the preceding example is $H(z) = -\ln(1 + z^{-1})$. In this case, the stability condition on $H(z)$ is not transparent and so would rarely be used. If an LTI system is lumped and causal, then its transfer function (the z-transform of $h[k]$, $k = 0, 1, 2, \ldots$) is a proper rational function of z. In this case, the stability condition on $H(z)$ is very simple and is widely used.

A transfer function can be written as $H(z)$ or $H(z^{-1})$. The stability conditions for both forms are different. Therefore, we must specify which form we are using. In this chapter, we use mainly $H(z)$.

Theorem 8.7 An LTIL discrete-time system with proper transfer function $H(z)$ is BIBO stable if and only if every pole of $H(z)$ has a magnitude less than 1 or, equivalently, lies inside the unit circle on the z-plane.

The proof of this theorem is similar to the proof of Theorem 8.2. This theorem also can be deduced from Theorem 8.2 using the transformation discussed in Section 5.2.2. The open left–s-plane is transformed by $z = e^{sT}$ into the interior of the unit circle on the z-plane. Thus, Theorem 8.7 follows from Theorem 8.2.

EXAMPLE 8.7.2

Consider the positive feedback system shown in Figure 8.21(a). Its impulse response is computed in (3.4) as

$$h(0) = 0, \qquad h[k] = a^k \qquad \text{for } k = 1, 2, \ldots$$

Because

$$\sum_{0}^{\infty} |h[k]| = \sum_{1}^{\infty} |a|^k = \begin{cases} \infty & \text{if } |a| \geq 1 \\ \dfrac{|a|}{1 - |a|} & \text{if } |a| < 1 \end{cases}$$

Figure 8.21 (a) Positive feedback system and (b) negative feedback system

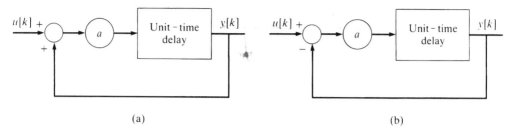

(a)   (b)

Theorem 8.6 implies that the positive feedback system is BIBO stable if and only if $|a| < 1$.

The positive feedback system can also be described by the difference equation

$$y[k + 1] - ay[k] = au[k]$$

which was discussed in Exercise 3.4.1. Thus, its transfer function is

$$H(z) = \frac{Y(z)}{U(z)} = \frac{a}{z - a}$$

It has one pole at a. Thus Theorem 8.7 implies that the system is BIBO stable if and only if $|a| < 1$. This is the same condition that was obtained by using Theorem 8.6.

EXERCISE 8.7.2

Use Theorems 8.6 and 8.7 to show that the negative feedback system in Figure 8.21(b) is BIBO stable if and only if $|a| < 1$. ■

EXAMPLE 8.7.3

Consider a system with discrete-time transfer function

$$H(z) = \frac{(z + 10)(z - 5)}{(z + 0.9)(z - 0.99)(z - 0.7 + j0.6)(z - 0.7 - j0.6)}$$

The two poles -0.9 and 0.99 have magnitudes smaller than 1. The magnitudes of the complex conjugate poles $0.7 \pm j0.6$ are both $\sqrt{0.49 + 0.36} = 0.92$. Thus, the system is BIBO stable.

EXAMPLE **8.7.4**

Consider an FIR filter of length N with impulse response $h[k]$, $k = 0, 1,$ $\ldots, N-1$ and $h[k] = 0$ for $k \geq N$. The filter is BIBO stable because

$$\sum_{k=0}^{\infty} |h[k]| = \sum_{k=0}^{N-1} |h[k]| < \infty$$

The same conclusion can also be reached from its transfer function:

FIR filter transfer function

$$H(z) = h(0) + h(1)z^{-1} + \cdots + h(N-1)z^{-N+1}$$
$$= \frac{h(0)z^{N-1} + h(1)z^{N-2} + \cdots + h(N-1)}{z^{N-1}}$$

All its poles $z = 0$ are located at the center of the unit circle. Thus, every FIR filter is BIBO stable.

EXERCISE 8.7.3

Are the following systems BIBO stable?

(a) $\dfrac{(z+1)}{(z+0.7+j0.8)(z+0.7-j0.8)}$

(b) $\dfrac{(z-2)}{(z-2)(z+0.2)z^4}$

(c) $\dfrac{(z-2)}{z^2(z-0.95)}$

[**ANSWERS:** (a) No, (b) yes, (c) yes.]

---

*8.8   **THE JURY TEST**

A.5

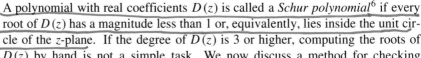

A polynomial with real coefficients $D(z)$ is called a *Schur polynomial*[6] if every root of $D(z)$ has a magnitude less than 1 or, equivalently, lies inside the unit circle of the z-plane. If the degree of $D(z)$ is 3 or higher, computing the roots of $D(z)$ by hand is not a simple task. We now discuss a method for checking whether or not a polynomial is Schur without solving for its roots. The method is called the *Jury test*.[7]

---

[6] In honor of Issai Schur (1875–1941), a German mathematician.

[7] Eliahu I. Jury (b. 1923), Sc.D. in electrical engineering from Columbia University; Professor Emeritus, University of California, Berkeley. Jury considers himself to be a lifetime graduate student. This author took a course from him at Berkeley.

Necessary conditions for a polynomial to be Schur.[8] We need the coefficients of the binomial expansion for $n = 1, 2, 3, \ldots$ listed in Table 8.2, for the following discussion. Consider the polynomial of degree 4

$$D(z) = z^4 + a_3 z^3 + a_2 z^2 + a_1 z + a_0 \qquad (8.28)$$

where the leading coefficient has been normalized to 1. Comparing the coefficients of (8.28) and those of $(z + 1)^4$, if any of the conditions

$$|a_3| < 4, \qquad |a_2| < 6, \qquad |a_1| < 4, \qquad |a_0| < 1 \qquad (8.29)$$

is not met, then the $D(z)$ is not Schur. More generally, a polynomial with leading coefficient 1 is not a Schur polynomial if the absolute value of any coefficient, except the leading coefficient, is equal to or larger than the corresponding coefficient in the binomial expansion of the same degree. To establish this assertion for (8.28), we factor $D(z)$ as

$$D(z) = (z + p_1)(z + p_2)(z + p_3)(z + p_4)$$

where p_i are the roots of $D(z)$. Then, we have

$$a_0 = p_1 p_2 p_3 p_4 \qquad (1 \text{ term})$$
$$a_1 = p_1 p_2 p_3 + p_1 p_2 p_4 + p_2 p_3 p_4 + p_1 p_3 p_4 \qquad (4 \text{ terms})$$
$$a_2 = p_1 p_2 + p_1 p_3 + p_1 p_4 + p_2 p_3 + p_2 p_4 + p_3 p_4 \qquad (6 \text{ terms})$$
$$a_3 = p_1 + p_2 + p_3 + p_4 \qquad (4 \text{ terms})$$

Table 8.2 *Coefficients of binomial expansions*

$(z + 1)$				1	1		
$(z + 1)^2$			1	2	1		
$(z + 1)^3$		1	3	3	1		
$(z + 1)^4$	1	4	6	4	1		
$(z + 1)^5$	1	5	10	10	5	1	

[8]It is suggested that the reader omit this part during the first reading and go directly to the part on necessary and sufficient conditions.

If $|p_i| < 1$, for $i = 1, 2, 3$, and 4, then the absolute value of every term in a_i is smaller than 1. Thus, we have the conditions in (8.29). The general case can be similarly established. This set of necessary conditions and some other tighter conditions can be found in Reference [23].

EXAMPLE 8.8.1

Consider

$$D(z) = 2z^3 + 7z^2 - 2z - 1.8$$

After normalizing the leading coefficient to 1, the second coefficient $7/2 = 3.5$ is larger than 3. Thus, it is not a Schur polynomial.

EXERCISE 8.8.1

Are the following polynomials Schur?

(a) $D_1(z) = z^5 - 6z^4 - 8z^3 + 4z^2 - 3z - 0.8$

(b) $D_2(z) = z^3 - 2z^2 + 2.5z - 0.8$

[**ANSWERS:** (a) No, (b) maybe.]   ■

Different necessary conditions for $D(z)$ of degree n and with a positive leading coefficient to be Schur are

$$D(1) > 0 \quad \text{and} \quad (-1)^n D(-1) > 0 \qquad \textbf{(8.30)}$$

For example, the polynomial

$$D(z) = z^3 + 2z^2 + z + 0.5$$

has $D(1) = 4.5$ and $(-1)^3 D(-1) = -0.5 < 0$, which violate (8.30). Thus, the polynomial is not Schur.

Necessary and sufficient conditions   For convenience of presentation, we will use a polynomial of degree 5 to develop a necessary and sufficient condition for the polynomial to be Schur. The general case can be similarly developed. Consider

$$D(z) = a_0 z^5 + a_1 z^4 + a_2 z^3 + a_3 z^2 + a_4 z + a_5 \qquad \text{with } a_0 > 0$$

$\ast$ $\ast$

Table 8.3 *The Jury table*

a_0	a_1	a_2	a_3	a_4	a_5	
a_5	a_4	a_3	a_2	a_1	a_0	$k_5 = a_5/a_0$
b_0	b_1	b_2	b_3	b_4	0	(1st a_i row) $- k_5$(2nd a_i row)
b_4	b_3	b_2	b_1	b_0		$k_4 = b_4/b_0$
c_0	c_1	c_2	c_3	0		(1st b_i row) $- k_4$(2nd b_i row)
c_3	c_2	c_1	c_0			$k_3 = c_3/c_0$
d_0	d_1	d_2	0			(1st c_i row) $- k_3$(2nd c_i row)
d_2	d_1	d_0				$k_2 = d_2/d_0$
e_0	e_1	0				(1st d_i row) $- k_2$(2nd d_i row)
e_1	e_0					$k_1 = e_1/e_0$
f_0	0					(1st e_i row) $- k_1$(2nd e_i row)

and real a_i. We form Table 8.3[9]. The first row is simply the coefficients of $D(z)$ arranged in the descending power of z. The second row is the reversal of the first row. We compute $k_5 = a_5/a_0$, or the ratio of the last elements of the first two rows. The first row subtracting the product of the second row and k_5 yields the first b_i row, as shown. Note that the last element must be zero and will be discarded in subsequent discussion. We then reverse the order of b_i and compute $k_4 = b_4/b_0$. The first b_i row subtracting the product of the second b_i row and k_4 yields the first c_i row. Proceeding in exactly the same manner, the table is completed as shown.

Theorem 8.8 (**The Jury test**) Every root of a polynomial of degree 5 with a positive leading coefficient has a magnitude less than 1 or, equivalently, lies inside the unit circle on the z-plane if, and only if, the five leading coefficients $\{b_0, c_0, d_0, e_0, f_0\}$ in Table 8.3 are all positive.

$\ast$

If $D(s)$ has degree n and with a positive leading coefficient, then $D(z)$ is Schur if and only if the subsequent n leading coefficients are all positive. If a zero or a negative leading coefficient appears in the table, we may stop the computation and conclude that the polynomial is not Schur. A proof of Theorem 8.8 can be found in Reference [4].

[9]This table was first presented by Jury in Reference [17], and so it is named after him.

EXAMPLE 8.8.2

Consider

$$D(z) = z^3 - 2z^2 + 2.5z - 0.8$$

We form

1	−2	2.5	−0.8
−0.8	2.5	−2	1

$$k_3 = \frac{-0.8}{1}$$

0.36	0	0.9	0
0.9	0	0.36	

$$k_2 = \frac{0.9}{0.36} = 2.5$$

−1.89

A negative leading coefficient appears. Thus, the polynomial is not a Schur polynomial.

EXAMPLE 8.8.3

Consider

$$D(z) = 2z^3 - 0.2z^2 - 0.24z - 0.08$$

We form

2	−0.2	−0.24	−0.08
−0.08	−0.24	−0.2	2

$$k_3 = \frac{-0.08}{2} = -0.04$$

1.9968	−0.2096	−0.248	0
−0.248	−0.2096	1.9968	

$$k_2 = \frac{-0.248}{1.9968} = -0.124$$

1.966	−0.236	0
−0.236	1.966	

$$k_1 = \frac{-0.236}{1.966} = -0.12$$

1.938	0

The three leading coefficients are all positive. Thus, $D(z)$ is a Schur polynomial.

EXERCISE 8.8.2

Check whether or not the following polynomials are Schur.

(a) $D_1(z) = 2z^3 - 0.2z^2 - 0.24z$

(b) $D_2(z) = 6z^4 + 4z^3 - 1.5z^2 - z + 0.1$

[**ANSWERS:** (a) Yes, (b) yes.]

*8.9   STEADY-STATE RESPONSE OF STABLE DISCRETE-TIME SYSTEMS

This section is the counterpart of Section 8.4 for the discrete-time case, and the discussion will be brief.

Theorem 8.9 Consider an LTIL discrete-time system with proper transfer function $H(z)$. If $H(z)$ is BIBO stable and if we apply the input $u[k] = \sin k\omega_0 T$, $k = 0, 1, 2, \ldots$, then the output $y[k]$ will approach

$$y_{ss}[k] := \lim_{k \to \infty} y[k] = A(\omega_0) \sin(k\omega_0 T + \theta(\omega_0)) \qquad (8.31)$$

where

$$A(\omega_0) := |H(e^{j\omega_0 T})| = [(Re\ H(e^{j\omega_0 T}))^2 + (Im\ H(e^{j\omega_0 T}))^2]^{1/2} \qquad (8.32a)$$

and

$$\theta(\omega_0) := \sphericalangle H(e^{j\omega_0 T}) = \tan^{-1} \frac{Im\ H(e^{j\omega_0 T})}{Re\ H(e^{j\omega_0 T})} \qquad (8.32b)$$

The response $y_{ss}[k]$ is called the *steady-state response*, and $H(e^{j\omega T})$ is called the *frequency response*.

PROOF The z-transform of

$$u[k] = \sin k\omega_0 T = \frac{1}{2j}(e^{jk\omega_0 T} - e^{-jk\omega_0 T})$$

is

$$U(z) = \frac{1}{2j}\left[\frac{z}{z - e^{j\omega_0 T}} - \frac{z}{z - e^{-j\omega_0 T}}\right]$$

The output is

$$Y(z) = H(z)U(z)$$

We expand $Y(z)/z$ by partial fraction expansion as

$$\frac{Y(z)}{z} = \frac{1}{2j} \left[\frac{1}{z - e^{j\omega_0 T}} - \frac{1}{z - e^{-j\omega_0 T}} \right] H(z)$$

$$= \frac{1}{2j} \left[\frac{H(e^{j\omega_0 T})}{z - e^{j\omega_0 T}} - \frac{H(e^{-j\omega_0 T})}{z - e^{-j\omega_0 T}} \right] + \text{(terms due to the poles of } H(z))$$

Thus, we have

$$Y(z) = \frac{1}{2j} \left[H(e^{j\omega_0 T}) \cdot \frac{z}{z - e^{j\omega_0 t}} - H(e^{-j\omega_0 T}) \frac{z}{z - e^{-j\omega_0 T}} \right]$$

$$+ z \text{ (terms due to the poles of } H(z)) \tag{8.33}$$

Because all poles of $H(z)$ lie inside the unit circle on the z-plane, their time responses all approach zero as $k \to \infty$, as shown in Figure 5.6. Thus, we have

$$Y_{ss}(z) = \frac{1}{2j} \left[H(e^{j\omega_0 T}) \frac{z}{z - e^{j\omega_0 T}} - H(e^{-j\omega_0 T}) \frac{z}{z - e^{-j\omega_0 T}} \right]$$

or

$$y_{ss}[k] = \frac{1}{2j} [H(e^{j\omega_0 T}) e^{jk\omega_0 T} - H(e^{-j\omega_0 T}) e^{-jk\omega_0 T}] \tag{8.34}$$

Let

$$H(e^{j\omega_0 T}) = A(\omega_0) e^{j\theta(\omega_0)}$$

where $A(\omega_0)$ and $\theta(\omega_0)$ are as defined in (8.32). Because all coefficients of $H(z)$ are assumed to be real, we have

$$H(e^{-j\omega_0 T}) = A(\omega_0) e^{-j\theta(\omega_0)}$$

(see Problem 8.29). Thus, (8.34) can be written as

$$y_{ss}[k] = \frac{1}{2j} A(\omega_0) [e^{j(k\omega_0 T + \theta(\omega_0))} - e^{-j(k\omega_0 T + \theta(\omega_0))}]$$

$$= A(\omega_0) \sin(k\omega_0 T + \theta(\omega_0))$$

This establishes the theorem.

EXAMPLE **8.9.1**

Consider a system with transfer function $H(z) = z/(z - 0.5)$. Let us find the steady-state response of the system excited by $u[k] = 3 \sin(2k - 0.1)$, for $k = 0$, 1, 2, ... The system is BIBO stable; therefore, Theorem 8.9 can be applied. We compute

$$H(e^{j\omega_0 T}) = H(e^{j2}) = \frac{e^{j2}}{e^{j2} - 0.5} = \frac{e^{j115°}}{1.29e^{j135°}} = 0.77e^{-j20°} = 0.77e^{-j0.35}$$

Thus, the steady-state response of the system is

$$y_{ss}[k] = 0.77 \times 3 \sin(2k - 0.1 - 0.35) = 2.31 \sin(2k - 0.45)$$

EXERCISE 8.9.1

Consider

$$H(z) = \frac{z - 2}{z + 0.5}$$

Compute its steady-state response excited by $u[k] = \sin 2kT$, $k = 0, 1, 2, \ldots$. The sampling period T is assumed to be 0.1 second.

0.1

[**ANSWER:** $y_{ss}[k] = 0.68 \sin(0.2k + 2.8)$.] ■

EXERCISE 8.9.2

Let $H(z)$ be BIBO stable. Use the final-value theorem to show that the steady-state response of the system excited by $u[k] = 1$, for $k = 0, 1, 2, \ldots$, is

$$y_{ss}[k] = \lim_{k \to \infty} y[k] = H(1) \tag{8.35}$$

■

In the proof of Theorem 8.9, we computed only the zero-state response. If the system is completely characterized by its transfer function, then the theorem still holds, even if initial conditions are nonzero. As in the continuous-time case, the response of an LTIL discrete-time system can be decomposed into the steady-state and transient responses. The steady-state response is excited exclusively by the applied input. The transient response is excited by nonzero initial conditions and/or an input. In other words, initial conditions will excite only transient responses but an input will excite both transient and steady-state responses. In the continuous-time case, the speed at which the response reaches steady state is determined mainly by the time constant or, equivalently, the smallest distance between the poles and the imaginary axis. The greater the distance, the faster the response reaches steady state. In the discrete-time case, the speed at which the response reaches steady state is also determined mainly by pole location. The closer the poles of $H(z)$ to the origin of the z-plane, the faster the total response reaches steady state. There is, however, one phenomenon in the discrete-time case that has no counterpart in the continuous-time case. If all poles of $H(z)$ are located at the origin of the z-plane, then the response excited by a step or sinusoidal sequence will reach steady state in a finite number of sampling periods. This is called the *deadbeat response*.

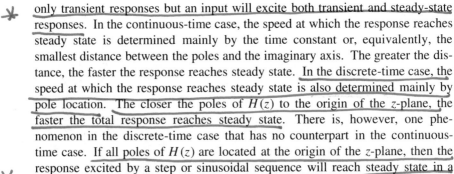

Theorem 8.10 Consider an LTIL discrete-time system with proper transfer function $H(z)$. If $H(z)$ is of the form $(b_n z^n + b_{n-1} z^{n-1} + \cdots + b_0)/z^n$ and if we apply the input $u[k] = 1$ $\{u[k] = \sin k\omega_0 T\}$, $k = 0, 1, 2, \ldots$, then we have, for $k \geq n$,

$$y[k] = H(1) \qquad \{y[k] = A(\omega_0) \sin(k\omega_0 T + \theta(\omega_0))\} \qquad (8.36)$$

where $A(\omega_0)$ and $\theta(\omega_0)$ are as defined in (8.32). This is called the deadbeat response.

In (8.31) and (8.35), the output will exactly equal the steady-state response only at $k = \infty$. If all poles of a transfer function are located at $z = 0$, then the output will equal the steady-state response after n sampling periods.

We will prove only the sinusoidal part of Theorem 8.10. The proof of Theorem 8.9 up to Equation (8.33) is directly applicable. In the present case, the last term of (8.33) is of the form

$$z\,(\text{terms due to the poles of } H(z))$$
$$= z \cdot \frac{c_{n-1} z^{n-1} + c_{n-2} z^{n-2} + \ldots + c_1 z + c_0}{z^n} \qquad (8.37)$$

where c_i depends on the initial conditions and $u[k]$. The inverse z-transform of (8.37) is identically zero for $k \geq n$ or the transient response will become identically zero for $k \geq n$. Thus, the response equals the steady-state response after n sampling periods. This establishes the theorem. This theorem has a very important application in the design of discrete-time control systems. If all poles of a system are designed to locate at the origin of the z-plane, then the transient will die out in a finite number of sampling periods and the design is called the deadbeat design.

*8.9.1   Amplitude and phase characteristics of $H(e^{j\omega T})$

If an LTIL discrete-time system with transfer function $H(z)$ is BIBO stable, then its steady-state response excited by sinusoids can be read out directly from $H(e^{j\omega T})$. Thus, $H(e^{j\omega T})$ is called the *frequency response* of the discrete-time system. The frequency response of a continuous-time system is defined on the imaginary axis $s = j\omega$ of the s-plane; whereas, the frequency response of a discrete-time system is defined on the unit circle $z = e^{j\omega T}$ of the z-plane, as shown in Figure 8.22. It is clear that $e^{j\omega T}$ is a periodic function of ω with period $2\pi/T$, that is,

$$e^{j\omega T} = e^{j(\omega + \frac{2\pi}{T})T}$$

for all ω. Thus, $H(e^{j\omega T})$ is also a periodic function of ω with period $2\pi/T$. For example, if

$$H(z) = \frac{z}{z - 0.9}$$

then

$$H(e^{j\omega T}) = \frac{e^{j\omega T}}{e^{j\omega T} - 0.9} = \frac{\cos \omega T + j \sin \omega T}{\cos \omega T - 0.9 + j \sin \omega T}$$

This is a complex-valued function of ω. For example, if $\omega T = \pi/4$, then

$$H(e^{j\pi/4}) = \frac{0.7 + j0.7}{-0.2 + j0.7} = \frac{1 \cdot e^{j45°}}{0.73 e^{j105.9°}} = 1.37 e^{-j60.9°}$$

If we compute more points, then the amplitude and phase of $H(e^{j\omega T})$ can be plotted as shown in Figures 8.23(a) and (b). These are called, respectively, the *amplitude* and *phase characteristics* of $H(z)$. They are plotted with respect to

Figure 8.22 *The unit circle on the z-plane*

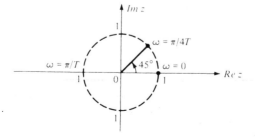

Figure 8.23   *Frequency response of a discrete-time system*

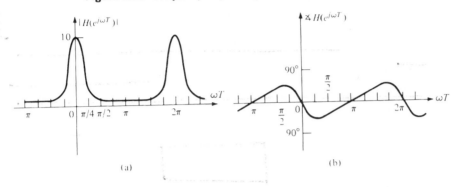

(a)   (b)

ωT; thus, the period is 2π. As was discussed in Chapter 7, the frequency range of discrete-time signals is $(-\pi/T, \pi/T]$. Thus, we need to plot the frequency response only in $(-\pi/T, \pi/T]$.

As in the continuous-time case, if all coefficients of $H(z)$ are real, then

$$H(e^{-j\omega T}) = H^*(e^{j\omega T}) \tag{8.38a}$$

$$|H(e^{-j\omega T})| = |H(e^{j\omega T})|$$

and

$$\measuredangle H(e^{-j\omega T}) = -\measuredangle H(e^{j\omega T}) \tag{8.38b}$$

That is, the amplitude characteristic is an even function of ω, and the phase characteristic is an odd function of ω. Because of (8.38), the frequency response of $H(z)$ is often plotted only from 0 to π/T, as in Figure 8.24, where the amplitude characteristics of ideal lowpass, bandpass, and highpass discrete-time filters are shown.

How to design discrete-time LTIL causal filters is discussed in texts on digital signal processing and will not be discussed here. We must again men-

Figure 8.24   *Amplitude characteristics of ideal lowpass, bandpass, and highpass filters*

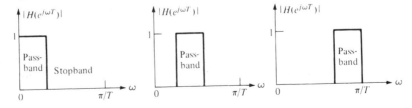

tion, however, two important points in filter design: *stability* and *steady state*. If a system is not BIBO stable, it cannot be used as a filter, even if it has the desired amplitude and phase characteristics. When designing filters, we consider mainly the steady-state response; the transient response is often disregarded. A good filter, however, should have a desired frequency response as well as an acceptable transient response. This is identical to the continuous-time situation.

EXERCISE 8.9.3

Plot the frequency response of the FIR system $H(z) = 1/z^2$. ■

*8.9.2   ## Response of stable discrete-time systems—general signals

The discussion in the preceding section applies only to sinusoidal sequences. We can extend it first to two-sided periodic signals, then to positive-time periodic signals, and finally to general signals. The process of extension is similar to that in the continuous-time case and will not be repeated. Instead, we will state only the results of the extension.

Consider a linear time-invariant lumped discrete-time system with the discrete transfer function $H(z)$. Let $u[k] = u(kT)$ and $y[k] = y(kT)$ be its input and output, and let $U(z)$ and $Y(z)$ be their z-transforms. Then, the zero-state response of the system can be described by

$$Y(z) = H(z)U(z) \tag{8.39}$$

This is a general equation, applicable regardless of whether $H(z)$ is BIBO stable or not and whether the frequency spectrum of $u[k]$ is defined or not. If $H(z)$ is BIBO stable, then its frequency response $H(e^{j\omega T})$ is defined. If $u[k]$ is discrete-time Fourier transformable, then its frequency spectrum $\bar{U}(\omega) := U(e^{j\omega T})$ is defined. Under these conditions, (8.39) can be written as

$$Y(e^{j\omega T}) = H(e^{j\omega T})U(e^{j\omega T}) \tag{8.40}$$

Thus, the frequency spectrum of the output is simply the product of the frequency response of the system and the frequency spectrum of the input. This extends Theorem 8.9 from sinusoidal signals to general signals and is the counterpart of $Y(j\omega) = H(j\omega)U(j\omega)$ in the continuous-time case. For example, consider the system shown in Figure 8.25. The system is an ideal lowpass filter. If the frequency spectrum of an input sequence is as shown, then its low frequency part will be amplified by 2 and its high frequency part will be completely eliminated at the output. This situation is similar to those in Figures 8.13 and 8.14.

Figure 8.25 *Discrete-time ideal lowpass filter*

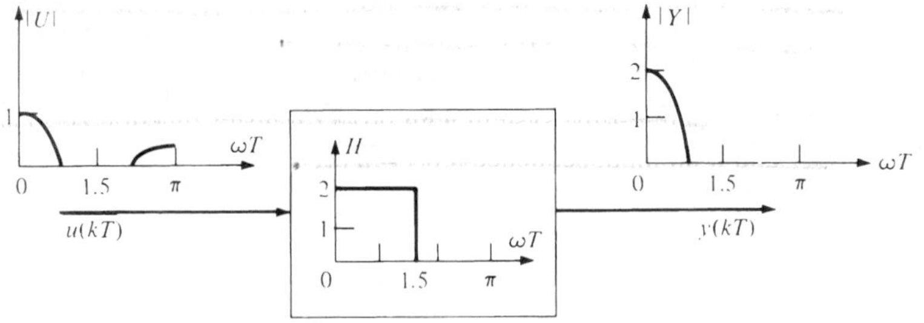

8.10   SUMMARY

1. A continuous-time or discrete-time system is BIBO stable if, for every bounded input, the zero-state response is bounded. An LTI distributed or lumped system is BIBO stable if, and only if, its impulse response is absolutely integrable or summable. The impulse response approaching zero does not imply the BIBO stability of the system (see Example 8.7.1).

2. An LTIL continuous-time system is BIBO stable if, and only if, every pole of its proper transfer function has a negative real part or, equivalently, lies inside the open left half−s-plane. An LTIL discrete-time system is BIBO stable if, and only if, every pole of its proper transfer function has a magnitude less than 1 or, equivalently, lies inside the unit circle on the z-plane.

3. A polynomial is called Hurwitz if all its roots lie inside the open left half−s-plane. A polynomial is called Schur if all its roots lie inside the unit circle of the z-plane. The Routh test can be used to determine whether or not a polynomial is Hurwitz; the Jury test can be used to determine whether or not a polynomial is Schur.

4. If a system is not BIBO stable, then the system will generally burn out or disintegrate. If a system with proper transfer function $H(s)$ $\{H(z)\}$ is BIBO stable, and if $u(t) = \sin \omega_0 t$ $\{u[k] = \sin \omega_0 kT\}$, then

$$y_{ss}(t) = \lim_{t \to \infty} y(t) = |H(j\omega_0)| \sin(\omega_0 t + \measuredangle H(j\omega_0)) \qquad \text{(8.41a)}$$

$$\{y_{ss}[k] = \lim_{k \to \infty} y[k] = |H(e^{j\omega_0 T})| \sin(\omega_0 kT + \measuredangle H(e^{j\omega_0 T}))\} \qquad \text{(8.41b)}$$

Thus, $H(j\omega)$ $\{H(e^{j\omega T})\}$ is called the frequency response of the system. If a system is not BIBO stable, the frequency response is not defined.

5. Equations (8.41a) and (8.41b) are fundamental in sinusoidal analysis and filter design. Stability and steady state are the two essential conditions for the validity of the equations.

6. If a continuous-time system with proper transfer function $H(s)$ is BIBO stable, and if an input $u(t)$ is Fourier transformable with frequency spectrum $U(j\omega)$, then the frequency spectrum $Y(j\omega)$ of the output is given by

$$Y(j\omega) = H(j\omega)U(j\omega)$$

If $H(s)$ is not BIBO stable, the output frequency spectrum may not be defined.

7. If a discrete-time system with proper transfer function $H(z)$ is BIBO stable, and if an input sequence $u[k] = u(kT)$ is discrete-time Fourier transformable with frequency spectrum $U(e^{j\omega T})$, then the frequency spectrum $Y(e^{j\omega T})$ of the output sequence is given by

$$Y(e^{j\omega T}) = H(e^{j\omega T})U(e^{j\omega T})$$

If $H(z)$ is not BIBO stable, the output frequency spectrum may not be defined.

PROBLEMS

8.1 Consider a system with transfer function $1/s(s + 1)$. Show that the zero-state response of the system excited by $u(t) = \sin at$, for $t \geq 0$ and for any real a, is bounded. Is the system BIBO stable?

8.2 Determine the BIBO stability of the systems with the following impulse responses.

(a) $h(t) = \begin{cases} \sin t & \text{for } 0 \leq t \leq 100 \\ 0 & \text{otherwise} \end{cases}$

(b) $h(t) = (t + 1)^{-1}$ for $t \geq 0$

(c) $h(t) = (t + 1)^{-3}$ for $t \geq 0$

8.3 Determine the BIBO stability of the systems with the following transfer functions.

(a) $H(s) = \dfrac{1}{(s^2 + 2s + 2)(s - 1)}$ *pole: +1 ∴ unstable*.

(b) $H(s) = \dfrac{s^2 - 1}{s^3 + s^2 - 2} = \dfrac{(s-1)(s+1)}{(s-1)(s+1-j)(s+1+j)} = \dfrac{s+1}{(s+1-j)(s+1+j)}$ *poles: -1±j ∴ STABLE*

(c) $H(s) = \dfrac{s^2 - 1}{s^3 + 3s^2 + 4s + 2} = \dfrac{(s-1)(s+1)}{(s+1)(s+1-j)(s+1+j)}$ ⟹ *poles: -1±j ∴ STABLE*

8.4 Is the network in Figure P8.4 BIBO stable?

8.5 Which of the following polynomials are Hurwitz polynomials?

(a) $s^5 + 3s^4 + s^2 + 2s + 10$

(b) $-s^4 - 2s^3 - 3s^2 - 4s - 1$

(c) $s^4 - 2s^3 + 3s^2 + 4s + 1$

(d) $s^4 + 3s^3 + 6s^2 + 5s + 3$

(e) $s^6 + 3s^5 + 7s^4 + 8s^3 + 9s^2 + 5s + 3$

Figure P8.4

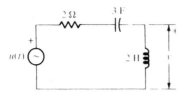

8.6 In Theorem 8.4, show that, if $u(t) = \cos \omega_0 t$, then

$$y_{ss}(t) = A(\omega_0) \cos(\omega_0 t + \theta(\omega_0))$$

8.7 Consider a system with transfer function $H(s) = (s-1)/(s^2 + 2s + 2)$. Find the steady-state responses excited by the following inputs.

(a) $u(t) = 5$

(b) $u(t) = \sin 2t$

(c) $u(t) = \sin(2t + \pi/4) + \cos 2t$

(d) $u(t) = 1 + \sin 2t + \cos 4t$

8.8 Find the response of

$$\dot{y}(t) + 2y(t) = u(t)$$

excited by $y(0-) = -2$ and $u(t) = 1$ for $t \geq 0$. Indicate the zero-input response and zero-state response. Indicate also the steady-state response and the transient response. Is it true that, if $y(0-) = 0$, then the transient response is zero? Is it true that, if $u(t) = 0$, then the steady-state response is zero?

8.9 What are the time constants of the following transfer functions?

(a) $\dfrac{s-2}{s^2 + 2s + 1}$

(b) $\dfrac{s-1}{(s+1)(s^2 + 2s + 2)}$

(c) $\dfrac{s^2 + 2s - 2}{(s^2 + 2s + 4)(s^2 + 2s + 10)}$

(d) $\dfrac{s+10}{s+1}$

(e) $\dfrac{s-10}{s^2 + 2s + 2}$

Do they all have the same time constant?

8.10 Consider the network shown in Figure P8.10. Plot its frequency response.

Figure P8.10

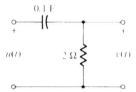

Use the plot to read out the steady-state response due to

$$u(t) = \sin 0.1t + \sin 1000t$$

What is the ratio of the amplitudes of the two components at the output? Is the filter a lowpass or a highpass filter?

8.11 Consider an LTIL system with impulse response

$$h(t) = e^{-t} \qquad \text{for } t \geq 0$$

What are the Fourier series of the outputs if the inputs are as follows?

(a) $1 + \cos t + \sin t + \cos 3t$, for all t

(b) $\sin t + \sin (t + \pi/4) + \cos (3t - \pi/4)$, for all t

(c) $\displaystyle\sum_{k=-\infty}^{\infty} \delta(t - k)$, for all t (when k is an integer)

(d) $|\sin \pi t|$, for all t

8.12 (a) Consider the input signal

$$u(t) = \cos t + \cos 1000t$$

for all t. What is the average power of $u(t)$? What is the radio of the average powers associated with each frequency?

(b) If $u(t)$ is applied to a system with transfer function

$$H(s) = \frac{5}{s + 5}$$

what is the average power of its output? What is the radio of the average powers associated with each frequency? Is the system a lowpass or a highpass filter?

(c) Repeat (b) for a system with transfer function

$$H(s) = \frac{s}{s + 5}$$

8.13 (a) Consider an LTI and lumped system with strictly proper transfer function $H(s)$. Let $h(t)$ be its impulse response. Show that

$$\int_0^\infty |h(t)| dt < \infty \qquad \text{if and only if} \qquad \int_0^\infty h^2(t) dt < \infty$$

That is, the system is BIBO stable if and only if $h(t)$ is an energy function. (Hint: Use the fact that $h(t)$ is a linear combination of exponentially decreasing functions.)

(b) Consider an LTIL system with impulse response $h(t)$. Show that, if the system is BIBO stable, every input with a finite energy will excite an output with a finite energy. Thus BIBO stability implies finite energy stability.

(c) In mathematical terminology, a system can be called an operator. An operator L assigns a unique output function $y(t)$ to every input function $u(t)$, denoted as $y(t) = Lu(t)$. For $u(t) \neq 0$, if the corresponding $y(t)$ is of the form

$$y(t) = Lu(t) = \alpha u(t) \qquad \text{for all } t$$

then $u(t)$ is called an eigenfunction of the operator L and α is an eigenvalue of L. What are the eigenvalue and eigenfunction of the system in (8.18)? What conditions on $h(t)$ are necessary in order for the statement to be valid? Does the statement hold if the input is applied from $t = 0$ on?

8.14 (a) Consider a signal $u(t)$ with the frequency spectrum shown in Figure P8.14(a). Show that the spectrum of the modulated signal $u_m(t) = u(t) \cos \omega_c t$ is as shown in Figure P8.14(b).

(b) To recover $u(t)$ from $u_m(t)$, we may multiply $u_m(t)$ with $\cos \omega_c t$ to yield

$$y(t) = u_m(t) \cos \omega_c t$$

Show that the frequency spectrum of $y(t)$ is as shown in Figure P8.14(c) and that $u(t)$ can be recovered from $y(t)$ using an ideal lowpass filter with gain 2 and cutoff frequency $W = 1.2\omega_c$. This demodulation scheme utilizes the same multiplier as in Figure 6.22(a). It is called a synchronous demodulation because the cosine signal in $y(t)$ must be locked in phase or synchronized with the cosine signal used in modulation. This synchronization is possible in control systems but is impractical in radio receivers. Thus, suppressed-carrier amplitude modulation is not used in radio transmission. A schematic diagram of this modulation and demodulation system is shown in Figure P8.14(d).

Figure P8.14

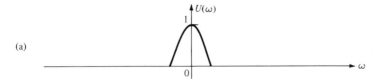

(a)

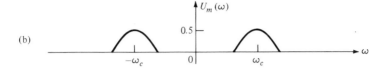

(b)

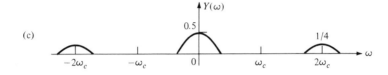

(c)

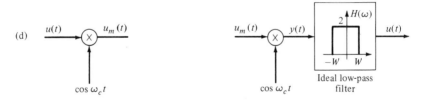

(d)

8.15 (a) Show that the following two transfer functions have the same amplitude characteristic

$$\frac{s^2 + 2s + 2}{(s + 1)(s^2 - 3s + 6)} \qquad \frac{s^2 - 2s + 2}{(s - 1)(s^2 + 3s + 6)}$$

Compare their pole and zero locations. Do they have the same phase characteristic?

(b) Find a BIBO stable transfer function that has the same amplitude characteristic as the one in (a).

8.16 Find the amplitude and phase characteristics of

$$H(s) = \frac{(s - 2)(s^2 - 2s + 2)}{(s + 2)(s^2 + 2s + 2)}$$

This is called an allpass filter. Can you give a reason for its name?

8.17 Consider the circuit shown in Figure P8.17. Find the transfer function $H(s)$ from u to y. Let $R = 10\ \Omega$, $L = 0.2\ H$, and $C = 5 \times 10^{-6}\ F$. Find $H(j\omega)$ for $\omega = 0$, ∞ and $\omega = 1/\sqrt{LC}$. Roughly sketch the amplitude characteristic of $H(s)$. Is it a lowpass, highpass, or bandpass filter? (This circuit is often called a *resonant* circuit.)

8.18 The transfer function

$$H(s) = \frac{b_2 s^2 + b_1 s + b_0}{s^2 + a_1 s + a_0}$$

is called a biquadratic transfer function. From its coefficients, it is possible to tell the types of filters. Verify the following:

(a) $\dfrac{b_0}{s^2 + a_1 s + a_0}$ lowpass

(b) $\dfrac{b_2 s^2}{s^2 + a_1 s + a_0}$ highpass

(c) $\dfrac{b_1 s}{s^2 + a_1 s + a_0}$ bandpass

(d) $\dfrac{s^2 - a_1 s + a_0}{s^2 + a_1 s + a_0}$ allpass

(e) $\dfrac{b_2 s^2 + b_0}{s^2 + a_1 s + a_0}$ notch or bandstop

An allpass filter has a magnitude characteristic equal to 1 for all ω; the magnitude characteristic of a bandstop filter is very small or equal to zero in a frequency band.

8.19 Show that, if all coefficients of $H(s)$ are real, then

$$H(-j\omega) = H^*(j\omega)$$
$$|H(-j\omega)| = |H(j\omega)|$$

and

$$\measuredangle H(-j\omega) = -\measuredangle H(j\omega)$$

Figure P8.17

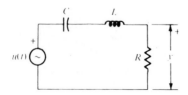

8.20 Is the system with the impulse sequence $h[k] = 1/k^2$, for $k \geq 1$, BIBO stable?

8.21 Which of the following polynomials are Schur polynomials?

(a) $z^3 + 4z^2 - 0.9$

(b) $z^3 - z^2 + 2z - 0.7$

(c) $2z^4 + 1.6z^3 + 1.92z^2 + 0.64z + 0.32$

(d) $z^4 - 0.6z^3 - 2z^2 - 1.2z - 0.8$

8.22 A Schur polynomial can also be tested by using the Routh test after the bilinear transformation

$$s = \frac{z-1}{z+1} \quad \text{or} \quad z = \frac{1+s}{1-s}$$

(a) Show that the transformation transforms the interior of the unit circle of the z-plane into the open left half of the s-plane. Is this a one-to-one mapping?

(b) Determine whether or not the polynomial

$$D(z) = z^3 - 0.3z^2 + 0.1z - 0.1$$

is Schur by first transforming it into the s-domain and then using the Routh test.

(c) Use the Jury test to check whether or not $D(z)$ is Schur.

(d) Which method, the one used in (b) or the one used in (c), is simpler?

8.23 Plot the frequency response of

$$H(z) = \frac{z-2}{z^2 - 0.25}$$

8.24 Find the steady-state responses of the system in Problem 8.23 excited by the following inputs.

(a) $2 + \sin 0.1k, \ k = 0,1,2,\ldots$

(b) $10 \sin 0.1k + \sin 4k, \ k = 0,1,2,\ldots$

8.25 Show (8.35) without using the final-value theorem.

8.26 Plot the frequency response of

$$H(z) = \frac{1 + z + z^2}{z^2}$$

Is it a lowpass, bandpass, or highpass filter?

Figure P8.27

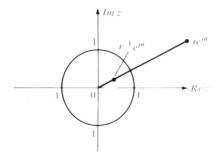

8.27 Show that the amplitude characteristics of the two transfer functions

$$\frac{1}{z - re^{j\theta}} \quad \text{and} \quad \frac{1}{r(z - r^{-1}e^{j\theta})} = \frac{1}{rz - e^{j\theta}}$$

are identical. The poles $re^{j\theta}$ and $r^{-1}e^{j\theta}$ are called reciprocal with respect to the unit circle, as shown in Figure P8.27.

8.28 Is the transfer function

$$H(z) = \frac{z - 0.8}{z^2 + 2z + 2}$$

BIBO stable? If not, can you find a BIBO stable $H_s(z)$ that has an amplitude characteristic identical to $H(z)$'s? (This process of stabilization is often used in filter design.)

8.29 Show that, if $H(z)$ has real coefficients, then

$$H(e^{-j\omega T}) = H^*(e^{j\omega T})$$
$$|H(e^{-j\omega T})| = |H(e^{j\omega T})|$$

and

$$\sphericalangle H(e^{-j\omega T}) = -\sphericalangle H(e^{j\omega T})$$

8.30 Prove the step input part of Theorem 8.10.

USING **MATLAB**

Although the Routh and Jury tests can be readily programmed, it is simpler and more informative to use the command `roots` to check BIBO stability or, more precisely, Hurwitz or Schur polynomials. For example, for the polynomial in Example 8.3.3, the command

```
roots([2 1 7 3 4 1.5])
```

yields

```
ans = -0.0363+1.6575i
      -0.0363-1.6575i
      -0.0202+0.8393i
      -0.0202-0.8393i
      -0.3871
```

Every root has a negative real part. Thus, the polynomial is Hurwitz.
   For the polynomial in Example 8.8.3, the commands

```
r = roots([2   -0.2   -0.24   -0.08]);
abs(r)
```

yield

```
ans = 0.5000
  0.2828
  0.2828
```

Every root has a magnitude less than 1. Thus, the polynomial is Schur. Thus, the BIBO stabilities of analog and digital systems can be easily checked using the command `roots`.
   Commands `freqs` and `freqz` can be used to compute frequency responses of analog and digital systems. For example, the frequency response of $H(s) = 10/(s + 10)$ in (8.15) can be obtained by typing

```
w=logspace(-1,2,200); (generates 200 points equally spaced
                       in [10⁻¹, 10²] = [0.1, 100])
n=10;d=[1 10];
y=freqs(n,d,w);
plot(w,abs(y),w,angle(y))
```

(MATLAB was used in the same way to compute frequency spectra in Chapter 6.) The frequency response of

$$H(z) = \frac{z - 2}{2z^3 - 0.2z^2 - 0.24z - 0.08} = \frac{0.5z - 1}{z^3 - 0.1z^2 - 0.12z - 0.04} \quad \text{(M8.1)}$$

can be obtained by typing

```
n=[0 0 0.5 -1];d=[1 -0.1 -0.12 -0.04];
[y,w]=freqz(n,d,200);
plot(w,abs(y),w,angle(y))
```

The command `freqz(n,d,200)` computes 200 points between 0 and π. If we want to compute 200 points between 0 and 2π, we use the command `freqz(n,d,200,'whole')`. It is important that `n` has the same length as `d` and that `d` has 1 as its leading coefficient. Actually, `n` and `d` are defined for the following negative-power form

$$H^-(z^{-1}) = \frac{n(1) + n(2)z^{-1} + n(3)z^{-2} + \cdots}{1 + d(2)z^{-1} + d(3)z^{-2} + \cdots}$$

as `n = [n(1) n(2) . . . ]` and `d = [1 d(2) d(3) . . . ]`. However, they can be formed directly from positive-power $H(z)$, if we require `n` and `d` to have the same length and `d` to have 1 as its leading coefficient. This situation is the same as in Chapter 7, where the command `freqz` was used to compute frequency spectra of sequences. Note that, if $H(s)$ and $H(z)$ are not BIBO stable, their frequency responses are not defined and the plots obtained by using the commands `freqs` and `freqz` are meaningless.

Consider now the processing of (8.14) by the filter in (8.15). The following commands will compute and plot the response of the filter.

```
t=0:0.01:10;
u = sin(t)+0.1*sin (100*t);
n=[10];d=[1 10];
y=lsim(n,d,u,t);
plot(t,y)
```

Note that command `lsim` stands for linear simulation; the first two arguments are the numerator and denominator of the system; the third argument is the input, and the last argument is the time. If the input is a unit step function, we may use command `step`. If the input is not a unit step function, then we must use `lsim`. Unfortunately, `lsim` is not available in the *Student Edition of MATLAB*.

Alternatively, we may use the command `conv` to compute the response of the filter. The impulse response of the filter with transfer function $10/(s + 10)$ is $h(t) = 10e^{-10t}$. Thus, the output of the filter is the convolution of the input $u(t)$ in (8.14) and the impulse response $h(t)$. Let us discuss the selection of the sampling period T. The period of sin $100t$ is $2\pi/100 = 0.0628$. Thus, the sampling period should be, at most, $0.0628/2 = 0.0314$ to meet the sampling theorem. We choose $T = 0.02$. The commands

```
t=0:0.02:10;
h=10*exp(-10*t);
u = sin(t)+0.1*sin(100*t);
y=0.02*conv(h,u);
plot(y)
```

yield the response shown in Figure M8.1(b). Note that only the first half of the plot is the convolution of h and u because the convolution is computed using the command fft and trailing zeros are padded to h and u (see Section 7.8). We type k = 0:0.02:20; u1 = sin(k) + 0.1*sin(100*k); plot(k,u1) to generate $u(t)$ in Figure M8.1(a) for comparison. We see that the filter suppresses the high-frequency signal. Thus, the filter is a lowpass filter. Note that we must multiply conv(h,u) by T to obtain y. See the discussion of using MATLAB in Chapter 3. In using conv(h,u), if we use more points by typing $t = 0:0.2:11$, then the message Student system limit on maximum variable size exceeded will appear. This is why we compute only up to $t = 10$ in Figure M8.1(b).

MATLAB PROBLEMS

Use MATLAB to compute the following.

M8.1 Which of the following polynomials are Hurwitz?

(a) $s^6 + 5s^5 + 13s^4 + 20s^3 + 19s^2 + 11s + 3$
(b) $s^4 + 1.4s^3 + 0.57s^2 + 0.116s + 0.0072$
(c) $20s^4 + 16s^3 + 19.2s^2 + 6.4s + 3.2$

[**ANSWERS:** (a), (b)]

Figure M8.1

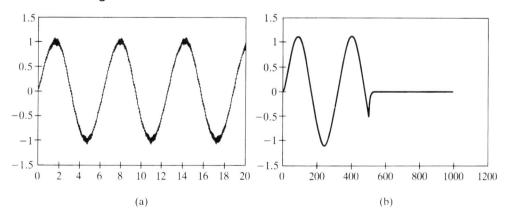

(a)   (b)

M8.2 Which of the polynomials in Problem M8.1 are Schur?

[**ANSWERS:** (b), (c)]

M8.3 Plot the frequency responses of the following continuous-time transfer functions

$$H_1(s) = \frac{s^2 + 2s + 2}{s^4 + 3s^3 + 6s^2 + 5s + 3} \quad \text{and} \quad H_2(s) = \frac{(s^2 - 2s + 2)(s - 3)}{(s^2 + 2s + 2)(s + 3)}$$

M8.4 Plot the frequency responses of the following discrete-time transfer functions

$$H_1(z) = \frac{z^2 + 2z + 2}{z^4 + 0.8z^3 + 0.96z^2 + 0.32z + 0.16} \quad \text{and} \quad H_2(z) = \frac{z^2 + 2z + 2}{2z^2 + 2z + 1}$$

Op-Amp Circuits, Model Reduction, and Asymptotic Stability

15

9.1   INTRODUCTION

Chapter 8 introduced the concept of BIBO stability and explained why systems designed to process signals must be stable. This chapter discusses further the importance of stability in practical systems.

Operational amplifier circuits, or op-amp circuits, are commonly introduced in first courses in electrical engineering. By using the idealized model, students in these courses can easily analyze and design many op-amp circuits. However, analyses and designs based on the idealized model may or may not hold for all circuits. This chapter will show that such analyses are valid only if the circuits are BIBO stable. Model reduction, which is widely employed in practice, will then be discussed by using the concept of BIBO stability and the concept of frequency spectrum of signals. This chapter will conclude with a study of the stability of the zero-input response and the Wien-bridge oscillator.

9.2   OPERATIONAL AMPLIFIERS

The operational amplifier is one of the most important circuit elements. It is built in integrated circuit form and it is small in size, inexpensive, and widely available. The operational amplifier is usually represented as shown in Figure 9.1(a) and modeled as shown in Figure 9.1(b). This is a simplified model of the operational amplifier. In reality, the op amp is rather complex and may contain over ten transistors. It has many more terminals than the three shown. There are terminals to be connected to power supplies, commonly +15 V and −15 V, and terminals for balancing. These terminals usually are not shown. The op amp in Figure 9.1(a) has two input terminals and one output terminal. The one

Figure 9.1   *Operational amplifier*

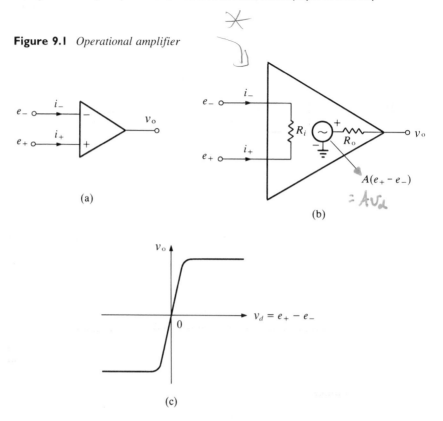

(a)

(b)

(c)

with the minus sign is called the inverting terminal and the one with the plus sign is called the noninverting terminal. Let e_- and e_+ denote the voltages applied at the inverting and noninverting terminals. Then, the output voltage v_o is a function of $v_d = e_+ - e_-$, called the *differential input voltage*. Their relationship generally is of the form shown in Figure 9.1(c). For $|v_d|$ small, typically less than 0.1 mV (millivolt $= 10^{-3}$ volt), the output voltage is a linear function of the differential voltage. For $|v_d|$ large, the output voltage saturates at a level about two volts below the supplied voltage. Some op-amps are designed to operate in the saturation region, such as in switching. If they are designed to operate in the linear region, then they can be used to build any LTIL continuous-time systems.

In this chapter, we will study op amps that operate only in the linear region. In this case, the output voltage and the differential voltage can be related by

$$v_o = A(e_+ - e_-)$$ (9.1)

where A is called the open-loop gain. The resistor R_i in Figure 9.1(b) is the input resistance, and R_o is the output resistance. R_i is generally very large, greater than 10^4 Ω, and R_o is very small, less than 50 Ω. The open-loop gain A is very large, usually over 10^5, at low frequencies.

Some assumptions are often made in analysis and design. If the output voltage is limited to the supplied voltage, say 15 volts, and if A is very large, say over 10^5, then $|e_+ - e_-| \leq 15/10^5 = 0.00015$. Thus, we assume $e_+ = e_-$ (that is, the voltages at the inverting and noninverting terminals equal each other), and the two terminals are said to be virtually short. Because R_i is very large, even if $e_+ \neq e_-$, the current $i_- = -i_+$ is practically zero. In other words, the two terminals are practically open. An op-amp with the short-circuit property

$$e_+ = e_- \qquad\qquad \textbf{(9.2)}$$

and the open-circuit property

$$i_- = -i_+ = 0 \qquad\qquad \textbf{(9.3)}$$

is called an *ideal* op amp. Therefore, an ideal op amp is often said to have $A = \infty$, $R_i = \infty$, and $R_o = 0$. With these two properties, the analysis and design of operational amplifier circuits can be greatly simplified.

Consider the circuit shown in Figure 9.2(a). The input signal v_i is connected to the noninverting terminal and the output is fed back into the inverting terminal. Thus, we have $v_i = e_+$ and $v_o = e_-$, and the short-circuit property implies

$$v_o = v_i \qquad\qquad \textbf{(9.4)}$$

Therefore, the output voltage always equals the input voltage, and the circuit is called the *negative-feedback voltage follower* or, simply, the *voltage follower*. This circuit can be used as an isolating amplifier and is also called a *buffer*.

Consider the circuit shown in Figure 9.2(b). Now, the output is fed back into the noninverting terminal and the input is applied to the inverting terminal. Because $v_i = e_-$ and $v_o = e_+$, we also have $v_o = v_i$. Thus, this circuit has the same input-output relationship as the one in Figure 9.2(a). If we use P-SPICE to simulate the responses of the two circuits excited by a sinusoid, they will yield identical results, as shown in Figure 9.2(c). Thus, the two circuits appear to be workable.

In practice, the negative-feedback voltage follower in Figure 9.2(a) is widely used, but the circuit in Figure 9.2(b) is never used. To see why, we build the circuits in Figure 9.2(a) and (b) using Fairchild μA 741 op amps. The response of the negative-feedback voltage follower in Figure 9.2(a) excited by a sinusoid is as shown in Figure 9.2(c); the output does follow the input signal. The response of the circuit in Figure 9.2(b) is as shown in Figure 9.2(d); it immediately reaches and then stays in the negative saturation region no matter what the magnitude of the sinusoid is. Therefore, the op-amp circuit does not have the property $v_o = v_i$. This shows that the analysis based on the short- and open-circuit properties may not always hold for actual circuits. This also shows

Figure 9.2 *Voltage followers*

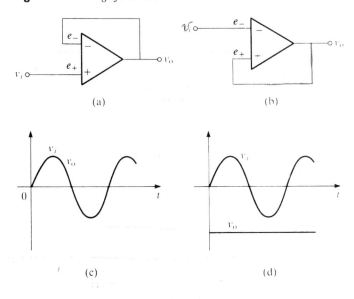

It is often said that negative feedback stabilizes a system and positive feedback destabilizes a system. Even though this statement happens to be true for the two circuits in Figures 9.2(a) and (b), it is not true in general, as the positive- and negative-feedback systems in Figure 8.4 show. Therefore, a proof of the stability of the two circuits in Figures 9.2(a) and (b) is needed. Before proceeding, let us discuss three more widely encountered op-amp circuits.

Consider the circuit shown in Figure 9.3, in which $Z_1(s)$ and $Z_2(s)$ denote two impedances. They could be resistors, capacitors, or RC combinations. We call $Z_2(s)$ the feedback impedance. Because the noninverting terminal is grounded, we have $e_+ = 0$ and, consequently, $e_- = 0$, following the short-circuit property. Thus, the current $i(t)$ passing through $Z_1(s)$ and the current $i_2(t)$ passing through $Z_2(s)$ are, respectively,

that computer simulation may yield erroneous results if models are not chosen properly.

Figure 9.3 *An op-amp circuit*

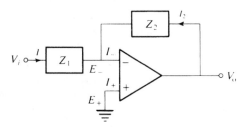

$$I(s) = \frac{V_i(s) - E_-(s)}{Z_1(s)} = \frac{V_i(s)}{Z_1(s)} \qquad I_2(s) = \frac{V_o(s)}{Z_2(s)}$$

The open-circuit property $i_- = 0$ implies $i(t) = -i_2(t)$ or

$$\frac{V_i(s)}{Z_1(s)} = -\frac{V_o(s)}{Z_2(s)}$$

which implies

$$H(s) := \frac{V_o(s)}{V_i(s)} = -\frac{Z_2(s)}{Z_1(s)} \tag{9.5}$$

Thus, the transfer function from v_i to v_o of the circuit in Figure 9.3 simply equals the ratio of the feedback impedance and the impedance connected to the inverting terminal.

Now, if $Z_1(s) = R_1$ and $Z_2(s) = R_2$, as shown in Figure 9.4(a), then

$$H(s) = -\frac{R_2}{R_1} \quad \text{(amplifier)} \tag{9.6}$$

It is an inverting amplifier with gain $-R_2/R_1$. If $Z_1(s) = R_1$ and $Z_2(s) = 1/Cs$, as shown in Figure 9.4(b), then

$$H(s) = -\frac{1}{R_1 Cs} \quad \text{(integrator)} \tag{9.7}$$

It is an integrator. If $Z_1(s) = 1/Cs$ and $Z_2(s) = R_2$, as shown in Figure 9.4(c), then

$$H(s) = -R_2 Cs \quad \text{(differentiator)} \tag{9.8}$$

Figure 9.4 *(a) Inverting amplifier, (b) integrator, and (c) differentiator*

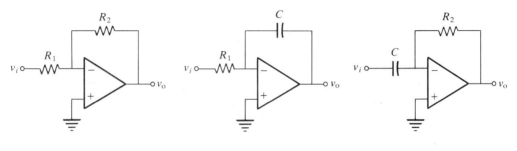

(a)   (b)   (c)

It is a differentiator. Therefore, amplifiers, integrators, and differentiators can be easily built using op amps, resistors, and capacitors.

The inverting amplifier and integrator are widely used in practice. Although the differentiator can be as easily built, the circuit in Figure 9.4(c), as will be shown later, may not be BIBO stable and its response may saturate no matter what input is applied. Therefore, the circuit cannot function as a differentiator.

To conclude this section, we mention that inductors are bulky in size and difficult to fabricate in integrated form. Because of these drawbacks, inductors are not used to build op-amp circuits; only resistors and capacitors are used.

9.3   BLOCK DIAGRAM OF OP-AMP CIRCUITS

In this section, we will develop block diagrams for the preceding op-amp circuits. The diagrams will be developed in the Laplace-transform domain, so all variables will be Laplace-transform variables. We will use capital letters to denote the Laplace transforms of the corresponding lower-case letters. For example, I_- and V_1 are the Laplace transforms of i_- and v_1. For convenience, the argument s will not be shown in the Laplace transform. In this section, the open-loop gain will be denoted by $A(s)$ or, simply, A.

In order to study the stability of op-amp circuits, we will use a more realistic model. Assume $I_- = I_+ = 0$, but do not assume $E_+ = E_-$. Equivalently, assume the input impedance R_i is infinite but the open-loop gain is finite. Thus, the short-circuit property does not hold and we must use

$$V_o = A(E_+ - E_-)$$

Now, reconsider the voltage follower in Figure 9.2(a). From the connection, we have

$$V_o = A(V_i - V_o)$$

which implies $(1 + A)V_o = AV_i$ or

$$H_n(s) := \frac{V_o}{V_i} = \frac{A}{1 + A} \qquad (9.9)$$

This is the transfer function from v_i to v_o of the voltage follower in Figure 9.2(a). It can be represented by the block diagram in Figure 9.5(a). Using (4.76), we can readily show that the transfer function of the negative-feedback block diagram in Figure 9.5(a) equals (9.9). Similarly, for the op-amp circuit in Figure 9.2(b), we have

$$V_o = A(V_o - V_i)$$

Figure 9.5 *Block diagrams for the circuits in Figures 9.2 and 9.3*

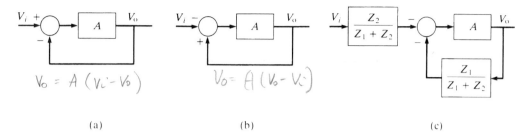

$$V_o = A(V_i - V_o)$$

(a)

$$V_o = A(V_o - V_i)$$

(b)

(c)

which implies

$$H_p(s) := \frac{V_o}{V_i} = -\frac{A}{1-A} \qquad (9.10)$$

Its block diagram is shown in Figure 9.5(b); it is a positive feedback system. Note that, if $A = \infty$, then both (9.9) and (9.10) become 1 and $v_o = v_i$. This is what we have in (9.4).

Now let us develop a block diagram for the circuit in Figure 9.3. Because the noninverting terminal is grounded, we have $E_+ = 0$ and

$$V_o = A(E_+ - E_-) = -AE_- \qquad (9.11)$$

Because $I_- = 0$, we have $I = -I_2$ or

$$\frac{V_i - E_-}{Z_1} = -\frac{V_o - E_-}{Z_2} \qquad (9.12)$$

which implies

$$Z_2 V_i - Z_2 E_- = -Z_1 V_o + Z_1 E_- \qquad (9.13)$$

Substituting (9.11) into (9.13) yields

$$Z_2 V_i = Z_2 E_- + Z_1 A E_- + Z_1 E_- = (Z_1 + Z_2 + AZ_1)E_-$$

or

$$E_- = \frac{Z_2}{Z_1 + Z_2 + AZ_1} V_i$$

Substituting this into (9.11) yields

$$V_o = \frac{-AZ_2}{Z_1 + Z_2 + AZ_1} V_i \tag{9.14}$$

Thus, the transfer function from v_i to v_o of the op-amp circuit in Figure 9.3 is

$$H(s) = \frac{V_o}{V_i} = \frac{-AZ_2}{Z_1 + Z_2 + AZ_1} \tag{9.15a}$$

$$= -\frac{Z_2}{Z_1 + Z_2} \frac{A}{1 + A\dfrac{Z_1}{Z_1 + Z_2}} \tag{9.15b}$$

This equation can be represented by the block diagram in Figure 9.5(c). Using (4.76), we can readily show that the transfer function from V_i to V_o in Figure 9.5(c) equals (9.15). Note that the block diagram in Figure 9.5(c) can be obtained directly from the circuit in Figure 9.3 using physical argument. If we ground the output ($V_o = 0$) and apply $V_i = 1$, then the voltage at the inverting terminal is $Z_2/(Z_1 + Z_2)$, which is the transfer function from the input V_i to the summer or adder. If we ground the input $V_i = 0$ and apply $V_o = 1$, then the voltage at the inverting terminal is $Z_1/(Z_1 + Z_2)$, which is the feedback transfer function in the block diagram. Note that (9.15) reduces to (9.5) if $A = \infty$.

9.4   STABILITY ANALYSIS

This section studies the BIBO stability of the op-amp circuits in Figures 9.2 and 9.4. The open-loop gain A is assumed to be a very large constant or an infinity in the idealized model. In reality, the gain is neither a constant nor an infinity. Figure 9.6 shows a typical frequency response of the Fairchild $\mu A741$, a widely available op amp. Such a response is obtained by measurement and is available from the manufacturer. We see that the gain A is roughly 4×10^5 at $\omega = 1$ Hz, but reduces to 10^2 at $\omega = 10^4$ Hz. Therefore, in a more realistic model, the open-loop gain A should be replaced by an open-loop transfer function $A(s)$. From the frequency response $A(j\omega)$ in Figure 9.6, we see that there are two corner frequencies at $8 \times 2\pi$ and $3 \times 10^6 \times 2\pi$ rad/s. Thus, the transfer function of the op amp can be shown to be

$$A(s) = \frac{4 \times 10^5}{(1 + \dfrac{s}{8 \times 2\pi})(1 + \dfrac{s}{3 \times 10^6 \times 2\pi})}$$

$$= \frac{37.86 \times 10^{13}}{(s + 50.24)(s + 18.84 \times 10^6)} \tag{9.16}$$

Figure 9.6 *Typical frequency response of Fairchild μA741 operational amplifier*

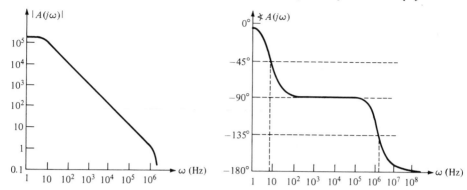

We shall use this $A(s)$ to study the BIBO stability of op-amp circuits.

Consider the voltage follower shown in Figure 9.2(a) and its block diagram in Figure 9.5(a). If $A(s)$ is given as in (9.16), then the transfer function from V_i to V_o is

$$
\begin{aligned}
H(s)_n &= \frac{A(s)}{1 + A(s)} = \frac{37.86 \cdot 10^{13}}{(s + 50.24)(s + 18.84 \cdot 10^6) + 37.86 \cdot 10^{13}} \\
&\approx \frac{37.86 \cdot 10^{13}}{s^2 + 18.84 \cdot 10^6 s + 37.86 \cdot 10^{13}} \\
&= \frac{37.86 \cdot 10^{13}}{(s + 9.42 \cdot 10^6 + j\,1.7025 \cdot 10^7)(s + 9.42 \cdot 10^6 - j\,1.7025 \cdot 10^7)}
\end{aligned}
\tag{9.17}
$$

This transfer function is clearly BIBO stable. If we apply a step input $v_i = a$, then its steady-state output, using (8.13), is

$$
v_o = H_n(0)a \approx \frac{37.86 \cdot 10^{13}}{37.86 \cdot 10^{13}} \cdot a = a
$$

The time constant of the transfer function is $1/(9.42 \cdot 10^6)$. Thus, the step response will reach the steady state in about $5/(9.42 \cdot 10^6)$ second, or less than 1 microsecond. For step inputs, then, the transfer function in (9.17) can be very well approximated by 1, and the analysis based on the open-circuit and short-circuit properties yields an acceptable result. This issue will be discussed further in the next section.

Consider now the op-amp circuit in Figure 9.2(b) and its block diagram in Figure 9.5(b). Its transfer function is

$$H_p(s) = -\frac{A(s)}{1 - A(s)} = \frac{-37.86 \cdot 10^{13}}{(s + 50.24)(s + 18.84 \cdot 10^6) - 37.86 \cdot 10^{13}}$$

$$\approx \frac{-37.86 \cdot 10^{13}}{s^2 + 18.84 \cdot 10^6 s - 37.86 \cdot 10^{13}} \qquad \textbf{(9.18)}$$

$$= \frac{-37.86 \cdot 10^{13}}{(s + 3.1038 \cdot 10^7)(s - 1.2198 \cdot 10^7)}$$

This transfer function has an unstable pole and is not BIBO stable. If we apply a unit step input $v_i = 1$, then the response becomes

$$V_o = \frac{-37.86 \cdot 10^{13}}{(s + 3.1038 \cdot 10^7)(s - 1.2198 \cdot 10^7)s}$$

$$= -\frac{-0.72}{s - 1.2198 \cdot 10^7} + \frac{0.72}{s + 3.1038 \cdot 10^7} + \frac{1}{s}$$

or

$$v_o(t) = -0.72e^{1.2198 \cdot 10^7 t} + 0.72e^{-3.1038 \cdot 10^7 t} + 1$$

for $t \geq 1$. Because of the positive real pole, the response approaches $-\infty$ as $t \to \infty$. In reality, the response will stay in the negative saturation region or the circuit will burn out. This is consistent with the response of the actual circuit shown in Figure 9.2(d). Thus, for the circuit in Figure 9.2(b), the analysis based on the ideal model yields an erroneous result and the circuit cannot be used as a voltage follower. For a different stability analysis of the circuit, see Reference [10], p. 326.

Consider the op-amp circuit in Figure 9.3 and its block diagram in Figure 9.5(c). Its transfer function is computed in (9.15) as

$$H(s) = \frac{V_o}{V_i} = \frac{-AZ_2}{(1 + A)Z_1 + Z_2} \qquad \textbf{(9.19)}$$

We will use this equation with A in (9.16) to study the stability of the op-amp circuits in Figure 9.4.

Inverting Amplifier The substitution of $Z_1 = R_1$, $Z_2 = R_2$, and (9.16) into (9.19) yields

$$H(s) = \frac{-\dfrac{37.86 \times 10^{13}}{(s + 50.24)(s + 18.84 \times 10^6)} \cdot R_2}{\left[1 + \dfrac{37.86 \times 10^{13}}{(s + 50.24)(s + 18.84 \times 10^6)}\right] R_1 + R_2} \qquad \textbf{(9.20)}$$

$$= \frac{-37.86 \cdot 10^{13} \cdot R_2}{(R_1 + R_2)s^2 + (R_1 + R_2) \cdot 18.84 \cdot 10^6 s + (R_1 + R_2) \cdot 94.65 \cdot 10^7 + 37.86 \cdot 10^{13} \cdot R_1}$$

The denominator of $H(s)$ is a polynomial of degree 2. A polynomial of degree 2 is Hurwitz if its coefficients are all positive. Because this is the case for any positive R_1 and R_2, the circuit in Figure 9.4(a) is BIBO stable.

Integrator We assume $Z_1 = R_1 = 1000\ \Omega$ and $Z_2 = 1/Cs = 1/10^{-6}s = 10^6/s$. The substitution of these and (9.16) into (9.19) yields

$$H(s) = \cfrac{-\cfrac{37.86 \times 10^{13}}{(s + 50.24)(s + 18.84 \times 10^6)} \cdot \cfrac{10^6}{s}}{\left[1 + \cfrac{37.86 \times 10^{13}}{(s + 50.24)(s + 18.84 \times 10^6)}\right] \times 10^3 + \cfrac{10^6}{s}}$$

This can be simplified as

$$H(s) = \frac{-37.86 \times 10^{16}}{s^3 + 18.84 \times 10^6 s^2 + 37.86 \times 10^{13} s + 94.65 \times 10^{10}} \tag{9.21}$$

To check its BIBO stability, we form the Routh table for the denominator of $H(s)$:

	s^3	1	$37.86 \cdot 10^{13}$	
$k_1 = 1/18.84 \cdot 10^6$	s^2	$18.84 \cdot 10^6$	$94.65 \cdot 10^{10}$	$[0\ \ 37.86 \cdot 10^{13}]$
$k_2 = 18.84 \cdot 10^6/37.86 \cdot 10^{13}$	s	$37.86 \cdot 10^{13}$		$[0\ \ 94.65 \cdot 10^{10}]$
	1	$94.65 \cdot 10^{10}$		

All entries in the table are positive; thus, the integrator circuit is BIBO stable.

Differentiator We assume $Z_1 = 1/Cs = 10^6/s$ and $Z_2 = R_2 = 1000\ \Omega$. The substitution of these and (9.16) into (9.19) yields

$$H(s) = \cfrac{-\cfrac{37.86 \times 10^{13}}{(s + 50.24)(s + 18.84 \times 10^6)} \times 1000}{\left[1 + \cfrac{37.86 \times 10^{13}}{(s + 50.24)(s + 18.84 \times 10^6)}\right] \cdot \cfrac{10^6}{s} + 1000}$$

which can be simplified as

$$H(s) = \frac{-37.86 \times 10^{16} s}{10^3 s^3 + 18.84 \times 10^9 s^2 + 19.79 \times 10^{12} s + 37.86 \times 10^{19}} \tag{9.22}$$

Its denominator has degree 3. We form the Routh table for the denominator as

$$k_1 = \frac{10^3}{18.84 \cdot 10^9} \qquad \begin{array}{c|cc} s^3 & 10^3 & 19.79 \cdot 10^{12} \\ s^2 & 18.84 \cdot 10^9 & 37.86 \cdot 10^{19} \\ s & -0.31 \cdot 10^{12} & \end{array} \qquad [0 \ -0.31 \cdot 10^{12}]$$

A negative number appears in the table, so the denominator of $H(s)$ is not Hurwitz and the differentiator circuit in Figure 9.4(c) is not BIBO stable. If we apply, for example, $\sin t$ to the input of Figure 9.4(c), its output will approach infinity as $t \to \infty$ rather than $d(\sin t)/dt = \cos t$. Therefore, the circuit cannot function as a differentiator. The stability of the differentiator depends on the specific C and R_2 used. For example, if $C = 1\ \mu F = 10^{-6}\ F$ and $R_2 = 1\ M\Omega = 10^6\ \Omega$, then the circuit in Figure 9.4(c) is BIBO stable (see Problem 9.6). Even if a stable differentiator can be built, differentiation will amplify high-frequency noise, as discussed in Figure 3.19. Therefore, differentiators are rarely used in practice. Note that the differentiator is not a basic element in the block diagrams discussed in Figures 3.25 and 3.26.

In conclusion, if an op-amp circuit is not BIBO stable, then the behavior of the circuit will differ drastically from the behavior predicted based on the idealized model. In such a case, the ideal model cannot be used in analysis and design.

9.4.1   Justification of using the ideal op-amp model

The analysis of an op-amp circuit that is not BIBO stable cannot use its idealized model. This raises the question: If an op-amp circuit is BIBO stable, can we use the idealized model? The answer is affirmative for a class of input signals, as will be discussed in this section.

Let $H_s(s)$ denote the transfer functions of an op-amp circuit using the idealized model and let $H(s)$ denote the transfer function using the more realistic model. We compare here the responses of these two transfer functions. Consider the inverting amplifier in Figure 9.4(a) with $R_1 = 1\ k\Omega$ and $R_2 = 10\ k\Omega$. The transfer function using the ideal model is

$$H_s(s) = -\frac{R_2}{R_1} = -10 \tag{9.23}$$

The transfer function in (9.20) for the more realistic model with $A(s)$ in (9.16) becomes

$$\begin{aligned} H(s) = \frac{V_o(s)}{V_i(s)} &= \frac{-3.44 \times 10^{14}}{s^2 + 18.84 \times 10^6 s + 94.65 \times 10^7 + 3.44 \times 10^{13}} \\ &\approx \frac{-3.44 \times 10^{14}}{s^2 + 18.84 \times 10^6 s + 3.44 \times 10^{13}} \\ &\approx \frac{-3.44 \times 10^{14}}{(s + 16.79 \times 10^6)(s + 2.05 \times 10^6)} \end{aligned} \tag{9.24}$$

It has two negative real poles. Now we compute their unit step responses. Let $\bar{v}_o(t)$ be the unit step response of $H_s(s)$. Then, we have

$$\bar{V}_o(s) = -10V_i(s) = -\frac{10}{s} \quad \text{or} \quad \bar{v}_o(t) = -10$$

for all $t \geq 0$. Let $v_o(t)$ be the unit step response of $H(s)$. Then, we have

$$V_o(s) = H(s)V_i(s) = \frac{-3.44 \times 10^{14}}{(s + 16.79 \times 10^6)(s + 2.05 \times 10^6)} \cdot \frac{1}{s}$$

$$\approx -\frac{1.39}{s + 16.79 \times 10^6} + \frac{11.39}{s + 2.05 \times 10^6} - \frac{10}{s}$$

which implies

$$v_o(t) = -1.39e^{-16.79 \times 10^6 t} + 11.39e^{-2.05 \times 10^6 t} - 10$$

The time constant of $H(s)$ in (9.24) is $1/(2.05 \times 10^6) = 0.49 \times 10^{-6}$ second. Therefore, the output $v_o(t)$ will reach the steady state -10 in five time constants, or 2.4×10^{-6} second. In conclusion, for step inputs, $\bar{v}_o(t)$ and $v_o(t)$ are essentially the same, so we can use the idealized model to study the circuit or, equivalently, to approximate the transfer function in (9.24) by $H_s(s) = -10$.

The inverting amplifier with transfer function in (9.24) can be simplified as -10 for step inputs. Can it be so simplified for any input signal? To answer this, consider $v_i(t) = \sin 10t$. We compute

$$H(j10) = \frac{-3.44 \times 10^{14}}{(j10 + 16.79 \times 10^6)(j10 + 2.05 \times 10^6)} \approx -10$$

Therefore, the steady-state response excited by $v_i(t) = \sin 10t$ is

$$v_{o,ss}(t) = -10\sin 10t$$

Furthermore, the response will reach the steady state almost instantaneously. Therefore, for $v_i(t) = \sin 10t$, the transfer function in (9.24) can still be simplified as -10.

Consider now the input $v_i(t) = \sin 10^{20}t$. If we use the idealized model of -10, then the output is

$$\bar{v}_o(t) = -10\sin 10^{20}t \quad \text{for } t \geq 0$$

However, because

$$H(j10^{20}) = \frac{-3.44 \times 10^{14}}{(j10^{20} + 16.79 \times 10^6)(j10^{20} + 2.05 \times 10^6)} \approx \frac{-3.44 \times 10^{14}}{-10^{40}} \approx 0$$

the steady-state output of (9.24) is

$$v_{o,ss}(t) = |H(j10^{20})|\sin(10^{20}t + \sphericalangle H(j10^{20})) \approx 0$$

which is quite different from $-10\sin 10^{20}t$. Thus, for the input signal $v_i(t) = \sin 10^{20}t$, we cannot use the idealized model.

The question then becomes: For what type of signals can we use the idealized model or, equivalently, can the transfer function in (9.24) be simplified as -10? To answer this question, we must plot the frequency response of (9.24), as shown in Figure 9.7. We see that, for $\omega < 10^6$ rad/s, the frequency response of (9.24) is essentially the same as that of -10. However, for $\omega > 10^6$, they are different. Therefore, we can conclude that, for $v_i(t) = \sin \omega_0 t$ with $\omega_0 < 10^6$ or, more generally, for signals with frequency spectra lying inside the frequency range between 0 and 10^6 rad/s, we can use the idealized model to study the inverting amplifier or, equivalently, the transfer function in (9.24) can be approximated by -10.

Now let us discuss the transfer functions of the integrator in Figure 9.4(b) using the idealized model and the more realistic model. If $R_1 = 1\,\mathrm{k\Omega}$ and $C = 10^{-6}\,F$, then the transfer function in (9.7) for the ideal model becomes

$$H_s(s) = -\frac{1}{R_1 Cs} = -\frac{1}{10^{-3}s} = \frac{-10^3}{s} \tag{9.25}$$

The transfer function using the more realistic model in (9.16) is computed in (9.21) and repeated in the following:

$$H(s) = \frac{-37.86 \times 10^{16}}{s^3 + 18.84 \times 10^6 s^2 + 37.86 \times 10^{13}s + 94.65 \times 10^{10}} \tag{9.26}$$

Now compare the responses of $H_s(s)$ and $H(s)$. If we apply a step input with magnitude 10 mV (10^{-2} V), then the output of $H_s(s)$ is

$$\bar{V}_o(s) = -\frac{10^3}{s} \cdot \frac{10^{-2}}{s} = -\frac{10}{s^2}$$

Figure 9.7 *The frequency responses of (9.23) and (9.24)*

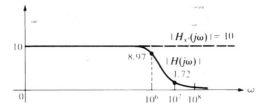

or, using Table 4.1,

$$\bar{v}_o(t) = -10t$$

which approaches infinity as $t \to \infty$. However, the output of $H(s)$ in (9.26) excited by the same input approaches

$$\lim_{t \to \infty} v_o(t) = H(0)10^{-2} = \frac{-37.86 \times 10^{16}}{94.65 \times 10^{10}} \cdot 10^{-2} = -4000$$

This is a finite number. Therefore, the step responses of (9.25) and (9.26) are quite different, and (9.26) cannot be approximated by (9.25) for step inputs.

To compare the two responses excited by $10^{-3} \cos t$, we calculate the output of $H_s(s)$ as

$$\bar{V}_o(s) = -\frac{10^3}{s} \cdot \frac{10^{-3}s}{s^2 + 1} = -\frac{1}{s^2 + 1}$$

This implies

$$\bar{v}_o(t) = -\sin t$$

for all t, which also can be obtained directly by integrating the input. To find the steady-state response of $H(s)$ excited by $10^{-3} \cos t$, we compute

$$H(j1) = \frac{-37.86 \cdot 10^{16}}{-j - 18.84 \cdot 10^6 + j37.86 \cdot 10^{13} + 94.65 \cdot 10^{10}} \approx \frac{-37.86 \cdot 10^{16}}{j37.86 \cdot 10^{13}}$$

$$= \frac{-10^3}{j} = 10^3 j = 10^3 e^{j90°}$$

Thus, the steady-state response, using (8.9b), is

$$v_{o,ss}(t) = 10^3 \cdot 10^{-3} \cos(t + 90°) = -\sin t$$

which is the same as $\bar{v}_o(t)$. For the input $10^{-3} \cos t$, the transfer function in (9.26) can be approximated very well by the simplified one in (9.25). Figure 9.8(a) shows the frequency responses of (9.25) and (9.26). We see that, except in the immediate neighborhood of $\omega = 0$, their frequency responses are almost identical. Therefore, for input signals that contain no DC parts, we can use the idealized model to study the integrator in Figure 9.4(b) or, equivalently, the transfer function in (9.26) can be simplified as (9.25).

The integrator in (9.25) is not BIBO stable; therefore, strictly speaking, its frequency response is not defined. However, for any bounded input that contains no DC parts, the output of the integrator will be bounded. Thus, the integrator can be considered to be BIBO stable for the class of signals that con-

Figure 9.8   *(a) The frequency responses of (9.25) and (9.26), (b) the frequency responses of (9.22) and −s/1000*

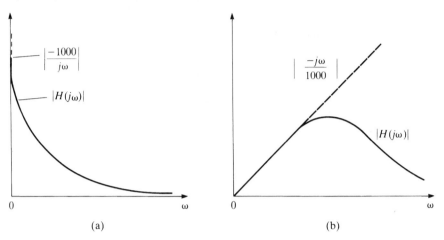

(a)   (b)

tain no DC parts. In practice, integrators are rarely used by themselves; however, they can be used to build any systems with proper transfer functions (see Figures 3.25 and 3.26 and Chapter 10). Therefore, the transfer function of the circuit in Figure 9.4(b) can be and is widely approximated by (9.7).

To summarize, the ideal op-amp model can be used only if the circuit is BIBO stable, and the analysis holds only for input signals whose frequency spectra lie in a limited range. Thus the concepts of stability and frequency spectrum are implicit in the use of the idealized model. Note that the plot of $H(s)$ in (9.22) along $s = j\omega$ is almost identical to that of the differentiator $-s/1000$ over a wide frequency range, as shown in Figure 9.8(b). However, because $H(s)$ is not BIBO stable, the responses of $H(s)$ and $-s/1000$ excited by any input will be entirely different and the differentiator circuit in Figure 9.4(c) cannot be analyzed using the idealized model.

*9.5   MODEL REDUCTION

In order to facilitate analysis and design, engineers often employ simplification and approximation when developing mathematical descriptions for systems. For example, almost all the linear and time-invariant models studied in this text were so obtained. Once linear and time-invariant models have been developed, their transfer functions may be simplified again, as was discussed in the preceding section. Here, we shall discuss how simplification and approximation are employed in the general case.

Consider a system with transfer function $H(s)$ and a class of input signals whose frequency spectra are limited to a frequency range B. The transfer function $H(s)$ can be simplified as $H_s(s)$ if

1. $H(s)$ is BIBO stable,

2. there exists a simplified transfer function $H_s(s)$ whose frequency response roughly equals that of $H(s)$ in B, and

3. the simplified transfer function is BIBO stable or at least BIBO stable for the class of input signals.

Note that the stability of $H(s)$ is essential and that the simplified transfer function $H_s(s)$ must be BIBO stable for the class of input signals. In some cases, simplifications are not unique. For example, the transfer function of degree 2 in (9.24) can be simplified as

$$H_1(s) = \frac{-3.44 \times 10^{14}}{s^2 + 18.84 \times 10^6 s + 3.44 \times 10^{13}} \approx \frac{-3.44 \times 10^{14}}{18.84 \times 10^6 s + 3.44 \times 10^{13}}$$

a transfer function of degree 1, or

$$H_0(s) \approx \frac{-3.44 \times 10^{14}}{3.44 \times 10^{13}} = -10$$

a transfer function of degree 0. The frequency ranges in which the simplifications hold, however, are different.

Accelerometer The accelerometer is a device that measures the acceleration of an object to which the device is attached. By integrating the acceleration signal twice, we obtain the velocity and position of the object. The device is a key element in the inertial navigation systems used on airplanes and ships. A simplified model of an accelerometer is shown in Figure 9.9(a). A block with mass m is connected to the box through two springs, as shown. The box is filled with oil to create viscous friction. The box is rigidly attached to, for example, an airplane. Let u and z be, respectively, the displacements of the box and the block relative to the inertia space. Because the block is floating inside the box, u and z are not necessarily equal. We define $y := z - u$. It is the displacement of

Figure 9.9 *(a) Accelerometer and (b) seismometer*

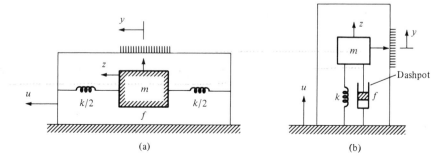

(a)   (b)

the block with respect to the box. This displacement y can be read from the scale on the box or be transformed into a voltage. The input of the accelerometer is the displacement u of the object, and the output is y. Let the spring constant for each spring be $k/2$ and the viscous friction coefficient be f. We now show that by choosing or designing m, f, and k properly, the output y can be approximately proportional to $\ddot{u}(t)$, the acceleration of the object.

The spring force acting on the block is $k[u(t) - z(t)]$. This force overcomes the viscous friction $f[\dot{z}(t) - \dot{u}(t)]$ and accelerates the block. Thus, we have

$$k[u(t) - z(t)] = m\ddot{z}(t) + f[\dot{z}(t) - \dot{u}(t)] \tag{9.27}$$

which can be written as, using $y = z - u$,

$$m\ddot{y}(t) + f\dot{y}(t) + ky(t) = -m\ddot{u}(t)$$

Assuming zero initial conditions and applying the Laplace transform yield

$$ms^2 Y(s) + fsY(s) + kY(s) = -ms^2 U(s)$$

Hence, the transfer function from u to y is

$$H(s) := \frac{Y(s)}{U(s)} = \frac{-ms^2}{ms^2 + fs + k} \tag{9.28}$$

This transfer function is BIBO stable for any positive m, f, and k. Typical amplitude characteristics of $H(s)$ are shown in Figure 9.10. They depend heavily on the poles of (9.28), which are

$$\frac{-f \pm \sqrt{f^2 - 4mk}}{2m}$$

Figure 9.10 *Frequency responses of (9.28)*

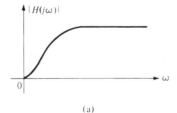

(a)

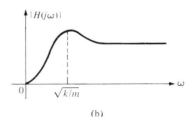

(b)

If $f^2 \geq 4mk$, the two poles of (9.28) are real and the amplitude characteristic does not have overshoot, as shown in Figure 9.10(a). If $f^2 < 4mk$, the two poles are complex conjugate and the amplitude characteristic has an overshoot, as shown in Figure 9.10(b). The smaller the viscous friction coefficient f, the larger the overshoot. The maximal value occurs approximately at $\omega_0 := \sqrt{k/m}$.

If the movement of an object is relatively smooth, like a commercial airline's, the frequency spectrum of the signal will be of low frequency. In this frequency range, if

$$|ms^2 + fs| \ll k$$

where $\ll$ denotes much smaller than, then $H(s)$ can be approximated as

$$H(s) = \frac{Y(s)}{U(s)} \approx \frac{-m}{k} s^2$$

or, in the time domain,

$$y(t) = -\frac{m}{k} \frac{d^2 u(t)}{dt^2} \tag{9.29}$$

The displacement or readout y is proportional to the acceleration. Thus, the device in Figure 9.9(a) can be used as an accelerometer.

Seismometer The seismometer is a device that measures the intensity of earthquakes. The basic structure of a seismometer is shown in Figure 9.9(b). It is basically the same as the structure of the accelerometer shown in Figure 9.9(a) except that the viscous friction is created with the use of a dashpot. The transfer function from u to y of the seismometer also equals

$$H(s) := \frac{Y(s)}{U(s)} = \frac{-ms^2}{ms^2 + fs + k} \tag{9.30}$$

If we choose m, f, and k so that

$$|ms^2| \gg |fs + k|$$

in the frequency spectrum of earthquakes, then we can approximate (9.30) by

$$H(s) \approx -\frac{ms^2}{ms^2} = -1$$

or

$$y(t) = -u(t)$$

Thus, the seismometer will register the intensity and duration of earthquakes. We see that the same transfer function can be designed to approximate either $-ms^2/k$ or -1. The accuracy of the approximation greatly depends on the frequency spectra of the input signals. It is suggested in Reference [15] that, in an accelerometer, the highest frequency component of input signals be smaller than $\sqrt{k/m}/2.5$ and, in a seismometer, the smallest frequency component of input signals be larger than $2.5\sqrt{k/m}$.

*9.5.1   Other reduction methods

One way to reduce a model is to find a simpler transfer function that matches the frequency response of the original transfer function, as discussed in the preceding section. Now, some other reduction methods will be presented.

Consider the systems shown in Figure 9.11. The system with transfer function $G_1(s) = 1/s(s^2 + 2s + 2)$ could be a motor driving a load such as an antenna and is called the plant. The system with transfer function $G_2(s) = 20/(s + 10)$ could be a power amplifier and is called the actuator. Because the pole of $G_2(s)$ is, compared to the poles of $G_1(s)$, much farther away from the imaginary axis, the response excited by the pole of $G_2(s)$ will die out much faster than the responses excited by the poles of $G_1(s)$. Therefore, the pole of $G_2(s)$ can be disregarded and $G_2(s)$ can be simplified as $20/10 = 2$. This type of simplification is widely used in engineering. For example, the transfer function of an amplifier is often modeled as a constant gain if its time constant is much smaller than the time constant of the system to which the amplifier is connected.

The preceding reduction is based essentially on the concept of dominant poles. If a transfer function $H(s)$ has a pole or a pair of complex conjugate poles very close to the imaginary axis, and the rest of the poles are far away from the imaginary axis, as shown in Figure 9.12, then the response of $H(s)$ is essentially governed by the pole or the pair of complex conjugate poles. Thus, they are called the *dominant poles*. For example, the transfer function

$$H(s) = \frac{3}{(1 + 0.01s)(1 + 100s/2525 + s^2/2525)(s + 2)} \qquad (9.31)$$

Figure 9.11 *Connection of two systems*

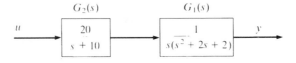

Figure 9.12 *Dominant poles*

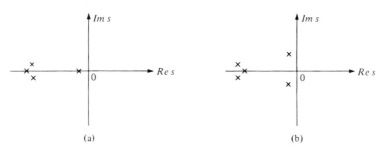

has poles at -100, $-50 \pm j5$, and -2. The pole -2 is the dominant pole. If we disregard the other poles, the transfer function in (9.31) can be simplified as

$$H_s(s) = \frac{3}{s+2} \tag{9.32}$$

The unit step responses of (9.31) and (9.32) are plotted in Figure 9.13(a). They are hardly distinguishable. Therefore, $H(s)$ in (9.31) can be approximated by $H_s(s)$ in (9.32). Note that, when simplifying, we must make sure that $H_s(0)$ equals $H(0)$, otherwise their steady-state responses will be different. Here is another example: The transfer function

$$H(s) = \frac{252500}{(s+100)(s^2+100s+2525)(s^2+2s+5)} \tag{9.33}$$

has poles at -100, $-50 \pm j5$, and $-1 \pm j2$. Because $-1 \pm j2$ are the dominant poles, we retain them in $H_s(s)$. Because $H(0) = 252500/(100 \cdot 2525 \cdot 5) = 0.2$, we require $H_s(0) = 0.2$. Thus, (9.33) can be simplified as

Figure 9.13 *Step responses*

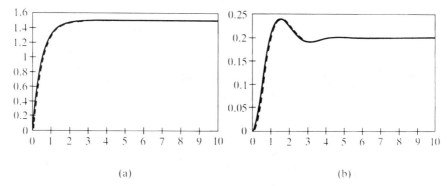

$$H_s(s) = \frac{1}{s^2 + 2s + 5} \tag{9.34}$$

The unit step responses of (9.33) and (9.34) are plotted in Figure 9.13(b). They are indeed very close.

The reduction procedure using dominant poles is not necessarily simple. If zeros appear in $H(s)$ or if poles are not clearly disjointed, then the procedure may not be applicable. In any case, model reduction often involves some engineering judgment. For instance, we never state precisely how close is close. Moreover, a reduction can appear to be rather arbitrary, and it may be in some sense. For example, using the dominant-pole concept, the transfer function in (9.24) can be reduced to −10, or

$$H_s(s) = \frac{-3.44 \times 10^{14}/16.79 \times 10^6}{s + 2.05 \times 10^6} = \frac{-20.5 \times 10^6}{s + 2.05 \times 10^6}$$

The ultimate test of the validity of a reduction must be the satisfactory performance of the resulting physical system.

Here is another type of reduction method: Consider a transfer function expressed as

$$H(s) = h_0 + h_1 s + h_2 s^2 + \cdots + h_n s^n + \cdots \tag{9.35}$$

Suppose we wish to find a transfer function of degree 2 of the form

$$H_s(s) = \frac{b_1 s + b_0}{a_2 s^2 + a_1 s + a_0} \tag{9.36}$$

to approximate $H(s)$. We could choose a_i and b_i to match h_i, $i = 0$, 1, 2, 3, and 4. This type of approximation is called *Pade's approximation*. This method will not always yield acceptable results, however. For example, the method may yield an unstable $H_s(s)$ from a stable $H(s)$. See Reference [30] for a method that resolves this difficulty.

Other reduction methods, such as matching moments, truncating continued fraction expansion, or minimizing an error between the responses of original and simplified transfer functions can be found in Reference [14].

*9.6   STABILITY OF ZERO-INPUT RESPONSES—MARGINAL AND ASYMPTOTIC STABILITIES

In this section, we will study the stability of zero-input responses. Chapter 8 introduced the concept of BIBO stability for zero-state responses and defined it in terms of the input and output signals. In the zero-input response, because the input is identically zero, the concept of BIBO stability is not applicable. The

Figure 9.14 *Stability of zero-input responses*

(a)   (b)

zero-input response is excited by nonzero initial conditions or initial state, so its stability must be defined accordingly. A system, or more precisely its zero-input response, is said to be *marginally stable* or *stable in the sense of Lyapunov*[1] (*stable i.s.L.*) if the response excited by every finite initial state is bounded. It is said to be *asymptotically stable* if the response excited by every finite initial state is bounded and approaches zero as $t \to \infty$. These definitions are applicable only to linear systems. For more general definitions that are applicable to both linear and nonlinear systems, see Reference [4].

Consider the system shown in Figure 9.14(a). If the ball rests at the bottom of the bowl (if the initial state is zero), it will remain at rest. If we move the ball to the location indicated by the dotted outline (if we generate a nonzero initial state), then the ball will oscillate inside the bowl. The amplitude of the oscillation will not increase to infinity. If there is no friction, the ball will continue to oscillate and will never stop. In this case, the system is marginally stable. If there is friction between the ball and the bowl, then the amplitude of the oscillation will decrease with time and the ball will eventually stop at the bottom of the bowl. In this case, the system is asymptotically stable.

Consider now the system shown in Figure 9.14(b). If the ball rests at the top, as shown, it will remain there forever if it is not disturbed. If we move the ball slightly away from the top (see the dotted outline) to generate a very small nonzero initial state, then the ball will roll far away from the top. So the system is not marginally stable.

Here is a different example to illustrate these concepts: Consider the network shown in Figure 9.15(a). Let the voltage across the 1 F capacitor be $y(t)$. Then, its current is $\dot{y}(t)$. The current passing through the inductor is $u(t) - \dot{y}(t)$. Consequently, the voltage across the 1 H inductor is $\dot{u}(t) - \ddot{y}(t)$, which equals $y(t)$. Thus, the network is described by the differential equation

$$y(t) = \dot{u}(t) - \ddot{y}(t)$$

or

$$\ddot{y}(t) + y(t) = \dot{u}(t)$$

[1] In honor of A. M. Lyapunov (1857–1918), a Russian mathematician.

Figure 9.15 *Networks*

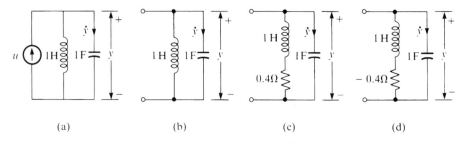

(a)   (b)   (c)   (d)

Now, if the input is zero, the network reduces to the one in Figure 9.15(b) and the equation becomes

$$\ddot{y}(t) + y(t) = 0 \tag{9.37}$$

The application of the Laplace transform to (9.37) yields

$$s^2 Y(s) - sy(0-) - \dot{y}(0-) + Y(s) = 0$$

which implies

$$Y(s) = \frac{sy(0-) + \dot{y}(0-)}{s^2 + 1}$$

Thus, the zero-input response is

$$y(t) = y(0-) \cos t + \dot{y}(0-) \sin t \tag{9.38}$$

For any finite initial conditions $y(0-)$ and $\dot{y}(0-)$, the response is a sinusoidal function and is clearly bounded. Thus, the network in Figure 9.15(b) is marginally stable. The response will not approach zero as $t \to \infty$, so the network is not asymptotically stable. Note that, for different initial conditions $y(0-)$ and $\dot{y}(0-)$, the amplitude of the sinusoid in (9.38) will be different. However, its frequency will always equal 1. The frequency is determined by the pure imaginary modes of the characteristic polynomial.

Consider the network shown in Figure 9.15(c). Let $y(t)$ be the voltage across the capacitor. Then, the current in the loop is $\dot{y}(t)$. The voltages across the resistor and inductor are, respectively, $0.4\dot{y}(t)$ and $\ddot{y}(t)$. Thus, we have

$$\ddot{y}(t) + 0.4\dot{y}(t) + y(t) = 0 \tag{9.39}$$

The application of the Laplace transform yields

$$s^2 Y(s) - sy(0-) - \dot{y}(0-) + 0.4sY(s) - 0.4y(0-) + Y(s) = 0$$

or

$$Y(s) = \frac{(s+0.4)y(0-) + \dot{y}(0-)}{s^2 + 0.4s + 1} = \frac{(s+0.4)y(0-) + \dot{y}(0-)}{(s+0.2)^2 + 0.96} \qquad \textbf{(9.40)}$$

If $y(0-) = 1$ and $\dot{y}(0-) = -1$, then

$$Y(s) = \frac{s + 0.4 - 1}{(s+0.2)^2 + (0.98)^2} = \frac{(s+0.2) - 0.82 \times 0.98}{(s+0.2)^2 + (0.98)^2}$$

and its inverse Laplace transform is

$$y(t) = e^{-0.2t} \cos 0.98t - 0.82 e^{-0.2t} \sin 0.98t$$

In general, the response excited by any finite initial conditions is a linear combination of $e^{-0.2t} \sin 0.98t$ and $e^{-0.2t} \cos 0.98t$, as was discussed in Section 4.6. Because of the exponentially decreasing term $e^{-0.2t}$, the response is bounded and approaches zero as $t \to \infty$. Thus, the network in Figure 9.15(c) is asymptotically stable.

The preceding analysis for the networks in Figure 9.15(b) and (c) was, in fact, unnecessary. Using physical reasoning, we can show that every network consisting of only LTI resistors, inductors, and capacitors is either marginally stable or asymptotically stable. Inductors and capacitors are elements with memory, and they can store energy in their magnetic and electric fields. They are called passive elements because the total energy that they provide to a network can never be larger than the initially stored energy. They are called lossless because they do not dissipate energy. Therefore, because the total energy of a network that consists of only capacitors and inductors will neither increase nor decrease, its zero-input response will approach neither infinity nor zero as $t \to \infty$ and the network is marginally stable. If a network also contains resistors that dissipate energy, then the total energy of the network will eventually approach zero. Thus, its zero-input response will approach zero as $t \to \infty$ and the network is asymptotically stable. In conclusion, *any linear time-invariant LC network is marginally stable and any linear time-invariant RLC network is asymptotically stable.*

So far, our discussion has implicitly assumed that resistors, inductors, and capacitors have positive resistance, inductance, and capacitance. Only under this assumption are they passive elements. Otherwise, they are active elements. If a network contains active elements, it is not always stable. For example, consider the network shown in Figure 9.15(d). It contains a resistor with negative resistance $-0.4 \, \Omega$. A negative resistor can be simulated or realized using operational amplifier circuits (see Problem 9.3). The network is described, as in (9.39), by

$$\ddot{y}(t) - 0.4\dot{y}(t) + y(t) = 0$$

which implies, as in (9.40),

$$Y(s) = \frac{(s - 0.4)y(0-) + \dot{y}(0-)}{(s - 0.2)^2 + (0.98)^2}$$

Thus, its zero-input response is a linear combination of the two modes $e^{0.2t} \cos 0.98t$ and $e^{0.2t} \sin 0.98t$. Because of the exponentially increasing factor $e^{0.2t}$, the two modes approach infinity as $t \rightarrow \infty$. Thus, the system in Figure 9.15(d) is neither marginally stable nor asymptotically stable.

With this preliminary, we are ready to discuss the general case. Consider a system described by the differential equation

$$D(p)y(t) = N(p)u(t) \tag{9.41}$$

where

$$D(p) := a_n p^n + a_{n-1} p^{n-1} + \cdots + a_1 p + a_0$$
$$N(p) := b_m p^m + b_{m-1} p^{m-1} + \cdots + b_1 p + b_0$$

and $p^i y(t) := y^{(i)}(t) := d^i y(t)/dt^i$, $i = 0, 1, 2, \ldots$. In this study of the zero-input response, we assume $u(t) = 0$, for all t, and so (9.41) reduces to

$$D(p)y(t) = 0 \tag{9.42}$$

The polynomial $D(p)$ or $D(s)$ is the characteristic polynomial, and the roots of $D(s)$ are the modes of the system. As discussed in Section 4.6, the zero-input response excited by any initial conditions is always a linear combination of the time responses of the modes. Thus, the stability of zero-input responses depends on the location of modes.

Theorem 9.1 An LTIL system described by (9.41) is marginally stable if, and only if, every root of $D(s)$ has a zero or negative real part and the root with a zero real part must be simple. The system is asymptotically stable if and only if every root of $D(s)$ has a negative real part or, equivalently, $D(s)$ is a Hurwitz polynomial.

A root λ is a simple root if it is the only root at λ. It is a repeated root if there are two or more roots located at λ. Theorem 9.1 differs from Theorem 8.2

in that it permits simple roots on the $j\omega$-axis. Simple roots on the $j\omega$-axis, such as $s = \pm j\omega_0$ and $s = 0$, yield $\sin \omega_0 t$ and step functions that are bounded in $[0, \infty)$. If the roots are repeated, then their time responses are $t\sin \omega_0 t$ or t that approaches infinity as $t \to \infty$. Therefore, marginal stability permits simple roots, but not repeated roots, on the imaginary axis. This essentially establishes Theorem 9.1.

Note that BIBO stability does not permit simple poles on the $j\omega$-axis. This is because we can always find a bounded input with the same poles to generate repeated poles in the Laplace transform of the output, and so the output will not be bounded.

EXERCISE 9.6.1

Consider the network shown in Figure 9.16(a). Is the network BIBO stable? If not, find a bounded input to excite an unbounded output. Is the network marginally stable? Is it asymptotically stable? How about the circuit in Figure 9.16(b)? Is it BIBO stable? marginally stable? asymptotically stable?

[**ANSWERS:** No, $\sin \sqrt{5}\, t$. Yes. No. Yes, yes, yes.]   ■

Asymptotic stability requires every zero-input response to be bounded and to approach zero, but marginal stability requires only boundedness. Therefore, asymptotic stability always implies marginal stability, but the converse is not true. Let us discuss the relationship between BIBO stability and asymptotic stability. Consider an LTIL system described by the differential equation

$$D(p)y(t) = N(p)u(t)$$

The system or, more precisely, the zero-input response is asymptotically stable if and only if every mode—that is, every root of $D(s)$—has a negative real part. The system or, more precisely, the zero-state response is BIBO stable if and only if every pole of its transfer function $H(s) = N(s)/D(s)$ has a negative real part. Because of the possible existence of common factors between $N(s)$ and $D(s)$, every pole is a mode but not every mode is a pole. Thus we conclude that

Figure 9.16 *Two networks*

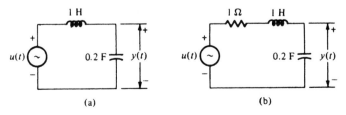

(a)   (b)

a system that is asymptotically stable is also BIBO stable, but a system that is BIBO stable is not necessarily asymptotically stable. An example illustrates.

EXAMPLE 9.6.1

Consider the system described by

$$\ddot{y}(t) - \dot{y}(t) - 2y(t) = \dot{u}(t) - 2u(t)$$

Its characteristic polynomial is

$$s^2 - s - 2 = (s - 2)(s + 1)$$

The system has two modes, 2 and −1. Because one mode does not have a negative real part, the system is not asymptotically stable. The transfer function of the system is

$$H(s) = \frac{s - 2}{s^2 - s - 2} = \frac{1}{s + 1}$$

The system has only one pole at −1, which has a negative real part. Thus, the system is BIBO stable.

If a system is completely characterized by its transfer function, then the set of modes equals the set of poles, as discussed in Section 4.10. In this case, when every pole has a negative real part, every mode does, too, and BIBO stability implies asymptotic stability. If a system is not completely characterized by its transfer function, then BIBO stability does not imply asymptotic stability as the preceding example illustrated. Note that if a system is not BIBO stable, then it is not asymptotically stable no matter the system is completely characterized by its transfer function or not.

*9.6.1   Discrete-time case

The discussion in the preceding section is applicable, with minor modification, to the discrete-time case. Consider an LTIL discrete-time system described by

$$D(p)y[k] = N(p)u[k]$$

where $D(p)$ and $N(p)$ are defined as in (5.50). The system or, more precisely, its zero-input response is marginally stable if the response excited by every set of finite initial conditions is bounded. It is asymptotically stable if the response is bounded and approaches zero as $k \to \infty$. The system is marginally stable if and only if every root of $D(z)$ has a magnitude less than or equal to 1 and the root with magnitude 1 must be simple. The system is asymptotically stable if and only if every root of $D(z)$ has a magnitude less than 1 or, equivalently, $D(z)$ is a Schur polynomial. These conditions are obtained from Theorem 9.1 by the mapping of $z = e^{Ts}$ or the mapping of the open left half–s-plane into the interior of the unit circle on the z-plane. All discussion for the continuous-time case applies directly to the discrete-time case and will not be repeated.

*9.7   MODEL OF OP-AMP CIRCUITS

Most systems are designed to be asymptotically stable and, consequently, BIBO stable. The only exceptions are oscillators, which are designed to be marginally stable. In order to study an oscillator, we will develop a block diagram for the op-amp circuit shown in Figure 9.17(a). This will be a generalization of the block diagram in Figure 9.5(c).

Consider the circuit in Figure 9.17(a). Assume $I_- = I_+ = 0$, but do not assume $E_- = E_+$. Capital letters are the Laplace transforms of the corresponding lower-case letters. The output of the circuit is fed back into the inverting terminal through impedance Z_2 and into the noninverting terminal through Z_4. The input v_1 is applied through impedance Z_1 to the inverting terminal, and the

Figure 9.17 *Op-amp circuit and its block diagram*

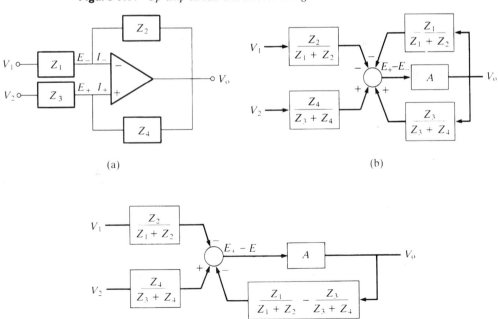

(a)

(b)

(c)

input v_2 is applied through impedance Z_2 to the noninverting terminal. Because $I_- = 0$ at the inverting terminal, we have

$$\frac{V_1 - E_-}{Z_1} = -\frac{V_o - E_-}{Z_2}$$

which implies

$$E_- = \frac{Z_2 V_1 + Z_1 V_o}{Z_1 + Z_2} = \frac{Z_2}{Z_1 + Z_2} V_1 + \frac{Z_1}{Z_1 + Z_2} V_o \qquad (9.43)$$

At the noninverting terminal, we have

$$\frac{V_2 - E_+}{Z_3} = -\frac{V_o - E_+}{Z_4}$$

which implies

$$E_+ = \frac{Z_4 V_2 + Z_3 V_o}{Z_3 + Z_4} = \frac{Z_4}{Z_3 + Z_4} V_2 + \frac{Z_3}{Z_3 + Z_4} V_o \qquad (9.44)$$

The substitution of (9.43) and (9.44) into the Laplace transform of (9.1) yields

$$V_o = A(E_+ - E_-) = A \left[\frac{Z_4}{Z_3 + Z_4} V_2 + \frac{Z_3}{Z_3 + Z_4} V_o - \frac{Z_2}{Z_1 + Z_2} V_1 - \frac{Z_1}{Z_1 + Z_2} V_o \right]$$

which can be grouped as

$$\left[1 + \frac{Z_1}{Z_1 + Z_2} A - \frac{Z_3}{Z_3 + Z_4} A \right] V_o = \frac{Z_4}{Z_3 + Z_4} A V_2 - \frac{Z_2}{Z_1 + Z_2} A V_1$$

Thus, we have

$$V_o = \frac{\dfrac{Z_4}{Z_3 + Z_4} A}{1 + \dfrac{Z_1}{Z_1 + Z_2} A - \dfrac{Z_3}{Z_3 + Z_4} A} V_2 - \frac{\dfrac{Z_2}{Z_1 + Z_2} A}{1 + \dfrac{Z_1}{Z_1 + Z_2} A - \dfrac{Z_3}{Z_3 + Z_4} A} V_1 \qquad (9.45)$$

This equation describes the op-amp circuit in Figure 9.17(a). Using this equation, we can develop the block diagram in Figure 9.17(b). The block diagram has two feedback paths: One feeds negatively into the adder through the transfer function $Z_1/(Z_1 + Z_2)$, the other feeds positively into the adder through the transfer function $Z_3/(Z_3 + Z_4)$. These two loops can be combined into one, as shown in Figure 9.17(c), with the feedback transfer function $[Z_1/(Z_1 + Z_2) - Z_3/(Z_3 + Z_4)]$. For the block diagrams in Figures 9.17(b) and (c), the output excited by v_1 is, using (4.76),

$$V_o = -\frac{Z_2}{Z_1+Z_2} \cdot \frac{A}{1 + A\dfrac{Z_2}{Z_1+Z_2} - A\dfrac{Z_3}{Z_3+Z_4}} V_1$$

The output excited by v_2 is, using again (4.76),

$$V_o = \frac{Z_4}{Z_3+Z_4} \cdot \frac{A}{1 + A\dfrac{Z_2}{Z_1+Z_2} - A\dfrac{Z_3}{Z_3+Z_4}} V_2$$

These equations become (9.45) following the additivity property of linear systems. Thus, the block diagrams in Figures 9.17(b) and (c) describe the op-amp circuit in Figure 9.17(a).

The block diagram in Figure 9.17(b) also can be obtained from the circuit in Figure 9.17(a) using physical argument. The block from v_o to the inverting terminal has the transfer function

$$\left.\frac{E_-}{V_o}\right|_{V_1=V_2=0} = \frac{Z_1}{Z_1 + Z_2}$$

It is the voltage at the inverting terminal due to $V_1 = 0$ (grounding the input v_1) and $V_o = 1$. The block transfer function from v_1 to the inverting terminal is the voltage at the inverting terminal due to $V_o = 0$ (grounding the output v_o) and $V_1 = 1$. The other two block transfer functions can be similarly obtained. Note that the block diagram in Figure 9.17(b) reduces to the one in Figure 9.5(c) if $V_2 = 0$, $Z_3 = 0$, and $Z_4 = \infty$.

As was discussed in Section 4.10.1, the loading problem must be considered when developing block diagrams. The block diagram in Figure 9.17(b) is developed from the mathematical equation describing the entire circuit in Figure 9.17(a); therefore, the loading problem has been taken care of in the block diagram.

EXAMPLE 9.7.1

Consider the op-amp circuit shown in Figure 9.18(a). Its block diagram can be readily obtained and is shown in Figure 9.18(b). Because the block transfer functions from V_1 and V_2 to the adder are the same, they can be combined as shown in Figure 9.18(c), where we have also reduced the loop to a single block by using (4.76). Thus, the transfer function from $(V_2 - V_1)$ to V_o is

Figure 9.18 *Difference amplifier and its block diagrams*

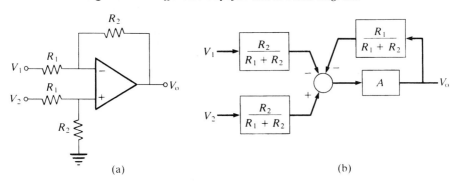

(a)   (b)

(c)

$$H(s) = \frac{R_2}{R_1 + R_2} \cdot \frac{A(R_1 + R_2)}{AR_1 + R_1 + R_2}$$

If $A = \infty$, the equation reduces to

$$H(s) = \frac{R_2}{R_1 + R_2} \cdot \frac{R_1 + R_2}{R_1} = \frac{R_2}{R_1} \qquad (9.46)$$

or

$$V_o = \frac{R_2}{R_1}(V_2 - V_1) \quad \text{or} \quad v_o = \frac{R_2}{R_1}(v_2 - v_1) \qquad (9.47)$$

Thus the circuit is called a *difference amplifier.*

---

*9.8   STABILITY ANALYSIS OF AN OSCILLATOR CIRCUIT

An oscillator is a device that will sustain, once it is excited, an oscillation without requiring that any further input be applied. The response of an oscillator is, therefore, a zero-input response. We argue that every LTIL oscillator must be marginally stable. If it is neither marginally stable nor asymptotically stable, the response will go to infinity and the device will saturate or burn out. If it is asymptotically stable, the response will eventually approach zero, and the oscillation will not be sustained. Therefore, an LTIL oscillator must be marginally stable but not asymptotically stable. What differentiates marginal stability from asymptotic stability is the existence of at least a pair of simple pure imaginary modes. The pure imaginary modes also determine the frequency of oscillation. This condition will be used to design an oscillator circuit in this section. Note that every oscillator must be connected to a power supply; otherwise the oscillation cannot be sustained. The power supply, however, is not considered as an input. This is similar to televisions: Their inputs are the signals emitted by television stations; power supplies are not considered as inputs.

Consider the circuit shown in Figure 9.19. The inverting terminal is connected to the input v_1 through resistor R_1 and to the output v_o through resistor R_2. The noninverting terminal is connected to the input v_2 through the parallel connection of resistor R and capacitor C and to the output through the series connection of the same R and C. If the circuit is properly designed, the circuit will maintain an oscillation even if v_1 and v_2 are zero or grounded. Thus the circuit with v_1 and v_2 grounded is called the *Wien-bridge oscillator*. To carry out an analysis of this circuit, we first assume $v_1 = 0$ and compute

$$Z_3 = \frac{R \cdot \dfrac{1}{Cs}}{R + \dfrac{1}{Cs}} = \frac{R}{RCs + 1} \tag{9.48a}$$

and

$$Z_4 = R + \frac{1}{Cs} = \frac{RCs + 1}{Cs} \tag{9.48b}$$

These are, respectively, the impedances of the parallel and series connections of R and C. The block diagram of this circuit is shown in Figure 9.20(a), which is

Figure 9.19 *Wien-bridge oscillator*

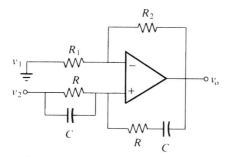

Figure 9.20   *Models of a Wien-bridge oscillator*

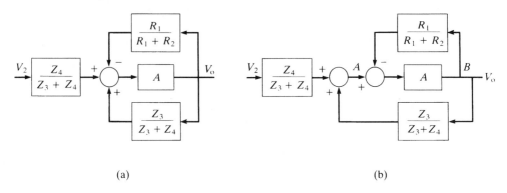

(a)   (b)

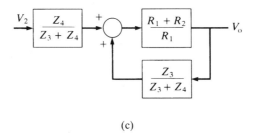

(c)

reduced from Figure 9.17(b). Because $v_1 = 0$, the input branch from v_1 to the summer is not drawn. If we redraw the block diagram as shown in Figure 9.20(b), then the transfer function from A to B shown can be computed as, using (4.76),

$$H_{AB} = \frac{A}{1 + A \cdot \dfrac{R_1}{R_1 + R_2}} = \frac{1}{\dfrac{1}{A} + \dfrac{R_1}{R_1 + R_2}}$$

which, if A is considered to be very large or infinite, reduces to

$$H_{AB} = \frac{R_1 + R_2}{R_1} \tag{9.49}$$

Thus, the two-loop feedback systems in Figures 9.20(a) and (b) can be reduced to the single-loop feedback system in Figure 9.20(c). The transfer function from v_2 to v_o is, using (4.77),

$$H(s) = \frac{V_o}{V_2} = \frac{Z_4}{Z_3 + Z_4} \cdot \frac{\dfrac{R_1 + R_2}{R_1}}{1 - \dfrac{R_1 + R_2}{R_1} \cdot \dfrac{Z_3}{Z_3 + Z_4}} \tag{9.50}$$

$$= \frac{Z_4(R_1 + R_2)}{R_1(Z_3 + Z_4) - (R_1 + R_2)Z_3} = \frac{Z_4(R_1 + R_2)}{R_1 Z_4 - R_2 Z_3}$$

Substituting (9.48) into (9.50) yields

$$H(s) = \frac{V_o}{V_2} = \frac{\dfrac{RCs + 1}{Cs}(R_1 + R_2)}{R_1\left[\dfrac{RCs + 1}{Cs}\right] - R_2\left[\dfrac{R}{RCs + 1}\right]}$$

$$= \frac{(R_1 + R_2)(RCs + 1)^2}{R_1(RCs + 1)^2 - R_2 RCs} = \frac{(R_1 + R_2)(RCs + 1)^2}{R_1(RCs)^2 + (2R_1 - R_2)RCs + R_1}$$

which implies

$$[R_1(RCs)^2 + (2R_1 - R_2)RCs + R_1]V_o = (R_1 + R_2)(RCs + 1)^2 V_2 \tag{9.51}$$

This algebraic equation, in the Laplace transform, describes the circuit in Figure 9.19. If the input v_2 is zero, the equation reduces to

$$[R_1(RCs)^2 + (2R_1 - R_2)RCs + R_1]V_o = 0$$

Thus, the characteristic polynomial of the circuit is

$$D(s) := R_1(RCs)^2 + (2R_1 - R_2)RCs + R_1 \tag{9.52}$$

The condition for $D(s)$ to have a pair of pure imaginary roots is

$$2R_1 = R_2 \tag{9.53}$$

Under this condition, $D(s)$ reduces to

$$D(s) = R_1[(RCs)^2 + 1] \tag{9.54}$$

and has a pair of pure imaginary roots at $\pm j/RC$. Therefore, the circuit or, more precisely, its zero-input response is marginally stable. The circuit has two modes $\pm j/RC$ and its zero-input response is of the form, as was discussed in Section 4.6,

$$v_o(t) = k_1 e^{jt/RC} + k_2 e^{-jt/RC} = \bar{k}_1 \sin\left(\frac{1}{RC}t + \bar{k}_2\right) \tag{9.55}$$

Thus, the frequency of oscillation is

$$\omega = \frac{1}{RC} \text{ rad/s} \tag{9.56}$$

Because of the simplicity in its structure and design, the Wien-bridge oscillator is widely used.

Although the frequency of oscillation is fixed by the circuit, the amplitude of oscillation depends on initial conditions and is not fixed. Different excitation yields different amplitude. In order to have a fixed amplitude of oscillation, the circuit in Figure 9.19 must be modified as follows: First, R_2 must be chosen to be slightly larger than $2R_1$ so that the characteristic polynomial in (9.52) has a pair of open right half-plane complex roots. Now the circuit is not stable and its zero-input response excited by, for example, thermal noise or a power-supply transient will grow without bound. Therefore, a nonlinear circuit, called a limiter, must be introduced to keep the amplitude at a fixed value. See Reference [28]. Thus, an actual Wien-bridge oscillator is more complex than the one shown in Figure 9.19. However, the preceding linear analysis does illustrate the basic design of oscillators.

*9.8.1   Loop gain and oscillation

There is an alternative method for analyzing Wien-bridge oscillators. Consider the feedback system shown in Figure 9.21(a). It has two blocks with transfer functions $\bar{A}(s)$ and $\beta(s)$. The product of $\bar{A}(s)$ and $\beta(s)$, or $L(s) := \bar{A}(s)\beta(s)$, is called the *loop transfer function* or the *loop gain*. In general, if

$$L(s) = \bar{A}(s)\beta(s) = 1 \tag{9.57a}$$

or, more precisely, if there exists a frequency ω_0 such that

$$L(s)\big|_{s=j\omega_0} = L(j\omega_0) = \bar{A}(j\omega_0)\beta(j\omega_0) = 1 \tag{9.57b}$$

then the system can sustain an oscillation with frequency ω_0. This is called the *loop gain condition* and is widely used in electronics texts to establish conditions of oscillation. This condition can be established intuitively as follows: Suppose the output is a sinusoid. If the loop gain is larger than 1, then the

Figure 9.21 *Feedback system*

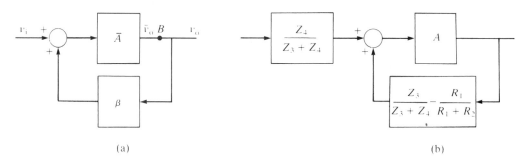

(a)   (b)

magnitude of the sinusoid will increase as the signal travels continuously around the loop. If the loop gain is smaller than 1, then the magnitude will decrease. Thus, if the loop gain is different from 1, a sustained oscillation cannot be maintained. On the other hand, if the loop gain is 1, the magnitude of the oscillation neither increases nor decreases as the signal travels around the loop and the oscillation is sustained. The condition also can be argued intuitively as follows: If we break the loop at B in Figure 9.21(a), then the signal $\bar{v}_o$ at the output of $\bar{A}$ is

$$\bar{v}_o = \bar{A}\beta v_o$$

If $\bar{A}\beta \neq 1$, then $\bar{v}_o \neq v_o$ and the signal v_o will change after reconnecting the loop. If $\bar{A}\beta = 1$, then $\bar{v}_o = v_o$ and v_o will not change after reconnecting the loop. Thus, the system can generate a sustained signal v_o. Although this argument is very loose, it is easy to understand.

Now let us use the loop gain to study the Wien-bridge oscillator. If we use the block diagram in Figure 9.20(c), then the condition of oscillation is

$$\frac{R_1 + R_2}{R_1} \cdot \frac{Z_3}{Z_3 + Z_4} = 1 \tag{9.58}$$

which can be simplified as

$$(R_1 + R_2)Z_3 - R_1(Z_3 + Z_4) = R_2 Z_3 - R_1 Z_4 = 0 \tag{9.59}$$

This is the denominator of the transfer function in (9.50). Substituting (9.48) into (9.59) yields

$$
\begin{aligned}
R_2 \frac{R}{RCs + 1} - R_1 \frac{RCs + 1}{Cs} &= \frac{R_2 RCs - R_1(RCs + 1)^2}{Cs(RCs + 1)} \\
&= \frac{-[R_1(RCs)^2 + (2R_1 - R_2)RCs + R_1]}{Cs(RCs + 1)} = 0
\end{aligned}
$$

which implies

$$R_1(RCs)^2 + (2R_1 - R_2)RCs + R_1 = 0 \tag{9.60}$$

This is the characteristic polynomial in (9.52). If $2R_1 = R_2$ and if $s = \pm j/RC$, then (9.60) and, consequently, the loop gain condition are met and the circuit will sustain an oscillation with frequency $1/RC$. This result is the same as the one obtained in the preceding section.

The block diagram in Figure 9.20(a) can also be reduced to the single-loop system shown in Figure 9.21(b). The loop gain condition then becomes

$$A\left[\frac{Z_3}{Z_3 + Z_4} - \frac{R_1}{R_1 + R_2} \right] = 1 \tag{9.61a}$$

If A is very large or infinite, then (9.61a) implies

$$\frac{Z_3}{Z_3 + Z_4} - \frac{R_1}{R_1 + R_2} = 0 \tag{9.61b}$$

or

$$(R_1 + R_2)Z_3 - R_1(Z_3 + Z_4) = R_2 Z_3 - R_1 Z_4 = 0$$

This is the same as (9.59). Thus, the loop gain condition in (9.61) yields the same result as the loop gain condition in (9.58), even though they appear to be quite different.

Now we discuss the relationship between the loop gain condition and the oscillation condition on the characteristic polynomial. Let

$$\bar{A}(s) = \frac{N_A(s)}{D_A(s)} \quad \beta(s) = \frac{N_\beta(s)}{D_\beta(s)}$$

Then, the transfer function from v_i to v_o in Figure 9.21(a) is

$$H(s) = \frac{V_o}{V_i} = \frac{\bar{A}(s)}{1 - \bar{A}(s)\beta(s)} = \frac{\dfrac{N_A(s)}{D_A(s)}}{\dfrac{D_A(s)D_\beta(s) - N_A(s)N_\beta(s)}{D_A(s)D_\beta(s)}}$$

$$= \frac{N_A(s)D_\beta(s)}{D_A(s)D_\beta(s) - N_A(s)N_\beta(s)} \tag{9.62}$$

Its denominator is the characteristic polynomial of the system. We see that the characteristic polynomial equals the numerator of the rational function

$$1 - L(s) = 1 - \bar{A}(s)\beta(s) = 1 - \frac{N_A(s)}{D_A(s)} \frac{N_\beta(s)}{D_\beta(s)} = \frac{D_A(s)D_\beta(s) - N_A(s)N_\beta(s)}{D_A(s)D_\beta(s)}$$

If there exists ω_0 such that $L(\pm j\omega_0) = 1$ or

$$1 - L(\pm j\omega_0) = 0$$

or, equivalently,

$$D(\pm j\omega_0) := D_A(\pm j\omega_0)D_\beta(\pm j\omega_0) - N_A(\pm j\omega_0)N_\beta(\pm j\omega_0) = 0$$

then the characteristic polynomial has a pair of pure imaginary roots. Thus, the loop gain condition in (9.57) only checks whether or not the characteristic polynomial has a pair of pure imaginary roots.

Consider a system with characteristic polynomial $D(s)$. Suppose $D(s)$ has a pair of pure imaginary roots $\pm j\omega_0$ and a real root α, or

$$D(s) = (s - \alpha)(s + j\omega_0)(s - j\omega_0)$$

The general form of the zero-input response of this system is

$$k_1 e^{\alpha t} + k_2 \sin(\omega_0 t + k_3)$$

If $\alpha < 0$, or if the root is in the open left half-plane, the zero-input response will approach a sinusoid. If $\alpha > 0$, or if the root is in the open right half-plane, then the response will approach infinity. Thus, in order to sustain an oscillation, the system must have a pair of pure imaginary modes and the remaining modes must be stable modes. The loop gain condition checks the existence of imaginary modes but does not check whether the remaining modes are stable. Thus, strictly speaking, the loop gain condition is only a necessary condition for the existence of oscillations.

If a system has only two modes, then the loop gain condition is also sufficient for the existence of an oscillation. This is the case for the Wien-bridge oscillator studied in this chapter. For systems with three or more modes, the loop gain condition will not guarantee the existence of an oscillation.

9.9   SUMMARY

1. Operational amplifier circuits are widely used in practice. Unlike passive RLC networks, which are always BIBO stable, circuits that contain op amps may not be BIBO stable. To study the stability of op-amp circuits, a more realistic open loop gain $A(s)$, such as the one in (9.16), must be used. Only if op-amp circuits are known to be BIBO stable can we use $A(s) = \infty$ for analysis and design.

2. The transfer functions of the op-amp circuit in Figure 9.2(b) and the differentiator in Figure 9.4(c) are, respectively, 1 and $-RCs$ if we use $A(s) = \infty$ in their analyses. However, because they are actually not BIBO stable, they cannot be used in practice and the analyses based on $A(s) = \infty$ are meaningless. Note that a positive feedback system is not always unstable and a negative feedback system is not always BIBO stable, as shown in Example 8.2.2 and Exercise 8.2.3.

3. The inverting amplifier and integrator in Figure 9.4(a) and (b) are BIBO stable and can be analyzed using the idealized model.

4. Model reduction is used in engineering to simplify analysis and design. However, model reduction holds only for input signals whose frequency spectra lie in a limited range. For example, the inverting amplifier with a constant gain holds only for signals with low-frequency spectra, and the integrator with transfer function $-1/RCs$ holds only for signals with no DC part.

5. Stability of the zero-input response and stability of the zero-state response of LTIL systems are often studied separately. BIBO stability is defined for the zero-state response, and marginal stability and asymptotic stability are defined for the zero-input response. If the zero-input response excited by every finite initial state is bounded, then the system is marginally stable. If, in addition, the response approaches zero, then the system is asymptotically stable.

6. The condition for a system to be asymptotically stable is that the system has no modes in the *closed* right half–s-plane. The condition for a system to be marginally stable is that the system has no modes in the *open* right half–s-plane and no repeated modes on the $j\omega$-axis. Simple modes on the $j\omega$-axis are permitted.

7. Asymptotic stability implies BIBO stability. A system can be BIBO stable without being asymptotically stable. However, if a system is completely characterized by its transfer function, then BIBO stability implies asymptotic stability.

8. A system with only one mode cannot perform as an oscillator. A system with two modes can perform as an oscillator if and only if the two modes are complex-conjugate pure imaginary modes or if and only if its loop gain equals 1. A system with two or more modes can perform as an oscillator if

and only if it is marginally stable. The loop gain condition only checks the existence of imaginary modes and, while necessary, is not sufficient for a system to perform as an oscillator.

PROBLEMS

9.1 Consider the op-amp circuit shown in Figure P9.1. Use the idealized model to show $v_0 = -(v_{i1} + v_{i2} + v_{i3})$. The circuit is called a summing circuit.

9.2 Use the idealized model to compute the transfer functions of the op-amp circuits shown in Figure P9.2.

9.3 (a) Consider the op-amp circuit shown in Figure P9.3. Use the idealized model to show

$$Z_{in} = \frac{V}{I} = -\frac{R_1}{R_2} Z$$

(b) Use the circuit to build a resistor with resistance -1 KΩ and a capacitor with capacitance -0.001 F.

9.4 Consider the integrator in Figure 9.4(b) with $R_1 C = 1$. If the input is $a\sin \omega_0 t$, what is its output? If the output is limited to ± 15 V and if $a = 1$, what is the frequency range in which the circuit will not saturate? If $\omega_0 = 10$ Hz, what is the range of a in which the circuit will not saturate?

9.5 Develop block diagrams for the op-amp circuits in Figure P9.2(b) and (c). If the open loop gain is given as in (9.16), what are their transfer functions? Are they BIBO stable?

Figure P9.1

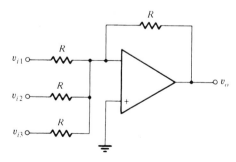

Figure P9.2

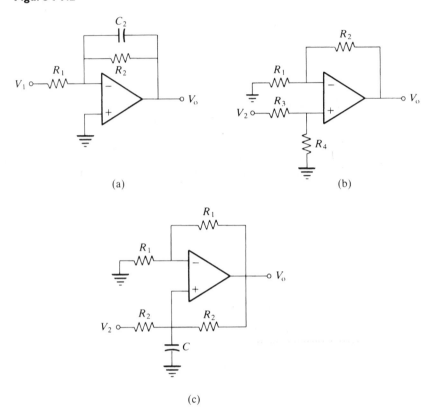

(a)

(b)

(c)

9.6 Show that the differentiator in Figure 9.4(c) with $A(s)$ in (9.16) is BIBO stable if $R_2 = 1\ \text{M}\Omega = 10^6\ \Omega$ and $C = 1\ \mu F = 10^{-6}\ F$.

9.7 Consider an amplifier with transfer function

$$H(s) = \frac{-10}{0.01s + 1}$$

(a) If the input $u(t)$ is $\sin 0.1t$, can $H(s)$ be simplified as -10?

(b) If the input $u(t)$ is $\sin 10^3 t$, can $H(s)$ be simplified as -10?

(c) Find the frequency range in which $H(s)$ can be simplified as -10.

(d) Repeat (a), (b), and (c) for

$$H(s) = \frac{-10}{-0.01s + 1}$$

9.8 The circuit in Figure P9.8 is sometimes called an integrator in industry.

Figure P9.3

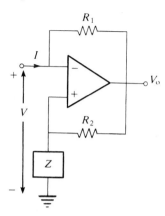

Find the values of R and C and the frequency range of input signals to justify such a name.

9.9 The circuit in Figure P9.9 is sometimes called a differentiator in industry. Find the values of R and C and the frequency range of input signals to justify such a name.

9.10 Consider the transfer function

$$H(s) = \frac{-s^2}{s^2 + (10^{-2} + 10^{-3})s + 10^{-5}}$$

Figure P9.8

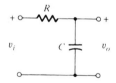

Figure P9.9

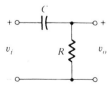

Plot its frequency response. For the input $u(t) = e^{-0.1t} \sin 10t$, can $H(s)$ be simplified as -1? Can it be simplified as

$$H_1(s) = \frac{-s}{s + (10^{-2} + 10^{-3})}$$

or

$$H_2(s) = \frac{-s^2}{s^2 + 10^{-5}}$$

9.11 Consider the transfer function

$$H(s) = \frac{-s^2}{10^{-5}s^2 + (10^{-2} + 10^{-3})s + 1}$$

Plot its frequency response. For the input $u(t) = e^{-0.1t}\sin t$, can $H(s)$ be simplified as $-s^2$? Can it be simplified as the following?

$$H_1(s) = \frac{-s^2}{(10^{-2} + 10^{-3})s + 1}$$

Or as this?

$$H_2(s) = \frac{-s^2}{10^{-5}s^2 + 1}$$

9.12 (a) Compute the unit step response of

$$H_1(s) = \frac{2}{(1 + 0.01s)(s^2 + 2s + 2)}$$

(b) Compute the unit step response of

$$H_2(s) = \frac{2}{s^2 + 2s + 2}$$

(c) Can the transfer function in (a) be simplified as in (b)? Can the approximation be used for any input? If not, find an input for which the approximation does not hold.

9.13 (a) Compute the unit step response of

$$H_1(s) = \frac{2(s + 1)}{(1 + 0.01s)(s^2 + 2s + 2)}$$

(b) Compute the unit step response of

$$H_2(s) = \frac{2(s+1)}{s^2 + 2s + 2}$$

(c) Can the transfer function in (a) be simplified as in (b)? Can the approximation be used for any input? If not, find an input for which the approximation does not hold.

9.14 (a) Plot the amplitude characteristics of

$$H_1(s) = \frac{1}{(1 - 0.0001s)(s + 1)} \quad \text{and} \quad H_2(s) = \frac{1}{(1 + 0.0001s)(s + 1)}$$

Are they the same?

(b) Compare their unit step responses.

9.15 The U-tube shown in Figure P9.15 can be used to measure pressure and is called a *manometer*. Suppose the transfer function from p to h is

$$G(s) = \frac{H(s)}{P(s)} = \frac{0.1}{0.01s^2 + 0.1s + 1}$$

(a) If $p(t) = a$, what is the steady-state response of h(t)? Is the steady-state height proportional to the applied pressure? (Calibrations are carried out for steady-state responses in most instruments.)

(b) What is the time constant of $G(s)$? Roughly, how many seconds will it take for its step response to reach the final height?

Figure P9.15

$p(t)$

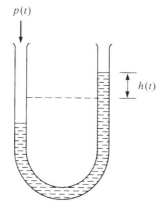

$h(t)$

   (c) Plot the amplitude characteristic of $G(s)$. If $p(t) = \sin \omega_0 t$, what can be the largest value of ω_0 with the manometer still yielding a correct reading within 10%?

9.16 Are the networks in Figure P9.16 marginally stable and asymptotically stable?

9.17 Study marginal stability, asymptotic stability, and BIBO stability of the systems described by the following.

   (a) $\ddot{y}(t) + 2\dot{y}(t) = \dot{u}(t) - u(t)$

   (b) $\ddot{y}(t) - 4y(t) = \dot{u}(t) - 2u(t)$

   (c) $(p^4 + 3p^3 + 6p^2 + 5p + 3)y(t) = (p - 2)u(t)$

9.18 Study marginal stability, asymptotic stability, and BIBO stability of the discrete-time systems described by the following.

   (a) $y[k + 2] + 0.2y[k + 1] = u[k + 1] - u[k]$

   (b) $y[k + 2] - 2.5y[k + 1] + y[k] = u[k + 1] - 2u[k]$

   (c) $y[k] - y[k - 2] = u[k] - 10u[k - 1]$

   (d) $(p^4 + 0.2p^3 - 0.3p^2 - 0.4p)y[k] = (p - 2)u[k]$

9.19 Consider the Wien-bridge oscillator shown in Figure 9.19. Let $R_1 = R = 1$ kΩ. Find R_2 and C so that the circuit will maintain an oscillation with frequency 1 kHz.

9.20 Use the short-circuit and open-circuit properties of the op amp to derive the oscillation conditions of the Wien-bridge oscillator. Do not use the block diagram in Figure 9.20.

Figure P9.16

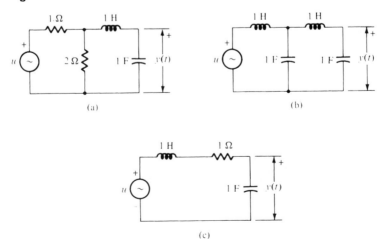

(a)

(b)

(c)

Figure P9.21

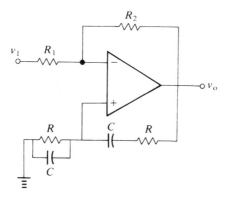

9.21 Consider the operational amplifier circuit shown in Figure P9.21. Develop a block diagram for the circuit and then derive the oscillation condition.

10

State-Variable Equations and Computer Simulations

50 ?

10.1 INTRODUCTION

3

The transfer function developed in the preceding chapters describes only the relationship between the input and output of LTIL systems; it is called the input-output description, or external description, of systems. In this chapter, we shall develop a description called the state-variable description, or internal description, of LTIL systems. Strictly speaking, this description is the same as the differential or difference equations discussed in Chapter 3. The difference is that the *high-order* differential or difference equations in Chapter 3 will now be written as sets of *first-order* differential or difference equations. By so doing, we can simplify our study.

The state-variable description is applicable only to lumped systems, systems with a finite number of state variables. For convenience of discussion, we will assume that the number of state variables is 3. Then, the continuous-time LTIL state-variable equation is of the form

$$\dot{x}_1(t) = a_{11}x_1(t) + a_{12}x_2(t) + a_{13}x_3(t) + b_1u(t)$$
$$\dot{x}_2(t) = a_{21}x_1(t) + a_{22}x_2(t) + a_{23}x_3(t) + b_2u(t) \qquad \textbf{(10.1a)}$$
$$\dot{x}_3(t) = a_{31}x_1(t) + a_{32}x_2(t) + a_{33}x_3(t) + b_3u(t)$$
$$y(t) = c_1x_1(t) + c_2x_2(t) + c_3x_3(t) + du(t) \qquad \textbf{(10.1b)}$$

where u and y are the input and output of the system; x_i, $i = 1, 2, 3$, are the state variables; a_{ij}, b_i, c_i, and d are constants; and $\dot{x}_i(t) = dx_i(t)/dt$. These equations are more often written in matrix form as

$$\dot{\mathbf{x}}(t) = \mathbf{A}\mathbf{x}(t) + \mathbf{b}u(t) \qquad \text{(state equation)} \qquad \textbf{(10.2a)}$$
$$y(t) = \mathbf{c}\mathbf{x}(t) + du(t) \qquad \text{(output equation)} \qquad \textbf{(10.2b)}$$

564

with

$$\mathbf{x} = \begin{bmatrix} x_1 \\ x_2 \\ x_3 \end{bmatrix}, \qquad \mathbf{A} = \begin{bmatrix} a_{11} & a_{12} & a_{13} \\ a_{21} & a_{22} & a_{23} \\ a_{31} & a_{32} & a_{33} \end{bmatrix}, \qquad \mathbf{b} = \begin{bmatrix} b_1 \\ b_2 \\ b_3 \end{bmatrix}$$

and

$$\mathbf{c} = [c_1 \ \ c_2 \ \ c_3]$$

The vector $\mathbf{x}$ is called the *state vector* or, simply, the *state*. If $\mathbf{x}$ has n state variables or is an $n \times 1$ vector, then $\mathbf{A}$ is an $n \times n$ square matrix, $\mathbf{b}$ is an $n \times 1$ column vector, $\mathbf{c}$ is a $1 \times n$ row vector, and d is a 1×1 scalar. $\mathbf{A}$ is called the *system matrix*, and d is called the *direct transmission* part. Equation (10.2a) describes the relationship between the input and state and is called the state equation. The state equation in (10.2a) consists of three first-order differential equations and is said to have dimension 3. The equation in (10.2b) relates the input, state, and output and is called the output equation. It is an algebraic equation; it does not involve the differentiation of $\mathbf{x}$. Thus if $\mathbf{x}(t)$ and $u(t)$ are known, the output $y(t)$ can be obtained from (10.2b) simply by multiplications and additions. State-variable equations are also called state-space equations because the state forms a linear space.

Before proceeding, we remark on the notation: Vectors are denoted by boldface lower-case letters, matrices are denoted by boldface capital letters, scalars are denoted by regular-face lower-case letters. In (10.2), $\mathbf{A}$, $\mathbf{b}$, $\mathbf{c}$, and $\mathbf{x}$ are boldface because they are either a vector or a matrix; u, y, and d are regular face because they are all scalars.

The discrete-time LTIL state-variable equation is of the form

$$\mathbf{x}[k + 1] = \mathbf{A}\mathbf{x}[k] + \mathbf{b}u[k] \qquad \text{(state equation)} \qquad \textbf{(10.3a)}$$
$$y[k] = \mathbf{c}\mathbf{x}[k] + du[k] \qquad \text{(output equation)} \qquad \textbf{(10.3b)}$$

where $k = 0, 1, 2, \ldots$. The state equation consists of a set of first-order *difference* equations. Otherwise, all discussion regarding (10.2) also applies to (10.3).

The transfer function describes only the zero-state response of a system. Thus, when we use the transfer function, we must assume that a system is initially relaxed. When using the state-variable equation, no such assumption is necessary. The equation is applicable even if the initial state is nonzero. The equation describes not only the output but also the state variables. Because the state variables reside inside the system, and are not necessarily accessible from the input and output terminals, the state-variable equation is also called the internal description. The state-variable description, then, is more general than the transfer function description. However, if a system is completely characterized by its transfer function (Section 4.10), the two descriptions are essentially the same.

Let us digress at this point to discuss briefly the history of transfer functions and state-variable equations. The transfer function (or, more generally, the Laplace transform) was introduced into electrical engineering in the 1920s. Since then, many methods using transfer functions have been developed to design circuits and control systems. For years, however, these methods were limited to single-input, single-output (SISO) systems and did not seem to be extendable to multiple-input, multiple-output (MIMO) systems. Because of this limitation and others, the state-variable approach was developed in the 1960s and began to dominate system analysis and design. This approach introduced the concepts of controllability and observability, which led to a better understanding of the structure of systems, and a number of new results, such as state feedback and state estimator. These new results reignited the interest in the transfer-function approach in the 1970s. By considering the transfer function as a ratio of two polynomials, it was possible to design state feedback and state estimator under the condition of complete characterization and to extend the results in the SISO case to the MIMO case. The equivalence of the state-variable equation and the transfer function became well established for LTIL systems, and both types of equations are now widely used.

The transfer function is not easily extendable to nonlinear or time-varying systems, however. For linear time-varying systems, the state-variable equations in (10.2) can be modified as

$$\dot{\mathbf{x}}(t) = \mathbf{A}(t)\mathbf{x}(t) + \mathbf{b}(t)u(t)$$
$$y(t) = \mathbf{c}(t)\mathbf{x}(t) + d(t)u(t) \tag{10.4}$$

where $\mathbf{A}, \mathbf{b}, \mathbf{c}$, and d are functions of time rather than constants. For nonlinear and time-varying systems, (10.2) can be modified as

$$\dot{\mathbf{x}}(t) = \mathbf{f}(\mathbf{x}(t), u(t), t)$$
$$y(t) = h(\mathbf{x}(t), u(t), t) \tag{10.5}$$

where $\mathbf{f}$ and h are nonlinear functions. For the discrete-time case, Equation (10.3) can be similarly modified. Thus, state-variable equations are more generally used in studying nonlinear and/or time-varying systems.

There is one more important reason for studying state-variable equations. Once the state-variable description of a system has been obtained, the system can be readily simulated on an analog or a digital computer. Such simulations will be studied later in this chapter.

10.2 EXAMPLES OF SETTING UP STATE-VARIABLE EQUATIONS

7

To develop state-variable equations for LTIL systems, we must first choose state variables. One way is to choose physical variables that associate with energy.[1] For example, the potential and kinetic energy of a mass are stored in its position and velocity. Therefore, position and velocity can be chosen as state variables for a mass. In RLC networks, capacitors and inductors are called the energy storage elements because they can store energy in their electric and magnetic fields. Therefore, all capacitor voltages and inductor currents are generally chosen as state variables for RLC networks. However, because resistors will not store energy (all energy is dissipated as heat), resistor voltages or currents are never chosen as state variables. Given this brief discussion, let us develop state-variable equations for LTIL systems.

EXAMPLE 10.2.1

(**Mechanical system**) Consider the mechanical system shown in Figure 3.20. For convenience, it is redrawn in Figure 10.1. Let k_1 be the spring constant and k_2 be the viscous friction coefficient. Then, as discussed in (3.73), the system is described by

$$m\ddot{y}(t) + k_2\dot{y}(t) + k_1 y(t) = u(t) \qquad (10.6)$$

where u is the applied force (input) and $y(t)$ is the displacement (output).

As noted before, because the potential energy and kinetic energy of a mass are stored in its position and velocity, position and velocity can be chosen as state variables of the system. Define

$$x_1(t) := y(t) \qquad (10.7a)$$

and

$$x_2(t) := \dot{y}(t) \qquad (10.7b)$$

We then have

$$\dot{x}_1(t) = \dot{y}(t) = x_2(t)$$

[1]The choice of state variables is not unique. State variables can also be chosen mathematically without regard to physical variables. See Reference [4].

Figure 10.1 *A mechanical system*

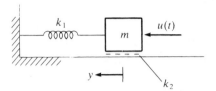

This relation follows from the definition of $x_1(t)$ and $x_2(t)$ and is independent of the system. Taking the derivative of $x_2(t)$ yields

$$\dot{x}_2(t) = \ddot{y}(t)$$

which, after substituting (10.6) and (10.7), becomes

$$\dot{x}_2(t) = \frac{1}{m}[-k_1 y(t) - k_2 \dot{y}(t) + u(t)]$$

$$= -\frac{k_1}{m} x_1(t) - \frac{k_2}{m} x_2(t) + \frac{1}{m} u(t)$$

These equations can be arranged in matrix form as

$$\begin{bmatrix} \dot{x}_1(t) \\ \dot{x}_2(t) \end{bmatrix} = \begin{bmatrix} 0 & 1 \\ -\dfrac{k_1}{m} & -\dfrac{k_2}{m} \end{bmatrix} \begin{bmatrix} x_1(t) \\ x_2(t) \end{bmatrix} + \begin{bmatrix} 0 \\ \dfrac{1}{m} \end{bmatrix} u(t) \qquad \textbf{(10.8a)}$$

$$y(t) = [1 \ \ 0] \begin{bmatrix} x_1(t) \\ x_2(t) \end{bmatrix} \qquad \textbf{(10.8b)}$$

This is a state-variable equation of dimension 2. Note that, for this example, the direct transmission part is zero.

A procedure for developing state-variable equations for RLC networks that may contain a voltage or current source is discussed next.

Procedure for developing state-variable equations for RLC networks

Step 1: Assign all capacitor voltages and all inductors currents as state variables. Write down capacitor currents and inductor voltages as shown in Figure 10.2. Note that currents flow from positive voltages to negative voltages.

Figure 10.2 *Branch characteristics of capacitor and inductor*

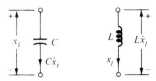

Step 2: Use Kirchhoff's voltage and/or current laws[2] to express every resistor's voltage and current in terms of state variables and, if necessary, the input.

Step 3: Use Kirchhoff's voltage and/or current laws to develop a state equation.

This procedure offers only a general guide. For a more detailed and specific procedure, see References [4]. We use an example to illustrate the procedure.

EXAMPLE 10.2.2

(**Electrical system**) Consider the RLC network shown in Figure 10.3. It consists of one resistor, one capacitor and two inductors. The input $u(t)$ is a voltage source and the output is chosen as the voltage across the 2-H inductor as shown.

Step 1: The capacitor voltage $x_1(t)$ and two inductor currents $x_2(t)$ and $x_3(t)$ are chosen as the state variables. The capacitor current is $2\dot{x}_1(t)$ and the inductor voltages are respectively $\dot{x}_2(t)$ and $2\dot{x}_3(t)$.

Step 2: The current passing through the 0.25-Ω resistor is clearly equal to $x_3(t)$. Thus the voltage across the resistor is $0.25x_3(t)$. The polarity of the voltage is specified as shown.

Step 3: From Figure 10.3, we see that the capacitor current $2\dot{x}_1$ is equal to $x_2(t) + x_3(t)$ or

$$\dot{x}_1(t) = 0.5x_2(t) + 0.5x_3(t)$$

The voltage across the 1-H inductor is, using the left-hand-side loop of the circuit,

$$\dot{x}_2(t) = u(t) - x_1(t)$$

[2]Gustave Robert Kirchhoff (1824–1887), a German physicist.

Figure 10.3   *Network with three state variables*

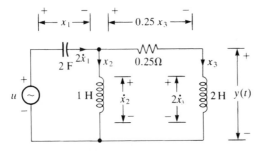

The voltage across the 2-H inductor is, using the outer loop of the circuit,

$$2\dot{x}_3(t) = u(t) - x_1(t) - 0.25x_3(t) \tag{10.9}$$

which implies

$$\dot{x}_3(t) = -0.5x_1(t) - 0 \cdot x_2(t) - 0.125x_3(t) + 0.5u(t)$$

These three equations can be arranged in matrix form as

$$\begin{bmatrix} \dot{x}_1 \\ \dot{x}_2 \\ \dot{x}_3 \end{bmatrix} = \begin{bmatrix} 0 & 0.5 & 0.5 \\ -1 & 0 & 0 \\ -0.5 & 0 & -0.125 \end{bmatrix} \begin{bmatrix} x_1 \\ x_2 \\ x_3 \end{bmatrix} + \begin{bmatrix} 0 \\ 1 \\ 0.5 \end{bmatrix} u(t) \tag{10.10a}$$

This state equation of dimension 3 describes the network. The output $y(t)$ is

$$y(t) = 2\dot{x}_3$$

which is not in the form of $\mathbf{cx} + du$. However, the substitution of (10.9) yields

$$y(t) = -x_1(t) - 0.25x_3(t) + u(t)$$

$$= [-1 \quad 0 \quad -0.25] \begin{bmatrix} x_1(t) \\ x_2(t) \\ x_3(t) \end{bmatrix} + u(t) \tag{10.10b}$$

This is the output equation.

EXERCISE 10.2.1

Find the state-variable descriptions of the networks shown in Figure 10.4 with the state variables shown.

Figure 10.4 *Two networks*

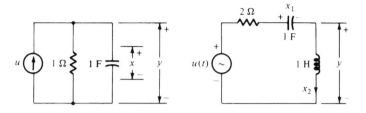

[**ANSWERS:** $\dot{x} = -x + u$, $y = x$;

$$\dot{\mathbf{x}} = \begin{bmatrix} 0 & 1 \\ -1 & -2 \end{bmatrix} \mathbf{x} + \begin{bmatrix} 0 \\ 1 \end{bmatrix} u, \qquad y = [-1 \ -2]\mathbf{x} + u.]$$

---

*EXAMPLE 10.2.3

(**Electromechanical system**) Consider the field-controlled DC motor studied in Figure 3.22 and repeated in Figure 10.5. The input voltage $u(t)$ is applied to the field circuit with resistance R_f and inductance L_f. The armature current i_a is kept constant. The generated torque $T(t)$ is proportional to i_f, written as $T(t) = k_t i_f(t)$. The torque is used to drive the load with moment of inertia J. The coefficient of the viscous friction between the motor shaft and bearing is f. The angular displacement of the load is θ. As was disscussed in (3.75), we have

$$T(t) = f\dot{\theta}(t) + J\ddot{\theta}(t) \tag{10.11}$$

Figure 10.5 *Field-controlled DC motor*

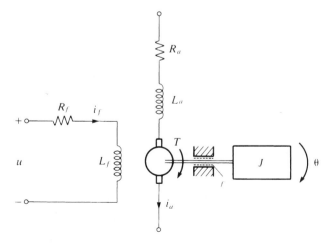

The state variables of the system are chosen as

$$x_1(t) := i_f(t)$$
$$x_2(t) := \theta(t)$$
$$x_3(t) := \frac{d}{dt}\theta(t) = \dot{\theta}(t)$$

From the field circuit, we have $u(t) = R_f i_f(t) + L_f \dot{i}_f(t)$, which implies

$$\dot{x}_1(t) = -\frac{R_f}{L_f}x_1(t) + \frac{1}{L_f}u(t)$$

By definition, we have

$$\dot{x}_2(t) = \dot{\theta}(t) = x_3$$

The substitution of (10.11) and $T(t) = k_t i_f(t) = k_t x_1(t)$ into $\dot{x}_3(t) = \ddot{\theta}(t)$ yields

$$\dot{x}_3(t) = \frac{1}{J}(T(t) - f\dot{\theta}(t)) = \frac{1}{J}k_t x_1(t) - \frac{f}{J}x_3(t)$$

These can be arranged in matrix form as

$$\dot{\mathbf{x}}(t) = \begin{bmatrix} -\dfrac{R_f}{L_f} & 0 & 0 \\ 0 & 0 & 1 \\ \dfrac{k_t}{J} & 0 & -\dfrac{f}{J} \end{bmatrix} \mathbf{x}(t) + \begin{bmatrix} \dfrac{1}{L_f} \\ 0 \\ 0 \end{bmatrix} u(t) \qquad (10.12)$$

If the angular displacement of the load is chosen as the output, then the output equation is

$$y(t) = [0 \ 1 \ 0]\mathbf{x}(t)$$

EXAMPLE 10.2.4

(**Hydraulic tanks**) Consider the system shown in Figure 3.23 and repeated in Figure 10.6. Define

$$q_i, q_1, q_2 = \text{rates of flow of liquid}$$
$$A_1, A_2 = \text{areas of the cross-section of tanks}$$
$$h_1, h_2 = \text{liquid levels}$$
$$R_1, R_2 = \text{flow resistances, controlled by valves}$$

Figure 10.6 *A liquid system*

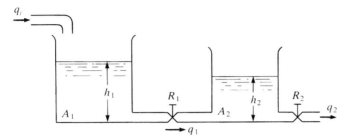

It is assumed that q_1 and q_2 are governed by

$$q_1 = \frac{h_1 - h_2}{R_1} \quad \text{and} \quad q_2 = \frac{h_2}{R_2} \tag{10.13}$$

They are proportional to relative liquid levels and inversely proportional to flow resistances. The changes of the liquid levels are governed by

$$A_1 dh_1 = (q_i - q_1)dt \tag{10.14a}$$

and

$$A_2 dh_2 = (q_1 - q_2)dt \tag{10.14b}$$

Let q_i and q_2 be the input and output of the system. The state variables of the system can be chosen as the liquid levels, that is, $x_i(t) := h_i(t)$, $i = 1, 2$. From (10.13) and (10.14), we have

$$\frac{dx_1(t)}{dt} = \dot{x}_1(t) = \frac{1}{A_1}\left[q_i - \frac{x_1 - x_2}{R_1} \right] = -\frac{1}{A_1 R_1}x_1 + \frac{1}{A_1 R_1}x_2 + \frac{1}{A}q_i$$

and

$$\dot{x}_2(t) = \frac{1}{A_2}(q_1 - q_2) = \frac{1}{A_2}\left[\frac{x_1 - x_2}{R_1} - \frac{x_2}{R_2} \right] = \frac{1}{A_2 R_1}x_1 - \frac{1}{A_2}\left[\frac{1}{R_1} + \frac{1}{R_2} \right]x_2$$

which can be expressed in matrix form as

$$\dot{\mathbf{x}}(t) = \begin{bmatrix} \dfrac{-1}{A_1 R_1} & \dfrac{1}{A_1 R_1} \\[2mm] \dfrac{1}{A_2 R_1} & \dfrac{-(R_1 + R_2)}{A_2 R_1 R_2} \end{bmatrix} \mathbf{x}(t) + \begin{bmatrix} \dfrac{1}{A} \\[2mm] 0 \end{bmatrix} q_i(t) \tag{10.15a}$$

The output equation is

$$q_2 = [0 \quad \frac{1}{R_2}]\mathbf{x}(t) \tag{10.15b}$$

This is the state-variable description of the liquid system.

10.3   SOLUTION OF STATE EQUATIONS—LAPLACE-TRANSFORM METHOD

Consider the linear time-invariant state-variable equation

$$\dot{\mathbf{x}}(t) = \mathbf{A}\mathbf{x}(t) + \mathbf{b}u(t) \tag{10.16a}$$
$$y(t) = \mathbf{c}\mathbf{x}(t) + du(t) \tag{10.16b}$$

The solution excited by the initial state $\mathbf{x}(0-)$ and input $u(t)$, $t \geq 0$, can be obtained by both analytical methods and computer simulations. In this and the next section, we will discuss analytical methods. A discussion of analog and digital computer simulations will then follow.

First, let us discuss the initial condition of $\mathbf{x}(t)$. If $u(t)$ is discontinuous, the discontinuity will appear at $y(t)$, as can be seen from (10.16b). However, because $\mathbf{x}(t)$ is an integration of $u(t)$, $\mathbf{x}$ is continuous at $t = 0$, even if $u(t)$ is discontinuous at $t = 0$. Thus, we have $\mathbf{x}(0-) = \mathbf{x}(0) = \mathbf{x}(0+)$. A discontinuity will appear at $\mathbf{x}(0)$ only if $u(t)$ contains an impulse at $t = 0$. This is unlikely in practice. Thus, we shall assume $\mathbf{x}(t)$ to be continuous at $t = 0$, and we will make no distinction between 0 and $0-$ in the remainder of this chapter.

When the Laplace transform is applied to a vector or matrix, it is to be applied to every entry of the vector or matrix. For example, if $\mathbf{x} = [x_1 \ x_2 \ x_3]'$, then

$$\mathcal{L}[\dot{\mathbf{x}}(t)] = \begin{bmatrix} \mathcal{L}[\dot{x}_1(t)] \\ \mathcal{L}[\dot{x}_2(t)] \\ \mathcal{L}[\dot{x}_3(t)] \end{bmatrix} = \begin{bmatrix} sX_1(s) - x_1(0) \\ sX_2(s) - x_2(0) \\ sX_3(s) - x_3(0) \end{bmatrix} = s\begin{bmatrix} X_1(s) \\ X_2(s) \\ X_3(s) \end{bmatrix} - \begin{bmatrix} x_1(0) \\ x_2(0) \\ x_3(0) \end{bmatrix}$$

$$= s\mathbf{X}(s) - \mathbf{x}(0)$$

Similarly, it can be verified that

$$\mathcal{L}[\mathbf{A}\mathbf{x}(t)] = \mathbf{A}\,\mathcal{L}[\mathbf{x}(t)] = \mathbf{A}\mathbf{X}(s)$$

The application of the Laplace transform to vectors or matrices is no different than its application in the scalar case. Recall that in the scalar case, the Laplace transform of lower-case letters are denoted by their capital letters. We follow the same convention for the vector case as evident from the preceding two equations.

The application of the Laplace transform to (10.16) yields

$$s\mathbf{X}(s) - \mathbf{x}(0) = \mathbf{A}\mathbf{X}(s) + \mathbf{b}U(s) \qquad \textbf{(10.17a)}$$
$$Y(s) = \mathbf{c}\mathbf{X}(s) + dU(s) \qquad \textbf{(10.17b)}$$

Since $s - \mathbf{A}$ is not defined, we cannot write $s\mathbf{X}(s) - \mathbf{A}\mathbf{X}(s)$ as $(s - \mathbf{A})\mathbf{X}(s)$. If we introduce a unit matrix $\mathbf{I}$ into $s\mathbf{X}(s)$, we have

$$s\mathbf{X}(s) - \mathbf{A}\mathbf{X}(s) = s\mathbf{I}\mathbf{X}(s) - \mathbf{A}\mathbf{X}(s) = (s\mathbf{I} - \mathbf{A})\mathbf{X}(s)$$

Thus, (10.17a) implies

$$(s\mathbf{I} - \mathbf{A})\mathbf{X}(s) = \mathbf{x}(0) + \mathbf{b}U(s)$$

which implies, in turn,

$$\mathbf{X}(s) = \underbrace{(s\mathbf{I} - \mathbf{A})^{-1}\mathbf{x}(0)}_{\text{zero-input response}} + \underbrace{(s\mathbf{I} - \mathbf{A})^{-1}\mathbf{b}U(s)}_{\text{zero-state response}} \qquad \textbf{(10.18)}$$

This is an algebraic equation. We see that the Laplace transform has transformed the differential equation in (10.16a) into the algebraic equation in (10.18). It also has revealed the decomposition of the response into the zero-state response and the zero-input response. The substitution of (10.18) into (10.16b) yields

$$\boxed{Y(s) = \underbrace{\mathbf{c}(s\mathbf{I} - \mathbf{A})^{-1}\mathbf{x}(0)}_{\text{zero-input response}} + \underbrace{[\mathbf{c}(s\mathbf{I} - \mathbf{A})^{-1}\mathbf{b} + d]U(s)}_{\text{zero-state response}}} \qquad \textbf{(10.19)}$$

This also reveals the decomposition into the zero-input and zero-state responses.

If $\mathbf{x}(0) = \mathbf{0}$, then (10.19) reduces to

$$Y(s) = [\mathbf{c}(s\mathbf{I} - \mathbf{A})^{-1}\mathbf{b} + d]U(s)$$

Thus, the transfer function from u to y of the system described by (10.16) is

$$\boxed{H(s) = \mathbf{c}(s\mathbf{I} - \mathbf{A})^{-1}\mathbf{b} + d} \qquad \textbf{(10.20)}$$

This is a fundamental equation that relates the state-variable equation and the transfer function.

EXAMPLE 10.3.1

To find the output of

$$\dot{\mathbf{x}}(t) = \begin{bmatrix} 0 & 1 \\ -2 & -3 \end{bmatrix} \mathbf{x}(t) + \begin{bmatrix} 0 \\ 1 \end{bmatrix} u(t) \tag{10.21}$$

$$y(t) = [1 \ -1]\mathbf{x}(t)$$

excited by $\mathbf{x}(0) = [2 \ -1]'$ and $u(t) = 1$ for $t \geq 0$, we compute

$$s\mathbf{I} - \mathbf{A} = s\begin{bmatrix} 1 & 0 \\ 0 & 1 \end{bmatrix} - \begin{bmatrix} 0 & 1 \\ -2 & -3 \end{bmatrix} = \begin{bmatrix} s & -1 \\ 2 & s+3 \end{bmatrix}$$

Its inverse, using (A.68), is

$$(s\mathbf{I} - \mathbf{A})^{-1} = \frac{1}{s(s+3)+2}\begin{bmatrix} s+3 & 1 \\ -2 & s \end{bmatrix} \tag{10.22}$$

Thus, we have

$$\mathbf{X}(s) = (s\mathbf{I} - \mathbf{A})^{-1}[\mathbf{x}(0) + \mathbf{b}U(s)] = \frac{1}{(s+2)(s+1)}\begin{bmatrix} s+3 & 1 \\ -2 & s \end{bmatrix}\left(\begin{bmatrix} 2 \\ -1 \end{bmatrix} + \begin{bmatrix} 0 \\ 1 \end{bmatrix}\frac{1}{s}\right)$$

$$= \frac{1}{(s+2)(s+1)}\begin{bmatrix} s+3 & 1 \\ -2 & s \end{bmatrix}\begin{bmatrix} 2 \\ -1 + \dfrac{1}{s} \end{bmatrix}$$

$$= \frac{1}{(s+2)(s+1)}\begin{bmatrix} 2(s+3) + \dfrac{-s+1}{s} \\ -4 - s + 1 \end{bmatrix}$$

$$= \begin{bmatrix} \dfrac{2s^2 + 5s + 1}{s(s+2)(s+1)} \\ \dfrac{-s-3}{(s+2)(s+1)} \end{bmatrix}$$

Only the output $y(t)$ is sought. It requires less computation to compute first $Y(s)$ and then its inverse than to compute first the two inverses of $X_1(s)$ and $X_2(s)$ and then $y(t)$. Therefore, we compute

$$Y(s) = [1 \ -1]\mathbf{X}(s) = \frac{2s^2 + 5s + 1}{s(s+2)(s+1)} - \frac{-s-3}{(s+2)(s+1)}$$

$$= \frac{3s^2 + 8s + 1}{s(s+2)(s+1)} = \frac{0.5}{s} + \frac{-1.5}{s+2} + \frac{4}{s+1}$$

Thus, the output is

$$y(t) = 0.5 - 1.5e^{-2t} + 4e^{-t} \qquad (10.23)$$

for $t \geq 0$.

EXERCISE 10.3.1

Find the output of

$$\dot{x}(t) = -2x(t) + 3u(t)$$
$$y(t) = -x(t)$$

excited by $x(0) = -1$ and $u(t) = 2$ for $t \geq 0$.

[**ANSWER:** $y(t) = -3 + 4e^{-2t}$.] ■

EXERCISE 10.3.2

Find $y(t)$ of

$$\dot{\mathbf{x}}(t) = \begin{bmatrix} 0 & 0 \\ 0 & -1 \end{bmatrix}\mathbf{x}(t) + \begin{bmatrix} 1 \\ 1 \end{bmatrix}u(t)$$
$$y(t) = [-1 \ \ 1]\mathbf{x}(t) + u(t)$$

excited by $\mathbf{x}(0) = \mathbf{0}$ and $u(t) = \cos t$, $t \geq 0$.

[**ANSWER:** $y(t) = 1.5 \cos t - 0.5 \sin t - 0.5e^{-t}$.] ■

EXERCISE 10.3.3

Find $y(t)$ of

$$\dot{\mathbf{x}}(t) = \begin{bmatrix} 0 & 1 \\ -0.5 & -1.5 \end{bmatrix}\mathbf{x}(t) + \begin{bmatrix} 0 \\ 1 \end{bmatrix}u(t) \qquad (10.24)$$
$$y(t) = [1 \ \ -1]\mathbf{x}(t)$$

excited by $\mathbf{x}(0) = [2 \ \ -1]'$ and $u(t) = 1$, for $t \geq 0$.

[**ANSWER:** $y(t) = 2 + 4e^{-t} - 3e^{-0.5t}$.] ■

*10.4   SOLUTION OF STATE EQUATIONS— TIME-DOMAIN METHOD

In this section, we shall discuss a method for computing the solution of state equations that does not use the Laplace transform. We discuss first a function of a square matrix.

Definition of e^{At} and its properties The solution of (10.16a) hinges on e^{At}, a function of **A**. Thus, we must first discuss this function. Consider the infinite series

$$e^{at} = 1 + ta + \frac{t^2}{2!}a^2 + \frac{t^3}{3!}a^3 + \cdots + \frac{t^n}{n!}a^n + \cdots \qquad (10.25)$$

where a is a constant. Because the factorial $n!$ increases, for large n, much faster than $(ta)^n$, the infinite series in (10.25) converges for all finite ta. Clearly, the rate of convergence depends on ta. For example, if $ta = 1000$, it will take 125 terms for (10.25) to converge to within one percent of its actual value. However, if $ta = 0.1$, then (10.25) will converge in two terms. Following (10.25), we define

$$e^{At} := \mathbf{I} + t\mathbf{A} + \frac{t^2}{2!}\mathbf{A}^2 + \frac{t^3}{3!}\mathbf{A}^3 + \cdots + \frac{t^n}{n!}\mathbf{A}^n + \cdots \qquad (10.26)$$

where $\mathbf{I}$ is the unit matrix of the same order as **A** and $\mathbf{A}^k := \mathbf{AA} \cdots \mathbf{A}$ (k times). For example, if

$$\mathbf{A} = \begin{bmatrix} 0 & 2 & 1 \\ 0 & 0 & -3 \\ 0 & 0 & 0 \end{bmatrix}$$

then

$$\mathbf{A}^2 = \mathbf{AA} = \begin{bmatrix} 0 & 2 & 1 \\ 0 & 0 & -3 \\ 0 & 0 & 0 \end{bmatrix}\begin{bmatrix} 0 & 2 & 1 \\ 0 & 0 & -3 \\ 0 & 0 & 0 \end{bmatrix} = \begin{bmatrix} 0 & 0 & -6 \\ 0 & 0 & 0 \\ 0 & 0 & 0 \end{bmatrix}$$

$$\mathbf{A}^3 = \mathbf{A}^2\mathbf{A} = \begin{bmatrix} 0 & 0 & -6 \\ 0 & 0 & 0 \\ 0 & 0 & 0 \end{bmatrix}\begin{bmatrix} 0 & 2 & 1 \\ 0 & 0 & -3 \\ 0 & 0 & 0 \end{bmatrix} = \begin{bmatrix} 0 & 0 & 0 \\ 0 & 0 & 0 \\ 0 & 0 & 0 \end{bmatrix}$$

and

$$\mathbf{A}^k = \mathbf{0} \qquad \text{for } k = 3, 4, 5, \ldots$$

Thus, we have

$$e^{At} = \mathbf{I} + t\mathbf{A} + \frac{t^2}{2!}\mathbf{A}^2 + \frac{t^3}{3!}\mathbf{A}^3 + \cdots$$

$$= \begin{bmatrix} 1 & 0 & 0 \\ 0 & 1 & 0 \\ 0 & 0 & 1 \end{bmatrix} + t\begin{bmatrix} 0 & 2 & 1 \\ 0 & 0 & -3 \\ 0 & 0 & 0 \end{bmatrix} + \frac{t^2}{2}\begin{bmatrix} 0 & 0 & -6 \\ 0 & 0 & 0 \\ 0 & 0 & 0 \end{bmatrix} = \begin{bmatrix} 1 & 2t & t - 3t^2 \\ 0 & 1 & -3t \\ 0 & 0 & 1 \end{bmatrix}$$

This example is chosen so that the infinite series terminates in a finite number of terms. For a general **A**, this will not happen and the computation of (10.26) by hand is not possible. If **A** is an $n \times n$ matrix, then $e^{\mathbf{A}t}$ defined in (10.26) is also an $n \times n$ matrix.

Let us consider some properties of $e^{\mathbf{A}t}$. If $t = 0$, (10.26) reduces to

$$e^{\mathbf{0}} = \mathbf{I} \tag{10.27}$$

where **0** is a square matrix with entries all equal to zero. The differentiation of (10.26) with respect to t yields

$$\frac{d}{dt} e^{\mathbf{A}t} = \mathbf{0} + \mathbf{A} + \frac{2t}{2!} \mathbf{A}^2 + \frac{3t^2}{3!} \mathbf{A}^3 + \cdots$$

$$= \mathbf{A}(\mathbf{I} + t\mathbf{A} + \frac{t^2}{2!} \mathbf{A}^2 + \cdots)$$

$$= (\mathbf{I} + t\mathbf{A} + \frac{t^2}{2!} \mathbf{A}^2 + \cdots)\mathbf{A}$$

which implies

$$\frac{d}{dt} e^{\mathbf{A}t} = \mathbf{A}e^{\mathbf{A}t} = e^{\mathbf{A}t}\mathbf{A} \tag{10.28}$$

Note that, in general, $\mathbf{AB} \neq \mathbf{BA}$, but $e^{\mathbf{A}t}\mathbf{A} = \mathbf{A}e^{\mathbf{A}t}$. Next, we show

$$e^{\mathbf{A}t}e^{-\mathbf{A}\tau} = e^{\mathbf{A}(t-\tau)} \tag{10.29}$$

By definition, we have

$$e^{\mathbf{A}t}e^{-\mathbf{A}\tau} = (\mathbf{I} + t\mathbf{A} + \frac{t^2}{2!} \mathbf{A}^2 + \cdots)(\mathbf{I} + \tau(-\mathbf{A}) + \frac{\tau^2}{2!} (-\mathbf{A})^2 + \cdots)$$

$$= \mathbf{I} + t\mathbf{A} - \tau\mathbf{A} + \frac{t^2}{2!} \mathbf{A}^2 - t\tau\mathbf{A}^2 + \frac{\tau^2}{2!} \mathbf{A}^2 + \cdots$$

$$= \mathbf{I} + (t - \tau)\mathbf{A} + \frac{(t - \tau)^2}{2!} \mathbf{A}^2 + \cdots$$

$$= e^{\mathbf{A}(t-\tau)}$$

This establishes (10.29). Using the same procedure, we can show, because $\mathbf{AB} \neq \mathbf{BA}$,

$$e^{\mathbf{A}t}e^{\mathbf{B}t} \neq e^{(\mathbf{A}+\mathbf{B})t}$$

If $t = \tau$, then (10.29) becomes, using (10.27),

$$e^{\mathbf{A}t}e^{-\mathbf{A}t} = e^{\mathbf{A} \cdot 0} = \mathbf{I}$$

which implies

$$(e^{\mathbf{A}t})^{-1} = e^{-\mathbf{A}t}$$

In general, computing the inverse of a matrix is complicated. However, the inverse of $e^{\mathbf{A}t}$ can be easily obtained by merely changing the sign of t or $\mathbf{A}$. This property and the properties in equations (10.27) through (10.29) are similar to those of the scalar exponential function e^{at}. For easy reference, they are listed here:

1. $e^{\mathbf{0}} = \mathbf{I}$

2. $\dfrac{d}{dt} e^{\mathbf{A}t} = \mathbf{A}e^{\mathbf{A}t} = e^{\mathbf{A}t}\mathbf{A}$

3. $e^{\mathbf{A}t} e^{-\mathbf{A}\tau} = e^{\mathbf{A}(t-\tau)}$

4. $(e^{\mathbf{A}t})^{-1} = e^{-\mathbf{A}t}$

5. $\mathcal{L}[e^{\mathbf{A}t}] = (s\mathbf{I} - \mathbf{A})^{-1}$

The last property will be established shortly.

Homogeneous equation Having introduced $e^{\mathbf{A}t}$, we are ready to study the solution of (10.16a). If $u = 0$, (10.16a) reduces to

$$\dot{\mathbf{x}}(t) = \mathbf{A}\mathbf{x}(t) \tag{10.30}$$

This is called the homogeneous equation. To claim that its solution excited by the initial state $\mathbf{x}(0)$ is given by

$$\mathbf{x}(t) = e^{\mathbf{A}t}\mathbf{x}(0) \tag{10.31}$$

we must show that (10.31) satisfies (i) the initial condition $\mathbf{x}(0)$ and (ii) the differential equation $\dot{\mathbf{x}}(t) = \mathbf{A}\mathbf{x}(t)$. Indeed, we have

$$\mathbf{x}(0) = e^{\mathbf{A}\cdot 0}\mathbf{x}(0) = e^{\mathbf{0}}\mathbf{x}(0) = \mathbf{I}\mathbf{x}(0) = \mathbf{x}(0)$$

so (10.31) does satisfy the initial condition. Using (10.28), we have

$$\frac{d}{dt}\mathbf{x}(t) = \frac{d}{dt}[e^{\mathbf{A}t}\mathbf{x}(0)] = [\frac{d}{dt}e^{\mathbf{A}t}]\mathbf{x}(0) = \mathbf{A}e^{\mathbf{A}t}\mathbf{x}(0) = \mathbf{A}\mathbf{x}(t)$$

which shows that (10.31) also satisfies the differential equation. Thus (10.31) is the solution of (10.30). If $u(t) = 0$, (10.18) reduces to

$$\mathbf{X}(s) = (s\mathbf{I} - \mathbf{A})^{-1}\mathbf{x}(0)$$

The application of the Laplace transform to (10.31) yields $\mathbf{X}(s) = \mathcal{L}[e^{\mathbf{A}t}]\mathbf{x}(0)$. Thus, we have

$$\mathcal{L}[e^{\mathbf{A}t}] = (s\mathbf{I} - \mathbf{A})^{-1}$$

EXERCISE 10.4.1

Show that the solution of

$$\dot{\mathbf{x}}(t) = \begin{bmatrix} 0 & 0 \\ -2 & 0 \end{bmatrix} \mathbf{x}(t)$$

excited by $\mathbf{x}(0) = [1 \ 0]'$ is

$$\mathbf{x}(t) = \begin{bmatrix} 1 & 0 \\ -2t & 1 \end{bmatrix} \begin{bmatrix} 1 \\ 0 \end{bmatrix} = \begin{bmatrix} 1 \\ -2t \end{bmatrix}$$

∎

EXERCISE 10.4.2

Show that the solution of

$$\dot{\mathbf{x}}(t) = \begin{bmatrix} -1 & 0 \\ 0 & 2 \end{bmatrix} \mathbf{x}(t)$$

due to $\mathbf{x}(0) = [3 \ -1]'$ is

$$\mathbf{x}(t) = \begin{bmatrix} e^{-t} & 0 \\ 0 & e^{2t} \end{bmatrix} \begin{bmatrix} 3 \\ -1 \end{bmatrix} = \begin{bmatrix} 3e^{-t} \\ -e^{2t} \end{bmatrix}$$

∎

Nonhomogeneous equation To obtain the solution of

$$\dot{\mathbf{x}}(t) = \mathbf{A}\mathbf{x}(t) + \mathbf{b}u(t) \tag{10.32}$$

excited by $\mathbf{x}(0)$ and $u(t)$, $t \geq 0$, we premultiply[3] (10.32) by $e^{-\mathbf{A}t}$ to yield

$$e^{-\mathbf{A}t}\dot{\mathbf{x}}(t) - e^{-\mathbf{A}t}\mathbf{A}\mathbf{x}(t) = e^{-\mathbf{A}t}\mathbf{b}u(t)$$

which implies

---

[3]By premultiplication, we mean multiplication from the left. In the matrix case, multiplications from the left and from the right are different because $\mathbf{AB} \neq \mathbf{BA}$.

$$\frac{d}{dt}[e^{-\mathbf{A}t}\mathbf{x}(t)] = e^{-\mathbf{A}t}\mathbf{b}u(t)$$

Its integration from 0 to t yields

$$\int_0^t \frac{d}{d\tau}[e^{-\mathbf{A}\tau}\mathbf{x}(\tau)]d\tau = \int_0^t e^{-\mathbf{A}\tau}\mathbf{b}u(\tau)d\tau$$

or

$$e^{-\mathbf{A}\tau}\mathbf{x}(\tau)\Big|_{\tau=0}^t = e^{-\mathbf{A}t}\mathbf{x}(t) - \mathbf{I}\mathbf{x}(0) = \int_0^t e^{-\mathbf{A}\tau}\mathbf{b}u(\tau)d\tau$$

which becomes, after the premultiplication with $e^{\mathbf{A}t}$,

$$\mathbf{x}(t) = \underbrace{e^{\mathbf{A}t}\mathbf{x}(0)}_{\substack{\text{zero-input}\\\text{response}}} + \underbrace{e^{\mathbf{A}t}\int_0^t e^{-\mathbf{A}\tau}\mathbf{b}u(\tau)d\tau}_{\substack{\text{zero-state}\\\text{response}}} \qquad (10.33)$$

This is the solution of (10.32). The substitution of (10.33) into (10.16b) yields

$$y(t) = \underbrace{\mathbf{c}e^{\mathbf{A}t}\mathbf{x}(0)}_{\substack{\text{zero-input}\\\text{response}}} + \underbrace{\mathbf{c}e^{\mathbf{A}t}\int_0^t e^{-\mathbf{A}\tau}\mathbf{b}u(\tau)d\tau + du(t)}_{\substack{\text{zero-state}\\\text{response}}} \qquad (10.34)$$

As in (10.18) and (10.19), the responses in (10.33) and (10.34) can be decomposed into the zero-input response and the zero-state response. The solutions in (10.33) and (10.34) hinge on the computation of $e^{\mathbf{A}t}$. In addition to the infinite series in (10.26), many methods for computing $e^{\mathbf{A}t}$ are available. See Reference [4]. It is important to note that Equations (10.33) and (10.34) are used mainly in the study of the general properties of (10.16). If we are interested in computing the responses of $\mathbf{x}(t)$ and $y(t)$ excited by some specific input and initial state, we can obtain them by direct integration, as will be discussed in following sections.

10.5   SOLUTION OF STATE EQUATIONS—DIGITAL COMPUTER COMPUTATION

The basic idea for the digital computer computation of

$$\dot{\mathbf{x}}(t) = \mathbf{A}\mathbf{x}(t) + \mathbf{b}u(t) \qquad (10.35)$$

is very simple and does not require any knowledge of the discussions in the preceding two sections. By definition, we have

$$\dot{\mathbf{x}}(t_0) = \lim_{\alpha \to 0} \frac{[\mathbf{x}(t_0 + \alpha) - \mathbf{x}(t_0)]}{\alpha}$$

The substitution of this into (10.35) yields

$$\mathbf{x}(t_0 + \alpha) - \mathbf{x}(t_0) = [\mathbf{A}\mathbf{x}(t_0) + \mathbf{b}u(t_0)]\alpha$$

or

$$\mathbf{x}(t_0 + \alpha) = (\mathbf{I} + \alpha\mathbf{A})\mathbf{x}(t_0) + \mathbf{b}u(t_0)\alpha \qquad (10.36)$$

If $\mathbf{x}(t_0)$ and $u(t_0)$ are known, then $\mathbf{x}(t_0 + \alpha)$ can be computed from (10.36). Using this equation repeatedly or recursively, the solution of (10.35) due to any $\mathbf{x}(0)$ and $u(t)$, $t \geq 0$, can be computed. For example, from the given $\mathbf{x}(0)$ and $u(0)$, we can compute

$$\mathbf{x}(\alpha) = (\mathbf{I} + \alpha\mathbf{A})\mathbf{x}(0) + \mathbf{b}u(0)\alpha$$

We then use this $\mathbf{x}(\alpha)$ and $u(\alpha)$ to compute

$$\mathbf{x}(2\alpha) = (\mathbf{I} + \alpha\mathbf{A})\mathbf{x}(\alpha) + \mathbf{b}u(\alpha)\alpha$$

Proceeding forward, $\mathbf{x}(k\alpha)$, for $k = 0, 1, 2, \ldots$, can be obtained. This procedure can be expressed in a programmatic format as

   DO 10 $k = 1, N$

10   $\mathbf{x}((k + 1)\alpha) = (\mathbf{I} + \alpha\mathbf{A})\mathbf{x}(k\alpha) + \mathbf{b}u(k\alpha)\alpha$

where N is the number of points to be computed. This can be easily programmed on a digital computer. Equation (10.36) is the simplest but least accurate method for computing (10.35).

When using (10.36), the problem of choosing α naturally arises. It is clear that, the smaller the α, the more accurate the result. However, the smaller the α, the greater the number of points that have to be computed for the same time interval. Therefore, α must be selected by compromising between acceptable accuracy and desired amount of computation. In actual programming, α may be chosen as follows: We choose an arbitrary α_0 and compute the response. We then repeat the computation by using $\alpha_1 = \alpha_0/2$. If the result of using α_1 is close to the one computed by using α_0, we stop because the result obtained by using α_0 is probably very close to the actual solution. If the result of using α_1 is quite different from the one obtained by using α_0, α_0 is not small enough and cannot be used. Whether or not α_1 is small enough cannot be answered at this point. Next, we choose $\alpha_2 = \alpha_1/2$ and repeat the computation. If the result using α_2 is close to the one obtained by using α_1, we may conclude that α_1 is small enough and stop the computation. Otherwise, we continue the process until the results of two subsequent computations are close.

EXAMPLE 10.5.1

Let us compute the output $y(t)$ of

$$\dot{\mathbf{x}}(t) = \begin{bmatrix} 0 & 1 \\ -0.5 & -1.5 \end{bmatrix} \mathbf{x}(t) - \begin{bmatrix} 0 \\ 1 \end{bmatrix} u(t) \tag{10.37}$$

$$y(t) = [1 \ -1]\mathbf{x}(t)$$

due to $\mathbf{x}(0) = [2 \ -1]'$ and $u(t) = 1$ for $t \geq 0$.

For this equation, (10.36) becomes

$$\begin{bmatrix} x_1(t_0 + \alpha) \\ x_2(t_0 + \alpha) \end{bmatrix} = (\begin{bmatrix} 1 & 0 \\ 0 & 1 \end{bmatrix} + \alpha \begin{bmatrix} 0 & 1 \\ -0.5 & -1.5 \end{bmatrix}) \begin{bmatrix} x_1(t_0) \\ x_2(t_0) \end{bmatrix} + \begin{bmatrix} 0 \\ 1 \end{bmatrix} \cdot 1 \cdot \alpha$$

which implies,

$$x_1((k + 1)\alpha) = x_1(k\alpha) + \alpha x_2(k\alpha)$$

$$x_2((k + 1)\alpha) = -0.5\alpha x_1(k\alpha) + (1 - 1.5\alpha)x_2(k\alpha) + \alpha$$

The output equation is

$$y(k\alpha) = x_1(k\alpha) - x_2(k\alpha)$$

We compute $y(t)$ from $t = 0$ to $t = 10$ seconds. Arbitrarily, we choose $\alpha_0 = 1$. We compute from $k = 0$ to $k = 10$. A FORTRAN program for this computation, where A stands for α, is as follows:

```
PROGRAM EULER
REAL X1, X2, Y
A = 1.
X1 = 2
X2 = -1
K = 0
DO 10, I = 1,10
K = K + 1
X1 = X1 + A*X2
X2 = -0.5*A*X1 + (1. - 1.5*A)*X2 + A
Y = X1 - X2
PRINT*, ´K = ´, ´Y = ´, Y
```

10 CONTINUE

   STOP

   END

The result is printed in Table 10.1 and plotted with +'s in Figure 10.7. We then repeat the computation using $A = 0.5$ and compute 20 points. The result is plotted with o's in Figure 10.7, but only ten points are printed in Table 10.1. The two results are quite different; therefore, we repeat the process for $A = 0.25$ and then for $A = 0.125$. The last two results are very close; therefore, we stop the computation.

   The exact solution of (10.37) was computed in (10.24) as

$$y(t) = 2 + 4e^{-t} - 3e^{-0.5t}$$

The computed result is very close to the exact solution.

   In Example 10.5.1, α is fixed for each run of computation and a number of runs may be required before a satisfactory solution is obtained. The method can be modified by using a variable α so that the computation can be completed in one run. To do so, we choose an α and compute the first point. We then use $\alpha/2$ to compute *again* the first two points. Note that the first point in α corresponds to the second point in $\alpha/2$. If these points are very close, we may proceed to the next point. Otherwise, we use $\alpha/4$ to compute again the first four points. Proceeding in this manner, a fairly accurate result can be obtained in one run.

   This example illustrates the basic idea of digital computer computation. The method used is the least accurate, and we recommend using it only for

Table 10.1 *Computation of (10.37) using four different values of* α.

	$\alpha = 1.0$	$\alpha = 0.5$	$\alpha = 0.25$	$\alpha = 0.125$	**Exact**
$T = 0.0$	$Y = 3.0000$	$Y = 3.0000$	$Y = 3.0000$	$Y = 3.0000$	$Y = 3.0000$
$T = 1.0$	$Y = 0.0000$	$Y = 1.3281$	$Y = 1.5099$	$Y = 1.5852$	$Y = 1.6519$
$T = 2.0$	$Y = 2.5000$	$Y = 1.4231$	$Y = 1.4146$	$Y = 1.4237$	$Y = 1.4377$
$T = 3.0$	$Y = 1.0000$	$Y = 1.6003$	$Y = 1.5579$	$Y = 1.5420$	$Y = 1.5297$
$T = 4.0$	$Y = 2.2500$	$Y = 1.7300$	$Y = 1.6990$	$Y = 1.6829$	$Y = 1.6673$
$T = 5.0$	$Y = 1.5000$	$Y = 1.8183$	$Y = 1.8017$	$Y = 1.7916$	$Y = 1.7807$
$T = 6.0$	$Y = 2.1250$	$Y = 1.8778$	$Y = 1.8708$	$Y = 1.8661$	$Y = 1.8605$
$T = 7.0$	$Y = 1.7500$	$Y = 1.9178$	$Y = 1.9162$	$Y = 1.9149$	$Y = 1.9130$
$T = 8.0$	$Y = 2.0625$	$Y = 1.9447$	$Y = 1.9457$	$Y = 1.9461$	$Y = 1.9463$
$T = 9.0$	$Y = 1.8750$	$Y = 1.9628$	$Y = 1.9649$	$Y = 1.9660$	$Y = 1.9672$
$T = 10.0$	$Y = 2.0313$	$Y = 1.9750$	$Y = 1.9773$	$Y = 1.9786$	$Y = 1.9799$

Figure 10.7   *Computation of (10.37) using four different values of* α

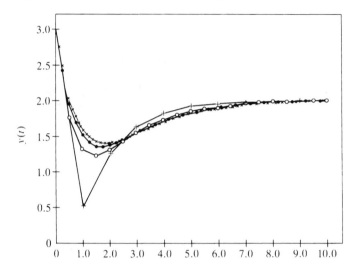

practice. For actual applications, we suggest using the existing programs, which will be discussed in Section 10.5.2.

Numerical integrations of $\dot{x}(t) = Ax(t)$

In this section, we will give a different derivation of (10.36) and discuss its relationship with the infinite power series in (10.26), or

$$x(t_0 + \alpha) = e^{A\alpha}x(t_0) = (I + \alpha A + \frac{\alpha^2}{2!}A^2 + \frac{\alpha^3}{3!}A^3 + \cdots)x(t_0) \quad \textbf{(10.38)}$$

If $u(t) = 0$, (10.36) becomes

$$x(t_0 + \alpha) = (I + \alpha A)x(t_0) \qquad \textbf{(10.39)}$$

A comparison of (10.38) and (10.39) reveals that (10.39) approximates $e^{A\alpha}$ by retaining the first two terms of its infinite power series. Therefore, it is the least accurate approximation.

Now consider a different derivation of (10.39). If $u(t) = 0$, then $\dot{x}(t) = Ax(t)$ implies

$$\int_{t_0}^{t_0 + \alpha} \dot{x}(t)dt = \int_{t_0}^{t_0 + \alpha} Ax(\tau)d\tau$$

or

$$\mathbf{x}(t_0 + \alpha) = \mathbf{x}(t_0) + \int_{t_0}^{t_0 + \alpha} \mathbf{A}\mathbf{x}(\tau)d\tau \tag{10.40}$$

In this integral equation, $\mathbf{x}(t_0)$ is known and $\mathbf{x}(t_0 + \alpha)$ is to be computed. Because $\mathbf{x}(\tau)$ is unknown for $\tau > t_0$, there is no way to carry out the integration. Therefore, we must proceed by approximation. If we assume

$$\mathbf{A}\mathbf{x}(t) = \mathbf{A}\mathbf{x}(t_0) \qquad \text{for } t_0 \leq t \leq t_0 + \alpha$$

as shown in Figure 10.8, then (10.40) becomes

$$\mathbf{x}(t_0 + \alpha) = \mathbf{x}(t_0) + \mathbf{A}\mathbf{x}(t_0)\int_{t_0}^{t_0 + \alpha} d\tau = (\mathbf{I} + \alpha\mathbf{A})\mathbf{x}(t_0) \tag{10.41}$$

This is the same as (10.39) and is called the *Euler method*.

A better approximation of the integration is to use the *trapezoidal method*, that is,

$$\int_{t_0}^{t_0 + \alpha} \mathbf{A}\mathbf{x}(\tau)d\tau = \frac{1}{2}\alpha[\mathbf{A}\mathbf{x}(t_0) + \mathbf{A}\mathbf{x}(t_0 + \alpha)] \tag{10.42}$$

However, this method cannot be used because $\mathbf{x}(t_0 + \alpha)$ is not known. One way to overcome this problem is to use the approximation of $\mathbf{x}(t_0 + \alpha)$ computed in (10.41). By so doing, (10.42) becomes

$$\mathbf{x}(t_0 + \alpha) = \mathbf{x}(t_0) + \frac{1}{2}\alpha[\mathbf{A}\mathbf{x}(t_0) + \mathbf{A}(\mathbf{I} + \alpha\mathbf{A})\mathbf{x}(t_0)]$$

$$= (\mathbf{I} + \alpha\mathbf{A} + \frac{\alpha^2}{2!}\mathbf{A}^2)x(t_0) \tag{10.43}$$

This approximation is called the *Heun method*. It is a combination of the Euler and trapezoidal methods. The Heun method uses the first three terms to approximate the infinite series of $e^{\mathbf{A}\alpha}$.

Figure 10.8 *Integrations of (10.40)*

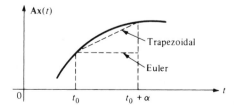

The fourth-order *Runge-Kutta method* is a widely used approximation in integration. Application of this method to (10.40) yields

$$\mathbf{x}(t_0 + \alpha) = (\mathbf{I} + \alpha\mathbf{A} + \frac{\alpha^2}{2!}\mathbf{A}^2 + \frac{\alpha^3}{3!}\mathbf{A}^3 + \frac{\alpha^4}{4!}\mathbf{A}^4)x(t_0) \qquad \textbf{(10.44)}$$

It uses the first five terms to approximate the infinite series of $e^{\mathbf{A}\alpha}$. For a more detailed discussion of the method, see Reference [19].

10.5.2   Using existing computer programs

Presently, of the many commercially available computer programs for personal computers and workstations, most contain programs for solving LTIL state-variable equations. For example, to use the command `step` in MATLAB to compute the unit step response of (10.37), we type

```
a = [0 1; -0.5 -1.5]; b = [0;1];
c = [1 -1]; d = 0;
step(a,b,c,d,1)
```

The result will appear on the screen. (For a discussion of the representation of matrices in MATLAB, see the last section of Appendix A.) The command `step` is developed for one or more inputs; the command `step(a,b,c,d,i)` computes the unit step response excited by the *i*th input. In our case, we have only one input, and we always set $i = 1$. The command automatically selects the integration step size and the time interval of computation. If we wish to select the integration step size and the time interval of computation ourselves, we may type

```
t = 0:0.01:10;
y = step(a,b,c,d,1,t);
plot(t,y)
```

This computes the output from $t = 0$ to $t = 10$ and prints the output every 0.01 second. If the input is other than the unit step function, then we may use command `lsim`, which stands for linear simulation. Unfortunately, `lsim` is not available in the *Student Edition of MATLAB*.

10.6   STATE-VARIABLE EQUATIONS AND BASIC BLOCK DIAGRAMS

The development of basic block diagrams for state-variable equations will be discussed in this section. A basic block diagram is any diagram built using only integrators, multipliers, and adders, (these elements were introduced in Figure 3.25). Note that an integrator is often denoted by s^{-1}. Given a state-variable equation, it is simple and straightforward to develop its basic block diagram, as Example 10.6.1 illustrates.

EXAMPLE **10.6.1**

Develop a basic block diagram for the state-variable equation

$$\dot{\mathbf{x}}(t) = \begin{bmatrix} 2 & 0.3 \\ 1 & -8 \end{bmatrix} \mathbf{x}(t) + \begin{bmatrix} 2 \\ 0 \end{bmatrix} u(t)$$

$$y(t) = [-2 \quad 3]\mathbf{x}(t)$$

or

$$\dot{x}_1(t) = 2x_1(t) + 0.3x_2(t) + 2u(t) \qquad \textbf{(10.45a)}$$
$$\dot{x}_2(t) = x_1(t) - 8x_2(t) \qquad \textbf{(10.45b)}$$
$$y(t) = -2x_1(t) + 3x_2(t) \qquad \textbf{(10.45c)}$$

The equation has two state variables. Thus, we need two integrators. The outputs of the integrators are assigned as $x_1(t)$ and $x_2(t)$, as shown in Figure 10.9. Their inputs are $\dot{x}_1(t)$ and $\dot{x}_2(t)$. Three amplifiers and one summer are used to generate the right-hand side of (10.45a) at the input of the first integrator, as in Figure 10.9(a). The right-hand side of (10.45b) is generated at the input of the second integrator. Finally, (10.45c) is used to generate $y(t)$ from x_1 and x_2. The resulting basic block diagram is shown in Figure 10.9(b).

Figure 10.9 *Development of the basic block diagram for (10.45)*

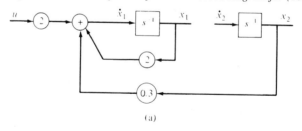

(a)

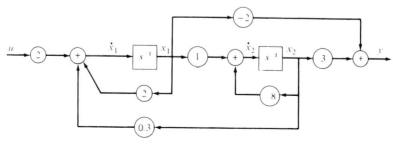

(b)

EXERCISE 10.6.1

Develop a basic block diagram for

$$\dot{\mathbf{x}}(t) = \begin{bmatrix} 2 & 1 \\ -1 & 0 \end{bmatrix} \mathbf{x}(t) + \begin{bmatrix} 2 \\ -3 \end{bmatrix} u(t)$$

$$y(t) = [1 \ 0]\mathbf{x}(t)$$

■

Developing a basic block diagram for any LTIL state variable equation is a straightforward task, as the preceding example showed. The next example illustrates that the converse is equally straightforward.

EXAMPLE **10.6.2**

Consider the basic block diagram shown in Figure 10.10. There are three integrators. The outputs of the three integrators are assigned as x_2, $-x_1$, and x_3. Note that this assignment is entirely arbitrary. Once the outputs are assigned, the inputs of the integrators become, respectively, $\dot{x}_2$, $-\dot{x}_1$, and $\dot{x}_3$. Now we can read the following equations directly from the diagram.

$$-\dot{x}_1 = x_2$$
$$\dot{x}_2 = 2x_2 - 3(-x_1) + x_3 + u$$
$$\dot{x}_3 = -x_1$$
$$y = -x_2 + 4x_3$$

Figure 10.10 *Basic block diagram*

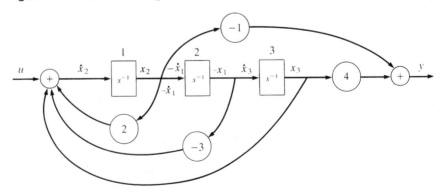

These equations can be expressed in matrix form as

$$\dot{\mathbf{x}} = \begin{bmatrix} 0 & -1 & 0 \\ 3 & 2 & 1 \\ -1 & 0 & 0 \end{bmatrix} \mathbf{x} + \begin{bmatrix} 0 \\ 1 \\ 0 \end{bmatrix} u$$

$$y = [0 \;\; -1 \;\; 4]\mathbf{x}$$

This state-variable equation describes the block diagram.

―――――――

EXERCISE 10.6.2

Consider again the basic block diagram in Figure 10.10. What is its state-variable equation if the outputs of the integrators 1, 2, and 3 are assigned as x_1, x_2, and x_3? What is its state-variable equation if the outputs of the integrators 1, 2, and 3 are assigned as x_3, x_2, and x_1?

$$[\textbf{ANSWERS: } \dot{\mathbf{x}} = \begin{bmatrix} 2 & -3 & 1 \\ 1 & 0 & 0 \\ 0 & 1 & 0 \end{bmatrix} \mathbf{x} + \begin{bmatrix} 1 \\ 0 \\ 0 \end{bmatrix} u, \; \dot{\mathbf{x}} = \begin{bmatrix} 0 & 1 & 0 \\ 0 & 0 & 1 \\ 1 & -3 & 2 \end{bmatrix} \mathbf{x} + \begin{bmatrix} 0 \\ 0 \\ 1 \end{bmatrix} u.]$$

$$y = [-1 \;\; 0 \;\; 4]\,\mathbf{x} \qquad\qquad y = [4 \;\; 0 \;\; -1]\mathbf{x} \qquad\qquad ■$$

The preceding example and exercise have shown that the assignment of state variables in a basic block diagram is not unique. Different assignments will yield different state-variable equations. These equations, however, are all equivalent. For a discussion of the equivalence of state-variable equations, see Reference [4].

10.6.1   Op-amp circuit implementation

The integrator, multiplier, and adder in a basic block diagram can be constructed using operational amplifier circuits. Figure 10.11(a) through (d) illustrates such construction where, except in (a), a grounded noninverting terminal is not shown. Therefore, once a basic block diagram has been developed for a state-variable equation, the equation can be built using operational amplifier circuits. For example, by replacing each element in Figure 10.9(b) with the corresponding op-amp circuit, we can implement (10.45) using operational amplifier circuits. Generally, however, this implementation would not be desirable because it would use an unnecessarily large number of components.

There is a method for implementing a state-variable equation without first drawing a basic block diagram. Consider the op-amp element shown in Figure 10.11(e). If the output is assigned as y, then its input must generate

$$-y = ax_1 + bx_2 + cx_3$$

Figure 10.11   *Implementation of basic elements*

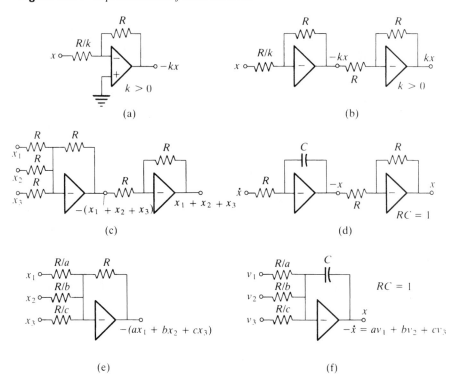

If the output is assigned as $-y$, then its input must generate

$$y = ax_1 + bx_2 + cx_3$$

Note that we have freedom in assigning the output as y or $-y$. Note, too, that the element acts as three multipliers and one adder. Similarly, the element in Figure 10.11(f) acts as one integrator, one adder, and three multipliers. The output of that element can be assigned as x or $-x$. If it is assigned as $-x$, then the input must generate

$$\dot{x} = av_1 + bv_2 + cv_3$$

We can use these two elements and inverters to implement (10.45). The equation in (10.45) has two state variables; thus, we need two integrators as in Figure 10.12. Rather arbitrarily, we assign the output of the first integrator (from left) as x_1 and the output of the second integrator as $-x_2$. Thus, we must generate $-\dot{x}_1$ at the input of the first integrator or

$$-\dot{x}_1 = -2x_1 - 0.3x_2 - 2u \tag{10.46}$$

Figure 10.12 *Implementations of (10.45)*

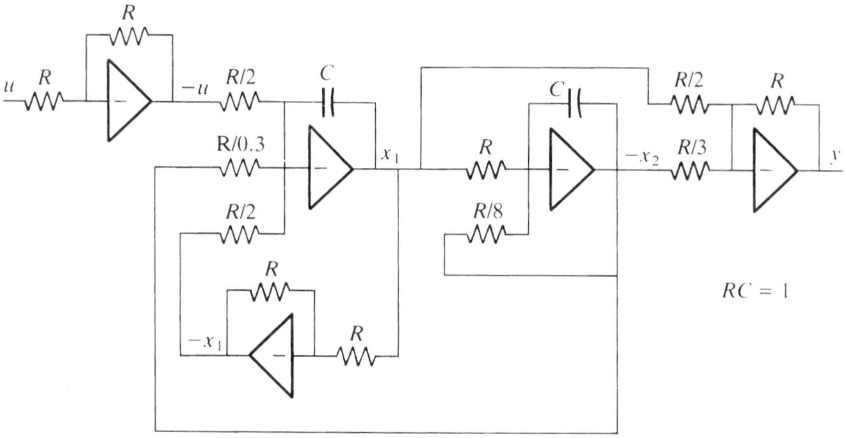

which is obtained from (10.45a) by reversing all the signs. Because the output of the first integrator is x_1, we use an inverter to change it to $-x_1$, as shown in Figure 10.12. We then generate the right-hand side of (10.46) at the input of the first integrator. Because the output of the second integrator is assigned as $-x_2$, we must generate $\dot{x}_2 = x_1 - 8x_2$ at the input of the second integrator, as shown. The rest is self-explanatory. This completes an op-amp circuit implementation of (10.45).

Although state-variable equations can be readily implemented using op-amp circuits, some difficulties may arise. For example, if op amps are powered by a ±15 volt power supply, the magnitude of all variables in the equation cannot exceed 13 (usually 2 volts below the power supply), or the circuit will saturate. Thus, magnitude scaling is required. This and other technical details are outside the scope of this text and will not be discussed. See Reference [8].

*10.7 DISCRETIZATION OF CONTINUOUS-TIME STATE EQUATIONS

Consider the continuous-time state-variable equation

$$\dot{\mathbf{x}}(t) = \mathbf{A}\mathbf{x}(t) + \mathbf{b}u(t) \tag{10.47a}$$
$$y(t) = \mathbf{c}\mathbf{x}(t) + du(t) \tag{10.47b}$$

On digital computer computation, we must discretize it into a discrete-time equation, such as

$$\mathbf{x}(t_0 + \alpha) = (\mathbf{I} + \alpha\mathbf{A})\mathbf{x}(t_0) + \mathbf{b}u(t_0)\alpha$$

Figure 10.13 *An input signal that is piecewise constant*

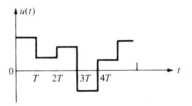

This equation, no matter how small the value of α, is only an approximation of (10.47a). In this section, we will discuss a discretization for a special class of inputs. The discretization will yield an exact solution of (10.47).

Let $u(t)$ be piecewise constant, as shown in Figure 10.13. That is,

$$u(t) = u(kT) \quad \text{for } kT \leq t < (k+1)T, \quad k = 0, 1, 2, \ldots \quad \textbf{(10.48)}$$

In other words, $u(t)$ changes value only at discrete sampling instants.[4] The solution of (10.47a) excited by the input $u(t)$ in (10.48) is, as derived in (10.33),

$$\mathbf{x}(t) = e^{\mathbf{A}t}\mathbf{x}(0) + e^{\mathbf{A}t}\int_0^t e^{-\mathbf{A}\tau}\mathbf{b}u(\tau)d\tau \quad \textbf{(10.49)}$$

This equation holds at every t, in particular at $t = kT$, $k = 0, 1, 2, \ldots$. We show that the equation can be reduced to a discrete-time equation at $t = kT$, $k = 0, 1, 2, \ldots$. At $t = (k+1)T$, (10.49) becomes

$$\mathbf{x}((k+1)T) = e^{\mathbf{A}(k+1)T}\mathbf{x}(0) + e^{\mathbf{A}(k+1)T}\int_0^{(k+1)T} e^{-\mathbf{A}\tau}\mathbf{b}u(\tau)d\tau$$

$$= e^{\mathbf{A}T}[e^{\mathbf{A}kT}\mathbf{x}(0) + e^{\mathbf{A}kT}(\int_0^{kT} e^{-\mathbf{A}\tau}\mathbf{b}u(\tau)d\tau + \int_{kT}^{(k+1)T} e^{-\mathbf{A}\tau}\mathbf{b}u(\tau)d\tau)] \quad \textbf{(10.50)}$$

$$= e^{\mathbf{A}T}[e^{\mathbf{A}kT}\mathbf{x}(0) + e^{\mathbf{A}kT}\int_0^{kT} e^{-\mathbf{A}\tau}\mathbf{b}u(\tau)d\tau] + e^{\mathbf{A}(k+1)T}\int_{kT}^{(k+1)T} e^{-\mathbf{A}\tau}\mathbf{b}u(\tau)d\tau$$

The term inside the brackets in (10.50) equals $\mathbf{x}(kT)$. Because $u(\tau)$ is equal to $u(kT)$ between kT and $(k+1)T$, after the change of variable $\alpha = (k+1)T - \tau$, the last term in (10.50) can be simplified as

$$[\int_{kT}^{(k+1)T} e^{\mathbf{A}(k+1)T}e^{-\mathbf{A}\tau}d\tau]\mathbf{b}u(kT) = [\int_T^0 e^{\mathbf{A}\alpha}(-d\alpha)]\mathbf{b}u(kT) = [\int_0^T e^{\mathbf{A}\alpha}d\alpha]\mathbf{b}u(kT)$$

---

[4]Inputs of this type occur if a digital computer and a zero-order hold are used to generate $u(t)$.

Thus, (10.50) becomes

$$\mathbf{x}((k+1)T) = e^{\mathbf{A}T}\mathbf{x}(kT) + [\int_0^T e^{\mathbf{A}\alpha}d\alpha]\mathbf{b}u(kT) \tag{10.51}$$

This is a discrete-time equation. Thus if an input changes value only at sampling instants kT, $k = 0, 1, 2, \ldots$, and if we want to compute the response only at kT, $k = 0, 1, 2, \ldots$, then the continuous-time equation in (10.47) can be replaced by

$$\mathbf{x}((k+1)T) = \overline{\mathbf{A}}\mathbf{x}(kT) + \overline{\mathbf{b}}u(kT)$$
$$y(kT) = \mathbf{c}\mathbf{x}(kT) + du(kT) \tag{10.52}$$

where

$$\overline{\mathbf{A}} := e^{\mathbf{A}T}$$
$$\overline{\mathbf{b}} := (\int_0^T e^{\mathbf{A}\tau}d\tau)\mathbf{b}$$

This is a discrete-time state-variable equation.

10.8   SOLUTION OF DISCRETE-TIME EQUATIONS

An LTIL discrete-time state-variable equation may arise from a continuous-time state-variable equation through discretization, as in the preceding section. It may also arise directly from systems that are inherently discrete-time, such as the inventory or population models discussed in Chapter 3. A general LTIL discrete-time state-variable equation is of the form

$$\mathbf{x}[k+1] = \mathbf{A}\mathbf{x}[k] + \mathbf{b}u[k] \tag{10.53a}$$

$$y[k] = \mathbf{c}\mathbf{x}[k] + du[k] \tag{10.53b}$$

where $\mathbf{x}$, u, and y are, respectively, the state, input, and output. If $\mathbf{x}$ is an $n \times 1$ vector, then $\mathbf{A}$ is an $n \times n$ square matrix, $\mathbf{b}$ is an $n \times 1$ column vector, $\mathbf{c}$ is a $1 \times n$ row vector, d is a 1×1 scalar, and the equation is said to be of dimension n. It is the counterpart of (10.16) in the discrete-time case. Equation (10.16a) consists of a set of first-order differential equations, whereas (10.53a) consists of a set of first-order difference equations. Equations (10.16b) and (10.53b) are identical; they are algebraic equations. As we shall see in this section, the study of (10.53) is considerably simpler than the study of (10.16).

The solution of (10.53a) excited by the initial state $\mathbf{x}[0]$ and the input sequence $u[k]$, $k = 0, 1, 2, \ldots$, can be obtained by direct substitution. Using $\mathbf{x}[0]$ and $u[0]$, we compute

$$\mathbf{x}[1] = \mathbf{A}\mathbf{x}[0] + \mathbf{b}u[0] \tag{10.54}$$

We then use $u[1]$ and $\mathbf{x}[1]$ to compute

$$\mathbf{x}[2] = \mathbf{A}\mathbf{x}[1] + \mathbf{b}u[1] \qquad (10.55)$$

Proceeding forward, $\mathbf{x}[k]$ and, consequently, $y[k]$ for any k can be computed.

EXERCISE 10.8.1

Compute $y[k]$, for $k = 0, 1, 2, 3, 4$, and 5, in

$$\mathbf{x}[k + 1] = \begin{bmatrix} 0 & 0 \\ 0 & 1 \end{bmatrix} \mathbf{x}[k] + \begin{bmatrix} 1 \\ 1 \end{bmatrix} u[k] \qquad (10.56)$$

$$y[k] = [1 \ \ -1]\mathbf{x}[k]$$

excited by $\mathbf{x}[0] = [1 \ \ -2]'$ and $u[k] = 1, k = 0, 1, 2, \ldots$.

[**ANSWERS**: 3, 2, 1, 0, −1, −2, ...] ∎

To develop a general solution of (10.53), we substitute (10.54) into (10.55) to yield

$$\mathbf{x}[2] = \mathbf{A}^2\mathbf{x}[0] + \mathbf{A}\mathbf{b}u[0] + \mathbf{b}[1]$$

We then compute

$$\mathbf{x}[3] = \mathbf{A}\mathbf{x}[2] + \mathbf{b}u[2] = \mathbf{A}^3\mathbf{x}[0] + \mathbf{A}^2\mathbf{b}u[0] + \mathbf{A}\mathbf{b}u[1] + \mathbf{b}u[2]$$

Proceeding forward, we can readily obtain

$$\mathbf{x}[k] = \mathbf{A}^k\mathbf{x}[0] + \mathbf{A}^{k-1}\mathbf{b}u[0] + \mathbf{A}^{k-2}\mathbf{b}u[1] + \cdots + \mathbf{A}\mathbf{b}u[k-2] + \mathbf{b}u[k-1]$$

$$= \underbrace{\mathbf{A}^k\mathbf{x}[0]}_{\substack{\text{zero-input} \\ \text{response}}} + \underbrace{\sum_{i=0}^{k-1}\mathbf{A}^{k-1-i}\mathbf{b}u[i]}_{\substack{\text{zero-state} \\ \text{response}}} \qquad (10.57)$$

and

$$\boxed{y[k] = \underbrace{\mathbf{c}\mathbf{A}^k\mathbf{x}[0]}_{\substack{\text{zero-input} \\ \text{response}}} + \underbrace{\sum_{i=0}^{k-1}\mathbf{c}\mathbf{A}^{k-1-i}\mathbf{b}u[i] + du[k]}_{\substack{\text{zero-state} \\ \text{response}}}} \qquad (10.58)$$

These two equations are the discrete counterparts of (10.33) and (10.34). We see that the responses can be decomposed into the zero-state responses and the zero-input responses. Equations (10.57) and (10.58) are used mainly in the study of the general properties of (10.53). They are *not* used in computing $\mathbf{x}[k]$ and $y[k]$. It is simpler to compute $\mathbf{x}[k]$ and $y[k]$ recursively from (10.53).

10.8.1   Application of the z-transform

Discrete-time state equations can be solved recursively by direct substitution. The solution, however, will not be in closed form. To obtain a closed-form solution, we employ the z-transform. The application of the z-transform to (10.53a) yields, using (5.18a),

$$z\,[\mathbf{X}(z) - \mathbf{x}[0]] = \mathbf{A}\mathbf{X}(z) + \mathbf{b}U(z)$$

which can be written as

$$z\mathbf{X}(z) - \mathbf{A}\mathbf{X}(z) = z\mathbf{x}[0] + \mathbf{b}U(z) \tag{10.59}$$

Note that $z\mathbf{X}(z) - \mathbf{A}\mathbf{X}(z) \neq (z - \mathbf{A})\mathbf{X}(z)$. However, we have $z\mathbf{X}(z) - \mathbf{A}\mathbf{X}(z) = (z\mathbf{I} - \mathbf{A})\mathbf{X}(z)$, where $\mathbf{I}$ is the unit matrix of the same order as $\mathbf{A}$. Thus, (10.59) implies

$$\mathbf{X}(z) = \underbrace{(z\mathbf{I} - \mathbf{A})^{-1}z\mathbf{x}[0]}_{\text{zero-input response}} + \underbrace{(z\mathbf{I} - \mathbf{A})^{-1}\mathbf{b}U(z)}_{\text{zero-state response}} \tag{10.60a}$$

and

$$\boxed{Y(z) = \underbrace{\mathbf{c}(z\mathbf{I} - \mathbf{A})^{-1}z\mathbf{x}[0]}_{\text{zero-input response}} + \underbrace{[\mathbf{c}(z\mathbf{I} - \mathbf{A})^{-1}\mathbf{b} + d\,]U(z)}_{\text{zero-state response}}} \tag{10.60b}$$

They are algebraic equations. The z-transform has transformed the difference equation in (10.53) into the algebraic equations in (10.60). It also has revealed the decomposition of the responses into the zero-state responses and the zero-input responses. If z is replaced by s, the zero-state part of (10.60) is identical to that of (10.19). The zero-input parts of (10.60) and (10.19) do not match, however; (10.60) has an extra z.

If the initial state $\mathbf{x}[0]$ is $\mathbf{0}$, then (10.60b) reduces to

$$Y(z) = [\mathbf{c}(z\mathbf{I} - \mathbf{A})^{-1}\mathbf{b} + d\,]U(z)$$

Thus, the sampled transfer function from u to y of the discrete-time system described by (10.53) is

$$\boxed{H(z) = \mathbf{c}(z\mathbf{I} - \mathbf{A})^{-1}\mathbf{b} + d} \qquad (10.61)$$

This is a fundamental equation relating the state-variable equation and discrete transfer function. It is identical to (10.20) if z is replaced by s.

EXAMPLE 10.8.1

How do we find the output of

$$\mathbf{x}[k+1] = \begin{bmatrix} 0 & 1 \\ -2 & -3 \end{bmatrix} \mathbf{x}[k] + \begin{bmatrix} 0 \\ 1 \end{bmatrix} u[k]$$

$$y[k] = [1 \;\; -1]\mathbf{x}[k]$$

excited by $\mathbf{x}[0] = [2 \;\; -1]'$ and $u[k] = 1$ for $k = 0, 1, 2, \ldots$?

This equation is similar to (10.21), and $(z\mathbf{I} - \mathbf{A})^{-1}$ is equal to (10.22) if s is replaced by z, that is,

$$(z\mathbf{I} - \mathbf{A})^{-1} = \frac{1}{z(z+3)+2} \begin{bmatrix} z+3 & 1 \\ -2 & z \end{bmatrix}$$

The substitution of $(z\mathbf{I} - \mathbf{A})^{-1}$ and $U(z) = z/(z-1)$ into (10.60a) yields

$$\mathbf{X}(z) = \frac{1}{(z+2)(z+1)} \begin{bmatrix} z+3 & 1 \\ -2 & z \end{bmatrix} \left(z \begin{bmatrix} 2 \\ -1 \end{bmatrix} + \begin{bmatrix} 0 \\ 1 \end{bmatrix} \frac{z}{z-1} \right)$$

$$= \frac{1}{(z+2)(z+1)} \begin{bmatrix} z+3 & 1 \\ -2 & z \end{bmatrix} \begin{bmatrix} 2z \\ -z + \dfrac{z}{z-1} \end{bmatrix}$$

Therefore, we have

$$Y(z) = [1 \;\; -1]\mathbf{X}(z) = \frac{1}{(z+2)(z+1)}[1 \;\; -1]\begin{bmatrix} z+3 & 1 \\ -2 & z \end{bmatrix}\begin{bmatrix} 2z \\ \dfrac{-z^2 + 2z}{z-1} \end{bmatrix}$$

$$= \frac{1}{(z+2)(z+1)}[z+5 \;\; 1-z]\begin{bmatrix} 2z \\ \dfrac{z^2 - 2z}{1-z} \end{bmatrix}$$

$$= \frac{2z(z+5) + (z^2 - 2z)}{(z+2)(z+1)} = \frac{3z^2 + 8z}{(z+2)(z+1)}$$

We expand $Y(z)/z$ by partial fraction expansion as

$$\frac{Y(z)}{z} = \frac{3z + 8}{(z + 2)(z + 1)} = \frac{k_1}{z + 2} + \frac{k_2}{z + 1}$$

with

$$k_1 = \left. \frac{3z + 8}{z + 1} \right|_{z=-2} = \frac{2}{-1} = -2$$

and

$$k_2 = \left. \frac{3z + 8}{z + 2} \right|_{z=-1} = \frac{5}{1} = 5$$

Thus, we have

$$Y(z) = \frac{-2z}{z + 2} + \frac{5z}{z + 1}$$

which implies, using Table 5.1,

$$y[k] = -2(-2)^k + 5(-1)^k, \qquad k = 0, 1, 2, \ldots$$

This is the output of the equation. The solution is in closed form.

EXERCISE 10.8.2

Find the closed-form output of

$$\mathbf{x}[k + 1] = \begin{bmatrix} 0 & 0 \\ 0 & 1 \end{bmatrix} \mathbf{x}[k] + \begin{bmatrix} 1 \\ 1 \end{bmatrix} u[k]$$

$$y[k] = [1 \;\; -1]\mathbf{x}[k]$$

excited by $\mathbf{x}[0] = [1 \;\; -2]'$ and $u[k] = 1$, for $k = 0, 1, 2, \ldots$. Check your result with the one computed for (10.56).

[**ANSWER:** $y[k] = 3 - k$.] ∎

10.8.2 ## Digital computer computation and basic block diagrams

The solution of an LTIL discrete-time state-variable equation can be obtained by repetitive substitution, as shown in (10.54) and (10.55). This procedure can be easily programmed on a digital computer. For example, a schematic computer program for solving (10.53) can be as simple as

$x[0], u[k]$

$k = 0$

DO 10 $I = 1, N$

$k = k + 1$

$x[k + 1] = Ax[k] + bu[k]$

$y[k] = cx[k] + du[k]$

PRINT $y[k]$

10   Continue

where N denotes the number of points to be computed. This program provides only the basic procedure; we must fill out some detail in actual programming. For example, if a program does not operate on matrices, then the matrix equation must be written out explicitly as scalar equations. In any case, the programming of LTIL discrete-time state-variable equations is simple and straightforward. For a specific example, see Example 10.5.1.

Now let us focus on the basic block diagram for an LTIL discrete-time state-variable equation. Such a block diagram is built using unit-time delay elements, multipliers, and adders, (these elements were introduced in Figure 3.6.) Note that the unit-time delay element is often denoted by z^{-1}. The procedure for developing basic block diagrams in the discrete-time case is identical to that in the continuous-time case discussed in Section 10.6. For example, consider

$$\mathbf{x}[k + 1] = \begin{bmatrix} 2 & 0.3 \\ 1 & -8 \end{bmatrix} \mathbf{x}[k] + \begin{bmatrix} 2 \\ 0 \end{bmatrix} u[k]$$

$$y[k] = [-2 \ \ 3]\mathbf{x}[k]$$

or

$$x_1[k + 1] = 2x_1[k] + 0.3x_2[k] + 2u[k]$$

$$x_2[k] = x_1[k] - 8x_2[k] \qquad\qquad (10.62)$$

$$y[k] = -2x_1[k] + 3x_2[k]$$

which is similar to (10.45). There are two state variables; therefore, we need two unit-time delay elements. We assign their outputs as $x_1[k]$ and $x_2[k]$. Then their inputs are $x_1[k + 1]$ and $x_2[k + 1]$. We use (10.62) to construct the basic block diagram shown in Figure 10.14. The diagram is identical to the one in Figure 10.9(b) if z^{-1} is replaced by s^{-1}.

As in the continuous-time case, it is also simple to develop a state-variable equation to describe a basic block diagram. For example, consider the diagram in Figure 10.15, which is the same as Figure 10.10 if integrators are replaced by unit-time delay elements. If we assign the outputs of the three delay elements

Figure 10.14 *Basic block diagram of (10.62)*

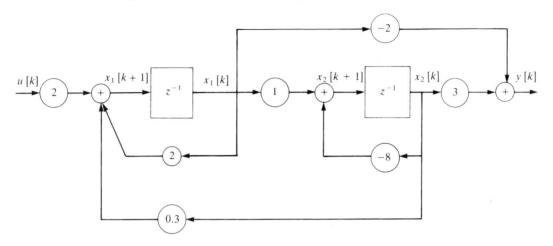

arbitrarily as $x_2[k]$, $-x_1[k]$, and $x_3[k]$ as shown, then their inputs are $x_2[k + 1]$, $-x_1[k + 1]$, and $x_3[k + 1]$. From the diagram, we can readily obtain

$$\mathbf{x}[k + 1] = \begin{bmatrix} 0 & -1 & 0 \\ 3 & 2 & 1 \\ -1 & 0 & 0 \end{bmatrix} \mathbf{x}[k] + \begin{bmatrix} 0 \\ 1 \\ 0 \end{bmatrix} u[k]$$

$$y[k] = [0 \; -1 \; 4]\mathbf{x}[k]$$

which is essentially the same as the one in Example 10.6.2.

As in the continuous-time case again, once a basic block diagram has been developed, the state-variable equation can be implemented using special hardware. Its implementation requires the use of shift registers as delay elements,

Figure 10.15 *Basic block diagram*

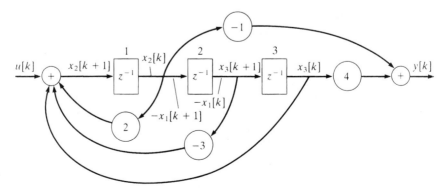

multipliers, and accumulators or adders. Generally, there are many ways to implement a basic block diagram. For example, the diagram in Figure 10.14 can be implemented using only one accumulator and only one multiplier, and all operations will be carried out sequentially. It also can be implemented using three accumulators and seven multipliers, and all operations will be carried out in parallel. The former implementation uses less hardware but takes a longer time to process; the latter uses much more hardware but is much faster. This is a topic outside the scope of this text.

10.9 CONTINUOUS-TIME REALIZATIONS

We discussed computer simulations and computations of state-variable equations in the preceding sections. Now we shall discuss the same topics for proper transfer functions. Suppose we want to compute the unit step response of

$$H(s) = \frac{2s^3 - s^2 + 3s + 10}{s^5 + 5s^4 + 11s^3 + 13s^2 + 8s + 2} =: \frac{N(s)}{D(s)} \quad \text{(10.63)}$$

If we use the procedure presented in Chapter 4, we must first compute the roots of $D(s)$ and then carry out partial fraction expansion. As was discussed at the end of Chapter 4, the roots of a polynomial are very sensitive to coefficient variations. Furthermore, partial fraction expansion is not simple to program. Therefore, the procedure in Chapter 4 is not the best way to carry out the computer computation of transfer functions. If we transform a transfer function into a state-variable equation, then all discussions in the preceding sections can be directly applied here. It turns out that this procedure yields more accurate results than direct computation on transfer functions. Thus, we will discuss first how to set up state-variable equations for transfer functions.

Given a transfer function $H(s)$, the problem of finding

$$\dot{\mathbf{x}}(t) = \mathbf{A}\mathbf{x}(t) + \mathbf{b}u(t) \quad \text{(10.64a)}$$
$$y(t) = \mathbf{c}\mathbf{x}(t) + du(t) \quad \text{(10.64b)}$$

so that the transfer function of (10.64) equals $H(s)$ or, equivalently, the problem of finding $\{\mathbf{A}, \mathbf{b}, \mathbf{c}, d\}$ so that

$$H(s) = \mathbf{c}(s\mathbf{I} - \mathbf{A})^{-1}\mathbf{b} + d \quad \text{(10.65)}$$

is called the *realization problem,* and (10.64) is called a realization of $H(s)$. The name is well justified, because $H(s)$ can be built or implemented through (10.64) using operational amplifier circuits. We will discuss a number of different realizations for the same $H(s)$. Before proceeding, note that $\mathbf{c}(s\mathbf{I} - \mathbf{A})^{-1}\mathbf{b}$ is a strictly proper rational function of s, that is,

$$\lim_{s \to \infty} \mathbf{c}(s\mathbf{I} - \mathbf{A})^{-1}\mathbf{b} = 0$$

Hence, from (10.65) we have

$$H(\infty) = d \qquad\qquad \textbf{(10.66)}$$

Thus, if a transfer function is strictly proper, the direct transmission part d of its realization is zero. If it is biproper, then d is nonzero. If we subtract $d = H(\infty)$ from $H(s)$, then

$$H_s(s) := H(s) - H(\infty) = H(s) - d \qquad\qquad \textbf{(10.67)}$$

is strictly proper. Therefore, every biproper transfer function can be easily decomposed into a constant and a strictly proper part.

10.9.1   Realizations of $b_0/D(s)$

Consider first a special class of transfer functions whose numerators are constants. Instead of discussing the general case, we will use the transfer function of degree 4

$$H(s) = \frac{b_0}{D(s)} = \frac{b_0}{a_4 s^4 + a_3 s^3 + a_2 s^2 + a_1 s + a_0}$$

where $a_4 \neq 0$, to illustrate the realization procedure. First, it is convenient to normalize the leading coefficient a_4 of $D(s)$ to 1 as

$$\frac{Y(s)}{U(s)} = H(s) = \frac{\bar{b}_0}{s^4 + \bar{a}_3 s^3 + \bar{a}_2 s^2 + \bar{a}_1 s + \bar{a}_0} =: \frac{\bar{b}_0}{D(s)} \qquad \textbf{(10.68a)}$$

where $\bar{a}_i = a_i/a_4$ and $\bar{b}_0 = b_0/a_4$. The corresponding differential equation is

$$y^{(4)}(t) + \bar{a}_3 y^{(3)}(t) + \bar{a}_2 y^{(2)}(t) + \bar{a}_1 y^{(1)} + \bar{a}_0 y(t) = \bar{b}_0 u(t)$$

or

$$y^{(4)}(t) = -\bar{a}_3 y^{(3)}(t) - \bar{a}_2 y^{(2)}(t) - \bar{a}_1 y^{(1)}(t) - \bar{a}_0 y(t) + \bar{b}_0 u(t) \quad \textbf{(10.68b)}$$

A basic block diagram of (10.68) is shown in Figure 10.16. Consider the four integrators connected in tandem. The input of the top integrator is assigned as $y^{(4)}(t)$, so the outputs of subsequent integrators are $y^{(3)}(t)$, $y^{(2)}(t)$, $y^{(1)}(t)$, and $y(t)$, as shown in Figure 10.16. The multipliers and summers are used to generate $y^{(4)}(t)$ in (10.68b). As we discussed in Figure 10.12, if we assign the output of each integrator as a state variable, then we can easily obtain a state-variable equation. Assigning the state variables shown in Figure 10.16, we obtain the following state-variable description of the block diagram:

Figure 10.16 *Basic block diagram of (10.68)*

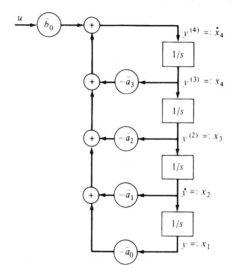

$$\begin{bmatrix} \dot{x}_1(t) \\ \dot{x}_2(t) \\ \dot{x}_3(t) \\ \dot{x}_4(t) \end{bmatrix} = \begin{bmatrix} 0 & 1 & 0 & 0 \\ 0 & 0 & 1 & 0 \\ 0 & 0 & 0 & 1 \\ -\bar{a}_0 & -\bar{a}_1 & -\bar{a}_2 & -\bar{a}_3 \end{bmatrix} \mathbf{x}(t) + \begin{bmatrix} 0 \\ 0 \\ 0 \\ \bar{b}_0 \end{bmatrix} u(t) \qquad (10.69)$$

$$y(t) = [1 \ 0 \ 0 \ 0]\mathbf{x}(t)$$

This is a realization of (10.68). The last row of the system matrix in (10.69) is made up of the coefficients of $\bar{D}(s)$ (after normalizing the leading coefficient) with signs reversed. The superdiagonal of **A** are all 1, and the rest are zeros. The column vector **b** is zero, except the last entry which equals the numerator without changing the sign. The row vector **c** is zero, except the first entry which is 1. Note that (10.69) can be obtained directly from (10.68); the basic block diagram in Figure 10.16 does not need to be plotted first.

EXERCISE 10.9.1

Find realizations of the following.

(a) $\dfrac{5}{2s^2 + 4s + 3}$

(b) $\dfrac{2}{s^3 + s - 1}$

(c) $\dfrac{3}{2s + 1}$

[ANSWERS: (a) $\dot{\mathbf{x}} = \begin{bmatrix} 0 & 1 \\ -1.5 & -2 \end{bmatrix} \mathbf{x} + \begin{bmatrix} 0 \\ 2.5 \end{bmatrix} u$, $y = [1 \ 0]\mathbf{x}$;

(b) $\dot{\mathbf{x}} = \begin{bmatrix} 0 & 1 & 0 \\ 0 & 0 & 1 \\ 1 & -1 & 0 \end{bmatrix} \mathbf{x} + \begin{bmatrix} 0 \\ 0 \\ 2 \end{bmatrix} u$, $y = [1 \ 0 \ 0]\mathbf{x}$;

(c) $\dot{x} = -0.5x + 1.5u$, $y = x$.]   ∎

Let us consider two more realizations of (10.68). If the outputs of the integrators in Figure 10.16 are chosen as x_1, x_2, x_3, and x_4, instead of as x_4, x_3, x_2, and x_1, then the state-variable equation description of Figure 10.16 becomes

$$\dot{\mathbf{x}} = \begin{bmatrix} -\bar{a}_3 & -\bar{a}_2 & -\bar{a}_1 & -\bar{a}_0 \\ 1 & 0 & 0 & 0 \\ 0 & 1 & 0 & 0 \\ 0 & 0 & 1 & 0 \end{bmatrix} \mathbf{x} + \begin{bmatrix} \bar{b}_0 \\ 0 \\ 0 \\ 0 \end{bmatrix} u \tag{10.70}$$

$$y = [0 \ 0 \ 0 \ 1]\mathbf{x}$$

This is a different realization. Because $\mathbf{c}(s\mathbf{I} - \mathbf{A})^{-1}\mathbf{b}$ is a 1×1 scalar, its transpose equals itself. Therefore, we have

$$\mathbf{c}(s\mathbf{I} - \mathbf{A})^{-1}\mathbf{b} = [\mathbf{c}(s\mathbf{I} - \mathbf{A})^{-1}\mathbf{b}]' = \mathbf{b}'(s\mathbf{I} - \mathbf{A}')^{-1}\mathbf{c}'$$

where the prime denotes the transpose. This means that if we use the transpose of $\mathbf{A}$ and interchange the roles of $\mathbf{b}$ and $\mathbf{c}$ after transposition, we will obtain a different realization. For example, if we use (10.70), then

$$\dot{\mathbf{x}} = \begin{bmatrix} -\bar{a}_3 & 1 & 0 & 0 \\ -\bar{a}_2 & 0 & 1 & 0 \\ -\bar{a}_1 & 0 & 0 & 1 \\ -\bar{a}_0 & 0 & 0 & 0 \end{bmatrix} \mathbf{x} + \begin{bmatrix} 0 \\ 0 \\ 0 \\ 1 \end{bmatrix} u \tag{10.71}$$

$$y = [\bar{b}_0 \ 0 \ 0 \ 0]\mathbf{x}$$

which is yet another realization of (10.68). In general, it is possible to obtain infinitely many realizations for $b_0/D(s)$.

Although we use the same $\mathbf{x}$ in (10.69), (10.70), and (10.71), they represent different variables in each equation. In (10.69), we have $x_1 = y$, $x_2 = \dot{y} = y^{(1)}$, $x_3 = y^{(2)}$, and $x_4 = y^{(3)}$. In (10.70), we have $x_1 = y^{(3)}$, $x_2 = y^{(2)}$, $x_3 = y^{(1)}$, and $x_4 = y$. In (10.71), the relationships between x_i and $y^{(i)}$ are more complex.

10.9.2   Realizations of N(s)/D(s)

To illustrate the realization of general proper transfer functions, we will use a transfer function of degree 4. Consider

$$H(s) = \frac{b_4 s^4 + b_3 s^4 + b_2 s^2 + b_1 s + b_0}{a_4 s^4 + a_3 s^3 + a_2 s^2 + a_1 s + a_0} =: \frac{N(s)}{D(s)} = \frac{Y(s)}{U(s)} \qquad \text{(10.72)}$$

where a_i and b_i are real constants and $a_4 \neq 0$. If $b_4 \neq 0$, the transfer function is biproper; if $b_4 = 0$, it is strictly proper. We will discuss two realizations: One that is used more often in signal processing, and one that is more common in control systems.

Method I   After dividing $N(s)$ and $D(s)$ by s^{-4}, we rewrite (10.72) as

$$(b_4 + b_3 s^{-1} + b_2 s^{-2} + b_1 s^{-3} + b_0 s^{-4}) U(s)$$
$$= (a_4 + a_3 s^{-1} + a_2 s^{-2} + a_1 s^{-3} + a_0 s^{-4}) Y(s)$$

which implies

$$Y(s) = \frac{1}{a_4} [(b_4 + b_3 s^{-1} + b_2 s^{-2} + b_1 s^{-3} + b_0 s^{-4}) U(s)$$
$$+ (-a_3 s^{-1} - a_2 s^{-2} - a_1 s^{-3} - a_0 s^{-4}) Y(s)] \qquad \text{(10.73)}$$

Verifying that the basic block diagram in Figure 10.17(a) generate $Y(s)$ in (10.73) is a straightforward task. That basic block diagram, however, is not desirable, because it uses an unnecessarily large number of integrators. We can modify it to cut that number in half. The basic block diagram in Figure 10.17(a) consists of two LTIL systems connected in tandem. Because the order of connection of LTIL systems can be reversed without affecting the input and output behavior (see Problem 10.30), the diagram can be modified as shown in Figure 10.17(b). The two chains of integrators are identical and can be combined to yield the basic block diagram in Figure 10.17(c). Now, if we choose the state variables as shown, we can readily obtain

$$\begin{bmatrix} \dot{x}_1 \\ \dot{x}_2 \\ \dot{x}_3 \\ \dot{x}_4 \end{bmatrix} = \begin{bmatrix} \dfrac{-a_3}{a_4} & \dfrac{-a_2}{a_4} & \dfrac{-a_1}{a_4} & \dfrac{-a_0}{a_4} \\ 1 & 0 & 0 & 0 \\ 0 & 1 & 0 & 0 \\ 0 & 0 & 1 & 0 \end{bmatrix} \mathbf{x} + \begin{bmatrix} \dfrac{1}{a_4} \\ 0 \\ 0 \\ 0 \end{bmatrix} u(t) \qquad \text{(10.74a)}$$

and

$$y = [b_3 \; b_2 \; b_1 \; b_0] \mathbf{x} + b_4 \dot{x}_1$$

The output equation contains the derivative of x_1. This can be eliminated by using the first equation of (10.74a) as

Figure 10.17 *Realization of (10.73)*

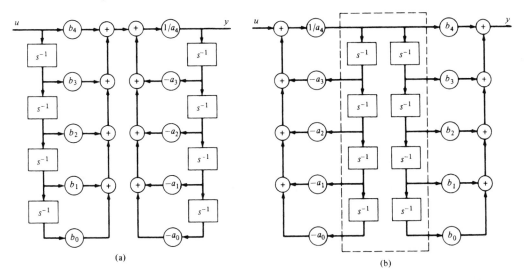

(a)   (b)

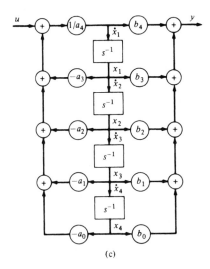

(c)

$$y = [b_3 \ b_2 \ b_1 \ b_0]\mathbf{x} + b_4[\frac{-a_3}{a_4} \ \frac{-a_2}{a_4} \ \frac{-a_1}{a_4} \ \frac{-a_0}{a_4}]\mathbf{x} + \frac{b_4}{a_4}u(t)$$

$$= [b_3 - b_4\frac{a_3}{a_4} \ b_2 - b_4\frac{a_2}{a_4} \ b_1 - b_4\frac{a_1}{a_4} \ b_0 - b_4\frac{a_0}{a_4}]\mathbf{x}(t) \quad \textbf{(10.74b)}$$

$$+ \frac{b_4}{a_4}u(t)$$

The set of two equations in (10.74) is a realization of (10.72).

EXERCISE 10.9.2

Find realizations for the following transfer functions.

$$\frac{-2s + 4}{2s^2 + 3s + 1} \qquad \frac{3s^2 - 2s + 4}{2s^2 + 3s + 1}$$

[**ANSWERS:** $\dot{x} = \begin{bmatrix} -1.5 & -0.5 \\ 1 & 0 \end{bmatrix} x + \begin{bmatrix} 0.5 \\ 0 \end{bmatrix} u$, $y = [-2 \ \ 4]x$;

$\dot{x} = \begin{bmatrix} -1.5 & -0.5 \\ 1 & 0 \end{bmatrix} x + \begin{bmatrix} 0.5 \\ 0 \end{bmatrix} u$, $y = [-6.5 \ \ 2.5]x + 1.5u$.] ∎

Method 2   In the second realization method, we must carry out two preliminary steps: (i) change $H(s)$ to strictly proper and (ii) normalize the leading coefficient of its denominator to 1. These are illustrated in an example.

EXAMPLE 10.9.1

Consider

$$H(s) = \frac{s^4 + 2s^3 - s^2 + 4s + 12}{2s^4 + 10s^3 + 20s^2 + 20s + 8} \tag{10.75}$$

Clearly, we have $H(\infty) = 1/2 = 0.5$. Thus, the direct transmission part d of any realization of $H(s)$ is $d = 0.5$. We compute

$$H_s(s) := H(s) - H(\infty)$$
$$= \frac{(s^4 + 2s^3 - s^2 + 4s + 12) - 0.5(2s^4 + 10s^3 + 20s^2 + 20s + 8)}{2s^4 + 10s^3 + 20s^2 + 20s + 8}$$
$$= \frac{-3s^3 - 11s^2 - 6s + 8}{2s^4 + 10s^3 + 20s^2 + 20s + 8}$$

It is strictly proper. We then normalize the equation, by dividing every entry by 2, as

$$H_s(s) = \frac{-1.5s^3 - 5.5s^2 - 3s + 4}{s^4 + 5s^3 + 10s^2 + 10s + 4} =: \frac{\overline{N}(s)}{\overline{D}(s)} \tag{10.76}$$

This completes the preliminary steps of realization.

Using the procedure in Example 10.9.1, we can write $H(s)$ in (10.72) as

$$H(s) = \frac{Y(s)}{U(s)} = H(\infty) + \frac{\bar{b}_3 s^3 + \bar{b}_2 s^2 + \bar{b}_1 s + \bar{b}_0}{s^4 + \bar{a}_3 s^3 + \bar{a}_2 s^2 + \bar{a}_1 s + \bar{a}_0} \qquad \textbf{(10.77)}$$

$$=: d + \frac{\bar{N}(s)}{\bar{D}(s)} =: d + H_s(s)$$

or

$$Y(s) = H(s)U(s) = dU(s) + H_s(s)U(s) =: Y_d(s) + Y_s(s)$$

where $Y_d(s)$ denotes the output due to the direct transmission part of $H(s)$, and where $Y_s(s)$ denotes the output due to the strictly proper part of $H(s)$. Because the direct transmission part can be easily obtained from $H(\infty)$, we will focus on the strictly proper part, that is,

$$H_s(s) = \frac{Y_s(s)}{U(s)} = \frac{\bar{N}(s)}{\bar{D}(s)} \qquad \textbf{(10.78)}$$

First, we introduce a new variable $w(t)$, whose Laplace transform $W(s)$ is defined by

$$W(s) = \frac{1}{\bar{D}(s)} U(s) \qquad \textbf{(10.79)}$$

The substitution of (10.79) into (10.78) yields

$$Y_s(s) = \bar{N}(s)W(s) \qquad \textbf{(10.80)}$$

Equation (10.79) is identical to (10.68a) if y is replaced by w and $\bar{b}_0$ is set to 1. Therefore, Figure 10.16 is directly applicable to (10.79) and is repeated in Figure 10.18 within the dotted lines. In the time domain, (10.80) becomes

$$y_s(t) = \bar{b}_3 w^{(3)}(t) + \bar{b}_2 w^{(2)}(t) + \bar{b}_1 w^{(1)}(t) + \bar{b}_0 w(t) \qquad \textbf{(10.81)}$$

Because $w^{(i)}$ is available in Figure 10.18, (10.81) can be easily generated. Finally, adding the direct transmission part d on the top completes the basic block diagram for the $H(s)$ in (10.77).

Depending on how the state variables are assigned, we will obtain different state-variable equations. If the state variables are chosen as shown, then we will obtain

Figure 10.18 *A basic block diagram of (10.77)*

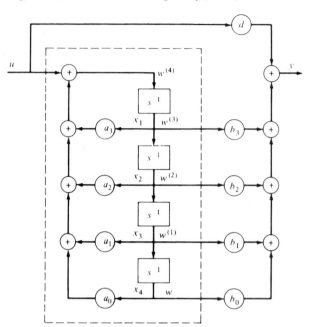

$$
\begin{bmatrix} \dot{x}_1(t) \\ \dot{x}_2(t) \\ \dot{x}_3(t) \\ \dot{x}_4(t) \end{bmatrix} = \begin{bmatrix} -\bar{a}_3 & -\bar{a}_2 & -\bar{a}_1 & -\bar{a}_0 \\ 1 & 0 & 0 & 0 \\ 0 & 1 & 0 & 0 \\ 0 & 0 & 1 & 0 \end{bmatrix} \mathbf{x} + \begin{bmatrix} 1 \\ 0 \\ 0 \\ 0 \end{bmatrix} u(t) \qquad \textbf{(10.82)}
$$

$$
y(t) = [\bar{b}_3 \ \bar{b}_2 \ \bar{b}_1 \ \bar{b}_0]\mathbf{x} + du(t)
$$

This is a realization of the $H(s)$ in (10.77). Note that (10.82) can be obtained directly from the coefficients of $H(s)$ in (10.77); a basic block diagram does not need to be drawn. The first row of the system matrix $\mathbf{A}$ can be read from the coefficients of $\bar{D}(s)$ with signs reversed. The subdiagonal of $\mathbf{A}$ all equal 1. All entries of the column vector $\mathbf{b}$ are zero, except the first entry, which is 1. The row vector $\mathbf{c}$ consists of all coefficients of $\bar{N}(s)$ *without* changing signs. An example illustrates.

EXAMPLE 10.9.2

Consider the transfer function in (10.75). It can be written as

$$
H(s) = 0.5 + \frac{-1.5s^3 - 5.5s^2 - 3s + 4}{s^4 + 5s^3 + 10s^2 + 10s + 4} =: d + \frac{\bar{N}(s)}{\bar{D}(s)} \qquad \textbf{(10.83)}
$$

Thus, it can be realized as

$$\dot{\mathbf{x}}(t) = \begin{bmatrix} -5 & -10 & -10 & -4 \\ 1 & 0 & 0 & 0 \\ 0 & 1 & 0 & 0 \\ 0 & 0 & 1 & 0 \end{bmatrix} \mathbf{x}(t) + \begin{bmatrix} 1 \\ 0 \\ 0 \\ 0 \end{bmatrix} u(t) \qquad (10.84)$$

$$y(t) = [-1.5 \ -5.5 \ -3 \ 4]\mathbf{x}(t) + 0.5u(t)$$

EXERCISE 10.9.3

Find realizations for the following transfer functions.

(a) $\dfrac{3s^2 - s + 2}{s^3 + 2s^2 + 1}$

(b) $\dfrac{4s^3 + 2s + 1}{2s^3 + 3s^2 + 1}$

(c) $\dfrac{5s + 1}{2s^2 + 3s + 2}$

[**ANSWERS:** (a) $\dot{\mathbf{x}} = \begin{bmatrix} -2 & 0 & -1 \\ 1 & 0 & 0 \\ 0 & 1 & 0 \end{bmatrix} \mathbf{x} + \begin{bmatrix} 1 \\ 0 \\ 0 \end{bmatrix} u, \quad y = [3 \ -1 \ 2]\mathbf{x};$

(b) $\dot{\mathbf{x}} = \begin{bmatrix} -1.5 & 0 & -0.5 \\ 1 & 0 & 0 \\ 0 & 1 & 0 \end{bmatrix} \mathbf{x} + \begin{bmatrix} 1 \\ 0 \\ 0 \end{bmatrix} u, \ y = [-3 \ 1 \ -0.5]\mathbf{x} + 2u;$

(c) $\dot{\mathbf{x}} = \begin{bmatrix} -1.5 & -1 \\ 1 & 0 \end{bmatrix} \mathbf{x} + \begin{bmatrix} 1 \\ 0 \end{bmatrix} u, \quad y = [2.5 \ 0.5]\mathbf{x}.$] ∎

The realization in Figure 10.17(c) is more often used in texts on digital signal processing; the realization in Figure 10.18 is more often used in texts on systems and control. The two resulting state-variable equations, however, are essentially the same. See Problem 10.18.

The realizations in Figures 10.17 and 10.18 are only two of infinitely many possible realizations of (10.72). Assigning different state variables will yield different realizations. Using the fact $\mathbf{c}(s\mathbf{I} - \mathbf{A})^{-1}\mathbf{b} = \mathbf{b}'(s\mathbf{I} - \mathbf{A}')^{-1}\mathbf{c}'$, we can again obtain different realizations, as in (10.70) and (10.71). A discussion of these different realizations would not introduce any new concept and so will be skipped here.

10.9.3 Minimal realization and complete characterization

Consider a proper rational function $H(s)$. While it is possible to find infinitely many realizations, these realizations may not have the same dimension. If we use these realizations to implement $H(s)$, then the numbers of integrators used will be different. It is, therefore, of practical interest to find a realization that will use the smallest possible number of integrators. Such a realization is called a *minimal dimensional realization*. This section discusses such realizations.

Given a proper transfer function $H(s) = N(s)/D(s)$, if we use the procedures discussed in the preceding section, then the dimension of the realization equals the degree of $D(s)$, denoted as deg $D(s)$. For example, the realization shown in (10.84), of the transfer function in (10.75) or (10.83), has dimension 4. Is it possible to obtain a realization with a smaller dimension? It turns out that the numerator and denominator of $H(s)$ in (10.75) have common factor $(s + 2)$. If we cancel out this common factor, then $H(s)$ in (10.75) or (10.83) becomes

$$H(s) = \frac{s^4 + 2s^3 - s^2 + 4s + 12}{2s^4 + 10s^3 + 20s^2 + 20s + 8} = \frac{s^3 - s + 6}{2s^3 + 6s^2 + 8s + 4}$$

$$= 0.5 + \frac{-1.5s^2 - 2.5s + 2}{s^3 + 3s^2 + 4s + 2} \tag{10.85}$$

From (10.85), we can obtain the realization

$$\dot{\mathbf{x}}(t) = \begin{bmatrix} -3 & -4 & -2 \\ 1 & 0 & 0 \\ 0 & 1 & 0 \end{bmatrix} \mathbf{x}(t) + \begin{bmatrix} 1 \\ 0 \\ 0 \end{bmatrix} u(t) \tag{10.86}$$

$$y(t) = [-1.5 \ -2.5 \ 2]\mathbf{x}(t) + 0.5u(t)$$

for the $H(s)$ in (10.83). The dimension of this realization is one smaller than that in (10.84). If we implement (10.86), the number of integrators used will be one smaller than the number used for (10.84). Thus, the realization in (10.86) is preferable to the one in (10.84).

Consider a proper rational function $H(s) = N(s)/D(s)$. If $N(s)$ and $D(s)$ have no common factor, the degree of $D(s)$ is defined as the degree of $H(s)$. Thus, the degree of $H(s)$, denoted as deg $H(s)$, equals the number of poles of $H(s)$. It can be shown that the smallest possible dimension of all realizations of $H(s)$ equals deg $H(s)$. These minimal dimensional realizations, or simply minimal realizations, are also called *irreducible realizations*. Realizations with dimensions larger than deg $H(s)$ are called *nonminimal* or *reducible realizations*.

EXAMPLE 10.9.3

Consider

$$H(s) = \frac{2s + 1}{(s + 2)(s + 3)} = \frac{2s + 1}{s^2 + 5s + 6} \tag{10.87}$$

The degree of $H(s)$ clearly equals 2. From the coefficients of $H(s)$, we can readily obtain the following realization:

$$\dot{\mathbf{x}} = \begin{bmatrix} -5 & -6 \\ 1 & 0 \end{bmatrix} \mathbf{x} + \begin{bmatrix} 1 \\ 0 \end{bmatrix} u$$

$$y = [2 \ 1]\mathbf{x}$$

It is of dimension 2 and is a minimal realization of (10.87).

If we write (10.87) as

$$H(s) = \frac{s(2s+1)}{s(s^2+5s+6)} = \frac{2s^2+s}{s^3+5s^2+6s+0}$$

then

$$\dot{\mathbf{x}} = \begin{bmatrix} -5 & -6 & 0 \\ 1 & 0 & 0 \\ 0 & 1 & 0 \end{bmatrix} \mathbf{x} + \begin{bmatrix} 1 \\ 0 \\ 0 \end{bmatrix} u(t)$$

$$y = [2 \ 1 \ 0]\mathbf{x}$$

is also a realization of $H(s)$. However, it is a nonminimal or reducible realization.

EXERCISE 10.9.4

Find two BIBO stable reducible realizations of dimension 4 and 5 for the $H(s)$ in (10.87). ■

The preceding example shows that the procedure for obtaining a minimal realization is very simple. All we need to do is to check first whether the denominator and numerator of $H(s)$ have any common factors. If they do, we cancel the common factors and then use the methods discussed in the preceding sections to find realizations for $H(s)$. The realizations will be minimal or irreducible. If we skip the step of checking for common factors, then the realization may or may not be minimal.

The concept of minimal realization can also be formulated from state-variable equations. Consider the state-variable equation[5]

$$\dot{\mathbf{x}} = \mathbf{A}\mathbf{x} + \mathbf{b}u \qquad (10.88)$$

$$y = \mathbf{c}\mathbf{x}$$

Let the dimension of $\mathbf{x}$ be n, denoted as dim $\mathbf{x} = n$. We compute the transfer function of (10.88) as

$$H(s) = \mathbf{c}(s\mathbf{I} - \mathbf{A})^{-1}\mathbf{b}$$

---

[5]We assume $d = 0$ to simplify the discussion. All discussion still holds if $d \neq 0$.

If deg $H(s) < \dim \mathbf{x} = n$, then the state-variable equation in (10.88) is said to be *reducible*. In this case, it is possible to find a smaller dimensional state-variable equation that has the same transfer function as the original equation in (10.88). Therefore, if deg $H(s) < \dim \mathbf{x}$, the equation is said to be reducible. On the other hand, if

$$\deg H(s) = \dim \mathbf{x}$$

then (10.88) is said to be *irreducible*. In this case, it is not possible to find a smaller dimensional state-variable equation that has the same transfer function as (10.88). ∎

EXAMPLE 10.9.4

Consider

$$\dot{\mathbf{x}} = \begin{bmatrix} 0 & 1 \\ 2 & 1 \end{bmatrix} \mathbf{x} + \begin{bmatrix} 1 \\ -1 \end{bmatrix} u \tag{10.89}$$

$$y = [1 \ -1]\mathbf{x}$$

We compute

$$s\mathbf{I} - \mathbf{A} = s\begin{bmatrix} 1 & 0 \\ 0 & 1 \end{bmatrix} - \begin{bmatrix} 0 & 1 \\ 2 & 1 \end{bmatrix} = \begin{bmatrix} s & -1 \\ -2 & s-1 \end{bmatrix}$$

and

$$(s\mathbf{I} - \mathbf{A})^{-1} = \frac{1}{s(s-1)-2}\begin{bmatrix} s-1 & 1 \\ 2 & s \end{bmatrix}$$

Thus, its transfer function is

$$H(s) = [1 \ -1]\frac{1}{(s+1)(s-2)}\begin{bmatrix} s-1 & 1 \\ 2 & s \end{bmatrix}\begin{bmatrix} 1 \\ -1 \end{bmatrix} = \frac{1}{(s+1)(s-2)}[1 \ -1]\begin{bmatrix} s-2 \\ 2-s \end{bmatrix}$$

$$= \frac{2(s-2)}{(s+1)(s-2)} = \frac{2}{s+1}$$

Clearly, we have deg $H(s) = 1 < \dim \mathbf{x} = 2$, so (10.89) is reducible. A realization of $H(s) = 2/(s+1)$ is

$$\dot{\mathbf{x}} = -x + u \qquad y = 2x \tag{10.90}$$

This has the same transfer function as (10.89). In other words, (10.89) can be reduced to (10.90).

EXAMPLE 10.9.5

Consider

$$\dot{\mathbf{x}} = \begin{bmatrix} 0 & 1 \\ 2 & 1 \end{bmatrix} \mathbf{x} + \begin{bmatrix} 0 \\ 1 \end{bmatrix} u \qquad (10.91)$$

$$y = [1 \ -1]\mathbf{x}$$

Its transfer function is

$$H(s) = [1 \ -1] \frac{1}{(s+1)(s-2)} \begin{bmatrix} s-1 & 1 \\ 2 & s \end{bmatrix} \begin{bmatrix} 0 \\ 1 \end{bmatrix} = \frac{1}{(s+1)(s-2)} [1 \ -1] \begin{bmatrix} 1 \\ s \end{bmatrix}$$

$$(10.92)$$

$$= \frac{1-s}{(s+1)(s-2)}$$

Clearly, we have deg $H(s) = \dim \mathbf{x} = 2$, so (10.91) is irreducible.

EXERCISE 10.9.5

Which of the following is irreducible?

(a) $\dot{\mathbf{x}} = \begin{bmatrix} 1 & 0 \\ 0 & -2 \end{bmatrix} \mathbf{x} + \begin{bmatrix} 0 \\ 1 \end{bmatrix} u$

$y = [0 \ 1]\mathbf{x}$

(b) $\dot{\mathbf{x}} = \begin{bmatrix} 1 & 0 \\ 0 & -2 \end{bmatrix} \mathbf{x} + \begin{bmatrix} 0 \\ 1 \end{bmatrix} u$

$y = [1 \ 0]\mathbf{x}$

[**ANSWER:** Both are reducible.]

The concept of irreducibility is closely related to the concept of complete characterization, which was discussed in Section 4.10. It can be shown that, if a state-variable equation is irreducible, then the system described by the state-variable equation is completely characterized by its transfer function. Thus, if a state-variable equation is irreducible, there is no loss of essential information when its transfer function is used in analysis and design.

If a state variable equation is irreducible, then it is also said to be controllable and observable. If a state-variable equation is reducible, then it is either uncontrollable or unobservable or both. Conditions to check controllability or observability of state-variable equations are available. The interested reader is referred to References [4] and [8].

10.10   TANDEM AND PARALLEL REALIZATIONS

One of the reasons for studying the realization problem is to simulate transfer functions on digital computers or to implement them using op-amp circuits. In the simulations or computations, we may encounter some error problems. For example, to represent $1/3 = 0.333\ldots$ on a digital computer, the number must be truncated. The multiplication of two b bit numbers will yield one $2b$ bit number. The $2b$ bit number must be approximated if it is to be represented by one b bit number. Thus, errors do occur in digital computation. Errors also occur in op-amp circuit realizations. This situation is actually much worse than the digital computation situation because the accuracy of op-amp circuits is very limited. Furthermore, their parameters may drift due to temperature variations and the drifting of power supply.

In computer simulation, then, it is natural to search for realizations on which the effect of parameter variations will be small. There are many possible ways to define the effect of parameter variations. For one such criterion, let $H(s)$ be a transfer function with poles λ_i, $i = 1, 2, \ldots, n$. If $H(s)$ can be realized exactly, then the poles of the transfer function of the realization should also equal λ_i, $i = 1, 2, \ldots, n$. However, due to inexact implementation or parameter variations, the poles of the realization may become $\bar{\lambda}_i$, $i = 1, 2, \ldots, n$. The error or deviation of each pole is $|\lambda_i - \bar{\lambda}_i|$. Thus,

$$J := \sum_{i=1}^{n} |\lambda_i - \bar{\lambda}_i|$$

is the total deviation of the poles due to parameter variations. A realization that has a smaller J is said to be less sensitive to parameter variations. Two realizations that are less sensitive than the ones discussed in Section 10.9.2 are presented here.

Tandem realization   In this realization, we factor a transfer function $H(s)$ as

$$H(s) = H_1(s)H_2(s) \cdots H_m(s)$$

where $H_i(s)$, $i = 1, 2, \ldots, m$, are proper transfer functions of degree 1 or 2 with real coefficients. Because we require coefficients of $H_i(s)$ to be real, complex-conjugate poles or zeros must be grouped together. We implement each $H_i(s)$ using the procedure discussed in the preceding sections, and then connect them in tandem as shown in Figure 10.19. The resulting equation is called a *tandem* or *cascade realization*. Example 10.10.1 illustrates.

Figure 10.19 *Tandem realization*

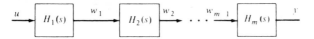

EXAMPLE 10.10.1

Consider

$$H(s) = \frac{(s-3)(s^2 - 2s + 3)}{(2s+1)(s+2)(s^2 + 2s + 2)} \qquad (10.93)$$

We group it as

$$H(s) = \frac{s-3}{2s+1} \cdot \frac{1}{s+2} \cdot \frac{2s^2 - 2s + 3}{s^2 + 2s + 2}$$

which can be written as

$$H(s) = \left[0.5 + \frac{-1.75}{s+0.5}\right] \cdot \frac{1}{s+2} \cdot \left[2 + \frac{-6s-1}{s^2 + 2s + 2}\right]$$

If we implement each term using Figure 10.18 and then connect them as shown in Figure 10.20, we will obtain a tandem realization of $H(s)$. If we assign the state variables as shown, then we have

$$\dot{x}_1 = -0.5x_1 + u$$
$$w_1 = -1.75x_1 + 0.5u$$
$$\dot{x}_2 = -2x_2 + w_1 = -2x_2 - 1.75x_1 + 0.5u$$

Figure 10.20 *Tandem realization of (10.93)*

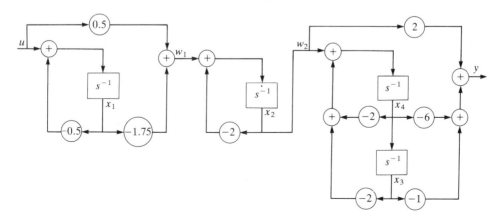

Proceeding forward, we can finally obtain

$$
\begin{bmatrix} \dot{x}_1 \\ \dot{x}_2 \\ \dot{x}_3 \\ \dot{x}_4 \end{bmatrix} = \begin{bmatrix} -0.5 & 0 & 0 & 0 \\ -1.75 & -2 & 0 & 0 \\ 0 & 0 & 0 & 1 \\ 0 & 1 & -2 & -2 \end{bmatrix} \begin{bmatrix} x_1 \\ x_2 \\ x_3 \\ x_4 \end{bmatrix} + \begin{bmatrix} 1 \\ 0.5 \\ 0 \\ 0 \end{bmatrix} u
$$

$$
y = [0 \ \ 2 \ \ -1 \ \ -6]\mathbf{x}
$$

Note that, if we group $(s - 3)$ with $(s + 2)$ rather than $(2s + 1)$, we will obtain a different tandem realization. Thus, tandem realizations are not unique.

EXERCISE 10.10.1

Find a tandem realization of the transfer function

$$
H(s) = \frac{s^3 - s + 6}{2s^3 + 6s^2 + 8s + 4} = \frac{(s + 2)(s^2 - 2s + 3)}{(s + 1)(2s^2 + 4s + 4)} \tag{10.94}
$$

∎

Parallel realization In this realization, we decompose a transfer function as

$$
H(s) = k_0 + \sum H_i(s)
$$

where $H_i(s)$ are strictly proper rational functions of degree 1 or 2 with real coefficients. We implement each $H_i(s)$ and then connect them in parallel as shown in Figure 10.21. The resulting equation is called a *parallel realization*. Example 10.10.2 illustrates.

EXAMPLE 10.10.2

Consider

$$
H(s) = \frac{(s - 3)(s^2 - 2s + 3)}{(2s + 1)(s + 2)(s^2 + 2s + 4)} \tag{10.95}
$$

We expand it, using partial fraction expansion, as

Figure 10.21 *Parallel realization*

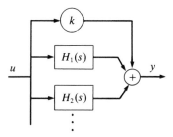

$$H(s) = k_0 + \frac{k_1}{2s + 1} + \frac{k_2}{s + 2} + \frac{k_3 s + k_4}{s^2 + 2s + 4} \qquad \textbf{(10.96)}$$

where

$$k_0 = H(\infty) = 0$$

$$k_1 = \left.\frac{(s - 3)(s^2 - 2s + 3)}{(s + 2)(s^2 + 2s + 4)}\right|_{s=-1/2} = \frac{(-3.5)(4.25)}{(1.5)(3.25)} = -3.05$$

$$k_2 = \left.\frac{(s - 3)(s^2 - 2s + 3)}{(2s + 1)(s^2 + 2s + 4)}\right|_{s=-2} = \frac{(-5)(11)}{(-3)(4)} = 4.58$$

To find k_3 and k_4, we select $s = 0$ and $s = 1$. Then, from (10.95) and (10.96), we have

$$\frac{(-3)(3)}{(2)(4)} = \frac{k_1}{1} + \frac{k_2}{2} + \frac{k_4}{4}$$

and

$$\frac{(-2)(2)}{(3)(3)(7)} = \frac{k_1}{3} + \frac{k_2}{3} + \frac{k_3 + k_4}{7}$$

which implies $k_4 = -1.46$ and $k_3 = -6.11$. If we use Figure 10.18 to implement each term in (10.96), then a parallel realization of (10.95) is as shown in Figure 10.22. If the state variables are chosen as shown, then we have

$$\dot{x}_1 = -0.5x_1 + u$$
$$\dot{x}_2 = -2x_2 + u$$
$$\dot{x}_3 = x_4$$
$$\dot{x}_4 = -4x_3 - 2x_4 + u$$
$$y = \frac{k_1}{2}x_1 + k_2x_2 + k_3x_3 + k_4x_4$$

They can be arranged in matrix form as

$$\dot{\mathbf{x}} = \begin{bmatrix} -0.5 & 0 & 0 & 0 \\ 0 & -2 & 0 & 0 \\ 0 & 0 & 0 & 1 \\ 0 & 0 & -4 & -2 \end{bmatrix} \mathbf{x} + \begin{bmatrix} 1 \\ 1 \\ 0 \\ 1 \end{bmatrix} u$$

$$y = [-1.53 \quad 4.58 \quad -6.11 \quad -1.46]\mathbf{x}$$

Figure 10.22 *Parallel realization of (10.95)*

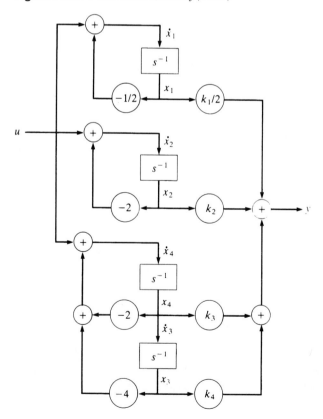

Unlike the tandem realization, the parallel realization is unique. Note that, if $H(s)$ has repeated poles, the parallel realization in Figure 10.21 must be modified. This is illustrated by an example.

EXAMPLE 10.10.3

Consider

$$H(s) = \frac{3}{s+2} + \frac{c_1}{s+1} + \frac{c_2}{(s+1)^2} + \frac{c_3}{(s+1)^3} + \frac{3s-5}{s^2+2s+4} \quad \text{(10.97)}$$

It has a repeated pole with multiplicity 3 at $s = -1$. Its parallel realization assumes the form in Figure 10.23(a). We see that terms associated with repeated poles are connected in tandem, while those associated with different poles are connected in parallel. Its basic block diagram is shown in Figure 10.23(b). The development of its state-variable equation is left as an exercise.

EXERCISE 10.10.2

Verify that the state-variable equation

$$\begin{bmatrix} \dot{x}_1 \\ \dot{x}_2 \\ \dot{x}_3 \\ \dot{x}_4 \\ \dot{x}_5 \\ \dot{x}_6 \end{bmatrix} = \begin{bmatrix} -2 & 0 & 0 & 0 & 0 & 0 \\ 0 & -1 & 0 & 0 & 0 & 0 \\ 0 & 1 & -1 & 0 & 0 & 0 \\ 0 & 0 & 1 & -1 & 0 & 0 \\ 0 & 0 & 0 & 0 & 0 & 1 \\ 0 & 0 & 0 & 0 & -4 & -2 \end{bmatrix} \mathbf{x} + \begin{bmatrix} 1 \\ 1 \\ 0 \\ 0 \\ 0 \\ 1 \end{bmatrix} u$$

$$y = [3 \ c_1 \ c_2 \ c_3 \ -5 \ 3]\mathbf{x}$$

describes the block diagram in Figure 10.23 or is a realization of (10.97). Note that the system matrix is a block diagonal matrix. ∎

EXERCISE 10.10.3

Find a parallel realization of

$$H(s) = \frac{s^3 - s^2 + 6}{(s+1)(2s^2 + 4s + 4)}$$

[**ANSWER:** $\dot{\mathbf{x}} = \begin{bmatrix} -1 & 0 & 0 \\ 0 & 0 & 1 \\ 0 & -2 & -2 \end{bmatrix} \mathbf{x} + \begin{bmatrix} 1 \\ 0 \\ 1 \end{bmatrix} u, \quad y = [-2 \ -2 \ -4]\mathbf{x} + 0.5u.$] ∎

Figure 10.23 *Parallel realization of (10.97)*

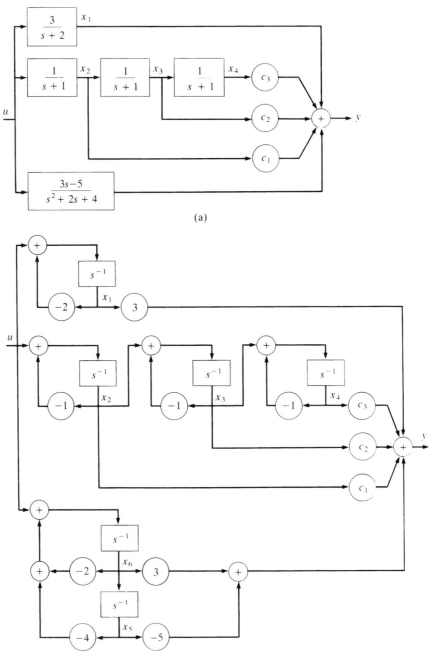

(a)

(b)

Parallel and tandem realizations are known to be less sensitive to parameter variations than are the realizations discussed in Section 10.9.2. It is not clear, however, which realization, parallel or tandem, is less sensitive. There is a so-called Jordan-form realization, which is the least sensitive to parameter variations. However, this realization may contain complex numbers. If no complex numbers are permitted, then the Jordan-form realization must be modified to become the parallel realization discussed in this section.

Here is a summary of the discussion in this section. When implementing an op-amp circuit or when building a special hardware with a small number of bits, we use the parallel or tandem realization to minimize the effect of parameter variations. However, if we are using a general purpose computer, because its accuracy is very good and the effect of parameter variations is usually negligible, it makes no difference which realization we use in simulation or computation.

10.11   DISCRETE-TIME REALIZATIONS

In this section, we will discuss the realization problem for discrete-time systems. All the results and discussions in Section 10.9 are directly applicable to the discrete-time case; therefore, the discussion here will be brief.

Consider the discrete-time transfer function

$$H(z) = \frac{b_4 z^4 + b_3 z^3 + b_2 z^2 + b_1 z + b_0}{a_4 z^4 + a_3 z^3 + a_2 z^2 + a_1 z + a_0} =: \frac{N(z)}{D(z)} \tag{10.98}$$

If we find a discrete-time state-variable equation

$$\mathbf{x}[k + 1] = \mathbf{A}\mathbf{x}[k] + \mathbf{b}u[k] \tag{10.99a}$$

$$y[k] = \mathbf{c}\mathbf{x}[k] + du[k] \tag{10.99b}$$

so that

$$H(z) = \mathbf{c}(z\mathbf{I} - \mathbf{A})^{-1}\mathbf{b} + d \tag{10.100}$$

then (10.99) is called a realization of (10.98). Equation (10.98) is the same as (10.72) if s is replaced by z. Using the same procedure that was shown in Figure 10.17, the basic block diagram in Figure 10.24(a) can be obtained for (10.98). If state variables are chosen as shown, then its state-variable equation is

Figure 10.24 *Two basic block diagrams for (10.98)*

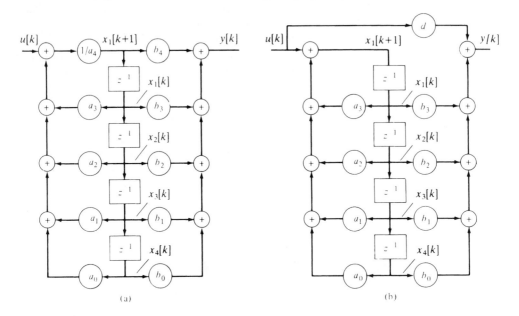

(a)   (b)

$$\mathbf{x}[k+1] = \begin{bmatrix} -a_3 & -a_2 & -a_1 & -a_0 \\ \dfrac{}{a_4} & \dfrac{}{a_4} & \dfrac{}{a_4} & \dfrac{}{a_4} \\ 1 & 0 & 0 & 0 \\ 0 & 1 & 0 & 0 \\ 0 & 0 & 1 & 0 \end{bmatrix} \mathbf{x}[k] + \begin{bmatrix} \dfrac{1}{a_4} \\ 0 \\ 0 \\ 0 \end{bmatrix} u[k] \qquad \textbf{(10.101a)}$$

$$y[k] = [b_3 \ b_2 \ b_1 \ b_0]\mathbf{x}[k] + b_4 x_1[k+1] \qquad \textbf{(10.101b)}$$

$$= [b_3 - b_4\frac{a_3}{a_4} \quad b_2 - b_4\frac{a_2}{a_4} \quad b_1 - b_4\frac{a_1}{a_4} \quad b_0 - b_4\frac{a_0}{a_4}]\mathbf{x}[k] + \frac{b_4}{a_4}u[k]$$

This is identical to (10.74). As in (10.77), we can write (10.98) as

$$H(z) = d + \frac{\bar{b}_3 z^3 + \bar{b}_2 z^2 + \bar{b}_1 z + \bar{b}_0}{z^4 + \bar{a}_3 z^3 + \bar{a}_2 z^2 + \bar{a}_1 z + \bar{a}_0} =: d + \frac{\bar{N}(z)}{\bar{D}(z)} \qquad \textbf{(10.102)}$$

Then, using the procedure used in (10.77) through (10.81), we can obtain the basic block diagram in Figure 10.24(b) for (10.102). The diagram is identical to the one in Figure 10.18 if s^{-1} is replaced by z^{-1}. If state variables are chosen as shown, then its state-variable equation is

$$\mathbf{x}[k+1] = \begin{bmatrix} -\bar{a}_3 & -\bar{a}_2 & -\bar{a}_1 & -\bar{a}_0 \\ 1 & 0 & 0 & 0 \\ 0 & 1 & 0 & 0 \\ 0 & 0 & 1 & 0 \end{bmatrix} \mathbf{x}[k] + \begin{bmatrix} 1 \\ 0 \\ 0 \\ 0 \end{bmatrix} u[k] \qquad \textbf{(10.103a)}$$

$$y[k] = [\bar{b}_3 \ \ \bar{b}_2 \ \ \bar{b}_1 \ \ \bar{b}_0]\mathbf{x}[k] + du[k] \qquad \textbf{(10.103b)}$$

which is identical to (10.82).

The parallel and tandem realizations discussed in Section 10.10 are also directly applicable to the discrete-time case. If $N(z)$ and $D(z)$ in (10.98) have a common factor, the realizations in (10.101) and (10.103) are not minimal realizations. If $N(z)$ and $D(z)$ have no common factors, then (10.101) and (10.103) are minimal realizations of (10.98).

Given the discrete-time state-variable equation

$$\mathbf{x}[k+1] = \mathbf{A}\mathbf{x}[k] + \mathbf{b}u[k]$$
$$y[k] = \mathbf{c}\mathbf{x}[k] \qquad \textbf{(10.104)}$$

we can compute its transfer function $H(z) = \mathbf{c}(z\mathbf{I} - \mathbf{A})^{-1}\mathbf{b}$. The degree of $H(z)$ is defined as the degree of its denominator after canceling all common factors between its denominator and numerator. If

$$\deg H(z) = \dim \mathbf{x}$$

the state-variable equation is said to be irreducible or controllable and observable. Under this assumption, there is no loss of essential information in using $H(z)$ to study (10.104), and (10.104) is said to be completely characterized by its transfer function. This is the same idea as in the continuous-time case, once again.

10.12 SUMMARY

1. The state-variable description of a system is more general than the transfer-function description because it describes not only the input and output but also the state variables inside the system, and because it describes the zero-state response as well as the zero-input response. The state-variable description can also be more easily extended to time-varying and nonlinear systems.

2. If an LTIL system is completely characterized by its transfer function, then the state-variable description and the transfer-function description are essentially the same.

3. The state-variable equation is very convenient to use in digital computer computation and op-amp circuit implementation. Although the transfer

function can be directly simulated on a computer, it is simpler to simulate it through its state-variable equation realization.

4. There are infinitely many ways to realize the proper transfer function $H(s) = N(s)/D(s)$. If $N(s)$ and $D(s)$ have common factors, then the resulting realization will not be irreducible. If we cancel the common factors, then the resulting realization will be irreducible. In this case, the number of integrators used in the realization will be minimal.

5. If we are using a general purpose computer, it probably makes no difference which realization we use in simulation or computation. However, if we are using a special hardware with a small number of bits, then we should use either a tandem or a parallel realization to minimize the effects of round-off errors.

PROBLEMS

10.1 Consider the pendulum system shown in Figure P10.1. Let the applied force u be the input and the angular displacement θ be the output. Assume that there is no air friction. Find a state-variable equation to describe the system. Is it a linear equation?

10.2 In Problem 10.1, if $\sin \theta$ and $\cos \theta$ are approximated as θ and 1 for θ small, what is the resulting equation? Is it a linear equation?

10.3 Find a state-variable equation to describe the network in Figure P10.3.

10.4 If the initial voltage of the capacitor in Figure P10.3 is 2 volts and the initial current of the inductor is 1 ampere, find the output $y(t)$ in Figure

Figure P10.1

Figure P10.3

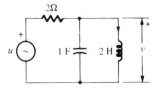

P10.3 due to the initial conditions and $u(t) = 1$ for $t \geq 0$. Use the state-variable equation obtained in Problem 10.3.

10.5 Repeat Problems 10.3 and 10.4 for the network in Figure P10.5 with the initial conditions shown and $u(t) = 1$ for $t \geq 0$.

10.6 Let $\mathbf{A} = \begin{bmatrix} 1 & 1 \\ 0 & 1 \end{bmatrix}$. Use (10.26) to show

$$e^{\mathbf{A}t} = \begin{bmatrix} e^t & te^t \\ 0 & e^t \end{bmatrix}$$

10.7 For the $\mathbf{A}$ and $e^{\mathbf{A}t}$ in Problem 10.6. Verify

$$e^{\mathbf{A}t} = \mathcal{L}^{-1}[(s\mathbf{I} - \mathbf{A})^{-1}]$$

10.8 Verify that, for any $\mathbf{A}$,

$$\mathcal{L}[e^{\mathbf{A}t}] \cdot (s\mathbf{I} - \mathbf{A}) = \mathbf{I} \quad \text{or} \quad \mathcal{L}^{-1}[e^{\mathbf{A}t}] = (s\mathbf{I} - \mathbf{A})^{-1}$$

10.9 For the $e^{\mathbf{A}t}$ in Problem 10.6, verify

$$(e^{\mathbf{A}t})^{-1} = e^{-\mathbf{A}t}$$

10.10 Show that the impulse response of (10.16) is, using (10.34),

$$h(t) = \mathbf{c}e^{\mathbf{A}t}\mathbf{b} + d\delta(t)$$

Show also

$$\mathcal{L}[h(t)] = \mathbf{c}(s\mathbf{I} - \mathbf{A})^{-1}\mathbf{b} + d$$

10.11 Find $y(t)$ in

$$\dot{\mathbf{x}} = \begin{bmatrix} 1 & 1 \\ 0 & 1 \end{bmatrix}\mathbf{x} + \begin{bmatrix} 0 \\ 1 \end{bmatrix}u$$

$$y = [1 \ -1]\mathbf{x}$$

Figure P10.5

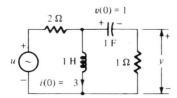

excited by $\mathbf{x}[0] = [1 \ 0]'$ and $u(t) = 1$ for $t \geq 0$. Solve it using the Laplace transform, and solve it directly in the time domain. Which method is simpler to use for this problem?

10.12 Draw basic block diagrams for the following equations.

(a) $\dot{\mathbf{x}} = \begin{bmatrix} 2 & 3 \\ -1 & -2 \end{bmatrix} \mathbf{x} + \begin{bmatrix} 2 \\ 1 \end{bmatrix} u$

$y = [1 \ 4]\mathbf{x} + 2u$

(b) $\dot{\mathbf{x}} = \begin{bmatrix} 0 & 0 & -3 \\ 1 & 0 & -2 \\ 0 & 1 & 1 \end{bmatrix} \mathbf{x} + \begin{bmatrix} 1 \\ 0 \\ -2 \end{bmatrix} u$

$y = [2 \ -2 \ 1]\mathbf{x}$

10.13 Assign state variables and then develop state-variable equations for the basic block diagrams shown in Figure P10.13.

10.14 Consider the equation in Problem 10.11. If $u(t)$ changes value only at $t = kT$, $k = 0, 1, 2, \ldots$, and $T = 0.1$ second, and if we are interested in only $x(t)$ and $y(t)$ at $t = kT$, what is its discretized state-variable equation?

Figure P10.13

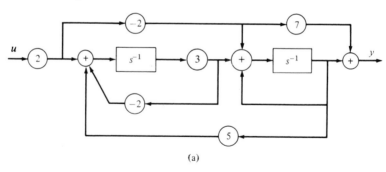

(a)

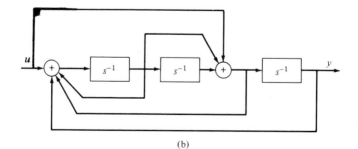

(b)

10.15 Find $y[k]$, for $k = 0, 1, 2, 3, 4$, in

$$\mathbf{x}[k + 1] = \begin{bmatrix} 1 & 1 \\ 0 & 1 \end{bmatrix} \mathbf{x}[k] + \begin{bmatrix} 0 \\ 1 \end{bmatrix} u[k]$$

$$y[k] = [1 \ -1]x[k]$$

excited by $\mathbf{x}[0] = [1 \ 0]'$ and $u[k] = 1$ for $k = 0, 1, 2, \ldots$. Compute $y[k]$ directly, and compute it using the z-transform. Which method is simpler to use for this problem?

10.16 Draw basic block diagrams for the following equations.

(a) $\mathbf{x}[k + 1] = \begin{bmatrix} 2 & 3 \\ -1 & -2 \end{bmatrix} \mathbf{x}[k] + \begin{bmatrix} 2 \\ 1 \end{bmatrix} u[k], \qquad y[k] = [1 \ 4]\mathbf{x}[k] + 2u[k]$

(b) $\mathbf{x}[k + 1] = \begin{bmatrix} 0 & 0 & -3 \\ 1 & 0 & -2 \\ 0 & 1 & 1 \end{bmatrix} \mathbf{x}[k] + \begin{bmatrix} 1 \\ 0 \\ -2 \end{bmatrix} u[k], \qquad y[k] = [2 \ -1 \ 1]\mathbf{x}[k]$

Compare your diagrams with the ones drawn for Problem 10.12.

10.17 Consider the basic block diagrams in Figure P10.13. Replace all integrators s^{-1} with unit-time delay elements z^{-1}. Assign state variables, and then develop discrete-time state-variable equations for the diagrams.

10.18 Consider the state-variable equations in (10.74) and (10.82). Show that, if $a_4 = 1$, they are identical. If $a_4 \neq 1$, are they still the same?

10.19 Refer to Figure 10.18. If the outputs of the integrators are assigned as $x_4, x_3, x_2,$ and x_1, instead of as $x_1, x_2, x_3,$ and x_4, what is its state-variable equation description?

10.20 Use (10.74) and (10.82) to find realizations for the following transfer functions.

(a) $\dfrac{2}{4s^3 + 2s^2 + 4s + 1}$

(b) $\dfrac{2s^3 + 1}{4s^3 + 2s^2 + 4s + 1}$

(c) $\dfrac{3s^2 + s + 1}{s^2 + 2s + 4}$

(d) $\dfrac{s^3 - 1}{s^4 + 1}$

10.21 Which of the realizations obtained in Problem 10.20 are minimal realizations?

10.22 What is the dimension of any minimal realization of $1/(s + 1)$? Find 2-, 3-, and 4-dimensional realizations for the transfer function.

10.23 What are the degrees of the following transfer functions? Find their minimal realizations, too.

(a) $\dfrac{s^2 + 2s + 1}{2s^2 + 3s + 1}$

(b) $\dfrac{s^2 - 1}{s^3 + 3s^2 + 5s + 3}$

10.24 Are the following equations irreducible? Which of them are completely characterized by their transfer functions?

(a) $\dot{\mathbf{x}} = \begin{bmatrix} 1 & 0 \\ 0 & 2 \end{bmatrix} \mathbf{x} + \begin{bmatrix} 0 \\ 1 \end{bmatrix} u$

$y = \begin{bmatrix} 1 & 0 \end{bmatrix} \mathbf{x}$

(b) $\dot{\mathbf{x}} = \begin{bmatrix} 1 & 1 & 0 \\ 0 & 1 & 0 \\ 0 & 0 & 1 \end{bmatrix} \mathbf{x} + \begin{bmatrix} 1 \\ 1 \\ 1 \end{bmatrix} u$

$y = \begin{bmatrix} 1 & 1 & 1 \end{bmatrix} \mathbf{x}$

(c) $\dot{\mathbf{x}} = \begin{bmatrix} 1 & 1 & 0 \\ 0 & 1 & 0 \\ 0 & 0 & 0 \end{bmatrix} \mathbf{x} + \begin{bmatrix} 1 \\ 1 \\ 1 \end{bmatrix} u$

$y = \begin{bmatrix} 1 & 1 & 1 \end{bmatrix} \mathbf{x}$

10.25 Consider the feedback system shown in Figure P10.25.

(a) Find its overall transfer function and then draw a basic block diagram for the feedback system.

(b) Find basic block diagrams for each block and then connect them to yield a basic block diagram for the feedback system.

Figure P10.25

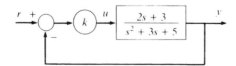

(c) Which diagram, the one drawn in (a) or the one drawn in (b), is more convenient to use if the value of k is to be changed?

10.26 Draw basic block diagrams for the systems shown in Figure P10.26.

10.27 Find tandem realizations for the following transfer functions.

(a) $\dfrac{(s + 3)^3}{2(s + 1)(s + 2)(2s^2 + 3s + 4)}$

(b) $\dfrac{3(s^2 - 2s + 2)(2s^2 + 2s + 1)}{(2s + 3)(s + 1)(s^2 + 3s + 4)}$

10.28 Find parallel realizations for the transfer functions in Problem 10.27.

10.29 Every problem from 10.19 to 10.28 can be restated for the discrete-time case. Try restating some of the problems and then solving them.

10.30 (a) Consider two LTIL systems with transfer functions $H_i(s)$, $i = 1, 2$. Will the transfer function of the tandem connection of H_1 followed by H_2 be the same as that of H_2 followed by H_1?

(b) The order of tandem connection is immaterial in LTIL systems. Is this true for linear time-varying systems? Consider the two systems described by $y_1(t) = du_1(t)/dt$ and $y_2(t) = tu_2(t)$.

Figure P10.26

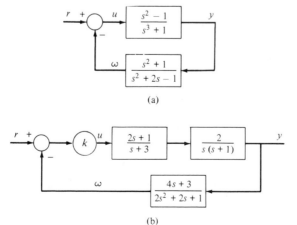

(a)

(b)

USING MATLAB

In MATLAB, a state-variable or state-space equation is represented by `[a,b,c,d]`, and a transfer function is represented by `[num,den]`. The command `ss2tf` computes the transfer function of a state-space equation. The command `tf2ss` finds a realization for a transfer function. For example, the equation in (10.21) is represented by

```
a=[0 1;-2 -3];b=[0;1];c=[1 -1];d=0
```

To find its transfer function, we type

```
[num,den]=ss2tf(a,b,c,d)
```

which yields

```
num = 0 -1.0000 1.0000
den = 1 3 2
```

Thus, we have

$$c\,(sI - a)^{-1}b + d = \frac{-s + 1}{s^2 + 3s + 2}$$

The command

```
step(a,b,c,d) or step(a,b,c,d,1)
```

generates the unit step response of (10.21) on the monitor. MATLAB automatically chooses the integration step size and the final time.

The discretization of a continuous-time state-space equation into a discrete-time equation can be carried out using command `c2d`. As shown in (10.52), the discretization affects only a and b. If we discretize (10.21) with sampling period $T = 0.1$, then the command

```
[da,db]=c2d(a,b,0.1)
```

yields

$$\mathrm{da} = \begin{bmatrix} 0.9909 & 0.0861 \\ -0.1722 & 0.7326 \end{bmatrix} \qquad \mathrm{db} = \begin{bmatrix} 0.0045 \\ 0.0861 \end{bmatrix}$$

Thus, the discretized equation of (10.21) with sampling period $T = 0.1$ is

$$\mathbf{x}[k + 1] = \begin{bmatrix} 0.9909 & 0.0861 \\ -0.1722 & 0.7326 \end{bmatrix} \mathbf{x}[k] + \begin{bmatrix} 0.0045 \\ 0.0861 \end{bmatrix} u\,[k]$$

$$y\,[k] = [1 \ \ -1]\mathbf{x}[k]$$

The *c* and *d* remain the same. The digital counterpart of step is dstep, which is not available in the student edition.

Realization finds a state-space equation for a transfer function and can be carried out by using tf2ss. Consider the transfer function in (10.75). The commands

```
n=[1 2 -1 4 12];d=[2 10 20 20 8];
[a,b,c,d]=tf2ss(n,d)
```

yield the realization in (10.84). Minimal realization (discussed in Section 10.9.3) can be carried out by using minreal, which is not available in the student edition. The commands ss2tf and tf2ss can be applied to discrete-time systems without any modification.

The command

```
step(num,den)
```

computes the unit step response of $H(s) = $ num/den. Internally, the command uses tf2ss to find a state-space equation and then compute the step response. Thus, in MATLAB, most computations for transfer functions are carried out using state-space realizations.

MATLAB PROBLEMS

Use MATLAB to carry out the following.

M10.1 Find the transfer functions of the state-variable equations in Exercises 10.3.1, 10.3.2, and 10.3.3.

M10.2 Find realizations for the transfer functions in Exercises 10.9.2 and 10.9.3.

Appendix A
Linear Algebraic Equations[1]

A.1   LINEAR ALGEBRAIC EQUATIONS FROM LTI MEMORYLESS SYSTEMS[2]

This appendix discusses linear algebraic equations. An example will illustrate how a set of linear algebraic equations arises in an LTI memoryless system: Consider the network shown in Figure A.1. It consists of five resistors, R_i, $i = 1, 2, 3, 4, 5$, and two current sources, u_i, $i = 1, 2$. It has no energy storage elements (inductors or capacitors) and is, therefore, a memoryless system. The problem is to find the node voltages v_i, $i = 1, 2, 3$, excited by the two constant current sources $u_1 = 2$ A (amperes) and $u_2 = 1$ A. If the directions of current i_j, $j = 1, 2, 3, 4, 5$, are chosen as shown, then we have, using Ohm's law,

$$i_1 = \frac{v_2 - v_1}{R_1}, \qquad i_2 = \frac{v_2}{R_2}, \qquad i_3 = \frac{v_3}{R_3}, \qquad i_4 = \frac{v_1 - v_3}{R_4}, \qquad i_5 = \frac{v_3 - v_2}{R_5}$$

Kirchhoff's current law states that the algebraic sum of all currents leaving (or entering) every node is identically zero. By choosing a current leaving a node as positive, we have, at node 1,

$$-i_1 + i_4 + u_1 = 0$$

or

$$-\frac{v_2 - v_1}{R_1} + \frac{v_1 - v_3}{R_4} + u_1 = 0$$

---

[1]The reader who has no prior knowledge of linear algebra may skip sections marked with an asterisk; there will be no loss of continuity.

[2]The reader who is not familiar with simple circuit analysis may skip this section. This section provides motivation for studying linear algebraic equations.

Figure A.1   *A memoryless system*

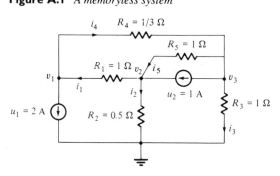

This can be simplified as

$$(\frac{1}{R_1} + \frac{1}{R_4})v_1 - \frac{1}{R_1}v_2 - \frac{1}{R_4}v_3 = -u_1$$

which becomes, after the substitution of $R_1 = 1\,\Omega$, $R_4 = 1/3\,\Omega$, and $u_1 = 2\,A$,

$$4v_1 - v_2 - 3v_3 = -u_1 = -2 \qquad \text{(A.1a)}$$

From $i_1 + i_2 - i_5 - u_2 = 0$ at node 2 and $i_3 - i_4 + i_5 + u_2 = 0$ at node 3, we can similarly obtain

$$-v_1 + 4v_2 - v_3 = u_2 = 1 \qquad \text{(A.1b)}$$
$$-3v_1 - v_2 + 5v_3 = -u_2 = -1 \qquad \text{(A.1c)}$$

The set of three equations in (A.1) is called a set of linear *algebraic* equations. In general, every LTIL memoryless system can be described by a set of linear algebraic equations.

EXERCISE A.1.1

Use the loop currents chosen to develop a set of linear algebraic equations to describe the network in Figure A.2.

[**ANSWER:** $2i_1 - i_2 = 10$, $-i_1 + 3i_2 = 0$.]   ■

The linear algebraic equations in (A.1) involve only real numbers. Mathematically speaking, what will be discussed in this appendix is equally applicable to the set of complex numbers and the set of rational functions with real or complex coefficients. For convenience, we will discuss only real numbers in Sections A.2 through A.9. The discussion will be extended to rational functions with real coefficients in Section A.10.

Figure A.2 *A network*

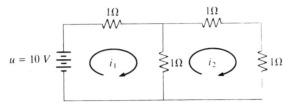

A.2 SOLVING LINEAR ALGEBRAIC EQUATIONS— GAUSSIAN ELIMINATION

The set of three equations in (A.1) has three unknowns v_i, $i = 1, 2, 3$. The problem is to find the v_i, $i = 1, 2, 3$, that simultaneously satisfies the three equations. This problem can be solved as follows: We add to (A.1b) the product of (A.1a) and (1/4), and add to (A.1c) the product of (A.1a) and (3/4) as

$$4v_1 - v_2 - 3v_3 = -2 \qquad \text{(A.1a)}$$
$$-v_1 + 4v_2 - v_3 = 1 \qquad \text{(A.1b)}$$
$$-3v_1 - v_2 + 5v_3 = -1 \qquad \text{(A.1c)}$$

to yield

$$4v_1 - v_2 - 3v_3 = -2 \qquad \text{(A.2a)}$$

$$\frac{15}{4}v_2 - \frac{7}{4}v_3 = \frac{1}{2} \qquad \text{(A.2b)}$$

$$-\frac{7}{4}v_2 + \frac{11}{4}v_3 = \frac{-5}{2} \qquad \text{(A.2c)}$$

where v_1 has been eliminated from the second and third equations. We then add to (A.2c) the product of (A.2b) and (7/15) to yield

$$4v_1 - v_2 - 3v_3 = -2 \qquad \text{(A.3a)}$$

$$\frac{15}{4}v_2 - \frac{7}{4}v_3 = \frac{1}{2} \qquad \text{(A.3b)}$$

$$\frac{29}{15}v_3 = \frac{-34}{15} \qquad \text{(A.3c)}$$

where v_2 has been eliminated from the last equation. This process is called *Gaussian elimination.*[3] We eliminate v_1 from (A.1b), and v_1 and v_2 from

---

[3]In honor of Carl Friedrich Gauss (1777–1855), a German mathematician. A variant of Gaussian elimination appeared about 250 B.C. in the Chinese work "Nine Chapters of Arithmetic." See Reference [1].

(A.1c). Once this process is completed, we then use (A.3c) to solve for v_3. The substitution of $v_3 = -34/29$ into (A.3b) yields

$$v_2 = \frac{4}{15}(\frac{1}{2} + \frac{7}{4} v_3) = \frac{-12}{29} \tag{A.4}$$

The substitution of v_3 and v_2 into (A.3a) yields

$$v_1 = \frac{1}{4}(-2 + v_2 + 3v_3) = \frac{-43}{29} \tag{A.5}$$

This process is called *back substitution*.

We see from this example that the procedure consists of two steps: elimination and back substitution. The procedure is very general and is applicable to linear algebraic equations with any number of unknown variables.

EXERCISE A.2.1

Find the solution of

$$2v_1 - v_2 = 10$$
$$-v_1 + 3v_2 = 0$$

[**ANSWER:** $v_1 = 6$, $v_2 = 2$.]   ∎

A.3   **MATRIX NOTATION AND OPERATION**

If the number of equations or the number of unknown variables is large, the writing of the set of equations becomes tedious and time-consuming. Therefore, it is desirable to introduce a more compact notation. This notation is also needed in order to fully develop the theory of linear algebraic equations.

The set of equations in (A.1) can be written as

$$\begin{bmatrix} 4 & -1 & -3 \\ -1 & 4 & -1 \\ -3 & -1 & 5 \end{bmatrix} \begin{bmatrix} v_1 \\ v_2 \\ v_3 \end{bmatrix} = \begin{bmatrix} -2 \\ 1 \\ -1 \end{bmatrix} \tag{A.6}$$

or simply

$$\mathbf{A}\mathbf{v} = \mathbf{b} \tag{A.7}$$

with

$$\mathbf{A} := \begin{bmatrix} 4 & -1 & -3 \\ -1 & 4 & -1 \\ -3 & -1 & 5 \end{bmatrix} \quad \mathbf{v} := \begin{bmatrix} v_1 \\ v_2 \\ v_3 \end{bmatrix} \quad \mathbf{b} := \begin{bmatrix} -2 \\ 1 \\ -1 \end{bmatrix} \tag{A.8}$$

This is called the matrix notation.[4]

A matrix $\mathbf{A}$ with m rows and n columns, called a matrix of order $m \times n$ or an $m \times n$ matrix, is an array of real numbers of the form

$$\mathbf{A} := \begin{bmatrix} a_{11} & a_{12} & \cdots & a_{1n} \\ a_{21} & a_{22} & \cdots & a_{2n} \\ \cdot & \cdot & \cdot & \cdot \\ \cdot & \cdot & \cdot & \cdot \\ \cdot & \cdot & \cdot & \cdot \\ a_{m1} & a_{m2} & \cdots & a_{mn} \end{bmatrix} =: \left[a_{ij} \right] \qquad \text{(A.9)}$$

The entry a_{ij} is located at the ith row and jth column. Therefore, the first subscript indicates row location and the second subscript indicates column location. If $m = n$, $\mathbf{A}$ is called an $n \times n$ matrix or a square matrix of order n. Clearly, the $\mathbf{A}$ in (A.8) is a 3×3 matrix, or a square matrix of order 3; $\mathbf{v}$ is a 3×1 matrix; and $\mathbf{b}$ is also a 3×1 matrix. An $n \times 1$ matrix is often called, more informatively, a *column* vector, and a $1 \times m$ matrix is called a *row* vector. We use boldface capital or lower-case letters to denote matrices or vectors. We also write (A.9) succinctly as $\mathbf{A} = [a_{ij}]$.

In the manipulation of matrices, the rules of addition, multiplication, and division are extensions of the corresponding operations in real numbers. Although there are some similarities, there are also some essential differences, as will be seen immediately.

Two matrices $\mathbf{A} = [a_{ij}]$ and $\mathbf{B} = [b_{ij}]$ are said to be equal if, and only if, they are of the same order and their corresponding entries are all equal; that is, $a_{ij} = b_{ij}$ for all i and j. Let $\mathbf{A}$ and $\mathbf{B}$ be two matrices of the same order. Then, the entries of their sum $\mathbf{C} := \mathbf{A} + \mathbf{B}$ are the sums of the corresponding entries of $\mathbf{A}$ and $\mathbf{B}$; that is, $c_{ij} = a_{ij} + b_{ij}$, for all i and j. For example, we have

$$\begin{bmatrix} 2 & 1 & -1 \\ 3 & -2 & 1 \end{bmatrix} + \begin{bmatrix} -2 & 1 & 0 \\ 1 & 0 & 1 \end{bmatrix} = \begin{bmatrix} 0 & 2 & -1 \\ 4 & -2 & 2 \end{bmatrix}$$

EXERCISE A.3.1

Compute the following, if they are defined.

(a) $[1 \ 0 \ -2] + [2 \ 3]$

(b) $\begin{bmatrix} 2 & 5 & -3 \\ 1 & -2 & 2 \end{bmatrix} + \begin{bmatrix} 2 & 3 & 4 \\ 1 & 9 & 9 \end{bmatrix}$

[**ANSWERS:** (a) Not defined, (b) $\begin{bmatrix} 4 & 8 & 1 \\ 2 & 7 & 11 \end{bmatrix}$.] ∎

---

[4]Recall that $\mathbf{A} := \mathbf{B}$ means that $\mathbf{A}$, by definition, is equal to $\mathbf{B}$ and $\mathbf{A} =: \mathbf{B}$ means that $\mathbf{B}$, by definition, is equal to $\mathbf{A}$.

Now let us discuss the multiplication of two matrices. Consider an $m \times n$ matrix $\mathbf{A}$ and a $q \times p$ matrix $\mathbf{B}$. If $n = q$, then the product of $\mathbf{A}$ and $\mathbf{B}$ is defined as

$$\underset{\substack{m \times p}}{\mathbf{C}} := \underset{\substack{m \times n}}{\mathbf{A}} \quad \underset{\substack{q \times p \quad (n = q)}}{\mathbf{B}} \tag{A.10}$$

where c_{ij} is defined as

$$c_{ij} := \begin{bmatrix} a_{i1} & a_{i2} & \cdots & a_{in} \end{bmatrix} \cdot \begin{bmatrix} b_{1j} \\ b_{2j} \\ \vdots \\ \vdots \\ b_{nj} \end{bmatrix} := a_{i1}b_{1j} + a_{i2}b_{2j} + \cdots + a_{in}b_{nj} \tag{A.11}$$

that is, c_{ij} is the sum of the product of the corresponding entries of the ith row of $\mathbf{A}$ and the jth column of $\mathbf{B}$.

EXAMPLE A.3.1

Consider

$$\begin{bmatrix} c_{11} & c_{12} & c_{13} \\ c_{21} & c_{22} & c_{23} \end{bmatrix} := \begin{bmatrix} 2 & 1 & 0 \\ 4 & -1 & 6 \end{bmatrix} \begin{bmatrix} 1 & 2 & 0 \\ -3 & -1 & 1 \\ 5 & 3 & 0 \end{bmatrix}$$

We have

$$c_{11} = \begin{bmatrix} 2 & 1 & 0 \end{bmatrix} \begin{bmatrix} 1 \\ -3 \\ 5 \end{bmatrix} := 2 \cdot 1 + 1 \cdot (-3) + 0 \cdot 5 = -1$$

$$c_{12} = \begin{bmatrix} 2 & 1 & 0 \end{bmatrix} \begin{bmatrix} 2 \\ -1 \\ 3 \end{bmatrix} := 2 \cdot 2 + 1 \cdot (-1) + 0 \cdot 3 = 3$$

Similarly, we have $c_{21} = 4 + 3 + 30 = 37$, $c_{22} = 8 + 1 + 18 = 27$, $c_{13} = 1$, and $c_{23} = -1$.

EXERCISE A.3.2

Compute

$$\begin{bmatrix} 2 & 3 \\ 1 & -1 \end{bmatrix} \cdot \begin{bmatrix} 2 & 2 \\ 1 & -1 \end{bmatrix}$$

[ANSWER: $\begin{bmatrix} 7 & 1 \\ 1 & 3 \end{bmatrix}$.]

EXERCISE A.3.3

Verify that the matrix equation in (A.6) gives the set of three equations in (A.1).
■

Consider again (A.10). If $n \neq q$, **AB** is not defined. It is possible that **AB** is defined, but **BA** is not. Even if both **AB** and **BA** are defined, they may not be of the same order. For example, if

$$\mathbf{A} = [2 \;\; 1 \;\; -1], \qquad \mathbf{B} = \begin{bmatrix} 1 \\ -2 \\ -4 \end{bmatrix}$$

then

$$\mathbf{AB} = [4] \quad \text{and} \quad \mathbf{BA} = \begin{bmatrix} 1 \\ -2 \\ -4 \end{bmatrix} [\,2 \;\; 1 \;\; -1\,] = \begin{bmatrix} 2 & 1 & -1 \\ -4 & -2 & 2 \\ -8 & -4 & 4 \end{bmatrix}$$

The former is a square matrix of order 1 or a scalar; the latter is a square matrix of order 3. Even if they are of the same order, generally they are not equal, that is,

$$\mathbf{AB} \neq \mathbf{BA}$$

Therefore, multiplications of matrices do not have the commutative property.

EXERCISE A.3.4

Show **AB** $\neq$ **BA**, where

$$\mathbf{A} = \begin{bmatrix} 1 & -2 \\ 1 & 1 \end{bmatrix} \quad \text{and} \quad \mathbf{B} = \begin{bmatrix} 3 & -1 \\ 1 & 0 \end{bmatrix}$$

■

For a square matrix $\mathbf{A} = [a_{ij}]$, the elements a_{ii}, $i = 1, 2, \ldots$, on the diagonal are called the diagonal elements or entries. A square matrix is called a diagonal matrix if all entries, except the diagonal entries, are zero. For example, the matrix

$$\begin{bmatrix} 2 & 0 & 0 \\ 0 & 0 & 0 \\ 0 & 0 & 3 \end{bmatrix}$$

is a diagonal matrix, often written as diag {2,0,3}. A diagonal matrix is called a *unit* or an *identity* matrix if its diagonal elements all equal 1. For example,

$$I_4 := \begin{bmatrix} 1 & 0 & 0 & 0 \\ 0 & 1 & 0 & 0 \\ 0 & 0 & 1 & 0 \\ 0 & 0 & 0 & 1 \end{bmatrix}$$

is the unit matrix of order 4. A unit matrix is often denoted by I without specifying its order. It is understood, however, that the order of I must be compatible with other matrices.

EXERCISE A.3.5

Let A be a 2×4 matrix. What are the orders of the two unit matrices in the following?

$$IA = AI = A$$

Verify the equation. This shows that a matrix remains unchanged after pre- or postmultiplication of a unit matrix.

[**ANSWERS:** 2, 4.] ∎

If c is a real number, called a scalar (1×1 matrix), then we define

$$c A := A c := \left[c a_{ij} \right]$$

For example, we have

$$3 \begin{bmatrix} 2 & 1 & -1 \\ 3 & -2 & 1 \end{bmatrix} = \begin{bmatrix} 6 & 3 & -3 \\ 9 & -6 & 3 \end{bmatrix}$$

and

$$-\begin{bmatrix} 2 & 1 & -1 \\ 3 & -2 & 1 \end{bmatrix} = \begin{bmatrix} -2 & -1 & 1 \\ -3 & 2 & -1 \end{bmatrix}$$

Note that, because $AB \neq BA$, the order of matrices cannot be changed. However, the position of a scalar can be moved to any position. For example, we have $ABc = AcB = cAB$, where c is a scalar.

EXERCISE A.3.6

Consider $c \cdot A$ where c is a scalar. If $c = 0$, then

$$0 \cdot A = 0$$

What are the order of the two zeros in the equation?

[**ANSWER:** The first zero is 1×1, the second zero is a matrix of the same order as **A**.] ∎

Although products of matrices do not commute, they do have the following associative property:

$$(A + B)C = AC + BC$$

and

$$A(B + C) = AB + AC$$

A square matrix $A = [a_{ij}]$ is called a *lower* triangular matrix if all elements above the diagonal are zero; it is called an *upper* triangular matrix if all elements under the diagonal are zero. If, in addition, all elements on the diagonal are 1, then the matrix is called a *unit* lower or upper triangular matrix. For example, the first matrix in

$$\begin{bmatrix} 2 & 0 & 0 \\ 0 & 0 & 0 \\ 1 & -2 & 3 \end{bmatrix}, \qquad \begin{bmatrix} 1 & -2 & 0 \\ 0 & 1 & 1 \\ 0 & 0 & 1 \end{bmatrix}$$

is a lower triangular matrix; the second matrix is a unit upper triangular matrix.

The division of two real numbers a and b can be written as a/b, ab^{-1}, or $b^{-1}a$. If we call b^{-1} the inverse of b, then the division becomes the multiplication of a and the inverse of b. In matrices, the notation A/B is not defined, and division must be defined through the concept of the inverse of matrices. This concept is introduced in the following discussion. It is defined only for square matrices.

A square matrix **A** is said to have an *inverse* if there exists a matrix, denoted by A^{-1}, such that

$$A^{-1}A = AA^{-1} = I$$

For example, consider the unit lower triangular matrix

$$A = \begin{bmatrix} 1 & 0 & 0 \\ -1 & 1 & 0 \\ 2 & 3 & 1 \end{bmatrix} \tag{A.12}$$

Let

$$M = \begin{bmatrix} 1 & 0 & 0 \\ m_{21} & 1 & 0 \\ m_{31} & m_{32} & 1 \end{bmatrix}$$

From

$$\mathbf{AM} = \begin{bmatrix} 1 & 0 & 0 \\ -1 & 1 & 0 \\ 2 & 3 & 1 \end{bmatrix} \begin{bmatrix} 1 & 0 & 0 \\ m_{21} & 1 & 0 \\ m_{31} & m_{32} & 1 \end{bmatrix} = \begin{bmatrix} 1 & 0 & 0 \\ -1 + m_{21} & 1 & 0 \\ 2 + 3m_{21} + m_{31} & 3 + m_{32} & 1 \end{bmatrix}$$

we see that, if $m_{21} = 1$, $m_{32} = -3$, and $m_{31} = -5$, then $\mathbf{AM} = \mathbf{I}$. We also have

$$\mathbf{MA} = \begin{bmatrix} 1 & 0 & 0 \\ 1 & 1 & 0 \\ -5 & -3 & 1 \end{bmatrix} \begin{bmatrix} 1 & 0 & 0 \\ -1 & 1 & 0 \\ 2 & 3 & 1 \end{bmatrix} = \begin{bmatrix} 1 & 0 & 0 \\ 0 & 1 & 0 \\ 0 & 0 & 1 \end{bmatrix}$$

Hence, the matrix $\mathbf{A}$ in (A.12) has an inverse given by

$$\mathbf{A}^{-1} := \mathbf{M} = \begin{bmatrix} 1 & 0 & 0 \\ 1 & 1 & 0 \\ -5 & -3 & 1 \end{bmatrix}$$

Using the same procedure, we can show that every upper or lower triangular matrix with nonzero diagonal elements has an inverse. Indeed, let

$$\mathbf{A} = \begin{bmatrix} a_{11} & a_{12} & a_{13} \\ 0 & a_{22} & a_{23} \\ 0 & 0 & a_{33} \end{bmatrix}$$

We assume

$$\mathbf{A}^{-1} := \mathbf{M} = \begin{bmatrix} m_{11} & m_{12} & m_{13} \\ m_{21} & m_{22} & m_{23} \\ m_{31} & m_{32} & m_{33} \end{bmatrix}$$

Then, we have

MA = AM

$$= \begin{bmatrix} a_{11}m_{11} & a_{11}m_{12} + a_{12}m_{22} + a_{13}m_{33} & a_{11}m_{13} + a_{12}m_{23} + a_{13}m_{33} \\ a_{22}m_{21} + a_{23}m_{31} & a_{22}m_{22} + a_{23}m_{32} & a_{22}m_{23} + a_{23}m_{33} \\ a_{33}m_{31} & a_{33}m_{32} & a_{33}m_{33} \end{bmatrix}$$

$$= \begin{bmatrix} 1 & 0 & 0 \\ 0 & 1 & 0 \\ 0 & 0 & 1 \end{bmatrix}$$

Equating the corresponding entries, we can readily obtain, under the assumption $a_{ii} \neq 0$, $i = 1, 2, 3$,

$$
\mathbf{A}^{-1} =
\begin{bmatrix}
\dfrac{1}{a_{11}} & \dfrac{-a_{12}}{a_{11}a_{22}} & m_{13} \\[2ex]
0 & \dfrac{1}{a_{22}} & \dfrac{-a_{23}}{a_{22}a_{33}} \\[2ex]
0 & 0 & \dfrac{1}{a_{33}}
\end{bmatrix}
$$

where

$$
m_{13} = \frac{a_{12}a_{23}}{a_{11}a_{22}a_{33}} - \frac{a_{13}}{a_{11}a_{23}}
$$

(Verify). This essentially establishes that every upper (or lower) triangular matrix with nonzero diagonal elements has an inverse, and the inverse is, again, an upper (or lower) triangular matrix. The statement does not hold if some of the diagonal elements are zero. For example, the diagonal matrix

$$
\begin{bmatrix} 1 & 0 \\ 0 & 0 \end{bmatrix}
$$

has no inverse because no m_{ij} exists to meet

$$
\begin{bmatrix} 1 & 0 \\ 0 & 0 \end{bmatrix}
\begin{bmatrix} m_{11} & m_{12} \\ m_{21} & m_{22} \end{bmatrix}
=
\begin{bmatrix} m_{11} & m_{12} \\ 0 & 0 \end{bmatrix}
=
\begin{bmatrix} 1 & 0 \\ 0 & 1 \end{bmatrix}
$$

EXERCISE A.3.7

Verify that there exists no m_{ij} to meet

$$
\begin{bmatrix} 1 & 1 \\ -1 & -1 \end{bmatrix}
\begin{bmatrix} m_{11} & m_{12} \\ m_{21} & m_{22} \end{bmatrix}
=
\begin{bmatrix} 1 & 0 \\ 0 & 1 \end{bmatrix}
$$

■

A square matrix that has an inverse is said to be *nonsingular*. Otherwise, it is said to be *singular*. If a matrix is not square, its inverse is not defined.[5]

---

[5] It is, however, possible to define something called the *pseudoinverse*.

Therefore, whenever a matrix is said to be nonsingular, the matrix is implicitly assumed to be square. Let **A** and **B** be nonsingular matrices. Then, we have

$$(\mathbf{AB})^{-1} = \mathbf{B}^{-1}\mathbf{A}^{-1}$$

This is a useful formula. It is important to note that

$$\mathbf{AB} = \mathbf{0}$$

does not imply **A** = **0** or **B** = **0**. For example, we have

$$\mathbf{AB} = \begin{bmatrix} 1 & 2 \\ 1 & 2 \end{bmatrix} \begin{bmatrix} 2 & 2 \\ -1 & -1 \end{bmatrix} = \begin{bmatrix} 0 & 0 \\ 0 & 0 \end{bmatrix} = \mathbf{0}$$

and **A** ≠ **0** and **B** ≠ **0**. However, if **A** is nonsingular, then **AB** = **0** does imply **B** = **0**. Indeed, the premultiplication of $\mathbf{A}^{-1}$ to both sides of **AB** = **0** yields

$$\mathbf{A}^{-1}\mathbf{AB} = \mathbf{IB} = \mathbf{B} = \mathbf{A}^{-1}\mathbf{0} = \mathbf{0}$$

This proves the assertion. More will be said later regarding nonsingular matrices and the computation of inverses.

A discussion of the concept of transpose concludes this section. A matrix is called the *transpose* of **A**, denoted by **A′**, if the *i*th *row* of **A** becomes the *i*th *column* of **A′**. Thus, if **A** is an $m \times n$ matrix, then **A′** is an $n \times m$ matrix. For example, if

$$\mathbf{A} = \begin{bmatrix} 2 & 3 & 1 \\ 4 & 0 & 1 \end{bmatrix} \quad \text{then} \quad \mathbf{A'} := \begin{bmatrix} 2 & 4 \\ 3 & 0 \\ 1 & 1 \end{bmatrix}$$

Thus, if **x** is an $m \times 1$ column vector, **x′** is a $1 \times m$ row vector. The vector

$$\mathbf{x} = \begin{bmatrix} -1 \\ 2 \\ 10 \\ 4 \end{bmatrix}$$

can be written as **x** = [−1 2 10 4]′ to save space.

EXERCISE A.3.8

Find the transposes of

$$\mathbf{A}_1 = \begin{bmatrix} 0 & -1 & 0 & 2 \\ 0 & 0 & 1 & -1 \\ 1 & 1 & 2 & 0 \end{bmatrix}, \quad \mathbf{A}_2 = \begin{bmatrix} 0 \\ 1 \\ 2 \end{bmatrix}, \quad \mathbf{A}_3 = \begin{bmatrix} 1 & -1 & 3 \\ -1 & 2 & 1 \\ 3 & 1 & 0 \end{bmatrix}$$

Which $\mathbf{A}_i$ has the property $\mathbf{A}_i' = \mathbf{A}_i$? (A matrix **A** with the property **A′** = **A** is said to be *symmetric*.) ∎

The following is a useful formula:

$$(\mathbf{AB})' = \mathbf{B}'\mathbf{A}'.$$

For example, we have

$$\left(\begin{bmatrix} 2 \\ 3 \end{bmatrix} [\,2\ \ 1\ \ 0\,] \right)' = \begin{bmatrix} 4 & 2 & 0 \\ 6 & 3 & 0 \end{bmatrix}' = \begin{bmatrix} 4 & 6 \\ 2 & 3 \\ 0 & 0 \end{bmatrix}$$

and

$$[\,2\ \ 1\ \ 0\,]' \begin{bmatrix} 2 \\ 3 \end{bmatrix}' = \begin{bmatrix} 2 \\ 1 \\ 0 \end{bmatrix} [\,2\ \ 3\,] = \begin{bmatrix} 4 & 6 \\ 2 & 3 \\ 0 & 0 \end{bmatrix}$$

They are indeed equal. We also have

$$(\mathbf{A} + \mathbf{B})' = \mathbf{A}' + \mathbf{B}'$$
$$(\mathbf{A}')' = \mathbf{A}$$

and

$$(\mathbf{A}^{-1})' = (\mathbf{A}')^{-1}$$

A.4 ELEMENTARY MATRICES AND GAUSSIAN ELIMINATION

Now we shall use matrix notations to develop the elimination process discussed in Section A.2. Consider an $m \times n$ matrix $\mathbf{A}$. The following operations on $\mathbf{A}$ are called elementary operations:

1. multiplication of a row (column) by a *nonzero* real number,
2. interchange of two rows (columns), and
3. addition of the product of one row (column) and a real number to another row (column).

These operations can be accomplished by using the *elementary matrices* of forms

$$\mathbf{E}_1 = \begin{bmatrix} 1 & 0 & 0 & 0 \\ 0 & 1 & 0 & 0 \\ 0 & 0 & c & 0 \\ 0 & 0 & 0 & 1 \end{bmatrix} \qquad \mathbf{E}_2 = \begin{bmatrix} 0 & 0 & 0 & 1 \\ 0 & 1 & 0 & 0 \\ 0 & 0 & 1 & 0 \\ 1 & 0 & 0 & 0 \end{bmatrix} \qquad \mathbf{E}_3 = \begin{bmatrix} 1 & 0 & 0 & 0 \\ 0 & 1 & 0 & 0 \\ 0 & 0 & 1 & 0 \\ 0 & d & 0 & 1 \end{bmatrix} \qquad \text{(A.13)}$$

where c and d are real numbers and $c \neq 0$. They are all square matrices of various orders and differ only slightly from the unit matrix. The first matrix differs from a unit matrix only by one element. The second matrix is equal to a unit matrix except that the ith and jth diagonal elements are moved to the (i, j)th and (j, i)th positions. The third matrix has an extra element in a unit matrix; the extra element can be located in any position. The subscript of $\mathbf{E}_i$ denotes only the type of operation; the order of $\mathbf{E}_i$ must be compatible with the matrix on which it operates. First we mention that all $\mathbf{E}_i$, $i = 1, 2, 3$, are nonsingular and their inverses are

$$
\mathbf{E}_1^{-1} = \begin{bmatrix} 1 & 0 & 0 & 0 \\ 0 & 1 & 0 & 0 \\ 0 & 0 & 1/c & 0 \\ 0 & 0 & 0 & 1 \end{bmatrix}, \qquad \mathbf{E}_2^{-1} = \mathbf{E}_2, \qquad \mathbf{E}_3^{-1} = \begin{bmatrix} 1 & 0 & 0 & 0 \\ 0 & 1 & 0 & 0 \\ 0 & 0 & 1 & 0 \\ 0 & -d & 0 & 1 \end{bmatrix}
$$

These can be directly verified by showing $\mathbf{E}_i^{-1}\mathbf{E}_i = \mathbf{E}_i\mathbf{E}_i^{-1} = \mathbf{I}$ (Verify). From $\mathbf{E}_1^{-1}$, we see the reason for requiring $c \neq 0$. We also see that the inverse of $\mathbf{E}_i$ has the same form as $\mathbf{E}_i$. Therefore, the inverses of elementary matrices are elementary matrices. Because every elementary matrix has an inverse, all elementary matrices are nonsingular.

The *pre*multiplication of $\mathbf{E}_i$ on $\mathbf{A}$ operates on the *rows* of $\mathbf{A}$. For example, the operation

$$
\mathbf{E}_2\mathbf{A} := \begin{bmatrix} 1 & 0 & 0 \\ 0 & 0 & 1 \\ 0 & 1 & 0 \end{bmatrix} \begin{bmatrix} 2 & 1 & 4 & 0 \\ 1 & 2 & 3 & 4 \\ 4 & 3 & 2 & 1 \end{bmatrix} = \begin{bmatrix} 2 & 1 & 4 & 0 \\ 4 & 3 & 2 & 1 \\ 1 & 2 & 3 & 4 \end{bmatrix}
$$

interchanges the second and third rows of $\mathbf{A}$. The operation

$$
\mathbf{E}_3\mathbf{A} := \begin{bmatrix} 1 & 0 & 0 \\ 0 & 1 & 0 \\ -2 & 0 & 1 \end{bmatrix} \begin{bmatrix} 2 & 1 & 4 & 0 \\ 1 & 2 & 3 & 4 \\ 4 & 3 & 2 & 1 \end{bmatrix} = \begin{bmatrix} 2 & 1 & 4 & 0 \\ 1 & 2 & 3 & 4 \\ 0 & 1 & -6 & 1 \end{bmatrix}
$$

adds the product of the first row of $\mathbf{A}$ and -2 to the third row. On the other hand, the *post*multiplication of $\mathbf{E}_i$ on $\mathbf{A}$ operates on the *columns* of $\mathbf{A}$. For example, the operation

$$
\mathbf{A}\mathbf{E}_2 := \begin{bmatrix} 2 & 1 & 4 & 0 \\ 1 & 2 & 3 & 4 \\ 4 & 3 & 2 & 1 \end{bmatrix} \begin{bmatrix} 1 & 0 & 0 & 0 \\ 0 & 0 & 0 & 1 \\ 0 & 0 & 1 & 0 \\ 0 & 1 & 0 & 0 \end{bmatrix} = \begin{bmatrix} 2 & 0 & 4 & 1 \\ 1 & 4 & 3 & 2 \\ 4 & 1 & 2 & 3 \end{bmatrix}
$$

interchanges the second and fourth columns. The operation

$$
\mathbf{A}\mathbf{E}_3 := \begin{bmatrix} 2 & 1 & 4 & 0 \\ 1 & 2 & 3 & 4 \\ 4 & 3 & 2 & 1 \end{bmatrix} \begin{bmatrix} 1 & 0 & 0 & 0 \\ 0 & 1 & 0 & 0 \\ -2 & 0 & 1 & 0 \\ 0 & 0 & 0 & 1 \end{bmatrix} = \begin{bmatrix} -6 & 1 & 4 & 0 \\ -5 & 2 & 3 & 4 \\ 0 & 3 & 2 & 1 \end{bmatrix}
$$

adds the product of the third column and -2 to the first column; the second, third, and fourth columns remain unchanged. Thus, $\mathbf{E}_i\mathbf{A}$ are called elementary row operations and $\mathbf{A}\mathbf{E}_i$ are called elementary column operations.

EXERCISE A.4.1

What are the effects of pre- and postmultiplication of the following elementary matrix on a matrix?

$$\begin{bmatrix} 1 & 0 & 0 \\ 0 & 1 & -2 \\ 0 & 0 & 1 \end{bmatrix}$$

Matrices like $\mathbf{E}_2$ are also called *permutation* matrices. They permute rows or columns. The permutation can be accomplished by simply interchanging indices, rather than explicitly multiplying $\mathbf{E}_2$ with a matrix. Therefore, $\mathbf{E}_2$ is not used in actual programming. It is introduced here to make our discussion complete.

In application, several $\mathbf{E}_3$ can be combined as

$$\begin{bmatrix} 1 & 0 & 0 & 0 \\ d_{21} & 1 & 0 & 0 \\ 0 & 0 & 1 & 0 \\ 0 & 0 & 0 & 1 \end{bmatrix} \begin{bmatrix} 1 & 0 & 0 & 0 \\ 0 & 1 & 0 & 0 \\ 0 & 0 & 1 & 0 \\ d_{41} & 0 & 0 & 1 \end{bmatrix} \begin{bmatrix} 1 & 0 & 0 & 0 \\ 0 & 1 & 0 & 0 \\ d_{31} & 0 & 1 & 0 \\ 0 & 0 & 0 & 1 \end{bmatrix} = \begin{bmatrix} 1 & 0 & 0 & 0 \\ d_{21} & 1 & 0 & 0 \\ d_{31} & 0 & 1 & 0 \\ d_{41} & 0 & 0 & 1 \end{bmatrix} =: \mathbf{M}_1 \quad \textbf{(A.14)}$$

(Verify). This means that the successive addition of the product of the first row and d_{i1} to the ith row for $i = 3$, 4, and 2 (the order is immaterial) can be accomplished by the one matrix $\mathbf{M}_1$ in (A.14). The inverse of $\mathbf{M}_1$ *happens* to be very simple, just reversing the signs of d_{i1} as

$$\mathbf{M}_1^{-1} = \begin{bmatrix} 1 & 0 & 0 & 0 \\ -d_{21} & 1 & 0 & 0 \\ -d_{31} & 0 & 1 & 0 \\ -d_{41} & 0 & 0 & 1 \end{bmatrix} \quad \textbf{(A.15)}$$

(Verify).

With this background, we can now use matrix notations to express the operations in (A.1) to (A.3). Consider (A.6) or

$$\begin{bmatrix} 4 & -1 & -3 \\ -1 & 4 & -1 \\ -3 & -1 & 5 \end{bmatrix} \mathbf{v} =: \mathbf{A}\mathbf{v} = \mathbf{b} := \begin{bmatrix} -2 \\ 1 \\ -1 \end{bmatrix} \quad \textbf{(A.16)}$$

Let

$$\mathbf{M}_1 = \begin{bmatrix} 1 & 0 & 0 \\ 1/4 & 1 & 0 \\ 3/4 & 0 & 1 \end{bmatrix}$$

This is an elementary matrix, formed from the first column of $\mathbf{A}$. The premultiplication of $\mathbf{M}_1$ to (A.16) yields

$$\begin{bmatrix} 4 & -1 & -3 \\ 0 & 15/4 & -7/4 \\ 0 & -7/4 & 11/4 \end{bmatrix} \mathbf{v} =: \mathbf{A}_1 \mathbf{v} = \mathbf{b}_1 := \begin{bmatrix} -2 \\ 1/2 \\ -5/2 \end{bmatrix} \tag{A.17}$$

where

$$\mathbf{A}_1 := \mathbf{M}_1 \mathbf{A} = \begin{bmatrix} 1 & 0 & 0 \\ 1/4 & 1 & 0 \\ 3/4 & 0 & 1 \end{bmatrix} \begin{bmatrix} 4 & -1 & -3 \\ -1 & 4 & -1 \\ -3 & -1 & 5 \end{bmatrix} = \begin{bmatrix} 4 & -1 & -3 \\ 0 & 15/4 & -7/4 \\ 0 & -7/4 & 11/4 \end{bmatrix} \tag{A.18a}$$

and

$$\mathbf{b}_1 := \mathbf{M}_1 \mathbf{b} = \begin{bmatrix} -2 \\ 1/2 \\ -5/2 \end{bmatrix} \tag{A.18b}$$

Thus, the operation of $\mathbf{M}_1$ has the same results as using the element $a_{11} = 4$ to eliminate all entries under it. The element 4 is, therefore, called the pivot of elimination or simply the *pivot*. We see that (A.17) is identical to (A.2). Let

$$\mathbf{M}_2 = \begin{bmatrix} 1 & 0 & 0 \\ 0 & 1 & 0 \\ 0 & 7/15 & 1 \end{bmatrix} \tag{A.19}$$

This is also an elementary matrix. The premultiplication of $\mathbf{M}_2$ to (A.17) yields

$$\begin{bmatrix} 4 & -1 & -3 \\ 0 & 15/4 & -7/4 \\ 0 & 0 & 29/15 \end{bmatrix} \mathbf{v} =: \mathbf{U}\mathbf{v} = \mathbf{b}_2 := \begin{bmatrix} 2 \\ 1/2 \\ -34/15 \end{bmatrix} \tag{A.20}$$

where

$$\mathbf{U} := \mathbf{M}_2 \mathbf{M}_1 \mathbf{A} = \begin{bmatrix} 4 & -1 & -3 \\ 0 & 15/4 & -7/4 \\ 0 & 0 & 29/15 \end{bmatrix}, \quad \mathbf{b}_2 := \mathbf{M}_2 \mathbf{M}_1 \mathbf{b} = \begin{bmatrix} 2 \\ 1/2 \\ -34/15 \end{bmatrix} \tag{A.21}$$

This step uses the pivot 15/4 to eliminate the element under it. Equation (A.20) is identical to (A.3). Therefore, Gaussian elimination can be carried out using matrix operations. It is important to note that all pivots must be nonzero in order for the elimination to be carried out to completion. Gaussian elimination transforms **A** into the upper triangular matrix **U** in (A.21). Thus, *Gaussian elimination actually achieves triangularization of a matrix.* Once **Av** = **b** is triangularized as **Uv** = **b**$_2$, the solution **v** can be obtained by back substitution, as in (A.3) through (A.5).

*A.4.1 Development of the algorithm

Now we are ready to discuss Gaussian elimination for the general case and to develop an algorithm. Consider

$$\mathbf{Av} = \mathbf{b} \tag{A.22}$$

where

$$\mathbf{A} = \begin{bmatrix} a_{11} & a_{12} & a_{13} & \cdots & a_{1n} \\ a_{21} & a_{22} & a_{23} & \cdots & a_{2n} \\ a_{31} & a_{32} & a_{33} & \cdots & a_{3n} \\ \cdot & & & & \\ \cdot & & & & \\ \cdot & & & & \\ a_{n1} & a_{n2} & a_{n3} & \cdots & a_{nn} \end{bmatrix}, \quad \mathbf{b} = \begin{bmatrix} b_1 \\ b_2 \\ b_3 \\ \cdot \\ \cdot \\ \cdot \\ b_n \end{bmatrix} =: \begin{bmatrix} a_{1,n+1} \\ a_{2,n+1} \\ a_{3,n+1} \\ \cdot \\ \cdot \\ \cdot \\ a_{n,n+1} \end{bmatrix}$$

Because the operations will be applied to **A** as well as to **b**, for convenience, we rename b_i as $a_{i,n+1}$. Let us assume $a_{11} \neq 0$. We compute $d_{i1} := a_{i1}/a_{11}$, $i = 2, 3, \ldots, n$, and

$$a_{ij}^{(1)} = a_{ij} - d_{i1}a_{1j}, \quad i = 2, 3, \ldots, n; j = 2, 3, \ldots, n + 1$$

Note that all $a_{i1}^{(1)}$ are not computed and are not needed in subsequent computation. Therefore, their memory locations can be used to store d_{i1}, for $i = 2, 3, \ldots, n$. This step is equivalent to the multiplication of the unit lower triangular matrix

$$\mathbf{M}_1 = \begin{bmatrix} 1 & 0 & 0 & \cdots & 0 \\ -d_{21} & 1 & 0 & \cdots & 0 \\ -d_{31} & 0 & 1 & \cdots & 0 \\ \cdot & & \cdot & & \cdot \\ \cdot & & & & \cdot \\ \cdot & & \cdot & & \cdot \\ -d_{n1} & 0 & 0 & \cdots & 1 \end{bmatrix} \tag{A.23}$$

to (A.22), which yields

$$\mathbf{A}_1 \mathbf{v} = \mathbf{b}_1 \tag{A.24}$$

where

$$
\mathbf{A}_1 := \mathbf{M}_1 \mathbf{A} =
\begin{bmatrix}
a_{11} & a_{12} & a_{13} & \cdots & a_{1n} \\
0 & a_{22}^{(1)} & a_{23}^{(1)} & \cdots & a_{2n}^{(1)} \\
0 & a_{32}^{(1)} & a_{33}^{(1)} & \cdots & a_{3n}^{(1)} \\
\cdot & \cdot & & & \\
\cdot & \cdot & & & \\
\cdot & \cdot & & & \\
0 & a_{n2}^{(1)} & a_{n2}^{(1)} & \cdots & a_{nn}^{(1)}
\end{bmatrix}, \qquad
\mathbf{b}_1 := \mathbf{M}_1 \mathbf{b} =
\begin{bmatrix}
a_{1,n+1} \\
a_{2,n+1}^{(1)} \\
a_{3,n+1}^{(1)} \\
\cdot \\
\cdot \\
\cdot \\
a_{n,n+1}^{(1)}
\end{bmatrix}
$$

Assuming that $a_{22}^{(1)} \neq 0$, we then compute $d_{i2} := a_{i2}^{(1)}/a_{22}^{(1)}$, $i = 3, 4, \ldots, n$, and

$$
a_{ij}^{(2)} = a_{ij}^{(1)} - d_{i2} a_{2j}^{(1)}, \qquad i = 3, 4, \ldots, n; j = 3, 4, \ldots, n+1
$$

Note that all $a_{i2}^{(2)}$ are not computed and are not needed in subsequent computation. Therefore, their memory locations can be used to store d_{i2}. This step is equivalent to the multiplication of the unit lower triangular matrix

$$
\mathbf{M}_2 =
\begin{bmatrix}
1 & 0 & 0 & \cdots & 0 \\
0 & 1 & 0 & \cdots & 0 \\
0 & -d_{32} & 1 & \cdots & 0 \\
\cdot & & \cdot & & \cdot \\
\cdot & & \cdot & & \cdot \\
\cdot & & \cdot & & \cdot \\
0 & -d_{n2} & 0 & \ldots & 1
\end{bmatrix}
\tag{A.25}
$$

to (A.24), which yields

$$
\mathbf{M}_2 \mathbf{A}_1 \mathbf{v} = \mathbf{b}_2 \tag{A.26}
$$

where

$$
\mathbf{A}_2 := \mathbf{M}_2 \mathbf{A}_1 =
\begin{bmatrix}
a_{11} & a_{12} & a_{13} & \cdots & a_{1n} \\
0 & a_{22}^{(1)} & a_{23}^{(1)} & \cdots & a_{2n}^{(1)} \\
0 & 0 & a_{33}^{(2)} & \cdots & a_{3n}^{(2)} \\
\cdot & \cdot & \cdot & & \\
\cdot & \cdot & \cdot & & \\
\cdot & \cdot & \cdot & & \\
0 & 0 & a_{n3}^{(2)} & \cdots & a_{nn}^{(2)}
\end{bmatrix}, \qquad
\mathbf{b}_2 := \mathbf{M}_2 \mathbf{b}_1 =
\begin{bmatrix}
a_{1,n+1} \\
a_{2,n+1}^{(1)} \\
a_{3,n+1}^{(2)} \\
\cdot \\
\cdot \\
\cdot \\
a_{n,n+1}^{(2)}
\end{bmatrix}
$$

Proceeding forward, we will finally obtain

$$
\mathbf{U}\mathbf{v} = \mathbf{b}_{n-1}
$$

where

$$
\mathbf{U} := \mathbf{M}_{n-1} \cdots \mathbf{M}_2 \mathbf{M}_1 \mathbf{A} =
\begin{bmatrix}
a_{11} & a_{12} & a_{13} & \cdots & a_{1n} \\
0 & a_{22}^{(1)} & a_{23}^{(1)} & \cdots & a_{2n}^{(1)} \\
0 & 0 & a_{33}^{(2)} & \cdots & a_{3n}^{(2)} \\
\cdot & \cdot & \cdot & & \cdot \\
\cdot & \cdot & \cdot & & \cdot \\
\cdot & \cdot & \cdot & & \cdot \\
0 & 0 & 0 & \cdots & a_{nn}^{(n-1)}
\end{bmatrix},
\tag{A.27}
$$

and

$$
\mathbf{b}_{n-1} := \mathbf{M}_{n-1} \cdots \mathbf{M}_2 \mathbf{M}_1 \mathbf{b} =
\begin{bmatrix}
a_{1,n+1} \\
a_{2,n+1}^{(1)} \\
a_{3,n+1}^{(2)} \\
\cdot \\
\cdot \\
\cdot \\
a_{n,n+1}^{(n-1)}
\end{bmatrix}
\tag{A.28}
$$

This completes the triangularization of $\mathbf{A}$. Because $a_{ij}^{(k)}$ are no longer needed after the computation of $a_{ij}^{(k+1)}$, $a_{ij}^{(k)}$ can be overwritten by $a_{ij}^{(k+1)}$. In other words, we may store $a_{ij}^{(k+1)}$ in the position of $a_{ij}^{(k)}$. By so doing, it becomes unnecessary to keep track of the superscripts and the algorithm can be simplified. For easy reference, this procedure is summarized as an algorithm. Note that the vector $\mathbf{b}$ was considered as the $(n + 1)$th column of $\mathbf{A}$; therefore, we use $\mathbf{A}_a$ to denote the $n \times (n + 1)$ matrix.

Algorithm A Let $\mathbf{A}_a = [a_{ij}]$ be an $n \times (n + 1)$ matrix. The following steps carry out Gaussian elimination:

1. $k = 1$

2. $d_{ik} = \dfrac{a_{ik}}{a_{kk}}$, $i = k + 1, k + 2, \ldots, n$

3. $a_{ij} \leftarrow a_{ij} - d_{ik} a_{kj}$, $i = k + 1, k + 2, \ldots, n$; $j = k + 1, k + 2, \ldots, n + 1$

4. $k \leftarrow k + 1$

5. If $k < n$, go back to step 2.

The notation $E \leftarrow F$ means that E is replaced by F. The entries a_{11}, a_{22}, $\ldots$, a_{nn} are called the pivot elements or, simply, the pivots of elimination. In this algorithm, all pivots are assumed to be nonzero. This restriction will be removed in the next section. At the end of the algorithm, we obtain $\mathbf{U}\mathbf{v} = \mathbf{b}_{n-1}$, where $\mathbf{U}$ and $\mathbf{b}_{n-1}$ are shown in (A.27) and (A.28) with superscripts removed. The solution $\mathbf{v} = [v_1 \ v_2 \ldots v_n]'$ can then be obtained as

$$v_i = \frac{1}{a_{ii}}\left(a_{i,n+1} - \sum_{j=i+1}^{n} a_{ij}v_j\right), \quad i = n, \, n-1, \ldots, 1 \qquad \text{(A.29)}$$

This is the procedure of back substitution.

EXERCISE A.4.2

Verify (A.29) for $n = 3$.   ■

*A.4.2   **Hand calculation**

In Gaussian elimination, we choose diagonal elements, if they are nonzero, as pivots and carry out elimination in the order of the first column, second column, and so forth. This algorithm is not convenient for hand calculation, however. For hand calculation, we modify the algorithm as follows: First, we search for a nonzero element in the first row to eliminate all elements under it. The nonzero element is chosen for convenience of computation. If there is no nonzero element in the first row, we move to the second row. We then search for a nonzero element in the second row to eliminate all elements under it. If there is no nonzero element in the second row, we move to the third row. We repeat the process until the last row. Example A.4.1 illustrates.

EXAMPLE **A.4.1**

Consider

$$\begin{bmatrix} 3 & 2 & 5 & 2 \\ 1 & 3 & 3 & 0 \\ 1 & 7 & 2 & 0 \\ 3 & 4 & 8 & 4 \end{bmatrix} \mathbf{v} = \begin{bmatrix} 2 \\ -1 \\ 2 \\ 0 \end{bmatrix}, \qquad \mathbf{M}_1 = \begin{bmatrix} 1 & 0 & 0 & 0 \\ 0 & 1 & 0 & 0 \\ 0 & 0 & 1 & 0 \\ -2 & 0 & 0 & 1 \end{bmatrix}$$

Because the last column has two zeros, if we choose the last element of the first row as the pivot, the computation can be reduced by two-thirds. The elimination is equivalent to $\mathbf{M}_1\mathbf{A}\mathbf{v} = \mathbf{M}_1\mathbf{b}$, and yields

$$\begin{bmatrix} 3 & 2 & 5 & 2 \\ 1 & 3 & 3 & 0 \\ 1 & 7 & 2 & 0 \\ -3 & 0 & -2 & 0 \end{bmatrix} \mathbf{v} = \begin{bmatrix} 2 \\ -1 \\ 2 \\ -4 \end{bmatrix}, \qquad \mathbf{M}_2 = \begin{bmatrix} 1 & 0 & 0 & 0 \\ 0 & 1 & 0 & 0 \\ 0 & -1 & 1 & 0 \\ 0 & 3 & 0 & 1 \end{bmatrix}$$

In the second row, if we choose the second element as the pivot, the computation will involve fractions. If we choose the first element, although it will require one additional elimination, the computation will involve only integers. Therefore, we choose the first element as the pivot to yield

$$\begin{bmatrix} 3 & 2 & 5 & 2 \\ 1 & 3 & 3 & 0 \\ 0 & 4 & -1 & 0 \\ 0 & 9 & 7 & 0 \end{bmatrix} \mathbf{v} = \begin{bmatrix} 2 \\ -1 \\ 3 \\ -7 \end{bmatrix}, \qquad \mathbf{M}_3 = \begin{bmatrix} 1 & 0 & 0 & 0 \\ 0 & 1 & 0 & 0 \\ 0 & 0 & 1 & 0 \\ 0 & 0 & 7 & 1 \end{bmatrix}$$

The pivot of the third row is chosen as -1. Using the pivot, we can obtain

$$\begin{bmatrix} 3 & 2 & 5 & 2 \\ 1 & 3 & 3 & 0 \\ 0 & 4 & -1 & 0 \\ 0 & 37 & 0 & 0 \end{bmatrix} \begin{bmatrix} v_1 \\ v_2 \\ v_3 \\ v_4 \end{bmatrix} = \begin{bmatrix} 2 \\ -1 \\ 3 \\ 14 \end{bmatrix}$$

This completes the elimination. The solution can then be obtained by back substitution as

$$37v_2 = 14 \quad \text{or} \quad v_2 = \frac{14}{37} = 0.3784$$

$$4v_2 - v_3 = 3 \quad \text{or} \quad v_3 = 4v_2 - 3 = \frac{56}{37} - 3 = -1.4865$$

and so forth.

This procedure differs from Gaussian elimination only in that pivots are not necessarily chosen on the diagonal. Consequently, the elimination is not necessarily carried out in the natural order. For hand calculation, however, using this modified procedure is considerably simpler than using Gaussian elimination.

EXERCISE A.4.3

Find the solution of

$$\begin{bmatrix} 3 & 0 & -1 \\ -2 & 2 & 1 \\ 1 & 1 & 0 \end{bmatrix} \mathbf{v} = \begin{bmatrix} 2 \\ 1 \\ 0 \end{bmatrix}$$

[**ANSWER:** $\mathbf{v} = [-3 \quad 3 \quad -11]$.]

A.5   GAUSSIAN ELIMINATION WITH PARTIAL PIVOTING

Gaussian elimination will break down if one of the pivots becomes zero. We can modify the method to take care of this problem. The modification is also needed in order to improve its numerical properties, as will be discussed in Section A.6.

In Gaussian elimination with partial pivoting, we first search for the element with the largest magnitude in the first column and bring it to the (1, 1) position by permutation of rows. We then use it as a pivot to eliminate all elements under it. After this step, we disregard the first column and first row and repeat the procedure for the remaining matrix. If all elements in a first column are zero, we skip the step of elimination and continue the process. Example A.5.1 illustrates.

EXAMPLE **A.5.1**

Consider the equation

$$\begin{bmatrix} 0 & 0 & 3 & -2 \\ -2 & 2 & 1 & 4 \\ 1 & -1 & -1.5 & 1 \\ 2 & -2 & 3 & -2 \end{bmatrix} \mathbf{v} = \begin{bmatrix} 4 \\ 2 \\ 0 \\ 8 \end{bmatrix}$$

Because $a_{11} = 0$, Gaussian elimination cannot be used. But we can use Gaussian elimination with partial pivoting. The elements with the largest magnitude in the first column are -2 and 2. Either one can be used as a pivot. We choose -2 and interchange the first two rows to yield

$$\begin{bmatrix} -2 & 2 & 1 & 4 \\ 0 & 0 & 3 & -2 \\ 1 & -1 & -1.5 & 1 \\ 2 & -2 & 3 & -2 \end{bmatrix} \mathbf{v} = \begin{bmatrix} 2 \\ 4 \\ 0 \\ 8 \end{bmatrix}$$

We then eliminate all elements under the pivot -2 by elementary row operations to yield

$$\begin{bmatrix} -2 & 2 & 1 & 4 \\ 0 & 0 & 3 & -2 \\ 0 & 0 & -1 & 3 \\ 0 & 0 & 4 & 2 \end{bmatrix} \mathbf{v} = \begin{bmatrix} 2 \\ 4 \\ 1 \\ 10 \end{bmatrix}$$

We repeat the process by disregarding the first column and first row and consider the subequation bounded by dotted lines. Because all entries in the first column of the submatrix are zero, no elimination is needed in this step. We then disregard its first column and first row and consider the subequation bounded by solid lines. We interchange its two rows to bring the largest element 4 to the pivotal position as

$$\begin{bmatrix} -2 & 2 & 1 & 4 \\ 0 & 0 & 3 & -2 \\ 0 & 0 & 4 & 2 \\ 0 & 0 & -1 & 3 \end{bmatrix} \mathbf{v} = \begin{bmatrix} 2 \\ 4 \\ 10 \\ 1 \end{bmatrix}$$

Elimination of the element under the pivot 4 yields

$$\begin{bmatrix} 2 & 2 & 1 & 4 \\ 0 & 0 & 3 & -2 \\ 0 & 0 & 4 & 2 \\ 0 & 0 & 0 & 3.5 \end{bmatrix} \mathbf{v} = \begin{bmatrix} 2 \\ 4 \\ 10 \\ 3.5 \end{bmatrix}$$

This completes Gaussian elimination with partial pivoting. After this step, the solution can be obtained by back substitution. Gaussian elimination with partial pivoting can always be carried out to completion. This is an important algorithm and is widely used in practice.

*A.5.1   Development of the algorithm

Now let us discuss Gaussian elimination with partial pivoting for the general case and develop an algorithm for it. Look at (A.22) again. Before carrying out the elimination, we search for the element with the largest magnitude in the first column, that is, we find α_1 such that

$$|a_{\alpha_1,1}| \geq |a_{i1}|, \qquad i = 1, 2, \ldots, n$$

We then interchange the indices of the first row and the α_1th row as

$$a_{\alpha_1,j} \longleftrightarrow a_{1j}, \qquad j = 1, 2, \ldots, n$$

This step brings the element with the largest magnitude to the pivotal (1, 1) position. Once this step is completed, we use the $\mathbf{M}_1$ in (A.23) to eliminate the first column. Because the new pivot has the largest magnitude, we have $|d_{i1}| \leq 1$ for $i = 2, 3, \ldots, n$. Next we search for α_2 such that

$$|a_{\alpha_2,2}^{(1)}| \geq |a_{i2}^{(1)}|, \qquad i = 2, 3, \ldots, n$$

and interchange the indices of the second row and the α_2th row to bring the element with the largest magnitude to the pivotal (2, 2) position. We then use $\mathbf{M}_2$ in (A.25) to carry out the elimination. Again, we have $|d_{i2}| \leq 1$ for $i = 3$, $4, \ldots, n$. Proceeding forward, if a new pivot is zero, no elimination in that column is necessary and we may set the corresponding $\mathbf{M}_i$ as $\mathbf{I}$ and proceed to the next step. This procedure, listed below as an algorithm, is developed for an $m \times n$ matrix. If the algorithm is used to solve $\mathbf{Av} = \mathbf{b}$, we append $\mathbf{b}$ to $\mathbf{A}$, as we did in Algorithm A.

Algorithm B (**Gaussian elimination with partial pivoting**) Let $\mathbf{A} = [a_{ij}]$ be an $m \times n$ matrix.

1. $k = 1$
2. Find $\alpha_k \geq k$ such that $|a_{\alpha_k,k}| \geq |a_{ik}|$, $i = k, k + 1, \ldots, m$. If $a_{\alpha_k,k} = 0$, go to step 6.
3. $a_{kj} \longleftrightarrow a_{\alpha_k,j}$, $j = k, k + 1, \ldots, n$
4. $d_{ik} = \dfrac{a_{ik}}{a_{kk}}$, $i = k + 1, k + 2, \ldots, m$
5. $a_{ij} \leftarrow a_{ij} - d_{ik}a_{kj}$, $i = k + 1, k + 2, \ldots, m; j = k + 1, k + 2, \ldots, n$
6. $k \leftarrow k + 1$
7. If $k < n$, go back to step 2.

In actual programming, the condition $a_{\alpha_k,k} = 0$ in step 2 must be replaced by $|a_{\alpha_k,k}| < \varepsilon$, where ε is a very small positive number, say, 10^{-10}. This is due to the fact that a zero on a digital computer may appear as a very small number because of truncation or round-off errors. The choice of ε depends on the software and computer used. How to choose ε is a difficult problem in digital computer computation.

Now, to modify (A.27) to include the permutations, let $\mathbf{P}_{i,j}$ be the permutation matrix that interchanges the ith and jth rows. Using this notation, $\mathbf{E}_2$ in (A.13) becomes $\mathbf{P}_{2,4}$. In general, $\mathbf{P}_{i,j}$ is a unit matrix with its ith and jth diagonal elements moved to the (i, j)th and (j, i)th positions. For example, if the order of $\mathbf{P}_{2,5}$ is five, then

$$\mathbf{P}_{2,5} = \begin{bmatrix} 1 & 0 & 0 & 0 & 0 \\ 0 & 0 & 0 & 0 & 1 \\ 0 & 0 & 1 & 0 & 0 \\ 0 & 0 & 0 & 1 & 0 \\ 0 & 1 & 0 & 0 & 0 \end{bmatrix} \tag{A.30}$$

Using $\mathbf{P}_{i,j}$, Algorithm B can be written as

$$\mathbf{U} = \mathbf{M}_{n-1}\mathbf{P}_{n-1,\alpha_{n-1}} \cdots \mathbf{M}_2\mathbf{P}_{2,\alpha_2}\mathbf{M}_1\mathbf{P}_{1,\alpha_1}\mathbf{A} =: \mathbf{NA} \tag{A.31}$$

where

$$\mathbf{N} := \mathbf{M}_{n-1}\mathbf{P}_{n-1,\alpha_{n-1}} \cdots \mathbf{M}_2\mathbf{P}_{2,\alpha_2}\mathbf{M}_1\mathbf{P}_{1,\alpha_1}$$

This differs from (A.27) only in the insertion of permutation matrices. If $\mathbf{A}$ is square, then the form of $\mathbf{U}$ in (A.31) is identical to the one in (A.27), that is, $\mathbf{U}$ is an upper triangular matrix. In Algorithm A, all pivots are assumed to be nonzero; therefore, all diagonal elements of the $\mathbf{U}$ in (A.27) are nonzero. In Algorithm B, pivots may become zero; therefore, the $\mathbf{U}$ in (A.31) may have zero diagonal elements. Note that, if a zero pivot appears in Algorithm B, the column is already zero and the step of elimination can be skipped. For any matrix $\mathbf{A}$, then, Algorithm B can always be carried out to completion.

Equation (A.31) can be used to compute the inverse of square matrices. If all pivots of $\mathbf{U}$ are nonzero, then its inverse $\mathbf{U}^{-1}$ exists and is also an upper triangular matrix. Then, we have

$$\mathbf{A}^{-1} = \mathbf{U}^{-1}\mathbf{N} \tag{A.32}$$

To show the validity of (A.32), we must show $(\mathbf{U}^{-1}\mathbf{N})\mathbf{A} = \mathbf{I}$ and $\mathbf{A}(\mathbf{U}^{-1}\mathbf{N}) = \mathbf{I}$. The premultiplication of $\mathbf{U}^{-1}$ to (A.31) yields the first equality. To show the second equality, we note that all $\mathbf{P}_{ij}$ and $\mathbf{M}_k$ are elementary matrices and their inverses exist. Therefore, the $\mathbf{N}$ in (A.31) has an inverse, and (A.31) implies $\mathbf{A} = \mathbf{N}^{-1}\mathbf{U}$ (Problem A.18). Thus, we have

$$\mathbf{A}(\mathbf{U}^{-1}\mathbf{N}) = \mathbf{N}^{-1}\mathbf{U}\mathbf{U}^{-1}\mathbf{N} = \mathbf{I}$$

This establishes (A.32). In conclusion, if we apply Algorithm B to a square matrix and if all pivots are nonzero, the matrix is nonsingular and its inverse is given by (A.32). If one or more pivots are zero, the matrix is singular and its inverse does not exist.

*A.6   THE CONDITION OF SETS OF EQUATIONS AND THE STABILITY OF ALGORITHMS

This section discusses an intrinsic property of linear algebraic equations and compares Gaussian elimination with and without partial pivoting. Consider the set of linear algebraic equations

$$\begin{aligned} 1.0001v_1 + 2v_2 &= 3.0001 \\ v_1 + 2v_2 &= 3 \end{aligned} \tag{A.33}$$

Its solution is $v_1 = 1$ and $v_2 = 1$. Suppose small perturbations or variations in coefficients occur in the second equation and (A.33) becomes

$$\begin{aligned} 1.0001v_1 + 2v_2 &= 3.0001 \\ v_1 + 1.999v_2 &= 3.0001 \end{aligned} \tag{A.34}$$

Its solution can be computed as $v_1 = 3.74965\ldots$ and $v_2 = -0.37496\ldots$. Thus, a very small change in coefficients generates a vastly different solution. In other words, the solution of (A.33) is very sensitive to coefficient perturbations and the set of equations is said to be *ill-conditioned*.

Now consider the following different set of equations

$$\begin{aligned} 0.0001v_1 + 2v_2 &= 2.0001 \\ v_1 + 2v_2 &= 3 \end{aligned} \tag{A.35}$$

Its solution is $v_1 = 1$ and $v_2 = 1$. We introduce small perturbations into the second equation to yield

$$0.0001v_1 + 2v_2 = 2.0001$$
$$v_1 + 1.999v_2 = 3.0001$$

(A.36)

Its solution can be computed as $v_1 = 1.0011\ldots$ and $v_2 = 0.99999\ldots$. Although the amounts of perturbations in (A.34) and (A.36) are the same, the solution of (A.36) is very close to the one of (A.35). Therefore, the solution of (A.35) is not sensitive to coefficient perturbations and the equation in (A.35) is said to be *well-conditioned*.

The conditions of (A.33) and (A.35) can be explained graphically. Equation (A.33) represents two straight lines, as shown in Figure A.3(a). Their intersection yields the solution. Because the intersection angle of the two lines is very small, a small change in either line will affect the point of intersection greatly. Therefore, the solution of (A.33) is very sensitive to coefficient variations and (A.33) is ill-conditioned. On the other hand, the intersection angle of the two straight lines of (A.35) is fairly large, as shown in Figure A.3(b). Therefore, the point of intersection is not sensitive to coefficient variations and (A.35) is well-conditioned.

Determining the condition of a set of linear algebraic equations is not a simple task. See Reference [32]. Whether an equation is well- or ill-conditioned is an intrinsic property of the equation. It is independent of the method used to solve it. Given a set of linear algebraic equations, we may solve it using Gaussian elimination with or without partial pivoting or using other methods. If no numerical errors are introduced in the computation, the result will be identical no matter which method is used. However, the situation changes completely if numerical errors occur in the computation.

Errors often occur in digital computer computations. For example, the number 5/3 cannot be represented exactly in any digital computer because of its

Figure A.3   *Conditions of (A.33) and (A.35)*

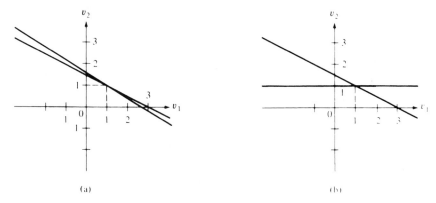

(a)   (b)

finite number of bits. The product $1.66667 \times 3.2568 = 5.4280108$ must be approximated by 5.42801 if only six significant digits are retained. Clearly, these errors, called truncation or round-off errors, must be accounted for in digital computer computations. A method of computation or an algorithm is said to be *numerically stable* if the effect of these errors on the solution is small; otherwise, it is said to be *numerically unstable*. Using Gaussian elimination (without pivoting) and Gaussian elimination with pivoting to solve the well-conditioned equation in (A.35) illustrates this concept.

Most digital computer computations are performed with floating-point arithmetic. Strictly speaking, floating-point numbers are represented in binary digits. For convenience, we shall use decimal digits. A number is represented as $\pm m \cdot 10^e$ with $0 \le m < 1$ in the floating-point mode. The integer e is called the *exponent*, and m is called the *mantissa*. For example, 2 is represented as 0.2×10^1, 0.02×10^2 or 0.002×10^3. In adding or subtracting two numbers, we must first set the exponents equal and then operate on the mantissas. In k-digit floating-point arithmetic, only k digits are retained in m by rounding off after every operation. For example, in 3-digit arithmetic, we have

$$0.210 \times 10^2 + 0.500 \times 10^4 = (0.0021 + 0.500) \times 10^4 = 0.502 \times 10^4$$

and

$$0.213 \times 10^2 \times 0.112 \times 10^3 = 0.023856 \times 10^5 = 0.239 \times 10^4$$

Given this preliminary discussion, we are ready to solve (A.35). We first rewrite it as

$$0.1 \times 10^{-3} v_1 + 0.2 \times 10 v_2 = 0.20001 \times 10 \qquad \text{(A.37a)}$$
$$0.1 \times 10 v_1 + 0.2 \times 10 v_2 = 0.3 \times 10 \qquad \text{(A.37b)}$$

Using 3-digit arithmetic, the application of Gaussian elimination (without partial pivoting) yields

$$0.100 \times 10^{-3} v_1 + 0.200 \times 10 v_2 = 0.200 \times 10$$
$$-0.200 \times 10^5 v_2 = -0.200 \times 10^5 \qquad \text{(A.38)}$$

The solution can be obtained by back substitution as $v_2 = 1$ and

$$0.100 \times 10^{-3} v_1 = 0.2 \times 10 - 0.2 \times 10 \times v_2 = 0$$

or $v_1 = 0$. This result is quite different from the actual solution $v_1 = v_2 = 1$. Therefore, Gaussian elimination (without partial pivoting) is not numerically stable.

Now let us try using Gaussian elimination with partial pivoting to solve (A.37). Using 3-digit arithmetic, we first interchange the order of the two equations in (A.37) as

$$0.1 \times 10v_1 + 0.2 \times 10v_2 = 0.3 \times 10 \tag{A.39a}$$
$$0.1 \times 10^{-3}v_1 + 0.2 \times 10v_2 = 0.20001 \times 10 \tag{A.39b}$$

The addition of the product of (A.39a) and (-10^{-4}) to (A.39b) yields

$$0.1 \times 10v_1 + 0.2 \times 10v_2 = 0.3 \times 10$$
$$0.2 \times 10v_2 = 0.2 \times 10 \tag{A.40}$$

The back substitution yields $v_2 = 1$ and

$$0.1 \times 10v_1 = 0.3 \times 10 - 0.2 \times 10 = 0.1 \times 10$$

or $v_1 = 1$. The result is identical to the actual solution $v_1 = v_2 = 1$.

This example shows that, using the same 3-digit arithmetic, Gaussian elimination yields an unacceptable result but Gaussian elimination with partial pivoting yields an excellent result. Thus, whereas Gaussian elimination (without partial pivoting) is not a numerically stable method, Gaussian elimination with partial pivoting is. The reason for the improvement can be explained roughly as follows: If $|a_{11}|$ in (A.22) is very small, then $d_{i1} = a_{i1}/a_{11}$ in (A.23) will be very large. Hence, if there are errors in $a_{ij}, j = 2, 3, \ldots, n$, using Gaussian elimination will greatly amplify the errors. However, if we carry out partial pivoting, then $|d_{i1}| = |a_{i1}/a_{11}| \leq 1$, for all i and the errors are reduced. Thus, Gaussian elimination with partial pivoting is a numerically stable method. A rigorous proof of the preceding assertions requires error analysis and can be found in Reference [32].

Other methods, such as Gaussian elimination with complete pivoting and Householder transformation, are numerically more stable than Gaussian elimination with partial pivoting. However, they require greater numbers of operations (addition and multiplication) and are more costly. Its simplicity makes Gaussian elimination with partial pivoting a more popular method for practical applications.

The condition of a set of equations and the stability of an algorithm are two independent concepts. Their possible combinations are given in Table A.1. By studying the table, we can conclude that, given a problem, well- or ill-conditioned, we should always use, if available, a stable method to solve it. If no stable method is available, we must exercise care in interpreting the computed result.

*A.6.1   Balancing

A process called scaling, balancing, or equalization is important in numerical computation. Consider the set of equations

$$0.1 \times 10v_1 + 0.2 \times 10^5v_2 = 0.20001 \times 10^5 \tag{A.41a}$$
$$0.1 \times 10v_1 + 0.2 \times 10v_2 = 0.3 \times 10 \tag{A.41b}$$

Table A.1 *Condition of Problems and Stability of Algorithms*

Problem	Unstable Algorithm	Stable Algorithm
Ill-conditioned	The result is generally unacceptable	The result will be as good as the data provided.
Well-conditioned	The result has to be carefully scrutinized. It may or may not be acceptable.	The result is generally good.

If we use 3-digit arithmetic and Gaussian elimination with partial pivoting to solve the equations, the result is $v_2 = 1$ and $v_1 = 0$. In fact, (A.41) is identical to (A.35) if (A.41a) is divided by 10^{-4}. Therefore, its actual solution is $v_1 = v_2 = 1$, and using Gaussian elimination with partial pivoting has not yielded a satisfactory solution for (A.41).

The largest coefficient associated with v_i in (A.41a) is 2×10^4; the largest coefficient associated with v_i in (A.41b) is 2. They differ by an order of 10^4. If the coefficients of $\mathbf{A}$ in $\mathbf{Av} = \mathbf{b}$ spread over a wide range, as in (A.41), the use of partial pivoting will not necessarily yield a good result. Therefore, the equations should be modified before they are solved. Note that an equation remains unchanged after the multiplication of a nonzero constant. For example, (A.41a) is equal to

$$0.1 \times 10^{-3} v_1 + 0.2 \times 10 v_2 = 0.20001 \times 10$$

Using this fact, we shall modify (A.22) before applying Gaussian elimination with partial pivoting. Let

$$a_{i,\max} = \max\{ |a_{ij}|, j = 1, 2, \ldots, n \}$$

It is the largest coefficient in magnitude in the ith row of $\mathbf{A}$. If $\{a_{1,\max}, a_{2,\max}, \ldots, a_{n,\max}\}$ are roughly equal or of the same order, we can apply Gaussian elimination with partial pivoting to solve the equation. If $\{a_{i,\max}, i = 1, 2, \ldots, n\}$ spread widely, we must first divide each equation by $a_{i,\max}$ so that the largest coefficients of all equations will be equal. This is the process called scaling, balancing, or equalization. After balancing, we can then use Gaussian elimination with partial pivoting, and the result we obtain will probably be improved. One way to avoid having to use balancing is to use the so-called *scaled* partial pivoting. See Reference [19].

A.7   LINEAR INDEPENDENCE AND RANK

In the preceding sections, we have discussed methods of solving linear algebraic equations without discussing whether or not solutions exist. In this and following sections, we will address this issue. Before proceeding, however, we need to introduce the concepts of linear independence and rank. Consider the equation $\mathbf{A}\mathbf{v} = \mathbf{b}$ in (A.7). Let $\mathbf{v} = [v_1 \; v_2 \; v_3]'$ and let $\mathbf{A} = [\mathbf{a}_1 \; \mathbf{a}_2 \; \mathbf{a}_3]$, where $\mathbf{a}_i$ denotes the ith column of $\mathbf{A}$, that is,

$$\mathbf{a}_1 = \begin{bmatrix} 4 \\ -1 \\ -3 \end{bmatrix}, \qquad \mathbf{a}_2 = \begin{bmatrix} -1 \\ 4 \\ -1 \end{bmatrix}, \qquad \mathbf{a}_3 = \begin{bmatrix} -3 \\ -1 \\ 5 \end{bmatrix}$$

Then, $\mathbf{A}\mathbf{v} = \mathbf{b}$ can be written as

$$\mathbf{a}_1 v_1 + \mathbf{a}_2 v_2 + \mathbf{a}_3 v_3 = \mathbf{b} \qquad \text{(A.42)}$$

Because v_i, $i = 1, 2, 3$, are scalars, they can be moved to the left-hand side of $\mathbf{a}_i$, and (A.42) becomes

$$v_1 \mathbf{a}_1 + v_2 \mathbf{a}_2 + v_3 \mathbf{a}_3 = \mathbf{b} \qquad \text{(A.43)}$$

Such an expression is called a linear combination. From this derivation, we see immediately that a solution $\mathbf{v}$ exists in $\mathbf{A}\mathbf{v} = \mathbf{b}$ if, and only if, $\mathbf{b}$ can be expressed as a linear combination of the columns of $\mathbf{A}$.

Consider a set of vectors $\mathbf{a}_i$, $i = 1, 2, \ldots, m$. If one of the vectors can be written as a linear combination of the others, then the set is said to be *linearly dependent*. If none of the vectors can be written as a linear combination of others, then the set is said to be *linearly independent*.

EXAMPLE A.7.1

Consider

$$[\, \mathbf{a}_1 \; \mathbf{a}_2 \; \mathbf{a}_3 \,] := \begin{bmatrix} 2 & -1 & 1 \\ 1 & 1 & 2 \\ 0 & 2 & 2 \end{bmatrix}$$

We have $\mathbf{a}_3 = \mathbf{a}_1 + \mathbf{a}_2$. Thus, the set of three vectors $\{\mathbf{a}_i, \; i = 1, 2, 3\}$ is linearly dependent.

EXAMPLE A.7.2

Consider the three vectors

$$\begin{bmatrix} 1 \\ 0 \\ 0 \end{bmatrix}, \quad \begin{bmatrix} 0 \\ 1 \\ 0 \end{bmatrix}, \quad \begin{bmatrix} 0 \\ 0 \\ 1 \end{bmatrix}$$

None of the vectors can be written as a linear combination of the other two vectors. Therefore, the three vectors are linearly independent.

Here is a formal definition of linear independence.

Definition A.1   A set of real vectors $\mathbf{a}_i$, $i = 1, 2, \ldots, n$, is said to be linearly dependent if there exist real numbers α_i, $i = 1, 2, \ldots, n$, not all zeros, such that

$$\alpha_1 \mathbf{a}_1 + \alpha_2 \mathbf{a}_2 + \ldots + \alpha_n \mathbf{a}_n = \mathbf{0} \tag{A.44}$$

If the only set of α_i for which (A.44) holds is $\alpha_i = 0$, $i = 1, 2, \ldots, n$, then the set is said to be linearly independent.

If not all α_i are zero in (A.44), then at least one α_i, say, $i = j$, is different from zero. Therefore, we can write (A.44) as

$$\mathbf{a}_j = \frac{-1}{\alpha_j}[\alpha_1 \mathbf{a}_1 + \cdots + \alpha_{j-1} \mathbf{a}_{j-1} + \alpha_{j+1} \mathbf{a}_{j+1} + \cdots + \alpha_n \mathbf{a}_n]$$

This shows that $\mathbf{a}_j$ is a linear combination of other vectors and the set is linearly dependent. If a set of vectors is linearly dependent, we can, generally, find many sets of α_i, not all zero, to meet (A.44). However, it is sufficient to find one set to conclude that the set is linearly dependent. Equation (A.44) always holds for $\alpha_i = 0$, $i = 1, 2, \ldots, n$. Therefore, to conclude that the set is linearly independent we must show that $\alpha_i = 0$, $i = 1, 2, \ldots, n$ is the *only* set of α_i for which (A.44) holds.

EXAMPLE A.7.3

The set of three vectors $\{\mathbf{a}_1, \mathbf{a}_2, \mathbf{a}_3\}$ with $\mathbf{a}_2 = \mathbf{0}$ is always linearly dependent because

$$\alpha_1 \mathbf{a}_1 + \alpha_2 \mathbf{a}_2 + \alpha_3 \mathbf{a}_3 = \mathbf{0}$$

with $\alpha_1 = 0$, $\alpha_2 = 1$, and $\alpha_3 = 0$.

EXERCISE A.7.1

Show that the set of one vector $\{\mathbf{a}\}$ is linearly independent if and only if $\mathbf{a} \neq \mathbf{0}$.

EXERCISE A.7.2

Is the following set of three vectors linearly independent?

$$\begin{bmatrix} 1 \\ 1 \\ 1 \end{bmatrix}, \quad \begin{bmatrix} 1 \\ 1 \\ 0 \end{bmatrix}, \quad \begin{bmatrix} 1 \\ 0 \\ 0 \end{bmatrix}$$

[**ANSWER:** Yes.]   ∎

The vectors $\mathbf{a}_i$ in Definition A.1 can be column or row vectors. Using the concept of linear independence, we can define the rank of matrices.

Definition A.2   The rank of a matrix is defined as the maximal number of linearly independent columns (or rows) in the matrix.

Because the concept of linear independence is defined for a set of vectors, this definition requires some explanation. Given a matrix, if the first column is a nonzero column, it is linearly independent by itself. If the set of the first two columns of the matrix is a dependent set, the second column can be written as a linear combination of the first column. We then delete the second column from subsequent consideration. If it is an independent set, we add the third column to the set. We repeat the process until the last column. Finally, we obtain a set that consists of only independent columns of the matrix. The number of the vectors in the set yields the rank of the matrix. The number of linearly independent columns can be shown to be equal to the number of linearly independent rows; therefore, either columns or rows can be used in the definition.

EXAMPLE A.7.4

Consider

$$\mathbf{A} = \begin{bmatrix} 1 & 0 & 0 & 2 & 4 \\ 0 & 1 & 0 & -1 & 3 \\ 0 & 0 & 1 & 3 & -3 \end{bmatrix} \tag{A.45}$$

The first three columns are, clearly, linearly independent. The fourth column can be expressed as a linear combination of the first three columns as

$$\begin{bmatrix} 2 \\ -1 \\ 3 \end{bmatrix} = 2 \begin{bmatrix} 1 \\ 0 \\ 0 \end{bmatrix} + (-1) \begin{bmatrix} 0 \\ 1 \\ 0 \end{bmatrix} + 3 \begin{bmatrix} 0 \\ 0 \\ 1 \end{bmatrix}$$

The fifth column can be similarly expressed. Thus, the maximal number of linearly independent columns in $\mathbf{A}$ is 3, and the rank of $\mathbf{A}$ is 3. The number of linearly independent rows can be shown to be equal to 3, too. Indeed, the equation

$$\beta_1[1 \ 0 \ 0 \ 2 \ 4] + \beta_2[0 \ 1 \ 0 \ -1 \ 3] + \beta_3[0 \ 0 \ 1 \ 3 \ -3]$$
$$= [\beta_1 \ \beta_2 \ \beta_3 \ (2\beta_1 - \beta_2 + 3\beta_3) \ (4\beta_1 + 3\beta_2 - 3\beta_3)] = [0 \ 0 \ 0 \ 0 \ 0]$$

implies $\beta_1 = 0$, $\beta_2 = 0$, and $\beta_3 = 0$. Therefore, the three rows of **A** are linearly independent. And, if we count the number of linearly independent rows, the rank of the matrix is again equal to 3.

Let **A** be an $m \times n$ matrix. Because the maximal number of linearly independent rows (columns) can never be larger than the total number of rows (columns), we have

$$\text{rank } \mathbf{A} \le \min\{m, n\} \tag{A.46}$$

If rank $\mathbf{A} = m$, the matrix **A** is said to have a full row rank. A necessary condition for **A** to have a full row rank is $n \ge m$. If $n < m$, then (A.46) implies rank $\mathbf{A} \le n$ and **A** cannot have a full row rank. In other words, in order to have a full row rank, **A** must have more columns than rows. Conversely, in order to have a full column rank, **A** must have more rows than columns. The matrix **A** in (A.45) has a full row rank but does not have a full column rank. A matrix, square or nonsquare, is said to have a full rank if it has a full column or row rank. If a square matrix has a full rank, then it has both a full column rank and a full row rank. For a nonsquare matrix, it is not possible to have a full column rank *and* a full row rank.

Consider the elementary matrix $\mathbf{E}_3$ in (A.13). The equation

$$\alpha_1 \begin{bmatrix} 1 \\ 0 \\ 0 \\ 0 \end{bmatrix} + \alpha_2 \begin{bmatrix} 0 \\ 1 \\ 0 \\ d \end{bmatrix} + \alpha_3 \begin{bmatrix} 0 \\ 0 \\ 1 \\ 0 \end{bmatrix} + \alpha_4 \begin{bmatrix} 0 \\ 0 \\ 0 \\ 1 \end{bmatrix} = \begin{bmatrix} \alpha_1 \\ \alpha_2 \\ \alpha_3 \\ d\alpha_2 + \alpha_4 \end{bmatrix} = \begin{bmatrix} 0 \\ 0 \\ 0 \\ 0 \end{bmatrix}$$

implies $\alpha_i = 0$, $i = 1, 2, 3, 4$. Thus, the four columns are linearly independent and the 4×4 matrix $\mathbf{E}_3$ has rank 4, or a full rank.

EXERCISE A.7.3

Show that the elementary matrices $\mathbf{E}_1$ and $\mathbf{E}_2$ in (A.13) each have a full rank. ■

In Section A.3, a square matrix was defined as nonsingular if its inverse exists. In the next section, we will show that every square matrix of full rank has an inverse and is, therefore, nonsingular.

*A.7.1   Computer computation of ranks

In this section, we shall modify Algorithm B to compute the rank of matrices. For convenience, we will assume **A** to be 5×8.

Suppose **A** becomes, after the first step of Algorithm B,

$$\begin{bmatrix} a_{11} & a_{12} & a_{13} & a_{14} & a_{15} & a_{16} & a_{17} & a_{18} \\ 0 & 0 & 0 & a_{24} & a_{25} & a_{26} & a_{27} & a_{28} \\ 0 & 0 & 0 & a_{34} & a_{35} & a_{36} & a_{37} & a_{38} \\ 0 & 0 & 0 & a_{44} & a_{45} & a_{46} & a_{47} & a_{48} \\ 0 & 0 & 0 & a_{54} & a_{55} & a_{56} & a_{57} & a_{58} \end{bmatrix} \tag{A.47}$$

At $k = 2$, because of $a_{i,2} = 0$ for $i \geq 2$, we choose 0 as a pivot and go to $k = 3$. Again, we have $a_{i3} = 0$ for $i \geq 3$. Therefore, we choose 0 as a pivot and go to $k = 4$. We then bring the larger of $|a_{44}|$ and $|a_{54}|$ to the (4, 4)th position and carry out elimination. Thus, the matrix **A** may become, after the completion of Algorithm B,

$$\begin{bmatrix} a_{11} & a_{12} & a_{13} & a_{14} & a_{15} & a_{16} & a_{17} & a_{18} \\ 0 & 0 & 0 & a_{24} & a_{25} & a_{26} & a_{27} & a_{28} \\ 0 & 0 & 0 & a_{34} & a_{35} & a_{36} & a_{37} & a_{38} \\ 0 & 0 & 0 & a_{44} & a_{45} & a_{46} & a_{47} & a_{48} \\ 0 & 0 & 0 & 0 & a_{55} & a_{56} & a_{57} & a_{58} \end{bmatrix} \tag{A.48}$$

We see that the pivots always appear on the (i, i)th position and that they may be zero or nonzero.

Now we shall make two modifications to the algorithm. The first modification is that all pivots will be chosen as nonzero and consequently can be normalized to 1. The normalization of pivots can be accomplished by simply dividing the row by the pivot element. For example, if we divide the first row of (A.47) by a_{11}, then the pivot becomes 1. The second modification is that, if no nonzero pivot exists in a column, we will search the columns on its right-hand side until we find a nonzero pivot. For the example in (A.47), at $k = 2$ of Algorithm B, we have $a_{i2} = 0$ for $i \geq 2$. Now, instead of going to $k = 3$, as in Algorithm B, we search the column $a_{i(2+1)}$ for $i \geq 2$. Because $a_{i,3} = 0$ for $i \geq 2$, we search the column $a_{i,4}$, $i \geq 2$. For the matrix in (A.47), suppose $a_{i,4}$, $i \geq 2$ are not all zero. We first carry out partial pivoting to bring the element with the largest magnitude to the (2, 4)th position and then carry out elimination. Proceeding forward, we finally obtain

$$\mathbf{U} = \begin{bmatrix} 1 & a_{12} & a_{13} & a_{14} & a_{15} & a_{16} & a_{17} & a_{18} \\ 0 & 0 & 0 & 1 & a_{25} & a_{26} & a_{27} & a_{28} \\ 0 & 0 & 0 & 0 & 1 & a_{36} & a_{37} & a_{38} \\ 0 & 0 & 0 & 0 & 0 & 0 & 1 & a_{48} \\ 0 & 0 & 0 & 0 & 0 & 0 & 0 & 0 \end{bmatrix} \tag{A.49}$$

We call this matrix a *staircase* matrix. It has the following features:

1. All nonzero rows appear in the upper part.
2. The left-most nonzero element of each nonzero row is a pivot. The position of the pivot of each row is on the right-hand side of the pivot of its previous row.
3. All entries under and on the left-hand side of every pivot are zero.

Note that every column has, at most, one pivot. Every unit upper triangular matrix is a staircase matrix.

Algorithm C Let $\mathbf{A} = [a_{ij}]$ be an $m \times n$ matrix. The algorithm transforms $\mathbf{A}$ into a staircase matrix.

1. $k = 1$
2. $l = 0$
3. Find $\alpha_k \geq k$ such that $|a_{\alpha_k, k+l}| = \max\{|a_{i,k+l}|, i = k, k+1, \ldots, m\}$. If $|a_{\alpha_k,k+l}| \leq \varepsilon$, $l \leftarrow l + 1$ and repeat step 3.
4. $a_{kj} \longleftrightarrow a_{\alpha_k, j}, j = k + l, k + l + 1, \ldots, n$
5. $a_{kj} \leftarrow \dfrac{a_{kj}}{a_{k,k+l}}, j = k + l, k + l + 1, \ldots, n$ (This step normalizes the pivot to 1.)
6. $a_{ij} \leftarrow a_{ij} - a_{ik}a_{kj}, i = k + 1, k + 2, \ldots, m; j = k + l + 1, k + l + 2, \ldots, n$
7. $k \leftarrow k + 1$
8. If $k < m$, go back to step 3.

Now consider the computer computation of the rank of a matrix $\mathbf{A}$. Using Algorithm C, we transform $\mathbf{A}$ into a staircase matrix $\mathbf{U}$. Because elementary row operations do not change the rank of a matrix, we have

$$\text{rank } \mathbf{A} = \text{rank } \mathbf{U}$$

Because of its staircase structure, the rank of $\mathbf{U}$ can easily be shown to be equal to its number of nonzero rows or its number of pivots. Therefore, once $\mathbf{A}$ is transformed into a staircase matrix, its rank can be readily determined.

EXAMPLE A.7.5

Consider a matrix that has been transformed, using Algorithm B, into

$$\begin{bmatrix} 1 & 3 & 0 & 1 \\ 0 & 0 & 1 & 2 \\ 0 & 0 & 1 & 3 \\ 0 & 0 & 0 & 0 \end{bmatrix}$$

It has two nonzero pivots, but it is not in a staircase form. Therefore, its rank is not necessarily equal to the number of nonzero pivots. If we use Algorithm C, the matrix can be transformed into

$$\begin{bmatrix} 1 & 3 & 0 & 1 \\ 0 & 0 & 1 & 2 \\ 0 & 0 & 0 & 1 \\ 0 & 0 & 0 & 0 \end{bmatrix}$$

It is in a staircase form, and it has three nonzero rows or three nonzero pivots. Therefore, the rank of the matrix is 3.

This example has shown that Algorithm B cannot be used to compute the rank of a matrix. However, it can be used to determine whether or not a *square* matrix has a full rank. In other words, if we apply Algorithm B to a square matrix and all pivots are nonzero, the matrix has a full rank. If one or more pivots are zero, the matrix does not have a full rank. However, we cannot tell what its rank is without additional computation, as Example A.7.5 illustrated.

EXAMPLE **A.7.6**

Consider

$$\mathbf{A} = \begin{bmatrix} 3 & -1 & 2 & 1 \\ 5 & 2 & 1 & 0 \\ -2 & -3 & 1 & 1 \\ 7 & 2 & -2 & 3 \\ 6 & 6 & -5 & 1 \end{bmatrix} \tag{A.50}$$

The application of Algorithm C to matrix **A** yields, using a personal computer,

$$\mathbf{U} = \begin{bmatrix} 1 & 0.2857 & -0.2857 & 0.4285 \\ 0 & 1 & -0.7666 & -0.3666 \\ 0 & 0 & 1 & -0.6744 \\ 0 & 0 & 0 & -3.0 \times 10^{-14} \\ 0 & 0 & 0 & 4.0 \times 10^{-14} \end{bmatrix}$$

The last two numbers in the last column are smaller than 10^{-10} and will be considered as zeros. Therefore, **U** has three nonzero rows and **A** has rank 3.

Digital computer computation of the rank of a matrix can be difficult. For example, the rank of the matrix

$$\begin{bmatrix} 7 & 14/3 \\ 3 & 2 \end{bmatrix} \tag{A.51}$$

is 1. However, if 14/3 is approximated as 4.66, then the matrix

$$\begin{bmatrix} 7 & 4.66 \\ 3 & 2 \end{bmatrix} \tag{A.52}$$

has rank 2. Thus, the rank of the matrix in (A.51) may change from 1 to 2 if there are perturbations in its elements or if there are round-off errors in the computation. Therefore, the computation of the rank is an ill-conditioned problem. Although Gaussian elimination with partial pivoting is numerically stable, care must be exercised in using it to compute the rank of matrices. The most reliable method for computing the rank of matrices is singular-value decomposition. For more detailed discussions of this method, see References [4] and [32].

The importance of staircase matrices lies not so much in their role in computing rank as in their application in system design. The staircase matrix can be transformed, by additional elementary row operations, into the echelon-form matrix. This form is very important in the design and identification of multivariable systems. See References [4] and [5].

*A.7.2   Hand calculation of ranks

The procedure presented in Section A.4.2 can be used to compute the rank of a matrix $\mathbf{A}$ by hand. At the end of the procedure, the rank of $\mathbf{A}$ is equal to the number of nonzero rows in the resulting matrix. This procedure is illustrated in the following example.

EXAMPLE **A.7.7**

Consider the matrix in (A.50) or

$$\mathbf{A} = \begin{bmatrix} 3 & -1 & 2 & 1 \\ 5 & 2 & 1 & 0 \\ -2 & -3 & 1 & 1 \\ 7 & 2 & -2 & 3 \\ 6 & 6 & -5 & 1 \end{bmatrix} \tag{A.53}$$

We choose $a_{14} = 1$ in the first row as a pivot and eliminate all entries under it to yield

$$\mathbf{A}_1 := \mathbf{M}_1 \mathbf{A} := \begin{bmatrix} 1 & 0 & 0 & 0 & 0 \\ 0 & 1 & 0 & 0 & 0 \\ -1 & 0 & 1 & 0 & 0 \\ -3 & 0 & 0 & 1 & 0 \\ -1 & 0 & 0 & 0 & 1 \end{bmatrix} \mathbf{A} = \begin{bmatrix} 3 & -1 & 2 & 1 \\ 5 & 2 & 1 & 0 \\ -5 & -2 & -1 & 0 \\ -2 & 5 & -8 & 0 \\ 3 & 7 & -7 & 0 \end{bmatrix}$$

Next, we choose $a_{23} = 1$ in the second row as a pivot and eliminate all entries under it to yield

$$\mathbf{A}_2 := \mathbf{M}_2\mathbf{A}_1 := \begin{bmatrix} 1 & 0 & 0 & 0 & 0 \\ 0 & 1 & 0 & 0 & 0 \\ 0 & 1 & 1 & 0 & 0 \\ 0 & 8 & 0 & 1 & 0 \\ 0 & 7 & 0 & 0 & 1 \end{bmatrix} \mathbf{A}_1 = \begin{bmatrix} 3 & -1 & 2 & 1 \\ 5 & 2 & 1 & 0 \\ 0 & 0 & 0 & 0 \\ 38 & 21 & 0 & 0 \\ 38 & 21 & 0 & 0 \end{bmatrix}$$

The third row is a zero row, and so we skip it or, equivalently, we set $\mathbf{M}_3 = \mathbf{I}$ and $\mathbf{A}_3 = \mathbf{M}_3\mathbf{A}_2 = \mathbf{A}_2$. We then choose a nonzero element in the fourth row and eliminate the entry under it to yield

$$\mathbf{U} := \mathbf{A}_4 := \mathbf{M}_4\mathbf{A}_3 := \begin{bmatrix} 1 & 0 & 0 & 0 & 0 \\ 0 & 1 & 0 & 0 & 0 \\ 0 & 0 & 1 & 0 & 0 \\ 0 & 0 & 0 & 1 & 0 \\ 0 & 0 & 0 & -1 & 1 \end{bmatrix} \mathbf{A}_3 = \begin{bmatrix} 3 & -1 & 2 & 1 \\ 5 & 2 & 1 & 0 \\ 0 & 0 & 0 & 0 \\ 38 & 21 & 0 & 0 \\ 0 & 0 & 0 & 0 \end{bmatrix}$$

The last row happens to be a zero row. We have computed $\mathbf{U} = \mathbf{M}_4\mathbf{M}_3\mathbf{M}_2\mathbf{M}_1\mathbf{A}$, where all $\mathbf{M}_i$ are elementary matrices. The matrix $\mathbf{U}$ has three nonzero rows; therefore, $\mathbf{A}$ has rank 3.

A.8   THE EXISTENCE OF SOLUTIONS

Having defined the concept of rank, we are ready to discuss the existence of solutions in linear algebraic equations. Consider

$$\mathbf{Av} = \mathbf{b} \tag{A.54}$$

where $\mathbf{A}$ is an $m \times n$ matrix, $\mathbf{v}$ is an $n \times 1$ vector, and $\mathbf{b}$ is an $m \times 1$ vector. The problem is to find $\mathbf{v}$ for a given $\mathbf{A}$ and $\mathbf{b}$. The number of equations in (A.54) is m, and the number of unknowns is n. In general, the number of equations can be larger than, equal to, or smaller than the number of unknowns.

Three situations may occur in (A.54):

1. Given $\mathbf{A}$ and $\mathbf{b}$, no $\mathbf{v}$ exists to meet $\mathbf{Av} = \mathbf{b}$. In this case, the equation is said to be inconsistent. For example, the equation

$$\begin{bmatrix} 2 & 2 & 4 \\ 1 & 1 & 2 \end{bmatrix} \begin{bmatrix} v_1 \\ v_2 \\ v_3 \end{bmatrix} = \begin{bmatrix} 2 \\ 0 \end{bmatrix}$$

is inconsistent, because $2(v_1 + v_2 + 2v_3) = 2$ and $v_1 + v_2 + 2v_3 = 0$ cannot hold simultaneously. Although the set has more unknowns than equations, it has no solution.

2. Given **A** and **b**, one and only one solution **v** exists to meet **Av** = **b**.

3. Given **A** and **b**, many solutions exist to meet **Av** = **b**.

A set of linear algebraic equations that meets item 2 or 3 is said to be consistent. In this section, we will discuss the condition for equations to be consistent. We will use [**A** : **b**] to denote the $m \times (n + 1)$ matrix with **b** appended to **A** as an additional column.

Theorem A.1 Equation (A.54) has a solution **v** if and only if

$$\text{rank } ([\mathbf{A} : \mathbf{b}]) = \text{rank } (\mathbf{A})$$

PROOF If there exists a **v** such that **Av** = **b**, then **b** can, as discussed in (A.43), be written as a linear combination of the columns of **A**. Consequently, **b** is linearly dependent on the columns of **A**. Thus, we have rank ([**A** : **b**]) = rank (**A**).

Conversely, if rank ([**A** : **b**]) = rank (**A**), then **b** can be written as a linear combination of the columns of **A**. Thus, there exists a **v** such that **Av** = **b**. This establishes the theorem.

EXAMPLE A.8.1

Consider

$$\mathbf{A}_1 \mathbf{v} := \begin{bmatrix} 1 & 1 \\ 1 & 1 \end{bmatrix} \begin{bmatrix} v_1 \\ v_2 \end{bmatrix} = \begin{bmatrix} 1 \\ 0 \end{bmatrix} \tag{A.55}$$

Because

$$\text{rank } \begin{bmatrix} 1 & 1 \\ 1 & 1 \end{bmatrix} = 1 \neq \text{rank } \begin{bmatrix} 1 & 1 & 1 \\ 1 & 1 & 0 \end{bmatrix} = 2$$

no solution exists in the equation.

Corollary A.1 Let **A** be a square matrix. Then, for *any* **b**, **Av** = **b** has a solution **v** if and only if **A** is nonsingular.

PROOF If **A** has a full row rank, the matrix [**A** : **b**] will have a full row rank no matter what **b** is. Hence, a solution will exist for any **b**. If **A** does not have a full row rank, we can always find a **b** to violate the rank condition in Theorem A.1. Therefore, the equation will not have a solution for that **b**. This establishes the corollary.

EXAMPLE **A.8.2**

Consider

$$\mathbf{A}_2 \mathbf{v} := \begin{bmatrix} 1 & 1 & 1 \\ 1 & 1 & 0 \\ 1 & 1 & 0 \end{bmatrix} \begin{bmatrix} v_1 \\ v_2 \\ v_3 \end{bmatrix} = \mathbf{b} \tag{A.56}$$

Clearly, $\mathbf{A}_2$ has rank 2, not a full rank. Therefore, the equation does not have a solution for every $\mathbf{b}$. For example, if $\mathbf{b} = [4\ 2\ 2]'$, then $\mathbf{v} = [1\ 1\ 2]'$ or $[2\ 0\ 2]'$ is a solution. However, if $\mathbf{b} = [0\ 2\ 3]'$, then (A.56) has no solution.

EXERCISE A.8.1

Consider (A.56). Under what condition on $\mathbf{b} = [b_1\ b_2\ b_3]'$ will a solution exist?

[**ANSWER:** $b_2 = b_3$.]   ■

To employ Theorem A.1 or its corollary requires computation of the rank of matrices. This is not a simple task. When given a set of linear equations, we shall proceed to solve it without first checking for the existence of solutions. At the end of the procedure, we will obtain either a solution or a set of inconsistent equations. If we obtain the former, the rank condition in Theorem A.1 is met; if we obtain the latter, it is not. Therefore, Theorem A.1 is not really used to solve linear algebraic equations. It is used for establishing theoretical results.

EXAMPLE **A.8.3**

Gaussian elimination can be used to solve the equation in (A.55). The premultiplication of the elementary matrix

$$\begin{bmatrix} 1 & 0 \\ -1 & 1 \end{bmatrix}$$

to (A.55) yields

$$\begin{bmatrix} 1 & 1 \\ 0 & 0 \end{bmatrix} \begin{bmatrix} v_1 \\ v_2 \end{bmatrix} = \begin{bmatrix} 1 \\ -1 \end{bmatrix}$$

which implies $v_1 + v_2 = 1$ and $0 \cdot v_1 + 0 \cdot v_2 = -1$. The latter equation can never hold. Thus, no solution exists in (A.55) and the condition in Theorem A.1 is not satisfied. Thus, Theorem A.1 can be easily checked at the end of the algorithm.

From the preceding discussion, we see that the existence of solutions depends only on the rank condition stated in Theorem A.1. It is independent of whether or not the number of unknowns is larger than the number of equations. A set of linear algebraic equations may have a solution even though its number of equations is larger than the number of unknowns. Similarly, a set of linear algebraic equations may not have a solution even though the number of unknowns is larger than the number of equations.

Before concluding this section, let us discuss the solution of $\mathbf{Av} = \mathbf{0}$. The equation $\mathbf{Av} = \mathbf{b}$ with $\mathbf{b} = \mathbf{0}$ is called a *homogeneous equation*. A solution $\mathbf{v}$ always exists in $\mathbf{Av} = \mathbf{0}$ because the condition rank $[\mathbf{A}]$ = rank $[\mathbf{A} : \mathbf{0}]$ is always satisfied. Clearly, $\mathbf{v} = \mathbf{0}$ is a solution no matter what $\mathbf{A}$ is. This solution is called a *trivial solution*. A nonzero $\mathbf{v}$ that satisfies $\mathbf{Av} = \mathbf{0}$ is called a *nontrivial solution*. A nontrivial solution may or may not exist in $\mathbf{Av} = \mathbf{0}$; its existence condition for square $\mathbf{A}$ is discussed below. The general case will not be discussed in this text.

Theorem A.2 Let $\mathbf{A}$ be a square matrix. Then, a nontrivial solution $\mathbf{v} \neq \mathbf{0}$ exists in $\mathbf{Av} = \mathbf{0}$ if and only if $\mathbf{A}$ is singular.

The matrix $\mathbf{A}$ is singular if and only if its columns are linearly dependent. Using this fact, the theorem can be readily established. This is left as an exercise.

EXAMPLE **A.8.4**

Consider

$$\begin{bmatrix} 1 & 2 & 5 \\ 0 & 1 & 1 \\ 0 & 0 & 0 \end{bmatrix} \begin{bmatrix} v_1 \\ v_2 \\ v_3 \end{bmatrix} = \mathbf{0}$$

The square matrix is in the staircase form; therefore, its rank is 2. Thus, the matrix does not have a full rank and is singular. We write the equation explicitly as

$$\begin{aligned} v_1 + 2v_2 + 5v_3 &= 0 \\ v_2 + v_3 &= 0 \end{aligned}$$

which imply $v_2 = -v_3$ and $v_1 = -3v_3$. Therefore, $\mathbf{v} = \begin{bmatrix} -3 & -1 & 1 \end{bmatrix}'$ or $\begin{bmatrix} -3v_3 & -v_3 & v_3 \end{bmatrix}'$ for any $v_3 \neq 0$ is a nontrivial solution.

EXERCISE A.8.2

Find a nontrivial solution, if one exists, in

$$\begin{bmatrix} 1 & 2 \\ -3 & -6 \end{bmatrix} \mathbf{v} = \mathbf{0}$$

[**ANSWER:** $[-2 \ \ 1]'$.]   ■

A.9   THE DETERMINANT AND RELATED TOPICS

The theory of determinants was developed much earlier than the theory of matrices. Its importance, however, has been diminishing in recent years. For example, the concept of the rank of a matrix was originally developed by using the determinant. However, it has become more convenient, conceptually and computationally, to define the rank by using the concept of linear independence of vectors. Because of this trend, the discussion of the theory of determinants in this section will not be exhaustive. Only those results useful in system analysis will be discussed.

Consider a square matrix $\mathbf{A} = [a_{ij}]$ of order n. Associated with each entry a_{ij}, we define a submatrix $\mathbf{A}_{ij}$ of $\mathbf{A}$ by deleting the ith row and jth column. Clearly, $\mathbf{A}_{ij}$ is a square matrix of order $(n-1)$. We use det $\mathbf{A}$ or $|\mathbf{A}|$ to denote the *determinant* of $\mathbf{A}$. It is defined recursively for $n = 1, 2, 3, \ldots,$ by using the formula

$$\det \mathbf{A} := |\mathbf{A}| := \sum_{i=1}^{n} (-1)^{i+j} a_{ij} |\mathbf{A}_{ij}| \qquad \text{(A.57a)}$$

for $j = 1$, or $2, \ldots,$ or n, or the formula

$$\det \mathbf{A} := |\mathbf{A}| := \sum_{j=1}^{n} (-1)^{i+j} a_{ij} |\mathbf{A}_{ij}| \qquad \text{(A.57b)}$$

for $i = 1$, or $2, \ldots,$ or n, with det $a := a$, for any 1×1 matrix a. The formula in (A.57) is often referred to as *Laplace's expansion*.[6] The expansion can be carried out along any row or any column. For $n = 2$, we have

$$\det \begin{bmatrix} a_{11} & a_{12} \\ a_{21} & a_{22} \end{bmatrix} = (-1)^{1+1} a_{11} \cdot \det a_{22} + (-1)^{1+2} a_{12} \cdot \det a_{21}$$

$$= a_{11} a_{22} - a_{12} a_{21} \qquad \text{(A.58)}$$

This equals the product of the entries along the solid line minus the product of the entries along the dotted line. For $n = 3$, we have

---

[6] In honor of French mathematician Pierre-Simon Laplace (1749–1827). He considered mathematics to be merely a kit of tools used to explain nature. The transform discussed in Chapter 4 also carries his name.

$$\det \begin{bmatrix} a_{11} & a_{12} & a_{13} \\ a_{21} & a_{22} & a_{23} \\ a_{31} & a_{32} & a_{33} \end{bmatrix} = (-1)^{1+1} a_{11} \cdot \det \begin{bmatrix} a_{22} & a_{23} \\ a_{32} & a_{33} \end{bmatrix} + (-1)^{1+2} a_{12} \cdot \det \begin{bmatrix} a_{21} & a_{23} \\ a_{31} & a_{33} \end{bmatrix}$$

$$+ (-1)^{1+3} a_{13} \cdot \det \begin{bmatrix} a_{21} & a_{22} \\ a_{31} & a_{32} \end{bmatrix}$$

$$= a_{11}(a_{22}a_{33} - a_{23}a_{32}) - a_{12}(a_{21}a_{33} - a_{23}a_{31})$$
$$+ a_{13}(a_{21}a_{32} - a_{22}a_{31})$$
$$= a_{11}a_{22}a_{33} + a_{12}a_{23}a_{31} + a_{13}a_{21}a_{32}$$
$$- (a_{11}a_{23}a_{32} + a_{12}a_{21}a_{33} + a_{13}a_{22}a_{31}) \qquad \text{(A.59)}$$

This equals the sum of the products of the entries along the three solid lines minus the sum of the products of the entries along the three dotted lines.

EXERCISE A.9.1

Compute the determinants of

$$\mathbf{A}_1 = \begin{bmatrix} 1 & -1 \\ 2 & 3 \end{bmatrix}, \qquad \mathbf{A}_2 = \begin{bmatrix} 1 & -1 & 2 \\ 2 & 1 & 1 \\ 3 & 0 & 2 \end{bmatrix}, \qquad \mathbf{A}_3 = \begin{bmatrix} -2 & 1 & 3 \\ 1 & 2 & 1 \\ 3 & 0 & -3 \end{bmatrix}$$

[**ANSWERS:** 5, −3, 0.] ∎

For $n \geq 4$, the formula in (A.57) is rarely used directly, except for matrices with many zero elements along some rows or columns. For example, consider

$$\det \begin{bmatrix} a_{11} & a_{12} & a_{13} & 0 \\ 0 & a_{22} & 0 & 0 \\ a_{31} & a_{32} & a_{33} & 0 \\ a_{41} & a_{42} & a_{43} & a_{44} \end{bmatrix} =: \det \mathbf{A} \qquad \text{(A.60)}$$

First, we expand it along the fourth column to obtain

$$\det \mathbf{A} = (-1)^{4+4} a_{44} \cdot \det \begin{bmatrix} a_{11} & a_{12} & a_{13} \\ 0 & a_{22} & 0 \\ a_{31} & a_{32} & a_{33} \end{bmatrix} =: a_{44} \det \mathbf{A}_{44}$$

We then expand $\mathbf{A}_{44}$ along its second row to obtain

$$\det \mathbf{A} = a_{44} \left\{ (-1)^{2+2} a_{22} \cdot \det \begin{bmatrix} a_{11} & a_{13} \\ a_{31} & a_{33} \end{bmatrix} \right\} = a_{44} a_{22}(a_{11}a_{33} - a_{13}a_{31})$$

This technique is often used in the hand calculation of determinants.

EXERCISE A.9.2

Show

$$
\det \begin{bmatrix} a_{11} & a_{12} & a_{13} & a_{14} \\ 0 & a_{22} & a_{23} & a_{24} \\ 0 & 0 & a_{33} & a_{34} \\ 0 & 0 & 0 & a_{44} \end{bmatrix} = a_{11}a_{22}a_{33}a_{44} \qquad \textbf{(A.61)}
$$

■

   In general, the determinants of any unit matrix, any unit lower triangular matrix, and any unit upper triangular matrix of any order are all equal to 1.

EXERCISE A.9.3

Show

$$
\det \begin{bmatrix} a_{11} & a_{12} & a_{13} & a_{14} \\ 0 & 0 & a_{23} & 0 \\ a_{31} & a_{32} & a_{33} & 0 \\ a_{41} & 0 & a_{43} & 0 \end{bmatrix} = a_{14}a_{23}a_{32}a_{41}
$$

■

EXERCISE A.9.4

Let $\mathbf{P}_{i,j}$ be the permutation matrix that interchanges the ith and jth rows. Show

$$
\det \mathbf{P}_{2,4} = \det \begin{bmatrix} 1 & 0 & 0 & 0 \\ 0 & 0 & 0 & 1 \\ 0 & 0 & 1 & 0 \\ 0 & 1 & 0 & 0 \end{bmatrix} = -1, \quad \det \mathbf{P}_{1,2} = \det \begin{bmatrix} 0 & 1 & 0 & 0 \\ 1 & 0 & 0 & 0 \\ 0 & 0 & 1 & 0 \\ 0 & 0 & 0 & 1 \end{bmatrix} = -1 \quad \textbf{(A.62)}
$$

In general, $\det \mathbf{P}_{i,j} = -1$ if $i \neq j$, and $\det \mathbf{P}_{i,i} = \det \mathbf{I} = 1$. ■

   Let $\mathbf{M}$ and $\mathbf{N}$ be any square matrices of the same order. Then, we have

$$
\det(\mathbf{MN}) = \det \mathbf{M} \cdot \det \mathbf{N} \qquad \textbf{(A.63)}
$$

This is an extremely useful formula. It is proved in References [18] and [20].
   It is sometimes convenient to partition a matrix into submatrices or to combine a number of matrices to form a composite matrix. For example, we may write

$$
\begin{bmatrix} \overline{\mathbf{A}}_{11} & \overline{\mathbf{A}}_{12} \\ \mathbf{A}_{21} & \mathbf{A}_{22} \end{bmatrix} = \begin{bmatrix} a_{11} & a_{12} & a_{13} & a_{14} \\ a_{21} & a_{22} & a_{23} & a_{24} \\ a_{31} & a_{32} & a_{33} & a_{34} \\ a_{41} & a_{42} & a_{43} & a_{44} \end{bmatrix} = \begin{bmatrix} \mathbf{A}_{11} & \mathbf{A}_{12} \\ \mathbf{A}_{21} & \mathbf{A}_{22} \end{bmatrix}
$$

where all $\overline{\mathbf{A}}_{ij}$ are 2×2 matrices bounded by the solid lines and all $\mathbf{A}_{ij}$ are bounded by the dotted lines. Clearly, $\mathbf{A}_{11}$ is a 1×1 matrix, $\mathbf{A}_{12}$ is a 1×3 matrix, $\mathbf{A}_{21}$ is a 3×1 matrix, and $\mathbf{A}_{22}$ is a 3×3 matrix. The rules of addition and multiplication discussed in Section A.3 are applicable to composite matrices. For example, consider the multiplication of a 2×4 matrix $\mathbf{A}$ and a 4×6 matrix $\mathbf{B}$. We may partition them and carry out multiplication as follows:

$$\begin{matrix} {\scriptstyle 1 \times 2 \ 1 \times 2} \\ \begin{bmatrix} \mathbf{A}_{11} & \mathbf{A}_{12} \\ \mathbf{A}_{21} & \mathbf{A}_{22} \end{bmatrix} \\ {\scriptstyle 1 \times 2 \ 1 \times 2} \end{matrix} \cdot \begin{matrix} {\scriptstyle 2 \times 2 \ 2 \times 4} \\ \begin{bmatrix} \mathbf{B}_{11} & \mathbf{B}_{12} \\ \mathbf{B}_{21} & \mathbf{B}_{22} \end{bmatrix} \\ {\scriptstyle 2 \times 2 \ 2 \times 4} \end{matrix} = \begin{matrix} {\scriptstyle 1 \times 2} \quad\quad {\scriptstyle 1 \times 4} \\ \begin{bmatrix} \mathbf{A}_{11}\mathbf{B}_{11} + \mathbf{A}_{12}\mathbf{B}_{21} & \mathbf{A}_{11}\mathbf{B}_{12} + \mathbf{A}_{12}\mathbf{B}_{22} \\ \mathbf{A}_{21}\mathbf{B}_{11} + \mathbf{A}_{22}\mathbf{B}_{21} & \mathbf{A}_{21}\mathbf{B}_{12} + \mathbf{A}_{22}\mathbf{B}_{22} \end{bmatrix} \\ {\scriptstyle 1 \times 2} \quad\quad {\scriptstyle 1 \times 4} \end{matrix}$$

The order of each submatrix is indicated. It is important that the orders of matrices in all operations be compatible. The rule of multiplication applied here is the same as that applied in (A.11).

Consider the matrices

$$\mathbf{M}_1 = \begin{bmatrix} \mathbf{I} & 0 \\ \mathbf{C} & \mathbf{A} \end{bmatrix}, \qquad \mathbf{M}_2 = \begin{bmatrix} \mathbf{I} & \mathbf{C} \\ 0 & \mathbf{A} \end{bmatrix}, \qquad \mathbf{M}_3 = \begin{bmatrix} \mathbf{A} & 0 \\ \mathbf{C} & \mathbf{I} \end{bmatrix}, \qquad \mathbf{M}_4 = \begin{bmatrix} \mathbf{A} & \mathbf{C} \\ 0 & \mathbf{I} \end{bmatrix}$$

where $\mathbf{A}$ is a square matrix, $\mathbf{I}$ is a unit matrix of any order, and $\mathbf{C}$ is an arbitrary matrix of compatible order. In computing the determinants of $\mathbf{M}_i$, $i = 1, 2, 3,$ and 4, if we expand $\mathbf{M}_i$ along the diagonal elements of $\mathbf{I}$, we can readily obtain

$$\det \mathbf{M}_i = \det \mathbf{A}$$

for $i = 1, 2, 3,$ and 4. This property can be extended as follows: Let $\mathbf{A}$ and $\mathbf{B}$ be square matrices of any order. Consider

$$\begin{bmatrix} \mathbf{A} & 0 \\ \mathbf{C} & \mathbf{B} \end{bmatrix} = \begin{bmatrix} \mathbf{A} & 0 \\ 0 & \mathbf{I} \end{bmatrix} \begin{bmatrix} \mathbf{I} & 0 \\ \mathbf{C} & \mathbf{B} \end{bmatrix}$$

Using $\det (\mathbf{MN}) = \det \mathbf{M} \cdot \det \mathbf{N}$, we have

$$\det \begin{bmatrix} \mathbf{A} & 0 \\ \mathbf{C} & \mathbf{B} \end{bmatrix} = \det \begin{bmatrix} \mathbf{A} & 0 \\ 0 & \mathbf{I} \end{bmatrix} \cdot \det \begin{bmatrix} \mathbf{I} & 0 \\ \mathbf{C} & \mathbf{B} \end{bmatrix} = \det \mathbf{A} \cdot \det \mathbf{B} \qquad \text{(A.64a)}$$

Similarly, we have

$$\det \begin{bmatrix} \mathbf{A} & \mathbf{C} \\ 0 & \mathbf{B} \end{bmatrix} = \det \mathbf{A} \cdot \det \mathbf{B} \qquad \text{(A.64b)}$$

This property is often used in computing the determinant of a large matrix.

*A.9.1 Computer Computation of Determinants

The determinant of a matrix can be computed using Laplace's expansion in (A.57). In this section, we will discuss the number of operations required in the computation. Because multiplication (including division) takes much more time than addition,[7] we will count only the number of multiplications.

Consider the determinant of the 3×3 matrix in (A.59). There are $3! = 6$ terms; each term requires $2 = (3 - 1)$ multiplications. Thus, for an $n \times n$ matrix, if we use (A.57) to compute its determinant, the number of multiplications is equal to $n!(n - 1)$. If some of the multiplications are combined, the number of multiplications can be smaller than $n!(n - 1)$, as can be seen from (A.59), but still larger than $n!$. For $n = 10$, $n!$ is equal to 3,628,800. Therefore, the use of (A.57) to compute the determinant of a large matrix is impractical.

Given a square matrix $\mathbf{A}$ of order n, we may apply Algorithm B to transform it into (A.31) or

$$\mathbf{U} = \mathbf{M}_{n-1}\mathbf{P}_{n-1,\alpha_{n-1}} \cdots \mathbf{M}_2\mathbf{P}_{2,\alpha_2}\mathbf{M}_1\mathbf{P}_{1,\alpha_1}\mathbf{A} =: \mathbf{N}\mathbf{A} \qquad \text{(A.65)}$$

where all $\mathbf{P}_{ij}$ are permutation matrices, all $\mathbf{M}_i$ are unit lower triangular matrices, and $\mathbf{U}$ is an upper triangular matrix. The application of (A.63) to (A.65) yields

$$\det \mathbf{U} = \cdots \det \mathbf{M}_2 \cdot \det \mathbf{P}_{2,\alpha_2} \cdot \det \mathbf{M}_1 \cdot \det \mathbf{P}_{1,\alpha_1} \cdot \det \mathbf{A}$$

Since $\det \mathbf{M}_i = 1$ for all i and $\det \mathbf{P}_{ij} = -1$ if $i \neq j$, as in (A.62), we have

$$\det \mathbf{U} = (-1)^k \det \mathbf{A}$$

or

$$\det \mathbf{A} = (-1)^k \det \mathbf{U} = (-1)^k \prod_{i=1}^{n} u_{ii} \qquad \text{(A.66)}$$

where u_{ii} are the diagonal elements of $\mathbf{U}$, and k is the total number of interchanges of rows or the number of $\mathbf{P}_{ij}$ with $i \neq j$.

Consider now the number of multiplications required in (A.66). Permutations of rows do not require multiplication; therefore, we consider only Gaussian elimination. For an $n \times n$ matrix, we require $(n - 1)$ divisions to obtain $\mathbf{M}_1$. The product of $\mathbf{M}_1$ and $\mathbf{A}$ requires $(n - 1)^2$ multiplications. Note that the first column of $\mathbf{M}_1\mathbf{A}$ can simply be set to zero without computation. Thus, the total number of multiplications (and divisions) in the first step is

$$(n - 1)^2 + (n - 1) = n(n - 1)$$

For convenience, $n(n - 1)$ is approximated as n^2. In the second step, the number of multiplications is $(n - 1)^2$. Proceeding forward, we conclude that (A.66) requires

---

[7]On personal computers, one multiplication may take 1 μs (microsecond = 10^{-6} seconds); division may take 2.9 μs; addition may take 0.4 μs, and subtraction may take 0.4 μs.

$$n^2 + (n-1)^2 + \cdots + 3^2 + 2^2 = \frac{n}{6}(n+1)(2n+1) - 1$$

number of multiplications. The number of multiplications in computing det $\mathbf{U}$ is $(n-1)$. Thus, the total number of multiplications in computing det $\mathbf{A}$ when using (A.66) is proportional to, for large n, $n^3/3$. It equals 333 for $n = 10$. It is 10,000 times more efficient than the method of Laplace's expansion! Thus, the latter method is not used for large n.

The matrix $\mathbf{A}$ is nonsingular or, equivalently, has a full rank if and only if all u_{ii} in (A.66) are different from zero. Thus, it follows from (A.66) that $\mathbf{A}$ is nonsingular if and only if det $\mathbf{A} \neq 0$. A square matrix $\mathbf{A}$ is said to be singular if det $\mathbf{A} = 0$ or, equivalently, it does not have a full rank.

A.9.2   The inverse of nonsingular matrices

This section introduces the formula for computing the inverse of a matrix using determinants. Let $\mathbf{A} = [a_{ij}]$ be a square matrix of order n. Let $\mathbf{A}_{ij}$ be the $(n-1) \times (n-1)$ submatrix of $\mathbf{A}$ obtained by deleting its ith row and jth column. We call $m_{ij} := \det \mathbf{A}_{ij}$ the *minor* and

$$c_{ij} := (-1)^{i+j} m_{ij} := (-1)^{i+j} \det \mathbf{A}_{ij}$$

the *cofactor* of $\mathbf{A}$ corresponding to a_{ij}. Then, we have

$$\mathbf{A}^{-1} = \frac{1}{\det \mathbf{A}} \operatorname{adj} \mathbf{A} \tag{A.67a}$$

and

$$\operatorname{adj} \mathbf{A} := \begin{bmatrix} c_{11} & c_{12} & \cdots & c_{1n} \\ c_{21} & c_{22} & \cdots & c_{2n} \\ \cdot & \cdot & & \cdot \\ \cdot & \cdot & & \cdot \\ \cdot & \cdot & & \cdot \\ c_{n1} & c_{n2} & \cdots & c_{nn} \end{bmatrix}' = \begin{bmatrix} c_{11} & c_{21} & \cdots & c_{n1} \\ c_{12} & c_{22} & \cdots & c_{n2} \\ \cdot & \cdot & & \cdot \\ \cdot & \cdot & & \cdot \\ \cdot & \cdot & & \cdot \\ c_{1n} & c_{2n} & \cdots & c_{nn} \end{bmatrix} \tag{A.67b}$$

where adj stands for the *adjoint* and the prime denotes the transpose. It is clear that (A.67a) is defined only if det $\mathbf{A} \neq 0$ or $\mathbf{A}$ is nonsingular. Formula (A.67) is used mainly for $n = 2$ and 3. For $n = 2$, we have $c_{11} = a_{22}$, $c_{12} = -a_{21}$, $c_{21} = -a_{12}$, $c_{22} = a_{11}$ and

$$\begin{bmatrix} a_{11} & a_{12} \\ a_{21} & a_{22} \end{bmatrix}^{-1} = \frac{1}{a_{11}a_{22} - a_{12}a_{21}} \begin{bmatrix} a_{22} & -a_{12} \\ -a_{21} & a_{11} \end{bmatrix} \tag{A.68}$$

Its correctness can be easily verified by showing

$$\begin{bmatrix} a_{11} & a_{12} \\ a_{21} & a_{22} \end{bmatrix} \begin{bmatrix} a_{11} & a_{12} \\ a_{21} & a_{22} \end{bmatrix}^{-1} = \begin{bmatrix} 1 & 0 \\ 0 & 1 \end{bmatrix}$$

The formula in (A.68) is very simple and worth remembering: Interchange the diagonal elements, change the signs of the off-diagonal elements (without changing positions), and divide the resulting matrix by the determinant.

EXERCISE A.9.5

Compute, if it exists,

$$\begin{bmatrix} 4 & 1 \\ -2 & 3 \end{bmatrix}^{-1} \qquad \begin{bmatrix} 1 & 4 \\ 3 & -2 \end{bmatrix}^{-1} \qquad \begin{bmatrix} 2 & 8 \\ 1 & 4 \end{bmatrix}^{-1}$$

[**ANSWERS:** $\dfrac{1}{14}\begin{bmatrix} 3 & -1 \\ 2 & 4 \end{bmatrix}$, $\begin{bmatrix} 1/7 & 2/7 \\ 3/14 & -1/14 \end{bmatrix}$, not defined.] ■

For $n = 3$, (A.67) becomes

$$\mathbf{A} := \begin{bmatrix} a_{11} & a_{12} & a_{13} \\ a_{21} & a_{22} & a_{23} \\ a_{31} & a_{32} & a_{33} \end{bmatrix}^{-1} = \frac{1}{\det \mathbf{A}} \begin{bmatrix} c_{11} & c_{12} & c_{13} \\ c_{21} & c_{22} & c_{23} \\ c_{31} & c_{32} & c_{33} \end{bmatrix} \qquad (A.69)$$

with

$$\begin{aligned} c_{11} &= a_{22}a_{33} - a_{23}a_{32}, & c_{12} &= -(a_{21}a_{33} - a_{23}a_{31}) \\ c_{13} &= a_{21}a_{32} - a_{22}a_{31}, & c_{21} &= -(a_{12}a_{33} - a_{13}a_{32}) \\ c_{22} &= a_{11}a_{33} - a_{13}a_{31}, & c_{23} &= -(a_{11}a_{32} - a_{12}a_{31}) \\ c_{31} &= a_{12}a_{23} - a_{13}a_{22}, & c_{32} &= -(a_{11}a_{23} - a_{13}a_{21}) \\ c_{33} &= a_{11}a_{22} - a_{12}a_{21} \end{aligned}$$

EXERCISE A.9.6

Verify

$$\begin{bmatrix} 2 & 1 & -1 \\ 3 & 0 & 1 \\ 0 & -1 & 1 \end{bmatrix}^{-1} = \frac{1}{3+2-3}\begin{bmatrix} 1 & -3 & -3 \\ 0 & 2 & 2 \\ 1 & -5 & -3 \end{bmatrix} = \frac{1}{2}\begin{bmatrix} 1 & 0 & 1 \\ -3 & 2 & -5 \\ -3 & 2 & -3 \end{bmatrix}$$

and show $\mathbf{AA}^{-1} = \mathbf{A}^{-1}\mathbf{A} = \mathbf{I}$. ■

If the order of a matrix is 4 or larger, the formula in (A.67) is rarely used. If a matrix $\mathbf{A}$ is triangular, then, in addition to using (A.67), we can solve $\mathbf{AA}^{-1} = \mathbf{I}$, as illustrated for the matrix in (A.12), to obtain its inverse. For computer computation, we may use Algorithm B to compute $\mathbf{U} = \mathbf{NA}$ as in (A.31) or (A.65). Then, the inverse of $\mathbf{A}$ is $\mathbf{U}^{-1}\mathbf{N}$, as in (A.32).

A.9.3 **Cramer's rule**

If $\mathbf{A}$ is square and nonsingular, the solution of $\mathbf{Av} = \mathbf{b}$ can also be obtained as

$$\mathbf{v} = \mathbf{A}^{-1}\mathbf{b} = \frac{1}{\det \mathbf{A}}[\text{adj } \mathbf{A}]\mathbf{b} \qquad (A.70)$$

Let $\mathbf{v} := [\, v_1 \, v_2 \ldots v_n \,]'$ and $\mathbf{b} := [\, b_1 \, b_2 \ldots b_n \,]'$, where the prime denotes the transpose. Let $\mathbf{A}_j(b)$ denote the matrix $\mathbf{A}$ with its jth column replaced by $\mathbf{b}$. Then, we have, for $i = 1, 2, \ldots, n$,

$$v_i = \frac{\det \mathbf{A}_i(b)}{\det \mathbf{A}} \qquad (A.71)$$

This is called *Cramer's rule*. It was first published by Gabriel Cramer (1704–1752) in 1750. Example A.9.1 illustrates its employment.

EXAMPLE **A.9.1**

Consider

$$\begin{bmatrix} 1 & 3 \\ 1 & 1 \end{bmatrix} \begin{bmatrix} v_1 \\ v_2 \end{bmatrix} = \begin{bmatrix} 2 \\ 0 \end{bmatrix}$$

Its solution is, using Cramer's rule,

$$v_1 = (\det \begin{bmatrix} 2 & 3 \\ 0 & 1 \end{bmatrix}) / (\det \begin{bmatrix} 1 & 3 \\ 1 & 1 \end{bmatrix}) = \frac{2}{1 - 3} = -1$$

and

$$v_2 = (\det \begin{bmatrix} 1 & 2 \\ 1 & 0 \end{bmatrix}) / (\det \begin{bmatrix} 1 & 3 \\ 1 & 1 \end{bmatrix}) = \frac{-2}{1 - 3} = 1$$

Based on this example, the use of Cramer's rule appears to be very simple. Unfortunately, if n is 3 or larger, it is not simple at all. Furthermore, it is applicable only if $\mathbf{A}$ is nonsingular. Thus, the use of Cramer's rule is rather limited.

*A.10 **EXTENSION TO RATIONAL FUNCTIONS**

In this section, we shall extend what we have studied so far for real numbers to rational functions. We will study rational functions with real coefficients exclusively; therefore, the qualification "with real coefficients" will be dropped. A rational function is, by definition, the ratio of any two polynomials. For example,

$$\frac{2}{1}, \quad \frac{s+1}{1}, \quad \frac{s+1}{s^2+2}, \quad \frac{1}{s^{10}+1}$$

are all rational functions. Because real numbers are polynomials of degree 0, the set of rational functions includes all polynomials and all real numbers.

The procedures for solving linear algebraic equations discussed in Section A.2 are directly applicable to equations with rational functions as coefficients. For example, consider

$$\frac{2s^2+2s+2}{s}I_1 - 2sI_2 = U \tag{A.72a}$$

$$-2sI_1 + \frac{4s^2+2s+1}{2s}I_2 = 0 \tag{A.72b}$$

where I_1 and I_2 are unknowns and U is given.[8] To eliminate I_1 from (A.72b), we add the product of (A.72a) and $s^2/(s^2+s+1)$ to (A.72b) to yield

$$(\frac{-2s^3}{s^2+s+1} + \frac{4s^2+2s+1}{2s})I_2 = \frac{s^2}{s^2+s+1}U$$

or

$$\frac{6s^3+7s^2+3s+1}{2s(s^2+s+1)}I_2 = \frac{s^2}{s^2+s+1}U \tag{A.73}$$

This completes the process of elimination. The solution can now be obtained by back substitution. From (A.73), we have

$$I_2 = \frac{2s(s^2+s+1)}{6s^3+7s^2+3s+1} \cdot \frac{s^2}{(s^2+s+1)}U = \frac{2s^3}{6s^3+7s^2+3s+1}U$$

The substitution of I_2 into (A.72b) or (A.72a) yields

$$I_1 = \frac{s(4s^2+2s+1)}{2(6s^3+7s^2+3s+1)}U \tag{A.74}$$

The procedure is identical to that for equations with real numbers.

EXERCISE A.10.1

Find the solutions of

---

[8]This equation arises from the network in Figure 2.5 with $R_1 = 2\,\Omega$, $R_2 = 1\,\Omega$, $C_1 = 0.5\,F$, $C_2 = 2\,F$, and $L_1 = 2\,H$. It is derived in Example 4.9.4.

$$sI_1 + \frac{1}{s+1}I_2 = 2$$

$$I_1 - \frac{1}{s}I_2 = -1$$

[**ANSWERS:** $I_1 = 1/s$, $I_2 = s + 1$.] ■

The concepts of determinants and inverses developed for real matrices are equally applicable to rational function matrices. For example, the inverse of

$$\mathbf{A} := \begin{bmatrix} \dfrac{2s+1}{2s} & \dfrac{-1}{2s} \\ \dfrac{-1}{2s} & \dfrac{6s^2+3}{2s} \end{bmatrix}$$

is, using (A.68),

$$\mathbf{A}^{-1} := \begin{bmatrix} \dfrac{2s+1}{2s} & \dfrac{-1}{2s} \\ \dfrac{-1}{2s} & \dfrac{6s^2+3}{2s} \end{bmatrix}^{-1} = \frac{1}{\Delta}\begin{bmatrix} \dfrac{6s^2+3}{2s} & \dfrac{1}{2s} \\ \dfrac{1}{2s} & \dfrac{2s+1}{2s} \end{bmatrix}$$

with

$$\Delta = \det \mathbf{A} = \frac{(2s+1)(6s^2+3)}{4s^2} - \frac{1}{4s^2} = \frac{12s^3 + 6s^2 + 6s + 2}{4s^2}$$

EXERCISE A.10.2

Find the inverses of

$$\begin{bmatrix} s & 1 \\ s & s \end{bmatrix} \qquad \begin{bmatrix} s^n & 1 \\ s^n - 1 & 1 \end{bmatrix}$$

[**ANSWER:** $\dfrac{1}{s^2 - s}\begin{bmatrix} s & -1 \\ -s & s \end{bmatrix}$, $\begin{bmatrix} 1 & -1 \\ 1 - s^n & s^n \end{bmatrix}$.] ■

Cramer's rule developed in Section A.9.3 is also directly applicable to rational function equations. A great deal more can be said regarding rational matrices, in particular, polynomical matrices. The interested reader is referred to Reference [4].

A.11   SUMMARY

1. An $m \times n$ matrix has m rows and n columns of elements. A square matrix has an equal number of rows and columns. A square matrix is called diagonal if all elements except the diagonal elements are zeros. A square matrix is called an upper triangular matrix if all entries below the diagonal are zeroes; it is called a lower triangular matrix if all entries above the diagonal are zeros. It is called a unit lower triangular matrix if, in addition, all diagonal elements are equal to one.

2. The sum of two matrices is not defined unless their orders are the same. The product of the two matrices $\mathbf{A}$ and $\mathbf{B}$ is not defined unless the number of columns of $\mathbf{A}$ is equal to the number of rows of $\mathbf{B}$. In this case, the resulting matrix $\mathbf{AB}$ has the same number of rows as $\mathbf{A}$ and the same number of columns as $\mathbf{B}$. In general, $\mathbf{AB}$ is different from $\mathbf{BA}$, and $\mathbf{AB} = \mathbf{0}$ implies neither $\mathbf{A} = \mathbf{0}$ nor $\mathbf{B} = \mathbf{0}$.

3. A square matrix $\mathbf{A}$ is nonsingular if there exists a matrix $\mathbf{A}^{-1}$ such that $\mathbf{AA}^{-1} = \mathbf{A}^{-1}\mathbf{A} = \mathbf{I}$.

4. Linear algebraic equations can be solved in two steps: elimination and back substitution. Although Cramer's rule can also be used, its applicability is limited and it is more computationally complex.

5. Gaussian elimination may break down and is not numerically stable. Gaussian elimination with partial pivoting can always be carried out to completion and is numerically stable. It is a very useful algorithm: It can solve a set of linear algebraic equations, It can compute the determinant much more efficiently than can Laplace's expansion, and it can compute the inverse of a square matrix more easily than can the inversion formula in (A.67).

6. When given a set of linear algebraic equations, we do not need to check for the existence of solutions first. (The existence condition requires the computation of ranks and is complicated.) Instead, we should immediately solve it, using a procedure such as Gaussian elimination with partial pivoting. At the end of the procedure, we will know whether or not a solution exists.

7. Developing a computer program to solve problems with few variables is not difficult. Developing a program to solve problems with many variables, for example, one thousand variables, however, may not be a simple task. The numerical stability of algorithms must be considered. The speed or the rate of convergence is also an important consideration. LINPACK and Eispack are two standard programs used in solving linear algebraic problems.

PROBLEMS

A.1 Solve the following set of equations.

$$2v_1 - v_2 + 3v_3 = 1$$
$$4v_1 - v_2 + v_3 = 3$$
$$v_1 - 2v_3 = 0$$

A.2 Compute the following.

$$\begin{bmatrix} 12 & -4 & 0 \\ 2 & 3 & 2 \end{bmatrix} + \begin{bmatrix} 1 & 5 & -9 \\ 4 & -5 & 7 \end{bmatrix}$$

$$[2 \ -3 \ 7] - 3 \cdot [-1 \ 5 \ -2]$$

$$\begin{bmatrix} 1 & 2 \\ -5 & 3 \end{bmatrix} + (-5) \cdot \begin{bmatrix} 3 & 0 \\ -2 & 3 \end{bmatrix}$$

A.3 Let

$$\mathbf{A} = \begin{bmatrix} 2 & -1 & 2 \\ 3 & 3 & 1 \end{bmatrix} \qquad \mathbf{B} = \begin{bmatrix} 1 & 0 \\ 0 & 1 \\ 1 & 1 \end{bmatrix}$$

Compute $\mathbf{AB}$ and $\mathbf{BA}$. Find $\mathbf{A}'$ and $\mathbf{B}'$, and then compute $\mathbf{A}'\mathbf{B}'$ and $\mathbf{B}'\mathbf{A}'$. Which equality, $(\mathbf{AB})' = \mathbf{A}'\mathbf{B}'$, or $(\mathbf{AB})' = \mathbf{B}'\mathbf{A}'$, do you have?

A.4 Given

$$\begin{bmatrix} a_{11} & a_{21} \\ a_{21} & a_{22} \end{bmatrix} \begin{bmatrix} d_1 & 0 \\ 0 & d_2 \end{bmatrix} = \begin{bmatrix} d_1 & 0 \\ 0 & d_2 \end{bmatrix} \begin{bmatrix} a_{11} & a_{21} \\ a_{21} & a_{22} \end{bmatrix}$$

in which $d_1 \neq d_2$. Under what condition on a_{ij} will the equality hold, or the two matrices commute? If $d_1 = d_2$, will the two matrices commute for all a_{ij}?

A.5 Find the inverse for each of the following triangular matrices.

$$\mathbf{A}_1 = \begin{bmatrix} 1 & 0 \\ 3 & -1 \end{bmatrix}, \qquad \mathbf{A}_2 = \begin{bmatrix} 1 & -1 & 2 \\ 0 & 1 & 1 \\ 0 & 0 & 1 \end{bmatrix}, \qquad \mathbf{A}_3 = \begin{bmatrix} 1 & -1 & 2 \\ 0 & 0.5 & 1 \\ 0 & 0 & 2 \end{bmatrix}$$

A.6 (a) Show $(\mathbf{A} + \mathbf{B})' = \mathbf{A}' + \mathbf{B}'$.

(b) Do we have $(\mathbf{A} + \mathbf{B})^{-1} = \mathbf{A}^{-1} + \mathbf{B}^{-1}$? If not, find counterexamples for scalar and 2×2 matrices.

A.7 Let $\mathbf{A}$ be a 4×4 matrix and let

$$\mathbf{U} = \mathbf{M}_3\mathbf{M}_2\mathbf{M}_1\mathbf{A}$$

where **U** is an upper triangular matrix and

$$
\mathbf{M}_1 = \begin{bmatrix} 1 & 0 & 0 & 0 \\ -d_{21} & 1 & 0 & 0 \\ -d_{31} & 0 & 1 & 0 \\ -d_{41} & 0 & 0 & 1 \end{bmatrix}, \qquad
\mathbf{M}_2 = \begin{bmatrix} 1 & 0 & 0 & 0 \\ 0 & 1 & 0 & 0 \\ 0 & -d_{32} & 1 & 0 \\ 0 & -d_{42} & 0 & 1 \end{bmatrix}, \qquad
\mathbf{M}_3 = \begin{bmatrix} 1 & 0 & 0 & 0 \\ 0 & 1 & 0 & 0 \\ 0 & 0 & 1 & 0 \\ 0 & 0 & -d_{43} & 1 \end{bmatrix}
$$

Compute $\mathbf{M} = \mathbf{M}_3 \mathbf{M}_2 \mathbf{M}_1$, and then compute $\mathbf{L} := \mathbf{M}^{-1}$.

A.8 In Problem A.7, verify that $\mathbf{M}_i^{-1}$ equals $\mathbf{M}_i$ with signs of d_{ij} reversed. Verify

$$
\mathbf{L} = \mathbf{M}_1^{-1} \mathbf{M}_2^{-1} \mathbf{M}_3^{-1} = \begin{bmatrix} 1 & 0 & 0 & 0 \\ d_{21} & 1 & 0 & 0 \\ d_{31} & d_{32} & 1 & 0 \\ d_{41} & d_{42} & d_{41} & 1 \end{bmatrix}
$$

From this, we conclude that, if **A** is triangularized as $\mathbf{U} = \mathbf{M}_3 \mathbf{M}_2 \mathbf{M}_1 \mathbf{A}$, then we can obtain

$$
\mathbf{L}\mathbf{U} := \mathbf{M}_1^{-1} \mathbf{M}_2^{-1} \mathbf{M}_3^{-1} \mathbf{U} = \mathbf{A}
$$

without any additional computation. This is called the LU decomposition of **A**, where **U** is an upper triangular matrix and **L** is a unit lower triangular matrix. Can any square matrix be factored into the LU decomposition? How about any nonsingular matrix? If **U** is not required to be lower triangular by including permutation matrices, can any square matrix be factored as $\mathbf{A} = \mathbf{L}\mathbf{U}$? In this case, what can you say about the diagonal elements of **U** if **A** is singular or nonsingular?

[**ANSWERS:** No; no; yes; at least one diagonal element is zero; all diagonal elements are nonzero.]

A.9 Express Problem A.1 in matrix notation and solve it using Gaussian elimination.

A.10 Repeat Problem A.9 using the procedure in Section A.4.2.

A.11 Solve the set of equations

$$
\begin{bmatrix} 2 & 0 & 0 \\ 3 & -1 & 0 \\ 2 & -3 & 4 \end{bmatrix} \begin{bmatrix} v_1 \\ v_2 \\ v_3 \end{bmatrix} = \begin{bmatrix} -2 \\ 0 \\ 3 \end{bmatrix}
$$

by first solving the first equation, then substituting the result into the second equation, and so forth. (This process is called *forward substitution*.)

A.12 Consider

$$\begin{bmatrix} a_{11} & 0 & 0 & \cdots & 0 \\ a_{21} & a_{22} & 0 & \cdots & 0 \\ \cdot & \cdot & \cdot & & \cdot \\ \cdot & \cdot & \cdot & & \cdot \\ \cdot & \cdot & \cdot & & \cdot \\ a_{n1} & a_{n2} & a_{n3} & \cdots & a_{nn} \end{bmatrix} \begin{bmatrix} v_1 \\ v_2 \\ \cdot \\ \cdot \\ \cdot \\ v_n \end{bmatrix} = \begin{bmatrix} a_{1,n+1} \\ a_{2,n+1} \\ \cdot \\ \cdot \\ \cdot \\ a_{n,n+1} \end{bmatrix}$$

Find a general formula, as in (A.29), to solve v_i, $i = 1, 2, \ldots, n$.

A.13 Show that, if a square matrix $\mathbf{A}$ is first factored as $\mathbf{LU}$, where $\mathbf{L}$ is a unit lower triangular matrix and $\mathbf{U}$ is an upper triangular matrix, as discussed in Problem A.8, then $\mathbf{Av} = \mathbf{b}$ can be solved by forward substitution and backward substitution. (Hint: Solve $\mathbf{Ly} = \mathbf{b}$ first, and then solve $\mathbf{Uv} = \mathbf{y}$.)

A.14 Solve the following equations using 4-digit floating-point arithmetic.

$$0.2 \times 10^{-5} v_1 + v_2 = 1.00002$$
$$v_1 + v_2 = 11$$

First use Gaussian elimination, and then use Gaussian elimination with partial pivoting. Compare your results.

A.15 Use 4-digit floating-point arithmetic and Gaussian elimination with partial pivoting to solve

$$0.2 \times 10 v_1 + 10^6 \times v_2 = 1.00002 \times 10^6$$
$$v_1 + v_2 = 11$$

Compare your result here with your results for Problem A.14. Is this result acceptable? Is this set of equations the same as the one in Problem A.14? (This problem shows that Gaussian elimination with partial pivoting may not yield satisfactory solutions if equations are not balanced.)

A.16 Compute the rank and determinant for each of the following matrices.

$$\begin{bmatrix} 2 & 0 & 1 \\ 1 & 3 & 2 \\ 1 & 0 & 1 \end{bmatrix} \qquad \begin{bmatrix} 2 & 0 & 1 \\ 1 & 3 & 2 \\ 3 & 3 & 3 \end{bmatrix}$$

Which matrix is nonsingular?

A.17 (a) Is it true that, if a set of linear algebraic equations has more unknowns than equations, a solution will always exist? If not, give a counterexample.

(b) Is it true that, if a set of linear algebraic equations has more equations than unknowns, no solution will exist? If not, give a counterexample.

A.18 Show that the inverse of $\mathbf{N}$ in (A.31) is given by

$$\mathbf{N}^{-1} = \mathbf{P}_{1,\alpha_1}^{-1}\,\mathbf{M}_1^{-1} \cdots \mathbf{P}_{n-1,\alpha_{n-1}}^{-1}\,\mathbf{M}_{n-1}^{-1}$$

and that (A.31) implies $\mathbf{A} = \mathbf{N}^{-1}\mathbf{U}$. Do you require $\mathbf{A}$ to be square in this equation?

A.19 Find the rank of

$$\begin{bmatrix} 2 & -1 & -1 & 2 & 1 \\ 4 & 1 & -1 & 2 & 4 \\ 8 & -1 & -3 & 6 & 6 \\ -4 & 5 & 1 & 10 & 1 \end{bmatrix}$$

using the procedure discussed in Section A.7.2.

A.20 Find the determinant of

$$\begin{bmatrix} 3 & 1 & 5 & 10 & 19 \\ 1 & 1 & 6 & -5 & 7 \\ 0 & 0 & 2 & 0 & 0 \\ 0 & 0 & 3 & 1 & 0 \\ 0 & 0 & 1 & 0 & 5 \end{bmatrix}$$

A.21 Show that the matrix

$$\begin{bmatrix} \mathbf{A} & \mathbf{0} \\ \mathbf{C} & \mathbf{B} \end{bmatrix}$$

is nonsingular if and only if $\mathbf{A}$ and $\mathbf{B}$ are nonsingular. Show also

$$\begin{bmatrix} \mathbf{A} & \mathbf{0} \\ \mathbf{C} & \mathbf{B} \end{bmatrix}^{-1} = \begin{bmatrix} \mathbf{A}^{-1} & \mathbf{0} \\ -\mathbf{B}^{-1}\mathbf{C}\mathbf{A}^{-1} & \mathbf{B}^{-1} \end{bmatrix}$$

A.22 Find the inverse for each of the following matrices.

$$\begin{bmatrix} 0 & -1 \\ 1 & 3 \end{bmatrix}, \qquad \begin{bmatrix} 2 & 1 & -1 \\ 1 & 1 & 0 \\ 0 & 2 & 3 \end{bmatrix}$$

A.23 Verify $(\mathbf{A}^{-1})' = (\mathbf{A}')^{-1}$ for the matrices in Problem A.22.

A.24 Consider a square matrix $\mathbf{A}$. We form the composite matrix

$$[\mathbf{A}\ \ \mathbf{I}]$$

and use elementary row operations to transform it into

$$[\mathbf{I}\ \ \mathbf{P}]$$

Show that $\mathbf{A}^{-1} = \mathbf{P}$.

A.25 Use the method in Problem A.24 to compute the inverses of the matrices in Problem A.22.

A.26 Use Cramer's rule to solve

$$\begin{bmatrix} 3 & 4 \\ 2 & 5 \end{bmatrix} \mathbf{v} = \begin{bmatrix} 1 \\ 5 \end{bmatrix}$$

A.27 Solve Problem A.1 using Cramer's rule.

A.28 Find the solution of

$$(s + 1)I_1 - 2I_2 = s$$
$$2I_1 + sI_2 = s + 1$$

The coefficients of the equations are all polynomials. Is the solution in polynomial form?

A.29 Consider $\mathbf{Av} = \mathbf{b}$. Let $\mathbf{v}_p$ be a solution, that is, $\mathbf{Av}_p = \mathbf{b}$. Let $\mathbf{v}_i$, $i = 1, 2, \ldots, m$, be nontrivial solutions of $\mathbf{Av} = \mathbf{0}$. Show that

$$\mathbf{v} = \mathbf{v}_p + \sum_{i=1}^{m} \alpha_i \mathbf{v}_i$$

for any α_i, is a solution of $\mathbf{Av} = \mathbf{b}$. (This is called the general solution. The general solution gives a parameterization of all solutions of $\mathbf{Av} = \mathbf{b}$.)

A.30 (a) Consider the network, called a bridge network, shown in Figure PA.30. The input u of the bridge is a voltage source and the output y is the voltage across A and B. Find the equation describing u and y.

(b) Find the conditions on R_i so that $y = 0$. (In this case, the bridge is said to be balanced.)

(c) If $R_1 = R_3 = 1$ ohm and $R_2 = 3$ ohms, what should be the value of R_4 for the bridge to be balanced? (This network can be used to measure an unknown resistor R_4 and be used to detect temperature changes if R_4 is a temperature-sensitive device.)

Figure PA.30

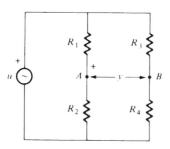

Figure PA.31

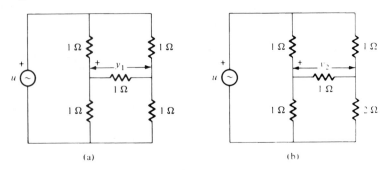

(a)   (b)

Figure PA.32

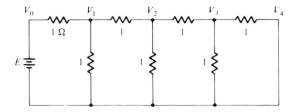

A.31 Consider the networks shown in Figure PA.31. Find the equations relating (a) u and y_1 and (b) u and y_2.

A.32 Consider the resistive network shown in Figure PA.32. Let V_i be the node voltages, as shown, with $V_0 = E$ and $V_4 = 0$. Show that the unknown voltages V_i, $i = 1, 2, 3$, can be solved from the following linear algebraic equation.

$$\begin{bmatrix} 3 & -1 & 0 \\ -1 & 3 & -1 \\ 0 & -1 & 3 \end{bmatrix} \begin{bmatrix} V_1 \\ V_2 \\ V_3 \end{bmatrix} = \begin{bmatrix} E \\ 0 \\ 0 \end{bmatrix}$$

USING **MATLAB**

In MATLAB, a matrix is represented row by row, separated by semicolons; entries of each row are separated by spaces or commas. For example, the following matrices

$$A = \begin{bmatrix} 4 & 1 & 1 \\ 2 & 4 & -1 \\ -3 & 2 & 5 \end{bmatrix} \qquad b = \begin{bmatrix} 1 \\ -2 \\ -4 \end{bmatrix} \qquad a = [2 \ 1 \ -1] \qquad d = [2]$$

are represented, respectively, by

```
A=[4 1 1;2 4 -1;-3 2 5];b=[1;-2;-4];
a=[2 1 -1];d=[2];
```

After each statement, if we press ENTER, the statement will be processed, stored, and displayed on the monitor. If we type a semicolon, the statement will be processed, stored, but not displayed. Because b has three rows, entries of b are separated by semicolons. The row vector a can also be represented as `[2, 1, -1]`, and the scalar d as `d = 2` without brackets.

The command in MATLAB

```
A'
```

yields the transpose of A as

```
ans =

    4            2           -3
    1            4            2
    1           -1            5
```

The command `a*b`, which is the product of a 1×3 and a 3×1 matrix, yields a 1×1 matrix or scalar 4 and the command

```
B=b*a
```

yields a 3×3 matrix as

```
B=
    2            1           -1
   -4           -2            2
   -8           -4            4
```

The determinant, rank, and inverse of a matrix can be obtained by typing `det(A)`, `rank(A)`, and `inv(A)` which yield, respectively, 97, 3, and

```
ans =
   0.2268      -0.0309      -0.0515
  -0.0722       0.2371       0.0619
   0.1649      -0.1134       0.1443
```

The solution of the linear algebraic equation $Av = b$ can be obtained by typing $A \backslash b$. Note that $A \backslash b$ (with back slash) yields the solution of $Av = b$ and A/c (with slash) yields the solution of $wA = c$. Both yield the same result if A, b, and c are scalars. For example, the solution of the equation in (A.6) can be obtained by typing

```
[4 -1 -3;-1 4 -1;-3 -1 5]\[-2;1;-1]
```

The answer is

```
ans =
  -1.4828
  -0.4138
  -1.1724
```

which is the same as $[-43/29; -12/29; -34/29]$ obtained in (A.3) through (A.5).

Note that computing the rank of matrices is an ill-conditioned problem because the result is very sensitive to the parameters. For example, the ranks of the two matrices

$$\begin{bmatrix} 1 & 3 \\ 1/3 & 1 \end{bmatrix} \qquad \begin{bmatrix} 1 & 3 \\ 0.333333333 & 1 \end{bmatrix}$$

are 1 and 2 on MATLAB. Note also that linear algebraic equations can be solved in many ways. For example, if we solve

$$\begin{bmatrix} 1 & 1 \\ 0 & 0 \end{bmatrix} v = \begin{bmatrix} 2 \\ 0 \end{bmatrix}$$

by typing

```
[1 1;0 0]\[2;0]
```

The result is

```
Warning: Matrix is singular to working precision
ans =
  ∞

  ∞
```

This result is clearly incorrect. Apparently, this equation is solved by computing the inverse of the matrix. If we modify the equation as

$$\begin{bmatrix} 1 & 1 & 0 \\ 0 & 0 & 0 \end{bmatrix} v = \begin{bmatrix} 2 \\ 0 \end{bmatrix}$$

and type

```
[1 1 0;0 0 0]\[2;0]
```

Then the result is [2;0;0]. This is a correct solution.

MATLAB PROBLEMS

MA.1 Use MATLAB to compute all Exercises in this appendix.

MA.2 Use MATLAB to solve Problems A.1, A.5, A.11, A.16, A.19, A.20, and A.22.

References

[1] C. B. Boyer, *A History of Mathematics*, New York: Wiley, 1968.

[2] C. F. Chen and L. S. Shieh, "A novel approach to linear model simplification," *Int. J. Control*, vol. 11, pp. 561–570, 1968.

[3] C. T. Chen, *One-dimensional Digital Signal Processing,* New York: Dekker, 1979.

[4] C. T. Chen, *Linear System Theory and Design,* New York: Holt, Rinehart & Winston, 1984.

[5] C. T. Chen, "Technique for identification of linear time-invariant multivariable systems," in *Advances in Control and Dynamic Systems*, vol. 27 (ed. C. T. Leondes), New York: Academic Press, l987.

[6] C. T. Chen, *System and Signal Analysis,* New York: Holt, Rinehart & Winston, 1989.

[7] C. T. Chen, "A model of op-amp circuits," *IEEE Trans. on Circuits and Systems*, vol. 38, no. 10, pp. 1233–1235, 1991.

[8] C. T. Chen, *Analog and Digital Control System Design: Transfer-function, State-space, and Algebraic Methods*, New York: Saunders, 1993.

[9] W. Cheney and D. Kincaid, *Numerical Mathematics and Computing*, Monterey, CA: Brooks/Cole, l980.

[10] L. O. Chua, C. A. Desoer, and E. S. Kuo, *Linear and Nonlinear Circuits*, New York: McGraw-Hill, 1987.

[11] G. C. Goodwin and K. S. Sin, *Adaptive Filtering Prediction and Control*, Englewood Cliffs, NJ: Prentice-Hall, 1984.

[12] S. Haykin, *Communication Systems*, 2nd ed., New York: John Wiley, 1983.

[13] R. G. Irvine, *Operational Amplifiers*, Englewood Cliffs, NJ:Prentice-Hall, 1981.

[14] M. Jamshidi, *Large Scale Systems: Modeling and Control*, New York: North Holland, 1983.

[15] C. D. Johnson, *Process Control Instrumentation Technology*, New York: John Wiley, 1977.

[16] E. I. Jury, *Theory and Application of the Z-transform Method*, New York: John Wiley, 1964.

[17] E. I. Jury, "A modified stability table for linear discrete system," *Proc. IEEE*, vol. 124, pp. 184–185, 1964.

[18] W. Kaplan, *Advanced Mathematics for Engineers*, Reading, MA: Addison-Wesley, 1981.

[19] J. T. King, *Introduction to Numerical Computation*, New York: McGraw-Hill, 1984.

[20] E. Kreyszig, *Advanced Engineering Mathematics*, 5th ed., New York: John Wiley, 1983.

[21] L. Ljung and T. Söderström, *Theory and Practice of Recursive Identification*, Cambridge, MA: MIT Press, 1982.

[22] W. A. Lynch and J. G. Truxal, *Introductory System Analysis*, New York: McGraw-Hill, 1961.

[23] M. Mansour, "Instability criteria of linear discrete systems," *Automatica,* vol. 2, pp. 167–178, 1965.

[24] J. F. Mahoney and B. D. Sivazlian, "Partial fractions expansion: A review of computational methodology and efficiency," *Journal of Computational and Applied Mathematics*, vol. 9, pp. 247–269, 1983.

[25] C. D. McGillem and G. R. Cooper, *Continuous and Discrete Signal and System Analysis*, 2nd ed., New York: Holt, Rinehart & Winston, 1984.

[26] A. V. Oppenheim and A. S. Willsky, *Signals and Systems,* Englewood Cliffs, NJ: Prentice-Hall, 1983.

[27] A. B. Rodriguez-Vazquez, J. L. Huertas, and L. O. Chua, "Chaos in a switched-capacitor circuit," *IEEE Trans. Circuits and Systems*, vol. CAS-32, pp. 1083–1085, 1985.

[28] A. S. Sedra and K. C. Smith, *Microelectronic Circuits*, 2nd ed., New York: Holt, Rinehart & Winston, 1987.

[29] B. Seo and C. T. Chen. "A relationship between the Laplace transfrom and Fourier transform," *IEEE Trans. on Automatic Control*, vol. AC–31, pp. 751, 1986.

[30] Y. Shamash, "Stable reduced-order models using Pade-type approximation," *IEEE Trans. on Automatic Control*, vol. AC–19, pp. 615–617, 1974.

[31] L. S. Shieh and M. J. Goldman, "A mixed Cauer form for linear system reduction," *IEEE Trans. Sys. Man. Cyb.*, vol. SMC–4, pp. 584–588, 1974.

[32] G. W. Stewart, *Introduction to Matrix Computations*, New York: Academic, 1973.

[33] G. W. Stroke, M. Halioua, V. Srinivasan, and M. Shinoda, "Retrieval of good images from accidentally blurred photographs," *Science*, vol. 189, pp. 261–263, 1975.

[34] J. G. Truxal, *Introductory System Engineering*, New York: McGraw-Hill, 1972.

[35] M. E. Van Valkenburg, *Analog Filter Design*, New York: Holt, Rinehart & Winston, 1982.

[36] A. H. Zemanian, *Distribution Theory and Transform Analysis*, New York: Dover, 1987.

Index